LabWindows/CVI Programming for Beginners

| VIRTUAL INSTRUMENTATION SERIES

Shahid F. Khalid
■ LabWindows/CVI Programming For Beginners

Mahesh L. Chugani • Abhay R. Samant • Michael Cerna
■ LabVIEW Signal Processing

Rahman Jamal • Herbert Pichlik
■ LabVIEW Applications

Barry Paton
■ Sensors, Transducers & LabVIEW

Jeffrey Travis
■ Internet Applications in LabVIEW

Lisa K. Wells • Jeffrey Travis
■ LabVIEW For Everyone

LabWindows/CVI Programming for Beginners

◆ Shahid F. Khalid

Prentice Hall PTR
Upper Saddle River, NJ 07458
www.phptr.com

Library of Congress Cataloging-in-Publication Data

Khalid, Shahid F.
   LabWindows/CVI programming for beginners / Shahid F. Khalid.
      p. cm. — (National Instruments virtual instrumentation series)
   Includes bibliographical references and index.
   ISBN 0-13-016512-3
   1. C (Computer program language)   2. Computer programming.   3. LabVIEW.
   4. Engineering instruments—Data processing.   I. Title.   II. Series.

QA76.73.C15 K485 2000
006—dc21
                                                                                                    99-086114

Editorial/Production Supervision: Joan L. McNamara
Acquisitions Editor: Bernard Goodwin
Editoral Assistant: Diane Spina
Marketing Manager: Lisa Konzelmann
Manufacturing Manager: Alexis R. Heydt
Cover Design: Talar Agasyan
Cover Design Direction: Jerry Votta
Series Design: Gail Cocker-Bogusz
Composition/Page Make-up: Pine Tree Composition, Inc.

© 2000 Prentice Hall PTR
Prentice-Hall, Inc.
Upper Saddle River, NJ 07458

Prentice Hall books are widely used by corporations and government agencies for training, marketing, and resale.

The publisher offers discounts on this book when ordered in bulk quantities.
For more information, contact:
   Corporate Sales Department
   Prentice Hall PTR
   One Lake Street
   Upper Saddle River, NJ 07458
   Phone: 800-382-3419; FAX: 201-236-7141
   Email (Internet): corpsales@prenhall.com

All rights reserved. No part of this book may be reproduced, in any form or by any means, without permission in writing from the publisher.

The author of this book has used his best effort in preparing this book. These efforts include the development, research, and testing of the theories and programs to determine their effectiveness. The author makes no warranty of any kind, expressed or implied, with regard to these programs or the documentation contained in this book. The author shall not be liable in any event for incidental or consequential damages in connection with, or arising out of the furnishing, performance, or use of these programs.

Author Contact
Email: shahid_khalid@hotmail.com
       shahid.f.khalid@boeing.com

Trademarks: CVI™, **LabVIEW**™ is the trademark of National Instruments Corporation. **Microsoft**™ is the trademark of Microsoft Corporation. **Word** ™, **Word 97**™, **Excel**™ are the trademarks of Microsoft Corporation. All other product names mentioned herein are the trademarks of their respective owners.

Printed in the United States of America

10 9 8 7 6 5 4 3 2

ISBN 0-13-016512-3

Prentice-Hall International (UK) Limited, *London*
Prentice-Hall of Australia Pty. Limited, *Sydney*
Prentice-Hall Canada Inc., *Toronto*
Prentice-Hall Hispanoamericana, S.A., *Mexico*
Prentice-Hall of India Private Limited, *New Delhi*
Prentice-Hall of Japan, Inc., *Tokyo*
Pearson Education Asia Pte. Ltd.
Editora Prentice-Hall do Brasil, Ltda., *Rio de Janeiro*

Dedicated to my family, who will probably never read this book

Contents

Contents vii
Illustrations xix
Tables xxxi
Preface xxxv

    What Is *LabWindows/CVI*?... xxxv
    Introduction... xxxvi
    Objectives of this Book xxxvii
    What You Need to Run
        LabWindows/CVI xxxix
    Conventions Used in this Book xl
    Acknowledgments.. xli

Foreword xliii

Getting Started 1

    What is *CVI*?.. 2
    Where is *CVI* used?... 3
    Starting *CVI* 5
    The First *CVI* Project 7
    Summary... 11

Basics of Creating the Graphical User Interface 13

    User Interface Overview 14
    CVI Programming Model.. 15
    Creating the First *CVI* Project.. 18

Analyzing the *CodeBuilder*-Generated Code 33
Overview of How to Develop a *CVI* Application 36
Creating the Second *CVI* Project 36
Callback Function and Generated Events 51
Summary 54

More Graphical User Interface 55

Using Timer Controls 56
The Anatomy of the Panel 69
Learning the Tool Bar 72
Creating and Using the Strip Charts 74
Project Window Icons 83
Observations 85
Summary 86

Enhancing the User Interface 87

Project Overview 88
Starting the Project 88
Creating and Using the Menu Bar 102
Generating the Source Code 109
Creating and Using Tool Bars 115
Adding a Text Box to the Project 122
Creating the Tab Order 132
Summary 134

Source Editor and Debugging Techniques 137

Source Editor Environment Overview 138
Exploring the Source Editor 140
    Source Editor Tool Bar 140
    Working with Function Panels 143

Interactive Execution Window	144
Building Your Project	146
Manipulating the Code	**149**
Setting Line Numbers and Selecting Text	149
Using the Code Selection Modes	150
Balancing Parentheses	152
Setting and Clearing Tags	153
Finding User Interface Objects (Controls)	154
Inserting C Constructs in Code	156
Comparing Source Files	158
Find, Search, and Replace Commands	160
Using the Debugger	**163**
Using Breakpoints	165
Exclude/Include Lines	169
Working with Variables, Arrays, and Strings	171
Watch Window	176
Memory Display Window	178
Summary	**179**

File Input/Output 181

What is File I/O?	182
Using Files	182
Overview of the Project	188
Creating the User Interface	189
Generating the Code	196
Pop-Up Panels	196
Confirm Pop-Up Panel	199
Message Pop-Up Panel	200
Generic Message Pop-Up Panel	200
Prompt Pop-Up Panel	203
File Select Pop-Up Panel	205
Install Pop-Up Function	205
Remove Pop-Up Function	205
Multi-File Select Pop-Up Panel	205
Directory Selection Pop-Up Panel	207
Graph Pop-Up Panels	208
Adding Code to the Project	211
Main Function	212
StartTestCB Function	212

WriteToFile Function 215
CheckFileName Function 219
Timer_2SecCB Function. 221
Timer_1SecCB Function. 222
StopTestCB Function 225
ClearTextBoxCB Function 226
ReadFileCB Function 228
RunExcelCB Function 228
ExplorerCB Function. 230
ExitPanelCB Function. 232
Summary. 233

List Boxes & Rings 235

Introduction 236
List Box Example 7-1 238
List Box Example 7-2 248
Using Ring Controls 257
Designing the Test
    Executive Project 262
Project7 Code Generation
    and Analysis. 270
        Main Function 271
        LoadTestNamesCB Function 271
        SelectTestsCB Function 275
        RemoveTestCB Function. 275
        SaveSeqCB Function 277
        LoadSeqCB Function 280
        ResetSeqCB Function. 281
        CopyTestCB Function. 283
        GoToTestExecCB Function. 284
        SelectTimesToRunCB Function. 286
        RunSequenceCB Function. 287
        ExitCB Function. 297
        ClearListBoxCB Function. 297
        ReturnCB Function. 297
        CheckFileName Function 297
        ExplorerCB Function. 297
Summary. 298

8

Creating Stand-Alone Executables and Distribution Disks 301

Creating Stand-Alone Executables
    and Distribution Disks Overview 302
Creating Stand-Alone Executables 303
Creating Distribution Disks 305
    Install DataSocket Support 307
    Install NI Reports Support 307
Error Handling for Stand-Alone Executables 314
Summary 315

9

Creating and Using Dynamic Link Libraries 317

Dynamic Link Libraries (DLL) Overview 318
Creating Dynamic Link Libraries 319
Creating DLLs without a User Interface 319
    Using the DLL 325
Creating DLLs with the User Interface 328
    Using the GUI DLL 336
Summary 340

10

External Compiler Support 341

External Compiler Support 342
Creating an Object File for Use in an External Compiler 342
Creating DLL Containing Callback Functions 354
Summary 364

11

GPIB Communications 365

GPIB Communication Introduction and History 366
GPIB Communication 367

Getting Started with GPIB 369
GPIB Communication Using the
    NI-488 Routines 370
Win32 Interactive Control Utility 388
NI Spy Utility 391
Using the IEEE-488.2 Standards 394
Using GPIB Instrument Drivers 408
Summary 426

12

RS-232 Serial Communication 427

Overview of Serial Communication 428
Connectors and Signals 429
Serial Communication Project 433
Configuring the RS-232 435
Error Checking Functions 443
Writing and Reading Data 444
    Sending Data Buffer 444
    Receiving Data Buffer 445
    Sending a Data Byte 447
    Reading a Data Byte 448
    Reading Data Using a
        Termination Byte 449
    Writing Data Files 450
    Reading Data Files 452
Sending and Receiving
    Data Packets 455
    Configuring *XModem* 455
    Sending Data Packets 457
    Receiving Data Packets 457
Serial Communication Utility Functions 459
    Flush Output Queue 459
    Flush Input Queue 460
    Queue Length 461
    Obtaining COM Port Status 462
    Break Function 463
Communicating with RS-485 465
Summary 466

LabWindows/CVI Installation 467

Installing *CVI* 468
Installing the Projects Directory 469

Project Window Environment 471

Starting the Project 472
Overview of the Project Window 473
    Using the File Menu 473
        New 473
        Open 475
        Save 476
        Save As 476
        Save All 476
        Auto Save Project 476
        Print 477
        Most Recently Closed Files 477
        Exit LabWindows/CVI 477
    Using the Edit Menu 477
        Add Files To Project 477
        Select All 478
        Exclude File From Build 479
        Remove File 479
        Move Item Up 479
        Move Item Down 479
    Using the View Menu 480
        Show Full Path Names 481
        Show Full Dates 481
        Sort By Date 481
        Sort By Name 481
        Sort By Pathname 481
        Sort By File Extension 481
        No Sorting 481
    Using the Build Menu 482
        Configuration 482
        Target Type 483
        Target Settings 484
        Compile File 488
        Mark File For Compilation 488

- Mark All For Compilation ... 488
- External Compiler Support ... 488
- Create Distribution Kit ... 488

Using the Run Menu ... 489
- Debug Project ... 489
- Continue ... 490
- Terminate Execution ... 490
- Break at First Statement ... 490
- Breakpoints ... 490
- Select External Process ... 490
- Execute ... 491
- Threads ... 492

Using the Instrument Menu ... 492
- Load ... 493
- Unload ... 493
- Edit ... 493

Using the Library Menu ... 494
- User Interface ... 495
- Advanced Analysis ... 495
- Easy I/O for Data Acquisition ... 496
- Data Acquisition ... 496
- VXI ... 496
- GPIB/GPIB 488.2 ... 496
- RS-232 ... 497
- VISA ... 499
- IVI ... 499
- TCP ... 500
- DataSocket ... 501
- DDE ... 502
- X Property (Unix Only) ... 502
- ActiveX Automation (Windows Only) ... 502
- Formatting and I/O ... 503
- Utility ... 504
- ANSI C ... 504

Using the Tools Menu ... 504
- Create ActiveX Automation Controller (Windows Only) ... 506
- Create IVI Instrument Driver ... 507
- Source Code Control ... 508

Using the Window Menu ... 508
- Cascade Windows ... 508
- Tile Windows ... 509
- Minimize All ... 509
- Close All ... 509
- Project ... 510
- Build Errors ... 510
- Run-Time Errors ... 510
- Source Code Control Errors ... 510

Contents

    Memory Display 510
    Variables 510
    Watch 511
    Interactive Execution 511
  Using the Options Menu 512
    Build Options 512
    Compiler Defines 515
    Include Paths 516
    Instrument Directories 516
    Run Options 517
    Command Line 518
    Environment 518
    Library Options 520
    Tools Menu Options 521
    Source Code Control Options 522
    Project Move Options 523
    Font 524
    Colors 525
  Using the Help Menu 525
Summary 525

User Interface Editor Environment 527

 File Menu 528
  New, Open, Save, Save As, Close, and Exit *LabWindows/CVI* 528
  Save Copy As 528
  Save All 529
  Add File to Project 529
  Read Only 529
  Print 529
 Edit Menu 529
  Undo and Redo 529
  Cut, Copy, and Paste 530
  Delete 530
  Cut Panel and Copy Panel 531
  Panel 531
  Control 531
  Tab Order 531
  Set Default Font 531
  Apply Default Font 531
  Control Style 531
 Create Menu 532

View Menu 533
    Find UIR Objects. 533
    Show/Hide Panels 536
    Bring Panel to Front 536
    Next Panel. 537
    Previous Panel 537
    Preview User Interface Header File. 537
Arrange Menu 537
Code Menu. 538
    Set Target File 538
    Generate 538
        Main Function 540
        All Callbacks 541
        Panel Callback. 541
        Control Callbacks 541
        Menu Callbacks 541
    View. 542
    Preferences 542
        Default Panel Events/Default Control Events. 543
        Always Append Code To End. 543
Run Menu 545
Library Menu 546
Tools Menu. 546
Window Menu 546
Options Menu. 546
    Operate Visible Panels 547
    Next Tool. 547
    Preferences 547
        Edit Color Preferences 547
        Preferences for New Panels 547
        Preferences for New Controls. 548
            More 549
    Save in Text Format 552
    Load from Text 552
Help Menu. 552

Formatting Functions   553

Introduction to Formatting and Scanning Functions 554
Formatting Functions 554
    Format Code Specifiers 559

Contents

    Formatting Modifiers 561
    Formatting Examples for the *Fmt* Function 565
        Integer to String 565
        Short Integer to String 569
        Inserting Characters Between Each Element of an Array 571
        Real to String in Floating Point Notation 572
        Real to String in Scientific Notation 574
        Integer and Real to String with Literals 576
        Concatenating Two Strings 577
        Appending to a String 577
        Converting Decimal Integer to
            Hexadecimal String 578
Scanning Functions 578
    Scan Functions — Format String 581
        String to String 581
        Converting String to Integer and String 586
        Converting Comma-Separated ASCII Numbers to Real Array 587
        Scanning Non-NUL Terminated Strings 587
        Converting Integer Arrays to Real Arrays 588
        Byte Swapping, with Integer Array
            to Real Array 588
        Converting Hexadecimal String
            to Decimal Integer 589
    ScanFile Function 590
    ScanIn Function 591
Status Functions 591
Using the Format Wizard 594

CVI Demo Programs 599

Bmath Demo Version 1.0 600
    Introduction 600
    Loading/Editing Instructions 601
    Using Parametric Run 602
    Iterative Evaluation 602
    Using the Plot Utility 603
    Using the Solver Utility 605
System Test Demo 605
    Introduction 605
    System Test Start Up Procedure 608
    How to Use Permission Levels 611
    System Test Window (Main Window) 611

How to Use the Molecules List Reordering Window. 614
System Self Tests Window 615
System Selective Testing Window 616
Loop/Select Window. 619
Sequencer Window 619
Test Builder Window 619
Last Test Window 621
Stop/Loop Window. 622
Vars Window 622

Bibliography   **625**
Index   **629**
The Author   **652**

Illustrations

Figure 1–1	Typical **Start** Button List under Windows	6
Figure 1–2	New **Project** Window	7
Figure 1–3	**Project** Window for `project1-1`	8
Figure 1–4	Project1–1 GUI	9
Figure 1–5	Project1–1 Run	10
Figure 2–1	*RunUserInterface/QuitUserInterface* Interaction	16
Figure 2–2	Project1–1 GUI	17
Figure 2–3	User Interface Toolbar	17
Figure 2–4	Events Generation	18
Figure 2–5	Configuration Menu	19
Figure 2–6	Create Panel Menu	20
Figure 2–7	Blank Panel	21
Figure 2–8	Parent/Child/Grandchild Panel	21
Figure 2–9	Panel Attributes Dialog Box	22
Figure 2–10	Creating the Command Button	23
Figure 2–11	Command Button Dialog Box	23
Figure 2–12	Attributes for RUN Command Button	24
Figure 2–13	First Project GUI	25
Figure 2–14	Default Control Callback Events	26
Figure 2–15	Default Panel Callback Events	26
Figure 2–16	Automatic Code Generation	27
Figure 2–17	Code Generation Dialog Box	28
Figure 2–18	Project2–1 *CodeBuilder* Skeleton Code	29
Figure 2–19	**Start** Command Button Callback Function	30
Figure 2–20	**Project** Window List	30
Figure 2–21	Project2–1 Build Configuration Sub-Menu	30
Figure 2–22	Executable File Creation Message	31
Figure 2–23	Project2–1 Run Menu	32
Figure 2–24	Project2–1 GUI	32
Figure 2–25	First Project Run	32
Figure 2–26	Edit Panel	37

Illustrations

Figure 2-27	Numeric Controls	38
Figure 2-28	Numeric Meter Attributes	39
Figure 2-29	Data Type Pull-Down Menu	40
Figure 2-30	Control Mode Pull-Down Menu	40
Figure 2-31	Range Values Dialog Box	41
Figure 2-32	Format and Precision Dialog Box	42
Figure 2-33	Show/Hide Parts Dialog Box	42
Figure 2-34	Text Attributes Dialog Box	43
Figure 2-35	Label Appearance Dialog Box	44
Figure 2-36	Voltmeter Control	44
Figure 2-37	Panel with Voltmeter and Command Buttons	46
Figure 2-38	*StartCB* Callback Function Listing	47
Figure 2-39	Project2-2 Code Listing	48
Figure 2-40	Project2-2 Run	50
Figure 3-1	Project2-2 GUI	57
Figure 3-2	Project3-1 GUI with the Timer	58
Figure 3-3	Timer Attributes Dialog Box	58
Figure 3-4	Timer Range Values Dialog Box	60
Figure 3-5	Format and Precision Dialog Box	61
Figure 3-6	Show/Hide Parts Dialog Box	61
Figure 3-7	*CodeBuilder*-Generated Code for Project3-1	62
Figure 3-8	Project3-1 **Project** Window	63
Figure 3-9	Code Listing for Project3-1	64
Figure 3-10	Project3-1 Run	67
Figure 3-11	*SetCtrlAttribute* Function Panel	68
Figure 3-12	Panel Attributes	69
Figure 3-13	Panel Coordinates and Dimensions	70
Figure 3-14	**Other Attributes** Panel	71
Figure 3-15	Edit Title Attributes	72
Figure 3-16	Editing Tool Bar	73
Figure 3-17	Project3-1 GUI with the Strip Chart	75
Figure 3-18	Strip Chart Attributes	75
Figure 3-19	Project3-2 GUI	76
Figure 3-20	Project3-2 Source Code Listing	77
Figure 3-21	Select Function Panel	79
Figure 3-22	*PlotStripChart* Function Panel	80
Figure 3-23	Project3-2 Run	83
Figure 3-24	**Project** Window Icons	84
Figure 4-1	New Project Creation	88
Figure 4-2	**User Interface Editor** Window	89
Figure 4-3	**Decoration** Boxes	90
Figure 4-4	**LEDs** in **Decoration** Boxes	91
Figure 4-5	Objects Aligning Menu	91
Figure 4-6	**LEDs** Aligned	92
Figure 4-7	*Coloring Tool*	92
Figure 4-8	Color Template	92

Illustrations   xxi

Figure 4–9	**LEDs** and **Binary Switches** after Alignment	93
Figure 4–10	Project4–1 GUI	95
Figure 4–11	Project4–1 Main Function	96
Figure 4–12	*StartCB* Callback Function	96
Figure 4–13	*StopCB* Callback Function	97
Figure 4–14	*ExitCB* Callback Function	97
Figure 4–15	*ResetCB* Callback Function	98
Figure 4–16	*Channel_CB* Callback Function	98
Figure 4–17	*Generate250LoopCB* Callback Function	100
Figure 4–18	TurnOnLED Callback Function	100
Figure 4–19	Project4–1 Run	102
Figure 4–20	Menu Bar List	103
Figure 4–21	Edit Menu Bar Dialog Box	103
Figure 4–22	Edit Menu Bar Modifier Key Choices	105
Figure 4–23	Edit Menu Bar Shortcut Key Choices	106
Figure 4–24	Edit Menu Commands (**Project** Window)	107
Figure 4–25	Project4–2 GUI	107
Figure 4–26	Menu Bar Items	109
Figure 4–27	Skeleton Code Generated by *CodeBuilder*	111
Figure 4–28	Project4–2 *Main* Function	112
Figure 4–29	*MenuStartCB* Callback Function	113
Figure 4–30	*MenuStopCB* Callback Function	114
Figure 4–31	*MenuResetCB* Callback Function	114
Figure 4–32	Project4–2 Run	115
Figure 4–33	**User Interface Editor** Tool Bar	116
Figure 4–34	Toolbar and Programmer's Toolbox Shown on the Instrument Menu	116
Figure 4–35	Project4–3 Header	117
Figure 4–36	Project4–3 *Main* Function	118
Figure 4–37	Accessing Icon Editor	121
Figure 4–38	Project4–3 Run	121
Figure 4–39	Creating a **Text Box** Control	123
Figure 4–40	**Text Box** Properties Dialog Box	123
Figure 4–41	Creating a String Control	124
Figure 4–42	Timer 1 sec. Properties Box	125
Figure 4–43	Project4–4 GUI	126
Figure 4–44	Control Callback Shortcuts	126
Figure 4–45	Generate_1secLoopCB Callback Function Skeleton Code	127
Figure 4–46	Project4–4 *Main* Function Code Differences	128
Figure 4–47	Additions to *Generate250LoopCB* Callback Function	129
Figure 4–48	Additions to *MenuStartCB* Callback Function	130
Figure 4–49	Additions to *MenuStopCB* Callback Function	131
Figure 4–50	Project4–4 Run	132
Figure 4–51	Tab Order Dialog Box	133
Figure 4–52	New Tab Order	134
Figure 5–1	Source Code Window	138
Figure 5–2	Menu Commands from Text Area	139

Illustrations

Figure 5–3	Menu Commands for Line Icons	140
Figure 5–4	Source Editor Tool Bar	140
Figure 5–5	Debugging Tool Bar Icons	143
Figure 5–6	Function Panel Insert Dialog Box	143
Figure 5–7	Matching Function Panel Names	144
Figure 5–8	Interactive Execution Window Run Demonstration	145
Figure 5–9	Inserting Popup Panels in IEW	146
Figure 5–10	IEW Run Using the Popup Panel	147
Figure 5–11	Insert Include Statements Dialog Box	148
Figure 5–12	Prototype Generated Header File	149
Figure 5–13	Character Select Mode Text Selection	150
Figure 5–14	Line Select Mode Text Selection	151
Figure 5–15	Column Select Mode Text Selection	152
Figure 5–16	Highlighted Text Using the Edit>>Balance Command	153
Figure 5–17	Code with Tags	154
Figure 5–18	Clear Tags Dialog Box	155
Figure 5–19	Find User Interface Object	155
Figure 5–20	**Insert Construct** Sub-Menu	156
Figure 5–21	*Switch* Construct	157
Figure 5–22	*Switch* Construct Created Using the **Insert Construct**	158
Figure 5–23	**Diff** Sub-Menu	159
Figure 5–24	**Match Criteria** Dialog Box	160
Figure 5–25	**Find** Dialog Box	160
Figure 5–26	Find Button Bar	161
Figure 5–27	**Replace** Dialog Box	161
Figure 5–28	**Replace Button Bar**	162
Figure 5–29	**Build Options** Dialog Box	164
Figure 5–30	Debugging Tool Bar Icons	164
Figure 5–31	Breakpoint Set	166
Figure 5–32	Execution Stopped at Breakpoint	167
Figure 5–33	**Breakpoint** Dialog Box	168
Figure 5–34	**Edit Breakpoint** Dialog Box	169
Figure 5–35	**Resolve All Excluded Lines** Dialog Box	170
Figure 5–36	Conditional Compilation	171
Figure 5–37	Excluded Lines Exit Dialog Box	171
Figure 5–38	Viewing Variables at the Breakpoint	172
Figure 5–39	Changing Values of the Variable	174
Figure 5–40	Array Display Window	175
Figure 5–41	Format in String Display Window	176
Figure 5–42	Add/Edit Watch Expression	177
Figure 5–43	Memory Display Window	178
Figure 6–1	Sample to Demonstrate Reading All the Lines from a File	186
Figure 6–2	Test Panel GUI with Top Decoration Box Controls	190
Figure 6–3	Test Panel GUI with Results Text Box	191
Figure 6–4	Test Panel GUI with Temperature Sensor	192
Figure 6–5	Test Panel GUI with Pressure Sensor Box	193

Illustrations

Figure 6–6	Test Panel GUI with Command Buttons	194
Figure 6–7	Completed Test Panel GUI	195
Figure 6–8	*CodeBuilder* Listing for Project6	197
Figure 6–9	Confirm Pop-Up Panel	199
Figure 6–10	Message Pop-Up Panel	201
Figure 6–11	Generic Message Pop-Up Panel	202
Figure 6–12	Prompt Pop-Up Panel	204
Figure 6–13	File Select Pop-Up Panel	207
Figure 6–14	Ygraph Pop-Up Panel	210
Figure 6–15	Xgraph Pop-Up Panel	211
Figure 6–16	Adding Code to *Main* Function	212
Figure 6–17	Project6 Main Function	213
Figure 6–18	StartTestCB Function Listing	214
Figure 6–19	WriteToFile Function Listing	216
Figure 6–20	OpenFile Function Panel	218
Figure 6–21	CheckFileName Function	220
Figure 6–22	Code Structure to Check File	221
Figure 6–23	Timer_2SecCB Function	221
Figure 6–24	Timer_1SecCB Function	223
Figure 6–25	StopTestCB Function	226
Figure 6–26	ClearTextBoxCB Function	227
Figure 6–27	ReadFileCB Function	229
Figure 6–28	RunExcelCB Function	231
Figure 6–29	ExplorerCB Function	232
Figure 6–30	ExitPanelCB Function	233
Figure 7–1	Creating a List Box	236
Figure 7–2	List Box	237
Figure 7–3	List Box Attributes	237
Figure 7–4	Label/Value Pair Control Box	238
Figure 7–5	Example 7–1 User Interface Resource	239
Figure 7–6	Listing of Example 7–1	241
Figure 7–7	String Sequence Structure	246
Figure 7–8	Example 7–2 GUI	249
Figure 7–9	List Box Options	250
Figure 7–10	Example 7–2 Listing	252
Figure 7–11	Formatted String Using Escape Sequences	256
Figure 7–12	Creating a Ring Control	258
Figure 7–13	Ring Control Demonstration Panel	259
Figure 7–14	Menu Ring Control Box	259
Figure 7–15	Ring Control Box	260
Figure 7–16	Vertical Level Slide and Knob	260
Figure 7–17	Meter and Horizontal Pointer Slide	261
Figure 7–18	Test Sequence Selection GUI	263
Figure 7–19	Test Executive GUI	265
Figure 7–20	Test Executive Command Buttons	268
Figure 7–21	Run Sequence GUI	269

Illustrations

Figure 7-22	Edit Label/Value Pairs	270
Figure 7-23	Main Function Source Code	272
Figure 7-24	LoadTestNamesCB Source Code	273
Figure 7-25	SelectTestsCB Source Code	276
Figure 7-26	RemoveTestCB Source Code	277
Figure 7-27	SaveSeqCB Source Code	278
Figure 7-28	LoadSeqCB Source Code	280
Figure 7-29	ResetSeqCB Source Code	282
Figure 7-30	CopyTestCB Source Code	284
Figure 7-31	GoToTextExecCB Source Code	285
Figure 7-32	SelectTimesToRunCB Source Code	286
Figure 7-33	RunSequenceCB Source Code	290
Figure 7-34	ExplorerCB Source Code	298
Figure 8-1	Set Configuration Menu	303
Figure 8-2	Target Settings Dialog Box	304
Figure 8-3	Create Distribution Disk Dialog Box	306
Figure 8-4	Edit File Group Dialog Box	308
Figure 8-5	Advanced Distribution Kit Options	310
Figure 8-6	Select Custom Script File Dialog Box	311
Figure 8-7	Build Progress Panel	312
Figure 8-8	Insert Disk Message	313
Figure 8-9	Test Exec Installation	313
Figure 9-1	SimpleDLL.c Code Listing	320
Figure 9-2	Setting Dynamic Link Library as Target Type	322
Figure 9-3	Target Settings for Dynamic Link Library Dialog Box	323
Figure 9-4	DLL Import Library Choices Dialog Box	323
Figure 9-5	Type Library Dialog Box	324
Figure 9-6	DLL Export Options	325
Figure 9-7	Creation of SimpleDLL.dll and .lib Files for Current Compiler	326
Figure 9-8	UseSimpleDLL.c Listing to Call the Export Functions	326
Figure 9-9	UseSimpleDLL Project Files	327
Figure 9-10	UseSimpleDLL Project Output	327
Figure 9-11	DLL User Interface Resource	328
Figure 9-12	GUIExportDLL Source Code Created by *CodeBuilder*	330
Figure 9-13	*Export Function* Callback Function	332
Figure 9-14	Code Fragment Added to *DLLMain*	332
Figure 9-15	Code Fragment Added to DLL_PROCESS_DETACH	333
Figure 9-16	GUIExportRoutine.c Listing	333
Figure 9-17	GUIExportHeader File	335
Figure 9-18	GUIExportDLL Project Window	335
Figure 9-19	GUIExport Create DLL Dialog Box	336
Figure 9-20	GUIExport.dll and GUIExport.lib Created	337
Figure 9-21	*UseGUIExport* Listing	337
Figure 9-22	UseGUIExportDLL Project Window	337
Figure 9-23	Functions Loaded in Main	338

Illustrations XXV

Figure 9–24	Values of a and b Swapped	338
Figure 9–25	Inside ExportFunction	339
Figure 9–26	Exported Functions Unloaded from DLLMain	339
Figure 10–1	Remove Lines from Project6 Main Function	344
Figure 10–2	Generate All Code Dialog Box	344
Figure 10–3	*CodeBuilder*-Created *WinMain* for Project6	345
Figure 10–4	*WinMain* Function Created for Project6	346
Figure 10–5	External Compiler Support Dialog Box	346
Figure 10–6	Microsoft Visual C++ File New Dialog Box	348
Figure 10–7	Microsoft Visual C++ Add to Project Menu Item	349
Figure 10–8	Microsoft Visual C++ Insert Files into Project	349
Figure 10–9	Microsoft Visual C++ Insert Library Files into Project	350
Figure 10–10	Microsoft Visual C++ Project Settings for Win32 Debug	351
Figure 10–11	Microsoft Visual C++ Project Settings Win32 Release	351
Figure 10–12	Microsoft Visual C++ Directory Options	352
Figure 10–13	Microsoft Visual C++ Project Build Menu	352
Figure 10–14	Microsoft Visual C++ ExtCallback Directory Listing	353
Figure 10–15	Microsoft Visual C++ ExtSupp Directory Listing	353
Figure 10–16	Project Run Using Microsoft Visual C++ as External Compiler	354
Figure 10–17	Completed DLL Creation Code	356
Figure 10–18	Create Dynamic Link Library for Project6DLL	357
Figure 10–19	Created Dynamic Link Library Files	358
Figure 10–20	DLL Import Library Choices	358
Figure 10–21	Importdll.c Source Code	359
Figure 10–22	Project Window for importdll.prj	359
Figure 10–23	Microsoft Visual C++ DLL File New Dialog Box	360
Figure 10–24	Microsoft Visual C++ Insert Files into DLL Project	360
Figure 10–25	Microsoft Visual C++ Insert Library Files into DLL Project	361
Figure 10–26	Microsoft Visual C++ DLL Project Settings for Win32 Debug	361
Figure 10–27	Microsoft Visual C++ DLL Project Settings Win32 Release	362
Figure 10–28	Microsoft Visual C++ Directory Options	363
Figure 10–29	Files in DLLCall Folder	363
Figure 10–30	Microsoft Visual C++ Project Build Menu	364
Figure 11–1	Typical GPIB System	368
Figure 11–2	Project11–1 User Interface	371
Figure 11–3	Project11–1 Code Listing	373
Figure 11–4	Accessing the GPIB Library Routines	382
Figure 11–5	GPIB/GPIB 488.2 Function Panel	382
Figure 11–6	Function Panel for *ibonl* Routine	383
Figure 11–7	Communication Command Buttons	385
Figure 11–8	Enter GPIB Device Command User Interface	386
Figure 11–9	Win32 Interactive Control Utility Start screen	389
Figure 11–10	Win32 Interactive Utility help ibdev	389
Figure 11–11	Win32 Interactive Utility ibdev data	390
Figure 11–12	NI Spy Main Window	392
Figure 11–13	Project11–1 Interactive Monitoring with NI Spy	392

Illustrations

Figure 11–14	Getting Communication Call Details with NI Spy	393
Figure 11–15	Project11–2 Code Listing	397
Figure 11–16	Enter Primary Address User Interface	406
Figure 11–17	Project11–3 Project Window	409
Figure 11–18	Project11–3 Code Listing	411
Figure 11–19	Project11–3 Run	425
Figure 12–1	Character Frame Format for Seven Data Bits, One Parity Bit, Two Stop Bits	429
Figure 12–2	Character Frame Format for Eight Data Bits, No Parity Bit, One Stop Bit	429
Figure 12–3	DB-9 (Male) Connector Pin Locations	431
Figure 12–4	DB-25 (Male) Connector Pin Locations	431
Figure 12–5	Project12 Main GUI	434
Figure 12–6	Serial Communication Configuration GUI	435
Figure 12–7	Eight Data Bits Using Odd Parity	436
Figure 12–8	Eight Data Bits Using Even Parity	437
Figure 12–9	RS-232 Configuration Source Code	439
Figure 12–10	Error Checking Code Fragment	443
Figure 12–11	*SendSerialCB* Function Source Code	444
Figure 12–12	*ReceiveSerialCB* Function Source Code	446
Figure 12–13	*SendByteCB* Function Source Code	447
Figure 12–14	*ReceiveByteCB* Function Source Code	448
Figure 12–15	*SendFileCB* Function Source Code	451
Figure 12–16	*ReadFileCB* Function Source Code	453
Figure 12–17	*FlushOutputQCB* Callback Function	459
Figure 12–18	*FlushInputQCB* Callback Function	460
Figure 12–19	*GetOutQLengthCB* Callback Function	462
Figure 12–20	*GetCOMStatusCB* Callback Function	463
Figure 12–21	*LoopEverySecondCB* Callback Function	464
Figure A–1	*CVI* Registration	469
Figure B–1	New Project Window	472
Figure B–2	Project Window File Menu	474
Figure B–3	Project Window New Sub-Menu	474
Figure B–4	Project Window Open Sub-Menu	475
Figure B–5	Project Window Edit Menu	478
Figure B–6	Add Files to Project Sub-Menu	478
Figure B–7	Project Window View Menu	480
Figure B–8	Project Window Build Menu	482
Figure B–9	Configuration Menu Commands	482
Figure B–10	Target Type Sub-Menu	483
Figure B–11	Target Settings Dialog Box	484
Figure B–12	Run with Output Directed to Windows Console	485
Figure B–13	Run with Output Directed to Standard Output	486
Figure B–14	Version Info Dialog Box	487
Figure B–15	Version Info Sub-Dialog Box	487
Figure B–16	Project Window Run Menu	489
Figure B–17	Breakpoints Dialog Box	491

Illustrations

xxvii

Figure B–18	Select External Process Dialog Box	491
Figure B–19	Threads Dialog Box	492
Figure B–20	Instrument Menu	493
Figure B–21	Project Window Library Menu	494
Figure B–22	User Interface Libraries	495
Figure B–23	Advanced Analysis Libraries	496
Figure B–24	Easy I/O for DAQ Menu	497
Figure B–25	VXI Utilities Library	498
Figure B–26	GPIB Function Library	498
Figure B–27	RS-232 Serial Port Library	499
Figure B–28	VISA Library	500
Figure B–29	IVI Library	500
Figure B–30	TCP Support Library	501
Figure B–31	Data Socket Library High-Level Functions	501
Figure B–32	DDE Support Library	502
Figure B–33	ActiveX Automation Library	503
Figure B–34	Formatting and I/O Library	504
Figure B–35	Utility Library	505
Figure B–36	ANSI C Library	505
Figure B–37	Tools Menu	506
Figure B–38	ActiveX Automation Server Choices	506
Figure B–39	Instrument Driver Development Wizard	507
Figure B–40	Select Instrument Driver Menu	508
Figure B–41	Window Menu	509
Figure B–42	Memory Display Window	511
Figure B–43	Options Menu	512
Figure B–44	Build Options	513
Figure B–45	Compiler Defines	515
Figure B–46	Include Paths	516
Figure B–47	Instrument Directories Dialog Box	517
Figure B–48	Run Options	517
Figure B–49	Command Line Dialog Box	518
Figure B–50	Environment Options	519
Figure B–51	Library Options	521
Figure B–52	Tools Menu Options	521
Figure B–53	Add/Edit Tolls Menu Dialog box	522
Figure B–54	Explorer Added to Tools Menu	522
Figure B–55	Source Code Control Options	523
Figure B–56	Project Move Options	523
Figure B–57	Select Project Font	524
Figure B–58	Color Option	524
Figure B–59	Help Menu	525
Figure C–1	User Interface Resource File Menu	528
Figure C–2	User Interface Editor Edit Menu	530
Figure C–3	Converting Control Style	532
Figure C–4	User Interface Editor Create Menu	533

Figure C-5	User Interface Editor View Menu	534
Figure C-6	Find UIR Objects Dialog Box	534
Figure C-7	Search By Ring Control	534
Figure C-8	Finding UIR Objects Using Prefix + Constant Box	535
Figure C-9	Successful Search Completion Dialog Box	536
Figure C-10	Show/Hide Panel Sub-Menu	537
Figure C-11	Arrange Menu	538
Figure C-12	Code Menu	538
Figure C-13	Set Target File Dialog Box	539
Figure C-14	Generate Sub-Menu	539
Figure C-15	Generate Code Dialog Box	540
Figure C-16	Generate Main Function Dialog Box	540
Figure C-17	Control Callbacks Shortcut	541
Figure C-18	Select Menu Bar Options	542
Figure C-19	Find UI Object Shortcut	543
Figure C-20	Preferences Menu Item	544
Figure C-21	Default Panel Events	544
Figure C-22	Default Control Events	545
Figure C-23	Run Menu	545
Figure C-24	Options Menu	546
Figure C-25	User Interface Editor Preferences Dialog Box	548
Figure C-26	Edit Panel/Other Attributes Dialog Box	549
Figure C-27	Control Text Style Dialog Box	550
Figure C-28	Label Text Style Dialog Box	550
Figure C-29	Other User Interface Editor Preferences	551
Figure D-1	*Fmt* Function Panel	555
Figure D-2	*Fmt* Function — Converting Integer to String	556
Figure D-3	*Fmt* Function — Converting Multiple Inputs to a String	556
Figure D-4	*Fmt* Function — Variables and Format Specifiers Correspondence	557
Figure D-5	*FmtFile* Function Panel	557
Figure D-6	Writing to File Using *Fmt* Function	558
Figure D-7	Writing to File Using *FmtFile* Function	558
Figure D-8	*FmtOut* Function Panel	559
Figure D-9	Integer to String	565
Figure D-10	Integer Hexadecimal Value to String	566
Figure D-11	Integer Octal Value to String	566
Figure D-12	Unsigned Integer Value to String	566
Figure D-13	Integer into String with Specified Width	567
Figure D-14	Integer into a String with Character Padding	567
Figure D-15	Integer to String with Specified Wdith Less than Source String's Width	568
Figure D-16	Reversing Byte Order	568
Figure D-17	Short Integer String	569
Figure D-18	Single-Byte Integer to String	569
Figure D-19	Short-Signed Integer to String	569
Figure D-20	Short-Signed Integer to String	570
Figure D-21	Short-Signed Integer to String Using Width Modifier	570

Illustrations

Figure D-22	Short-Signed Integer to String Using Width and Padding Modifiers	570
Figure D-23	Short Integer to String with Specified Width Less than Source String Width	571
Figure D-24	Inserting Specified Character Between Each Element of Array	571
Figure D-25	Specified Character Insertion Demo Using Interactive Window	572
Figure D-26	Floating-Point Number to String	572
Figure D-27	Floating-Point Number to String Using the Precision Modifier	573
Figure D-28	Floating-Point Number to String with More Precision Digits than in Source	573
Figure D-29	Floating-Point Number to String Using Width Modifier	573
Figure D-30	Floating-Point Number to String Specifying Width *Modifier* Less than Floating-Point Width	574
Figure D-31	Small Floating-Point Number to String	574
Figure D-32	Floating-Point Number into String with Scientific Notation	574
Figure D-33	Floating-Point Number to String Using Scientific Notation and Precision Modifier	575
Figure D-34	Floating-Point Number into String Using Scientific Notation Specifying Precision Modifier and Number of Digits in Exponent	575
Figure D-35	Floating-Point Number to String Using Scientific Notation and Specifying Precision and Width Modifiers plus Number of Digits in Exponent	575
Figure D-36	Floating-Point Number to String Using Scientific Notation and Specifying Too Small Width Modifier	576
Figure D-37	Floating-Point and Integers with Literal Characters to a String	576
Figure D-38	Concatentate Two Strings	577
Figure D-39	Appending Characters to a String	577
Figure D-40	Decimal Integer to Hexadecimal Integer	578
Figure D-41	*Scan* Function Panel	579
Figure D-42	*ScanFile* Function Panel	580
Figure D-43	*ScanIn* Function Panel	580
Figure D-44	Source String to Target String	582
Figure D-45	Blank and Tabs in String	582
Figure D-46	Converting All Characters Up to LineFeed in String	582
Figure D-47	String Conversion with White Space Character	583
Figure D-48	Convert String Up to Terminator Character	583
Figure D-49	Convert String Up to a Number	584
Figure D-50	Convert String with Modifiers	584
Figure D-51	Discarding Terminators from Source String	584
Figure D-52	Converting Selected Bytes from Source	585
Figure D-53	Appending Source to Target String	585
Figure D-54	Breaking an Alpha-Numeric String	586
Figure D-55	Erroneous Integer Converting	586
Figure D-56	Converting Comma-Separated ASCII Numbers to Real Numbers Array	587
Figure D-57	Null-Terminating Communication Data	588
Figure D-58	Converting Array of Integers to Array of Floating-Point Numbers	588
Figure D-59	Swapping Byte-Order and Integer to Real Values	589
Figure D-60	Hexadecimal String to Decimal Integer Using %i Format String	589
Figure D-61	Hexadecimal String to Decimal Integer Using %x Format String	589
Figure D-62	Converting Comma-Separated from a File	590
Figure D-63	Example of Using *FmtOut* and *ScanIn*	591
Figure D-64	GetFmtErrNdx() Function	592
Figure D-65	Use of GetFmtErrNdx Function	592
Figure D-66	Error Index	593

Illustrations

Figure D-67	NumFmtdBytes() Function	593
Figure D-68	Example of NumFmtdBytes()	594
Figure D-69	Format Wizard Selection	595
Figure D-70	Format Wizard Main GUI	595
Figure D-71	Format Wizard Type Pull-Down Menu	596
Figure D-72	Format Wizard Edit Item Dialog Box	596
Figure D-73	Format Wizard Completed Function Call	597
Figure E-1	Bmath — Main Menu	600
Figure E-2	Functions Window	601
Figure E-3	Parametric Run Window	603
Figure E-4	Param. Plot Window	604
Figure E-5	Solver Window	604
Figure E-6	Permissions Window	609
Figure E-7	System Test Window (Main Window)	609
Figure E-8	System Self Tests Window	610
Figure E-9	System Selective Testing Window	612
Figure E-10	Test Builder Window	613
Figure E-11	Molecules List Reordering Window	614
Figure E-12	Loop/Select Window	617
Figure E-13	Sequencer Window	617
Figure E-14	Vars Window	618
Figure E-15	Last Test Window	621
Figure E-16	Stop/Loop Window	621

Tables

Table 2-1	*LoadPanel* Function	34
Table 2-2	*DisplayPanel* Function	34
Table 2-3	*RunUserInterface* Function	35
Table 2-4	*QuitUserInterface* Function	35
Table 2-5	*SetCtrlVal* Function	49
Table 2-6	Control Callback Function Events	52
Table 2-7	Panel Callback Function Events	53
Table 3-1	Continuous Events	56
Table 3-2	*GetCtrlVal* Function	66
Table 3-3	*PlotStripChart* Function	82
Table 3-4	Memory Usage	86
Table 4-1	*SetCtrlAttribute* Function	99
Table 4-2	*GetCtrlAttribute* Function	101
Table 4-3	*LoadMenuBar* Function	113
Table 4-4	*Toolbar_New* Function Arguments	119
Table 4-5	*Toolbar_InsertItem* Function Arguments	120
Table 4-6	*InsertTextBoxLine* Function	130
Table 6-1	*OpenFile* Function	183
Table 6-2	*WriteFile* Function	184
Table 6-3	*ReadFile* Function	185
Table 6-4	*ReadLine* Function	187
Table 6-5	*WriteLine* Function	187
Table 6-6	*GetFileInfo* Function	188
Table 6-7	*ConfirmPopup* Function	200
Table 6-8	*MessagePopup* Function	201
Table 6-9	*GenericMessagePopup* Function	202
Table 6-10	*PromptPopup* Function	204
Table 6-11	*FileSelectPopup* Function	206
Table 6-12	*InstallPopup* Function	207
Table 6-13	*RemovePopup* Function	208

Tables

Table 6–14	*MultiFileSelectPopup* Function	209
Table 6–15	*DirSelectPopup* Function	210
Table 6–16	*MakePathname* Function	218
Table 6–17	*CopyString* Function	224
Table 6–18	*DefaultCtrl* Function	227
Table 7–1	*GetNumListItems* Function	244
Table 7–2	*GetCtrlIndex* Function	244
Table 7–3	*GetValueFromIndex* Function	245
Table 7–4	*GetLabelFromIndex* Function	246
Table 7–5	*InsertListItem* Function	247
Table 7–6	*DeleteListItem* Function	248
Table 7–7	*GetNumCheckedItems* Function	254
Table 7–8	*IsListItemChecked* Function	255
Table 7–9	*GetIndexFromValue* Function	257
Table 7–10	*SavePanelState* Function	279
Table 7–11	*RecallPanelState* Function	281
Table 7–12	*SetPanelPos* Function	287
Table 7–13	*FindPattern* Function	293
Table 7–14	*SplitPath* Function	294
Table 7–15	*CompareStrings* Function	296
Table 9–1	*LoadPanelEx* Function	331
Table 11–1	*ibonl* Routine	381
Table 11–2	*ibdev* Routine	384
Table 11–3	*ibln* Routine	385
Table 11–4	*ibwrt* Routine	387
Table 11–5	*ibcmd* Routine	387
Table 11–6	*ibrd* Routine	388
Table 11–7	IEEE–488.2 Commands	395
Table 11–8	*FindLstn* Routine	405
Table 11–9	*Send* Routine	406
Table 11–10	*SendCmds* Routine	407
Table 11–11	*hp3478a_config* Function	424
Table 12–1	DB-9 Connector Pin Signals	430
Table 12–2	DB-25 Connector Pin Signals	430
Table 12–3	DB-9 to DB-25 Adapter Connections	432
Table 12–4	DB-9 to DB-9 Null Modem Pin Connections	433
Table 12–5	DB-25 to DB-25 Null Modem Pin Connections	434
Table 12–6	*OpenComConfig* Function	440
Table 12–7	*SetCTSMode* Function	441
Table 12–8	*SetXMode* Function	442
Table 12–9	*SetComTime* Function	442
Table 12–10	*ComWrt* Function	445
Table 12–11	*ComRd* Function	446
Table 12–12	*ComWrtByte* Function	448
Table 12–13	*ComRdByte* Function	449

Tables

Table 12–14	*ComRdTerm* Function	450
Table 12–15	*ComFromFile* Function	452
Table 12–16	*ComToFile* Function	454
Table 12–17	*XModemConfig* Function	456
Table 12–18	*XModemSend* Function	458
Table 12–19	*XModemReceive* Function	458
Table 12–20	*FlushOutQ* Function	460
Table 12–21	*FlushInQ* Function	461
Table 12–22	*GetOutQLen* Function	462
Table 12–23	*GetComStat* Function	463
Table 12–24	*ComBreak* Function	464
Table 12–25	*ComSetEscape* Function	465
Table D–1	Formatcode Specifier	560
Table D–2	Integer Modifiers (`%i, %d, %x, %o, %c`)	561
Table D–3	Floating-Point Modifiers (`%f`)	562
Table D–4	String Modifiers (`%s`)	564

Preface

What Is *LabWindows/CVI*?

LabWindows/CVI™ is an integrated American National Standards Institute (ANSI) C environment (C is a computer programming language) developed by National Instruments Corporation™ and designed primarily for engineers and scientists creating virtual instrumentation applications. *Virtual instrumentation* refers to the combination of hardware and software elements that provides you, the user, with complete flexibility in designing and controlling the elements of stand-alone instruments from your computer system. You have the choice of designing the instrument's functionality. *LabWindows/CVI* helps you leverage the power of the computer to create flexible, reusable, and inexpensive measurement applications that outperform traditional test and measurement methods.

LabWindows/CVI is a programming environment that has been widely adopted throughout industry, academia, and research labs as the standard for data acquisition and instrument control software. It is a powerful and flexible instrumentation software system that embraces the traditional programming methodologies and enables you to create structured code, and it features an easy-to-use programming environment.

LabWindows/CVI includes all the tools necessary for data analysis and presentation of the results on the Graphical User Interface (GUI, pronounced "gooee") on your computer screen. It comes with a complete set of integrated input/output (I/O) libraries, analysis routines, and user interface tools that aid you in rapid application development, reducing the time that would be required by other conventional programming environments.

The integrated environment consists of code generation, prototyping tools, and libraries. There are input/output libraries, user interface libraries, and statistical and mathematical analysis libraries. *LabWindows/CVI* supports a comprehensive Advanced Analysis Library for time/frequency analysis, curve-fitting, digital filters, integration and differentiation functions, statistical functions, linear equations solutions, and many more. This releases you from seeking an outside application package to perform the analysis. Also included are signal processing, instrument driver creation, and ActiveX support, multithreading, DataSocket, OpenGL libraries, to name a few.

LabWindows/CVI's forte lies in building virtual instrumentation systems with General Purpose Interface Bus (GPIB), VME (Versa-Modular Eurocard) extensions for Instrumentation (VXI), PCI (Peripheral Component Interconnect) extensions for instrumentation (PXI), serial interface communication (RS-232), Transmission Control Protocol/Internet Protocol (TCP/IP) based devices with plug-in data acquisition (DAQ) boards without spending too much effort to create the applications. It combines an interactive, easy-to-use development approach with the programming power and flexibility of compiled ANSI C code.

The full potential of *LabWindows/CVI* is used to automate test systems, bench top experiments, DAQ monitoring projects, verification tests and measurements, process monitoring, and controlling systems. This integrated development environment gives full credibility to the National Instruments® slogan: "The Software is the Instrument" by enabling the user to create *virtual instruments* on the personal computer and to communicate with the real instruments via the communication interfaces.

The *LabWindows/CVI* environment supports open software architecture enabling you to reuse existing programs within its environment. If you are programming in C using your preferred environment, *LabWindows/CVI* complements your existing efforts and streamlines your future development. You can incorporate standard ANSI C source code, object files, and dynamic link libraries (DLLs) within *LabWindows/CVI*. You also have the flexibility to use the instrumentation libraries from *LabWindows/CVI* within the C/C++ compilers such as Microsoft Visual C++, Borland C++, Borland C++ Builder, Symantec C, or WATCOM C/C++ with which you may be more familiar.

For simplicity, from this point onwards *CVI* will refer to *LabWindows/CVI* and Windows will refer to Windows 2000, Windows NT version 4.0 Service Pack 3, Windows 98, Windows 95, or Windows 3.1. The differences between *CVI* using different operating systems will be explained in the text as necessary.

Introduction

Why write a book on *CVI*? The answer is simple. There is not a single book on *CVI* in the market at the time of this writing. *CVI* does come with an excellent set of manuals (both bound and on-line), which enables you to get started. Any packaged software in the market always has a couple of books written to give you a different approach from the manuals.

There are about a dozen good books written on LabVIEW®, a graphical programming language, called G, in which you can program in block diagram notation. LabVIEW was invented by National Instruments Corporation©, which is also the creator of *CVI*. Weeklong courses conducted by National Instruments Corporation teach both these programming tools. I have personally taken the National Instruments sponsored courses for both *CVI* and LabVIEW, which give the student very useful hands-on experience. During these courses, the students often asked the instructor for the name of some good books on the subject from which they could learn more. The instructor could always recommend a couple of good LabVIEW books but none for *CVI*.

This is the first book on *CVI*, and its aim is to teach you how to get started quickly and to create *CVI* projects. I am assuming you are unfamiliar with using *CVI*. If you are familiar with *CVI*, this book can serve as an adequate reference to refresh your memory and to glance over some of the examples on a particular topic.

Since *CVI* is based on C, it is recommended that before you try to learn *CVI*, you should know C. The bibliography section contains the names of a few good books on C.

Preface xxxvii

Objectives of this Book

The purpose of *LabWindows/CVI Programming for Beginners* is to serve as a tutorial to help you, a willing *CVI* learner, get started quickly with *CVI* to develop your instrumentation and analysis applications. Almost every aspect of the *CVI* programming environment has been introduced or pointed out in the *CVI* manuals. *CVI* contains many capabilities, and this book does not provide a comprehensive guide. The goal of the book, however, is to give you enough information to provide a foundation on which to build.

This book uses a systematic approach to teaching *CVI*. Every facet of creating the *CVI* projects is explained in detail. The chapters are arranged in an order that facilitates learning *CVI*. Detailed examples are included where necessary. The prototypes of the *CVI* library functions are explained when introduced for the first time in the project. These prototypes may look very similar to ones given in the manuals. They are included in this book for your convenience, sparing you the task of finding them among the alphabetical listings of innumerable functions spread over half a dozen *CVI* manuals or searching in Online Help.

The format of the book is designed with a beginner in mind. Each chapter contains an overview explaining the intent of the chapter. Where applicable, (a) *CVI* project(s) is/are created explaining the chapter's features, showing the necessary steps starting from a blank panel, adding objects to the panel to create the GUI, and incorporating the callback functions to the objects to execute the code. You are given a hands-on experience in compiling, building, and running the projects. You can appreciate the ease with which *CVI* can create and run the applications.

The complete running project with the code and the user interface resource file are listed/shown in the book and included on the CD-ROM distributed with the book. The projects supplied with this book on the CD-ROM have been thoroughly tested and executed numerous times for any possible bugs in the programs.

This book will best serve its purpose if you can create the project as you go over the examples in this book, though you can run the project even if you choose not to build the complete project.

The important items and the shortcuts at the end of each section are highlighted. At the conclusion of each chapter, summaries of the important features covered in that chapter are outlined.

Here is a short description of the chapters:

Chapter 1, "Getting Started," introduces you to the working environment of *CVI*, including some of its prominent capabilities and its uses in the industry and academia, and provides step-by-step basics for getting *CVI* up and running using a real-world application example with an explanation of the versatility of the *CVI* project.

Chapter 2, "Basics of Creating the Graphical User Interface," takes you through a systematic procedure of adding controls to the GUI. The *CVI* event-driven programming and the use of callback functions is explained so the user can control the program's execution. You are shown how to create the code using the *CVI CodeBuilder* for compiling, linking, and running the first *CVI* projects.

Chapter 3, "More Graphical User Interface," gives details of using additional features of the **User Interface Editor** that are not covered in the previous chapter. Additional user interface controls are discussed here. You are shown how to add and use the timer control and strip charts.

Chapter 4, "Enhancing the User Interface," covers the creation of decoration boxes to enhance the GUI to add menu bars, toolbars, and text boxes to increase the projects' functionality. The projects created in this chapter are executed using the menu bar items and the toolbar icons, instead of the command buttons, to give you familiarity with the creation and use of menus and toolbars.

Chapter 5, "Source Editor Debugging Techniques," are discussed here. This chapter provides the fundamentals of the **Source Code Editor** and the use of the *Debugger*. Many *CVI* features are discussed in detail, including details on how to make coding and debugging easier, to manipulate the code in the **Source Code Editor** window, and to effectively learn to use the *Debugger*.

Chapter 6, "File Input/Output," discusses the creation of various input and output files and the opening of files for writing and reading in ASCII (American Standard Code for Information Interchange) or binary format. Almost all the related file I/O library functions are discussed, and a project is created demonstrating the use of these library functions. Different types of *pop-up* panels are explained. Their purpose and creation using the function panels is demonstrated through creating the project(s).

Chapter 7, "List Boxes and Rings," describes the use and the features of the list boxes in detail by means of examples and includes a simple Test Executive project. Examples of different uses of list boxes are shown and the differences between list boxes and text boxes are explained. Examples are also given to show you how to enhance the displayed data on the list box. The creation and use of rings is discussed, including the different types of rings and their various uses.

Chapter 8, "Creating Stand-Alone Executables and Distribution Disks," describes the process of creating the stand-alone *CVI* executable to enable the user to install and run the project on a computer devoid of the *CVI* environment.

Chapter 9, "Creating and Using Dynamic Link Libraries," introduces you to the concepts and the creation of DLLs. The created DLLs are exported and run from the calling *CVI* project.

Chapter 10, "External Compiler Support," introduces you to the *CVI* features of using the external compiler. This is a useful topic for users who are comfortable using the compiler of their choice but who may want to take advantage of the ease of building GUIs and using *callback* functions, instrument drivers, and function panels in the *CVI* environment. Two ways to achieve this functionality are introduced by means of project examples: first, using and creating the *callback* function object file by using the external compiler, and secondly, by creating the DLL in *CVI* and loading and executing the project from the external compiler. Projects are created demonstrating both methods.

Chapter 11, "GPIB Communications," introduces the features and protocol associated with communicating with the GPIB-based instruments. The GPIB library functions are discussed in detail and demonstrated by means of projects.

Chapter 12, "RS-232 Serial Communication," deals with the communication with instruments using the RS-232 interface. Various hardware and software aspects of RS-232 are discussed. The use of the RS-232 library functions are demonstrated and described by means of a project.

Appendix A, "Installing *CVI*," contains the step-by-step procedure for installing *CVI* and explains the folder (directory) structure and purpose of the files.

Appendix B, "**Project** Window Environment," introduces you to *CVI* environment for the **Project** window with an emphasis on explaining the menus of the **Project** window. This appendix gives the user the basic information regarding the **Project** window, enabling you to create, code, compile, build, and run the projects.

Appendix C, "**User Interface Editor** Environment," explains the menus and menu commands enabling you to understand the **User Interface Editor.**

Appendix D, "Formatting Functions," explains the functions to translate or reformat data items into other forms. A variety of examples is included in this chapter to assist you in understanding this topic.

Appendix E, "*CVI* Demo Programs," describes two sophisticated demo programs written in *CVI* that are included on the CD-ROM. The system test application enables the user to configure the test parameters and test functions to run the system tests on specific hardware. Another program consists of a mathematical application that analyzes functions parametrically. These demos are included with limited functionalities for you to envision the power of the more complex features and uses of *CVI*. This appendix and the accompanying demo programs are written by Yaakov Ben-Ami, who is also the co-author of the forthcoming advanced book.

You should note that this text is for beginners and some of the more advanced topics are not covered here but are instead included in the forthcoming advanced book. You are made aware of these exclusions at the appropriate place in the book and are asked to refer to the manual(s) for further reading.

What You Need to Run *LabWindows/CVI*

CVI versions prior to 5.5 use the Windows 95/98/NT/3.1 operating systems on personal computers (PC) and the Solaris operating system on Sun SPARC stations. *CVI* version 5.5 runs on PCs using Windows 95 and 98, Windows NT 4.0 with Service Pack 3, and Windows 2000. *CVI* 5.5 no longer supports the Unix operating system or Windows 3.1.

The discussion and examples in this book are limited to the Windows environment though relative differences for Sun users will be mentioned as appropriate.

Installing and running *CVI* on Windows requires a Pentium 90 or faster processor, a minimum of 50 MB of free hard disk space, and at least 16 MB of RAM. A 800 by 600 resolution (or higher) video adapter is recommended. To install versions prior to *CVI* 5.5 on SPARC stations, you will need 12 MB of free disk space, a minimum of 32 MB disk swap space, and 23 MB of main memory.

To get really effective results when creating and displaying GUIs, a seventeen-inch or larger Super VGA monitor and a high-resolution video card, which can support a resolution of at least 1024 by 768 pixels, are recommended. The GUIs included on the CD-ROM with this book are designed using a 1024 by 768 pixels resolution on the monitor. The GUIs have been tested to work on seventeen-inch to twenty-one-inch monitors, though in some cases they may appear disproportionate. Since you will be following step-by-step instructions for creating the GUIs, you can re-size the panels and objects so they are compatible with your monitor's resolution and size.

As with any Windows programming task, the use of a mouse or trackball is a great convenience, and is a recommended device to use with *CVI*. In this book, only mouse-related instructions are given for accessing objects or selecting menu commands. The keyboard commands will only be used to show the alternate capabilities in controlling and running the projects, as needed.

Preface

It is assumed that you are familiar with the Windows environment on the PC or the Solaris operating environment on Sun workstations.

See Appendix A for the *CVI* installation and setup procedure.

Conventions Used in this Book

The conventions used in this book are similar to the conventions used in National Instruments™ *CVI* manuals. Conventions similar to those of the manuals are purposely adopted for this book because you will have to refer to the manuals to check for various functions or for more information. To avoid confusion, you are not introduced to a new set of conventions.

The following conventions are used in this book.

<>	The names of keyboard keys are enclosed between the angle brackets, e.g., <Ctrl> is the Control key and <F5> is the Function 5 key.
-	A hyphen between two or more keys within angle parentheses means to press these keys simultaneously, e.g., <Ctrl-Alt> means to hold down the Control key and press the Alternate key at the same time.
>>	This means to follow the nested menu commands, or dialog boxes, one after the other, e.g., Library>>UserInterface>>Panels>>DefaultPanel.
bold	Bold text denotes the names of menus, menu commands, arguments, dialog buttons, or user-created items.
bold italics	Bold, italicized text denotes caution, a warning, or any kind of important message.
italics	Italicized text denotes program functions, emphasis, cross-references, or text from which you supply the appropriate word or value.
`monospace`	Monospace text is used for illustrating exactly what should be entered from the keyboard, for any literal examples of programming, for file names and extensions, and for folder (directory) names.
`monospace bold`	Bold monospace text represents text and messages that the computer automatically prints to the screen.

At various places in the text, I have referenced the *CVI* manuals for more information on certain topics. Some information from the manuals may be included in the Online Help and may no longer exist in the form of a paper bound manual(s). It is not possible for me to give you the appropriate reference for all the manuals or Online Help used for different versions

Preface

of *CVI* software. Therefore, textual references to manuals implies either the paper manual(s) listed in the bibliography or the manuals available from Online Help.

In particular, for *CVI* version 5.5, *LabWindows/CVI User Interface Reference Manual* no longer exists as a separate entity, as parts of this manual have been moved to the *LabWindows/CVI User Manual* and other parts to Online Help. When references are made to this manual, the information can be obtained from either of these sources if you are using *CVI* 5.5. However, for versions of *CVI* prior to 5.5, the reference is to the manual listed in the bibliography. For *CVI* 5.5 version the *LabWindows/CVI Standard Libraries Reference Manual* is available from Online Help.

At various places in the text, certain *CVI* features being available for Windows or Unix are mentioned as appropriate. Note that this refers only to *CVI* versions prior to 5.5.

This book explains the concepts and features of *CVI* and is not specific to a certain version of *CVI* release. Some of the menus and submenus shown in various chapters and appendices may not look the same as they do in your current version of *CVI* because the menus in this text are based on *CVI* version 5.5. You should have little problem using the earlier version(s) of *CVI* with comparable menu and submenu commands.

Acknowledgments

Many people contributed to the creation of this book. Writing a book is an experience one never forgets. On the surface it may appear that the author is the one who creates the book, but in actuality there are numerous people behind the scenes whose contributions, suggestions, edits, comments, and reviews make all this possible. I would like to acknowledge the people who helped me in this effort.

I am fortunate to have had the company and friendship of Yaakov BenAmi, my mentor in *CVI,* from the days we worked together on projects at Boeing. Yaakov taught me many of the intricacies of this language. I am much indebted to him for reviewing the book so thoroughly, painfully scrutinizing each and every word and giving me his invaluable advice on material arrangements and technical, linguistic, and programming style. Yaakov has also written Appendix E, "CVI Demo Programs" and the two demo programs included on the CD-ROM.

Jana Lasseter deserves my special appreciation for reviewing the book thoroughly for errors, omissions, arrangement, and textual style. She went over the *CVI* projects in the text, painfully recreating all of them and checking for accuracy and functionality. Jana's priceless suggestions took me a long way in improving the quality of this text.

I appreciate Patricia Correa's never-ending help in editing the book for numerous grammar, syntax, and structure errors. Without her help, I would still be grasping for better words to make the text more understandable. I appreciate Edward Nielsen's help in reviewing parts of the book as well.

I thank my daughter Farah, whose creative and artistic ideas were used to the utmost in the design of the front cover of the book. Also thanks to my daughter Lubna, who spent time in editing parts of the book.

I appreciate my son Omar's help in building a computer system of exorbitant speed and memory that enabled me to write this book more productively. I am also thankful to him for retrieving many chapters of this text that were lost due to file corruption errors.

I would like to thank my family for their understanding of the numerous hours required in writing this book. In particular, I thank my wife, Uzma, who could not understand why I had to set the alarm so I could wake up early on weekends and work on this book. After seeing the results, I hope she understands.

Ravi Marawar, Academic Program Manager at National Instruments, deserves special mention and my extreme gratitude. He was instrumental in creating a group of *LabWindows/CVI* experts at National Instruments to review the book and guide me when the book was still in its infancy. This group reviewed the book and provided technical support and invaluable suggestions for making this book a reality.

Chuck Boecking's help in reviewing this book also needs a very special mention. Chuck, as the *LabWindows/CVI* Product Marketing Manager at National Instruments, did a thorough review of the book. Chuck made very useful suggestions for restructuring the book to aid comprehension by approaching the text from a beginner's perspective and improving the overall quality of this book. My thanks and appreciation go out to him.

I specially acknowledge the efforts of Nigel D'Souza, Senior Group Manager at National Instruments for his invaluable technical review of the text and for supplying me with much-needed information on a timely basis to complete this book. His effort on writing an impressive Foreword is deeply appreciated.

The help of numerous technical support engineers at National Instruments who supplied me with useful information at various stages of writing this book is very much appreciated.

I appreciate Joan McNamara's effort at Prentice Hall PTR Production Department for producing the book on time despite many obstacles.

Finally, thanks to the staff at Prentice Hall, led by Bernard Goodwin and supported by Diane Spina and Lisa Konzelmann, for their help and support from the early days of the inception of this book to its publication.

Foreword

The LabWindows/CVI product line has a long history of innovative features that have allowed thousands of engineers to easily develop sophisticated virtual instrumentation applications in a standard programming language. The LabWindows/CVI product started in 1988 as LabWindows for DOS. From these early beginnings to the present, the LabWindows development group has tried to deliver a highly productive environment in which engineers can build virtual instrumentation solutions on standard PC hardware and operating systems. To achieve this we incorporated features such as interactive execution to allow engineers to work with instruments without worrying about the structure of a complete program. These features have kept us challenged over the years as we have provided ever-increasing power in the LabWindows/CVI environment and libraries. In LabWindows/CVI 5.5, we have even allowed users to execute and debug multithreaded code from the interactive window. And even after all these years, we know of no other product that allows you to interactively execute ANSI C code.

It is exciting to see all of the LabWindows/CVI users build their applications faster and better than they could with traditional development environments. LabWindows/CVI users are always providing us with feedback that helps us better understand their needs. By seriously considering the input from our customers, we continuously add new built-in features to further increase the productivity of LabWindows/CVI users.

Over the years, we have seen many operating systems from DOS to Windows 2000. Along with these operating systems came new technologies that have made the computer more powerful and easier to use. In this constantly changing world of software, we are excited to make the latest technologies more accessible to our customers. It is also fun to help our customers take advantage of technologies such as TCP/IP, ActiveX, Web, and multithreading by providing libraries that make it easy to use the full power of these technologies.

Even as we deliver the latest technologies, we kept the software investments of LabWindows users in mind. We have worked hard to provide an easy upgrade path for our users to move to newer operating systems and technologies. It gives us a lot of satisfaction to hear about the many users who have successfully moved their application from LabWindows on DOS to LabWindows/CVI on Windows NT.

We have endeavored to make LabWindows/CVI easy to use and at the same time provide power through flexibility in the environment and libraries. This book fits in well with these goals. It adds to LabWindows/ CVI's ease of use of by going beyond the tutorial included with LabWindows/CVI. Shahid shows you how to make the best use of the many features and functions in LabWindows/CVI through concrete examples. If you are new to LabWindows/CVI, you will find that this book covers all of the topics necessary to help you quickly become productive with LabWindows/CVI.

We hope that this book will be the beginning of your long relationship with LabWindows/CVI, which will continue to evolve with new technologies and operating systems.

Nigel D'Souza
LabWindows/CVI Group Manager
National Instruments

GETTING STARTED

Chapter Highlights
- What is *CVI*?
- Where is *CVI* Used?
- Starting *CVI*
- The First *CVI* Project
- Summary

This chapter gives you an overview of *CVI*, including some of its prominent features and the operating environment. The main emphasis of this chapter is to show you some of the real-world applications presently being used in industry and academia and to explain the purpose and usefulness of *CVI*. You are shown how to access and run the first *CVI* project (application) to enable you to get familiar with the look and feel of *CVI*.

What Is *CVI*?

Before you get started, you would probably like to know what exactly *CVI* is and why you need to use it. *CVI (C for Virtual Instruments)* is a programming environment in which you create applications via a user-friendly graphical user interface (GUI) editor to interactively design *virtual instruments*. You do not use complicated graphics calls to build your user interface in *CVI*. You control your application using the full functionality of ANSI C and library functions.

Using the graphical user interface editor, you can select from a wide variety of controls designed specifically for instrumentation: knobs, meters, light emitting diodes (LEDs), gauges, dials, graphs, strip charts, and many more. These controls can be refined to customize their appearance and operation for your particular application.

In addition to the instrumentation controls, *CVI* lets you add menus, toolbars, text boxes, list boxes, and various kinds of ring controls. You can add sliders, command buttons, binary switches, check boxes, scroll bars, and decoration boxes to enhance the appearance of your GUI.

The power of *CVI* does not stop here. There are a myriad of built-in library functions ranging from standard library functions to special purpose libraries, such as data acquisition, General Purpose Interface Bus (GPIB), serial instrument control, Transmission Control Protocol/Internet Protocol (TCP/IP) communication, data analysis, data presentation, and data storage.

CVI was designed to minimize the development effort required to create engineering and scientific instrumentation applications and includes many timesaving features, which are shown below:

- **Function Panel Concept**—Function panels enable you to select the correct number, order, type, and attributes of arguments in the library function templates. The function panel lets you select from among the many attributes available for the function arguments to speed your application development. The library functions can be inserted into the source code via the function panels by just pressing a key.

- **Automatic Skeleton Code Generation**—This feature is a time saver as it creates the "skeleton code" in C from your GUI. "Skeleton code" is a program boiler-plate created from the GUI with the structure of the *main* and callback functions using the proper syntax.

- **Tight Integration Between Compiler and Linker**—*CVI* automatically inserts header files required for its internal libraries.

- **Run-Time Error Checking**—*CVI* prevents you from overwriting memory locations and catches errors that other environments may not.
- **Interactive Execution of Library Functions**—Function panels and code fragments can be executed interactively before including these functions in the source code.
- **Compatibility and Portability**—*CVI* supports ANSI C and is available on several platforms. All *CVI* applications are upward compatible to newer releases of this tool. Code developed on Windows (Win32) environments can be accessed from other Win32 development environments.
- **Open System Architecture**—*CVI* supports open software architecture, enabling you to reuse your existing programs within the *CVI* environment. You can incorporate standard ANSI C source code, object files, and dynamic link libraries (DLLs) with *CVI*. You can use the full power of *CVI* instrumentation libraries and GUI creation tools in a general purpose compiler of your choice from Microsoft, Borland, WATCOM or Symantec.
- **Add-On Enhancements**—Instead of building your own applications from scratch, *CVI* has a number of add-on packages to perform various specialized functions that suit your needs. These packages include *Test Executive*, *TestStand*, *SQL Toolkit*, *Statistical Process Control*, *Image Processing*, *PID* (proportional-integral-derivative) *Control*, and many more.

Using the strength of the *CVI* environment, you can markably increase your productivity. Programs that took weeks or months to write using the conventional programming languages can be completed in hours using *CVI* because it is specifically designed to take measurements, analyze data, and display the results in a user-preferred format.

Where Is *CVI* Used?

The convenience of design and ease of use has made *CVI* the product of choice in many industries, such as the automobile, aerospace, defense, energy, telecommunications, insurance, chemical, and test equipment industries, as well as many more.

The automobile industry uses *CVI* to test the Cadillac DeVille and DeVille Concours models for premium sound quality. The system used is called the

Automotive Audio System Tester (AAST), which conducts all the testing, data analysis, and storage of data for later use. The AAST includes microphones, a signal generator, a PC with *CVI* instrumentation software, an AT-DSP2200 board, and an AT-GPIB instrument controller board. This system can test operation and sound quality of eleven power amplifiers and eleven speakers.

The aerospace industry is using *CVI* to detect intermittent faults of 757 aircraft flap/slat controls during thermal and vibration cycling. Commercial Avionics Systems—Irving Operations (CAS-IO), a division of Boeing Defense and Space Group, has designed an automatic test system called the Boeing Mature Programs Environmental Stress Screening (ESS) Monitor. The ESS Monitor is a PC-based automatic test system using VXI and GPIB instrumentation, including the National Instruments VXIpc-486™ Model 550 embedded controller and *CVI* development software. The test runs constantly initiating the internal Built In Test Equipment (BITE) of a 757 Flap/Slat electronics module while subjecting it to thermal and vibration cycling. All the necessary input signals are provided, the BITE is activated, and the test results read. The module data is read back in a standard eight-bit (ASCII) format over the RS-232 serial bus. There are plans to develop the tests for 737, 747, 757 and 767 modules.

CVI is used in many facets of the defense industry and national security organizations. In particular, it is used successfully in the testing of scaled models of aircraft, missile, and space systems. These models are tested in conditions from vacuum pressure to many fold pressures and from low speeds to speeds up to Mach 6.

The U.S. Department of Energy uses *CVI* to investigate and map the conditions of the outer walls of the storage tanks containing radioactive waste materials using DAQ boards to control a remote robotics systems. These containers, which are 30 ft. tall and 80 ft. in diameter, are buried underground in a four foot layer of concrete. The only access to the outside of these underground tanks is through a very narrow space where the pipes feed the tanks. *CVI* was able to easily integrate the computer systems and create a GUI with operator interface menus that duplicate the indicators and controls of the instrument. Other control systems were used previously, but *CVI* proved to be superior for performing motion control.

In the world of developers of test systems, True Diagnostic Systems (TDS) uses *CVI* to test high-density matrix switching systems. These test systems use a mixture of analog and digital systems as well as GPIB and VXI instruments. Using *CVI* as a programming language for TDS test systems has made a major improvement in its product line.

At the Radar and Counter Measures (RCM) division of Thomson-CSF in France, *CVI* is being used for testing and controlling the performance characterization of microwave transmitter/receiver modules, which are part of active electronic-scanning antennas used for airborne radar. The software controls the GPIB instruments and VXI-bus instruments and is used to develop increasingly complex units under test designs. The decision to use *CVI* was driven by the quality of GPIB and VXI instrument drivers available in the *CVI* Instrument Driver Library, its wide range of GUI Library functions, its mathematical and analysis functions, and its basis in industry-standard programming languages. C programming language remains the choice of test instrument manufacturers because it easily controls the hardware in use.

In the telecommunications industry, Alcatel used *CVI* to find the best possible position to build a new radio network antenna for clear signals in that area. *CVI* was used to solve this problem. With the help of the Global Position System (GPS), a spectrum analyzer, and a laptop computer running *CVI*, the Alcatel engineers drove around the new site and measured the signal strength from a proposed transmitter location.

CVI is used in testing deep-sea diving suits and high-energy heart defibrillators and in performing battle-tank audio analysis. It is also used to reduce the downtime on fiber spinning machines, to test rail-vehicle propulsion logic printed circuit boards, and to teach pharmacology to students in universities as well as for many additional real-world uses.

CVI is not limited to use with instruments only. It can also be used as a stand-alone tool to perform database manipulations, mathematical modeling, advanced analysis, algorithm testing, or whatever you need to achieve programmatically using exotic GUIs.

I hope that by now your interest in learning *CVI* has been aroused. So let us charge ahead to run your first *CVI* project!

Starting *CVI*

The easiest way to run *CVI* is to double-click on the desktop icon that you created after you installed the software as explained in Appendix A. If you prefer not to create a desktop icon, you can run *CVI* from the **Programs** option by selecting the **Start** button in Windows. See Figure 1–1 for an example of what you might see.

To run the program, select **CVI IDE.** You will see a copyright screen telling you the version of the *CVI* software and to whom it is licensed. After

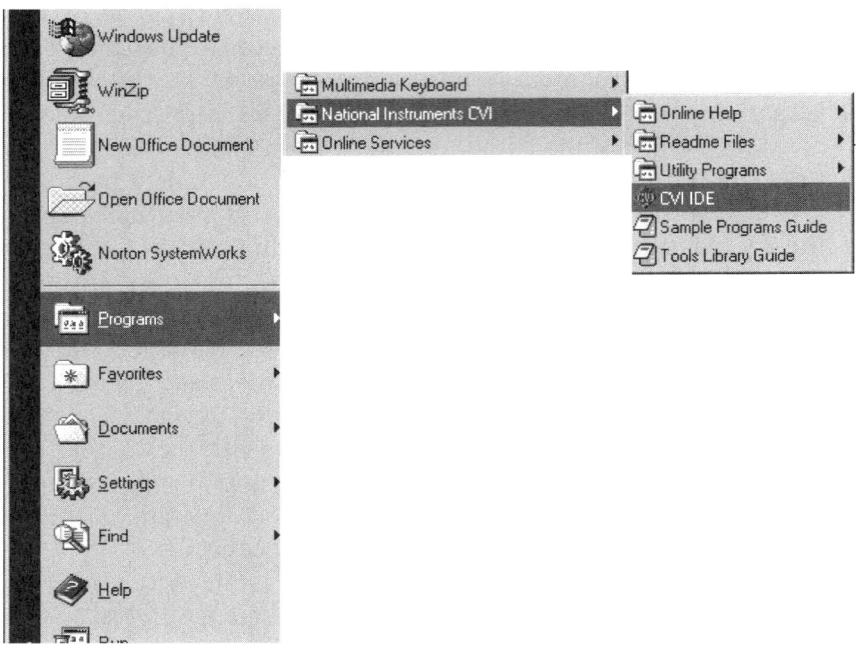

Figure 1-1
Typical **Start** Button List under Windows

that, you see a screen like the one in Figure 1–2. This is called the **Project** window.

A **Project** window is used to open a new or previously created project and to compile, build, and run the project. It also saves the files, sets the various options, creates stand-alone executables and DLLs, and performs debugging. From the **Project** window, you can setup the *CVI* environment and numerous other features that you will see as you build the projects in later chapters.

Projects are *CVI* applications that you build and execute. A project consists of various types of files that will be discussed as we go along. All the projects are built by starting with the *CVI* **Project** window.

If you have built and saved any projects previously, the *CVI* **Project** window will open with the last project files displayed and the project title on the **Title Bar**. If you are starting a new project, you will see a blank **Project** window template with no files listed, as shown in Figure 1–2, with `Untitled1.prj` shown on the title bar. `Untitled1.prj` is the default

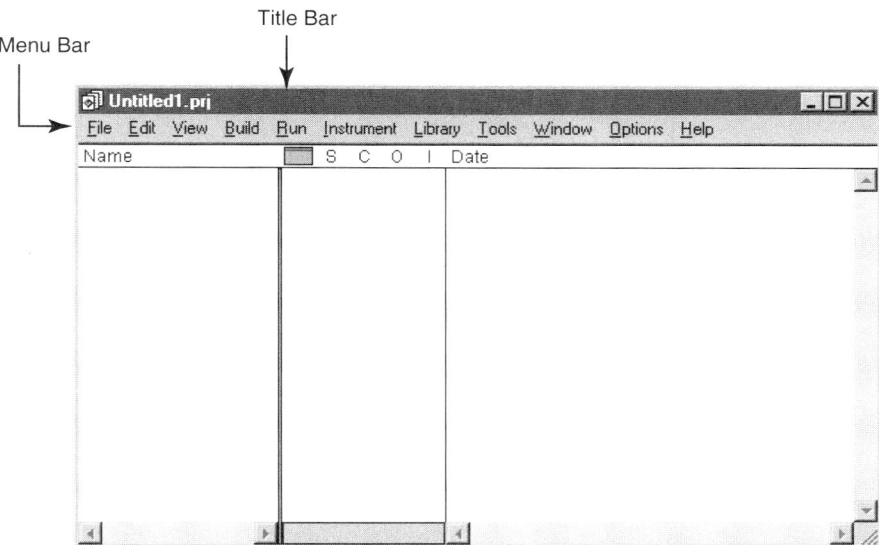

Figure 1–2
New **Project** Window

project name when a new project has not been saved with a user-supplied name.

This new **Project** window lets you start building the *CVI* applications.

The First *CVI* Project

To run your first project, select **File>>Open>>Project (*.prj)** from the **Project** window. In the folder on the hard disk where you installed the project files from the CD-ROM, find and double click on project 1-1.prj. Depending on your folder settings option, you may not see the file extension; instead you may see a description of the file. The files associated with this project are listed in Figure 1–3 in the **Project** window list.

Notice that the pathname of the project is displayed in the **Title Bar** of the **Project** window. Throughout the text, your pathnames may be different and will indicate where you installed the project files.

Select **Run>>Execute project1–1.exe,** and the application will start to run. The GUI for this project, as shown in Figure 1–4, will be displayed.

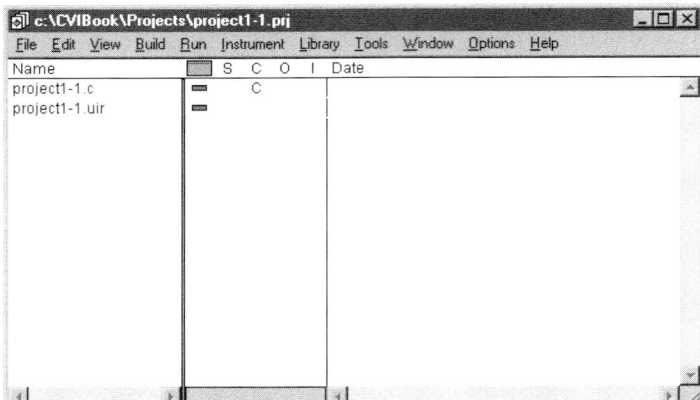

Figure 1–3
Project Window for `project1-1`

Before you execute this project, let us discuss the project's functionalities.

This project is used for monitoring the data from a rocket engine controller. The data monitored consists of voltages from two power supplies (five volts DC (direct current) and fifteen volts DC), open position for four engine valves, and pressures for gas generation and combustion chamber. Other data can be monitored for a rocket engine, but for this demonstration, the above data monitoring will suffice. Data can be acquired through a variety of communication interfaces: a data acquisition card, GPIB, RS-232, or other available instrumentation. For this demonstration, the simulated data is read from an ASCII file and run to demonstrate some of the capabilities of *CVI*.

The top sub-panel labeled **POWER SUPPLY** in Figure 1–4 records the data on a strip-chart for either the five-volt or the fifteen-volt power supply. Select which power supply to monitor by checking either the **5 Vdc** or the **15 Vdc** supply box. You can change the lower and upper limits of the voltages by clicking on the increment/decrement arrows to the left side of these control boxes. Move the mouse over the **PS Timer** knob (PS is short for Power Supply). Hold down the left mouse button over the indicator line on the knob and move it to set the time interval at which you want the data displayed. The same can be accomplished by clicking on the increment/decrement arrows on the left side of the box that is located below the timer knob.

To start monitoring, click the red **ON/OFF** command button on this sub-panel. The command button will be pushed in and its color will change to

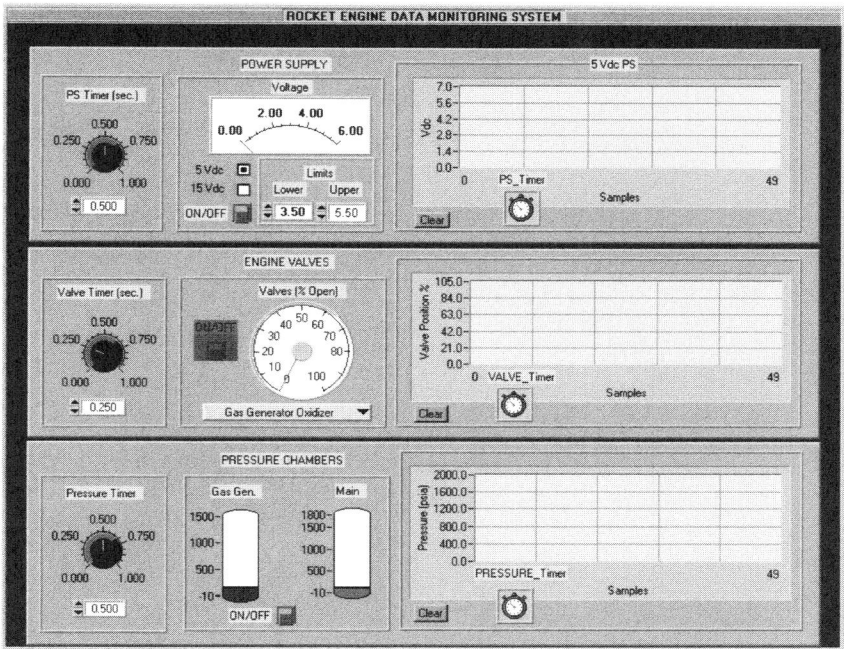

Figure 1–4
Project1–1 GUI

green. The data will be plotted on the strip-chart at the time interval that you selected on the **PS Timer** control. A snapshot of the run is shown in Figure 1–5. The values for the lower and upper limits are also plotted on the same strip-chart. Click the **ON/OFF** command button again. It will pop up to **OFF** position, its color will become gray, and the data plotting will be stopped. Now change the time interval on the **PS Timer** to 0 to make it run at the fastest speed allowed by your system (1 millisecond for Windows). Click the **ON/OFF** command button again and notice the increase in speed with which the data is displayed on the meter and plotted on the strip-chart.

Notice that when the 15 Vdc supply is selected, the meter's display maximum limit changes from 6 to 17. The maximum scale on the Y-axis of the voltage strip-chart also changes accordingly. The lower and upper limits default values are also changed to 8.0 and 14.25 for the 15 Vdc power supply. The label on the strip-chart changes to **15 Vdc PS** to indicate the power supply voltages being plotted. These labels and values are changed programmatically.

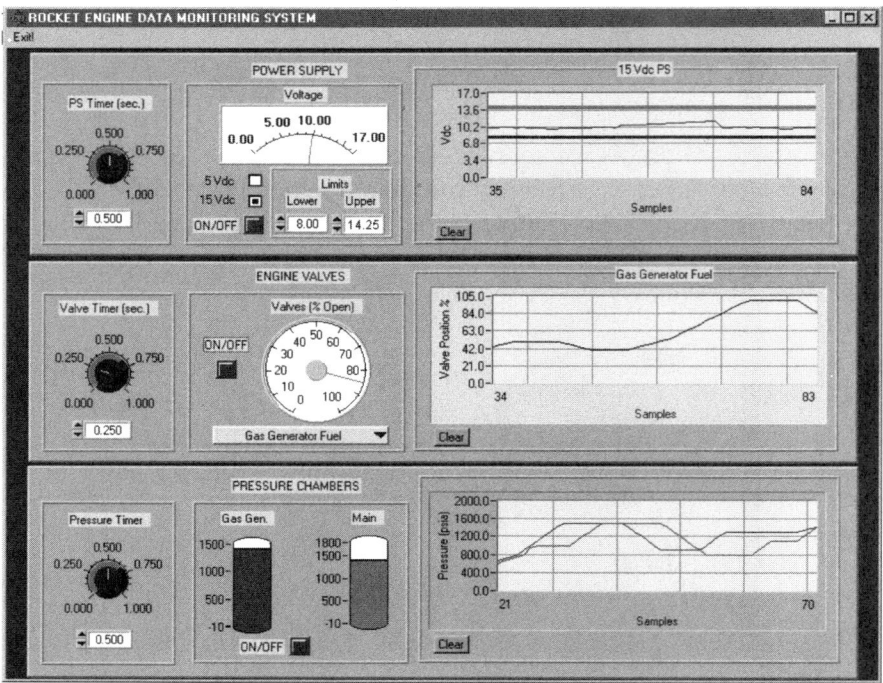

Figure 1–5
Project1–1 Run

To clear the strip-chart, click the **Clear** button below the strip-chart.

The middle sub-panel is labeled **ENGINE VALVES**. From this sub-panel, you can monitor the engine valve data and display it on the strip chart. Click on the pull-down menu shown below the valve gauge, and you will have an option to select one of the four engine valves. The selected valve name is placed in the strip chart title. Set the time interval on the **Valve Timer** knob and click the **ON/OFF** button to run the selected engine valve data. Notice that this **ON/OFF** command button is disabled until the engine valve is selected from the pull-down menu. The valve gauge will also move as each data value indicating the percentage open position of the valve is generated. The strip chart is plotted as before. As with the **POWER SUPPLY,** you can stop the monitoring and change the timer interval to make the monitoring run slower or faster. You can select a different valve from the pull-down menu and display and observe its simulated data.

The lower sub-panel is labeled **PRESSURE CHAMBERS.** In this sub-panel, the two pressure chambers are monitored and their data displayed simultaneously on the strip chart by clicking on the **ON/OFF** command button. The data is also "written" to the pressure tank controls to graphically depict the changes for each pressure value. The **Pressure Timer** knob can be varied appropriately to vary the display speed of the data.

To exit the project, click on the **Exit!** command on the menu, which is located on the top left corner of the menu bar.

Summary

This chapter introduced you to *CVI* and its many uses in real-world applications. It got you started with *CVI* and showed you how to run your first *CVI* project. A data acquisition application was run and its many features explained. The purpose of the first project was to show you some of the capabilities of *CVI* and enable you to get familiar with the way this tool can be used for your own needs.

Basics of Creating the Graphical User Interface

Chapter Highlights
- User Interface Overview
- *CVI* Programming Model
- Creating the First *CVI* Project
- Analyzing the *CodeBuilder*-Generated Code
- Overview of How to Develop a *CVI* Application
- Creating the Second *CVI* Project
- Callback Function and Generated Events
- Summary

This chapter will teach you how to develop projects (applications) in *CVI*. You will proceed systematically through the process of building a project, so you can understand the power and ease with which you can add objects to your GUI and control program flow by means of callback functions. The *CVI* event-driven programming methodology is explained, and the use of callback functions is introduced. In this chapter, you will learn how to create GUIs and the skeleton code from the GUIs using the *CVI CodeBuilder* program. You will learn to compile, build, and execute the applications.

User Interface Overview

The user interface is an integral part of *CVI*. It originally appeared for Disk Operating System (DOS) for the personal computer in January 1989. It offered one of the first ways to graphically represent measurement data on the Microsoft platform. For this reason, *LabWindows* was superior to almost all common development environments. When Windows 3.0 came about in March 1994, *LabWindows* morphed into what is known today as *LabWindows/CVI*. Again, the user interface continued to be the focal point for creating interactive measurement applications. Even today, with Microsoft development tools, *CVI* continues to excel because it combines rapid user interface development with built-in measurement display capabilities not available in any other development packages.

CVI consists of the tools needed to make a sophisticated, professional-looking, user-friendly application. The strength of the GUI lies in its easy to use drag-and-drop features. These features will help you to create a versatile data measurement system, analysis, and display package that simplifies, customizes, and enhances your measurement application. *CVI* has a very large set of user interface tools that will enable you to present your data in various eye-catching formats.

The **User Interface Editor** is part of the *CVI* environment that enables you to create effective GUIs. The versatility of your GUI is further enhanced since you can customize it to select only the features of the measurement instrument(s) your application needs. You can choose the communication interface, configure the device, and send data to and receive data from the device to perform the functions applicable to your requirements.

The flexibility of the **User Interface Editor** enables you to select the user interface tools from an available large set. You have a choice, for example, of selecting instruments from a pull-down menu or from the communication interface from the ring control. You can use switches, command buttons, and numeric controls that allow you to send and receive data and perform limit checking on the data. You have the choice of adding menus and toolbars to enhance the capabilities of your user interface.

After acquiring your instrumentation data, you can analyze the data using many of the Advanced Analysis Library functions, and you can save it to a file for later analysis and/or display it in a tabular format with data elements in different colors. If you desire, you can make great presentations of the acquired data by plotting it as a graph, creating a histogram, or displaying your output on a strip chart as you receive the data.

What needs to be emphasized here is that *CVI* is a measurement tool that can communicate easily with your instrumentation via a user-defined GUI through which you can command, control, and communicate with the actual instrument(s).

CVI Programming Model

Before you start to create a user interface, it is important to understand the event-driven programming model used by *CVI*. Sequential programs run in sequence from top to bottom, mostly irrespective of the events generated by the "outside world" (except for some user inputs or real-time interrupts). After the program has run to the end, the program stops. Certain user operations on the GUI generate events. These operations could be selecting a control, entering a value in a control box, selecting a menu item, pressing a key, selecting the panel (the area on which the controls are placed), or moving or re-sizing the panel. A left click of the mouse will generate a different event than the right click, which is different from a double-click.

CVI generates event-driven code. Using event-driven code allows you to execute only the code linked to the control by means of a program function called the callback function, a user-defined function that performs certain task(s) when the event is generated. *Only the code in the callback function is executed in response to the events generated.* The program returns to what it was doing before the event trigger and waits for another event to occur to invoke its callback function. Even-driven programming is what makes *CVI* programs fundamentally different from sequential programs.

There are two *CVI* library functions that process the events: *RunUserInterface* and *QuitUserInterface*. *RunUserInterface* is a *CVI* library function that processes and controls the events generated when you select an active control. This function is usually the last statement in the main function. The *RunUserInterface* function continuously monitors the user interface for events and makes sure that these events are passed to the proper callbacks through the operating system, keeping track of the control object that induced this event.

QuitUserInterface is a library function that works in conjunction with the *RunUserInterface* library function to stop the processing of events. The *RunUserInterface* function returns the value passed to the *QuitUserInterface* function as shown in the code fragment shown in Figure 2–1.

It is important to understand that the *QuitUserInterface* library function does not terminate your application; it disables the *RunUserInterface* function. If there are other statements in the main function after *RunUserInterface*, then those statements are executed. For example, if you have a *printf* state-

```
int returnvalue = RunUserInterface (void);
int status = QuitUserInterface (int returnvalue);
```

Figure 2–1
RunUserInterface/QuitUserInterface Interaction

ment after the *RunUserInterface* function, it will be executed. However, the event processing is stopped by the *QuitUserInterface* function.

The *RunUserInterface* function calls the *ProcessSystemEvents* function that actually monitors and updates the user interface. Normally, the *ProcessSystemEvents* is not explicitly called in the code. If a certain section of the code is taking a long time to process and you want another event to be handled immediately, you should make an explicit call to this function. It is recommended that you use this function sparingly as it may interfere with normal event processing of the *RunUserInterface* function.

Let us understand the event generation by looking at some of the command buttons for `project1-1.prj` that was shown in Chapter 1, "Getting Started." Load `project1-1.prj` as before by double clicking on `project1-1.prj` from the `Projects` folder. The list of the files included in this project is displayed in the **Projects** window. Double click on `project1-1.uir` file. The GUI shown in Figure 2–2 is displayed in the **User Interface Editor.**

Observe the toolbar shown in the top left corner below the menu in Figure 2–2. The toolbar is shown in detail in Figure 2–3 and consists of four toolbar icons.

These toolbar icons will be explained in Chapter 4, "Enhancing the User Interface." For now, click on the toolbar icon at the extreme left, the **Operating Tool.** The icon will become highlighted when selected. Using this tool, you can click on active controls to observe the sequence of events displayed in the top right corner of the GUI.

Use the **Operating Tool** to click the **OFF** button on one of the sub-panels. Notice that when you click on an object that has an associated callback function, the events that are associated with that control are generated. You can observe the events being generated by looking at the top right corner (below the menu bar) of the GUI. The event change is shown in Figure 2–4 when the

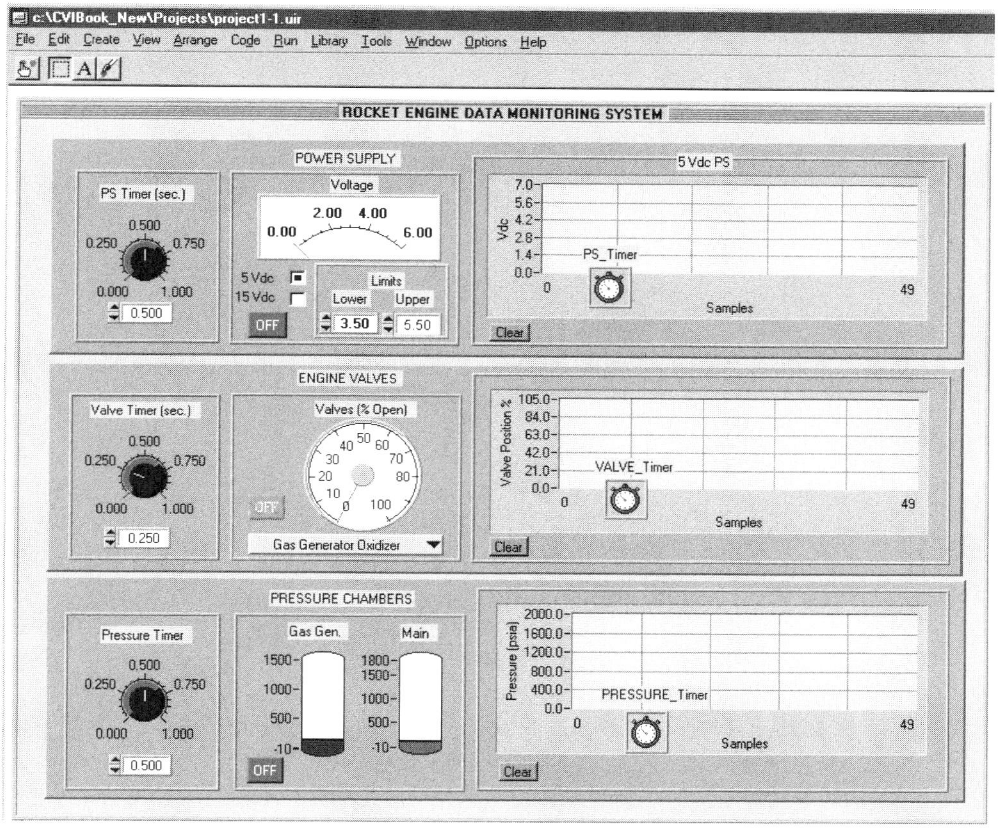

Chapter 2 • Basics of Creating the Graphic User Interface

Figure 2–2
Project1–1 GUI

OFF command button in the **POWER SUPPLY** panel is selected by left clicking the mouse.

A callback function is called three times. First, the callback function is called to process the EVENT_GOT_FOCUS event. This event is generated when the control is made the active control. (To make the control active, you

Figure 2–3
User Interface Toolbar

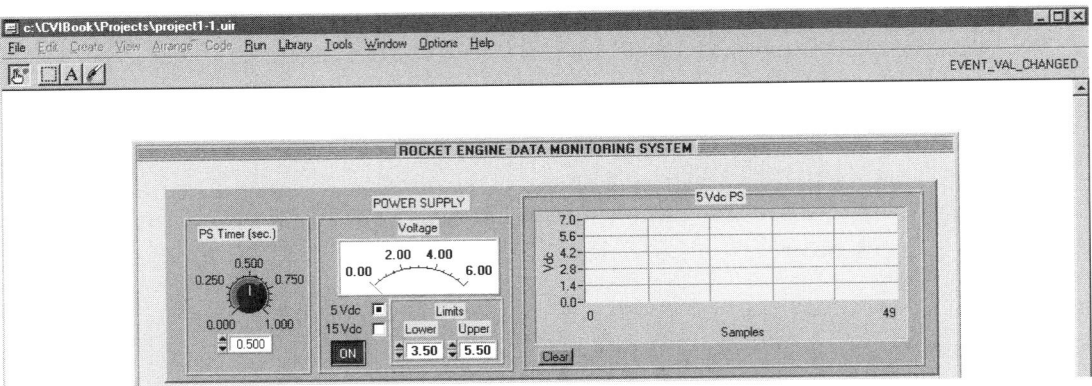

Figure 2–4
Events Generation

can perform a mouse click on the control or use the <Tab> key on the keyboard to move to the control).

The second time, the *callback* function is called to process the EVENT_LEFT_CLICK if the left mouse button is clicked.

The third time, the callback function is called to process EVENT_COMMIT event. This is generated when you click on a control with the left mouse button and release it while still on top of the control. This event is also generated when you select a control and press <Enter>. This is valid only if the control mode is set to **Hot** or **Validate.** The control modes will be explained below in the "Creating the Second *CVI* Project" section.

EVENT_LOST_FOCUS event is generated when the control that was previously active is no longer active. This happens when another control is made active.

In this GUI, the timer knobs have no callback function associated with them, and when you left click on these knobs you get only EVENT_LEFT_CLICK, EVENT_LOST_FOCUS, and EVENT_GOT_FOCUS events. If you change the value of the knob using the mouse, the event EVENT_VAL_CHANGED is generated, and the EVENT_COMMIT is generated when you release the mouse button.

Creating the First *CVI* Project

Let us create the first *CVI* project. Click on the *CVI* icon on the desktop. If you did not create the *CVI* icon on the desktop as explained in Appendix A, select **Start>>Programs>>National Instruments CVI** and then click on the

CVI IDE icon as shown in Figure 1–1. The **Project** window is displayed. Note that all the projects shown in this book are compatible with the Windows operating system and run on a personal computer. Unix users can run the same code with a few modifications; the differences between the two operating systems are explained where relevant in the book.

To create a new project, select **New** from the **File** menu and **Project** (*.prj) from the submenu. A dialog box will ask you to verify if you want to unload the current project (if any). Select **Yes**. If you are using *CVI* for the first time and did not create a project, this message will not appear.

Figure 2–5 is displayed showing a typical configuration of the *CVI* environment. Select **OK** for the default settings.

As mentioned before, *CVI* consists of three environments from which you can create or modify the project. To reiterate, these environments are the **Project** window, the **User Interface Editor,** and the **Source Code Editor.**

The name of the current, open *CVI* environment is conveniently located in the title bar of the window. The **Project** window is indicated by the .prj extension appended to the file name, the **User Interface Editor** window by the .uir (user interface resource) extension, and the **Source Code Editor** window by the .c extension.

Note that throughout this text the terms user interface resource and GUI have the same meaning and will be used interchangeably.

To create a GUI, you need to open the **User Interface Editor.** From the **Project** window, click on the **File>>New** menu and select **User Interface**

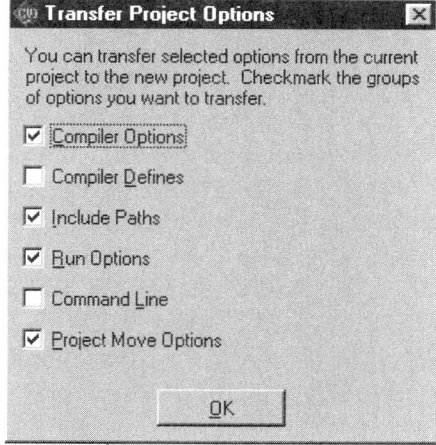

Figure 2–5
Configuration Menu

(***.uir**). This opens a blank template in the **User Interface Editor** with a default name as `Untitled1.uir` in the template header.

Select **Create** in the **User Interface Editor**. The menu in Figure 2–6 is displayed.

Select **Panel** from the submenu to create a panel. An **Untitled Panel** appears as shown in Figure 2–7.

A panel is where you place the various control objects to create the GUI. It is a workspace onto which you can add, move, and re-size objects, additional panels, or menu or tool bars. Panels that contain other panels are called parent panels and the panels within them are called child panels. Panels within child panels are called grandchild panels. An example of the parent/child/grandchild panel relationship is shown in Figure 2–8.

You cannot drag a child panel outside its parent panel. If you shrink a panel, a child panel might be partially or completely hidden in the shrunken panel view.

Now, double-click on the panel you created in Figure 2–7 to display the properties (attributes) box for the panel. A dialog box, similar to that shown in Figure 2–9, will be displayed. Every control on the *CVI* has a properties (attributes) box associated with it. Via these attribute boxes, you can set the properties for that control. Notice that the **Constant Name** is entered as `PANEL` by default. Change it to `FIRST_PNL` for this project. You have a choice

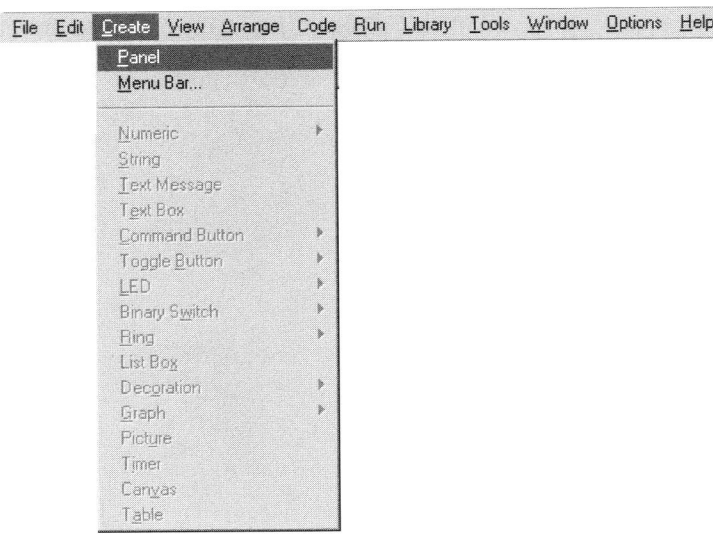

Figure 2–6
Create Panel Menu

Figure 2–7
Blank Panel

of entering any **Constant Name** you wish, but with no spaces between the characters and no more than 10 characters. In the **Panel Title** box enter FIRST PROJECT. This will be the title displayed on the panel when you close the properties box by selecting **OK**.

To place a command button on the panel, select **Create>>Command Button>>Square Command Button** as shown in Figure 2–10.

A command button with the default name OK will be created on the panel. To change the command button name and to assign a callback function to it, double-click on the command button. A command button attributes box, shown in Figure 2–11 is displayed.

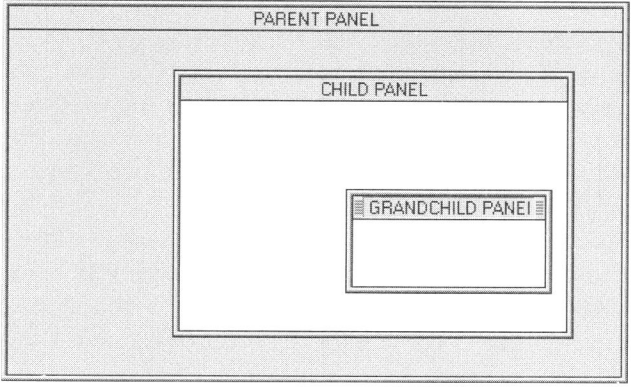

Figure 2–8
Parent/Child/Grandchild Panel

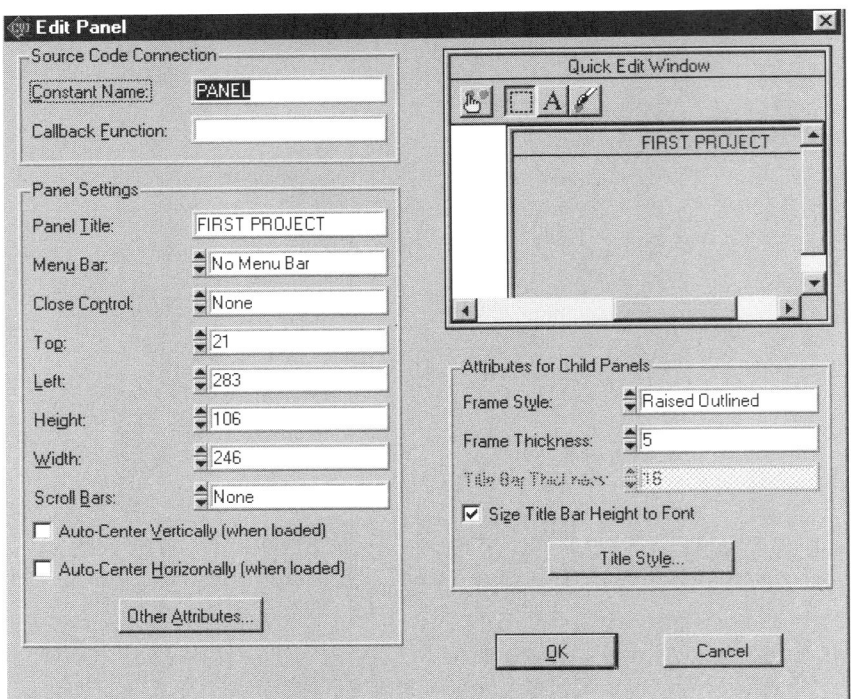

Figure 2–9
Panel Attributes Dialog Box

The command button properties box is displayed with the default **Constant Name** of COMMANDBUTTON and the label name as OK. Overwrite the **Constant Name** with RUN, and enter RunCB in the **Callback Function** dialog box to give it a more meaningful name. The **Constant Name** for the control should contain no spaces and is limited to a maximum length of 20 characters. According to *CVI* convention, the **Constant Name** is in all uppercase letters.

For the command button label, replace OK with RUN in the **Label** box shown in the **Label Appearance** group in the lower right-hand side of the properties box as shown in Figure 2–12. You can chose to type the label using uppercase or lowercase letters. Select OK to save the settings and return to the panel.

When you want to refer to this command button, you enter its name in the **Constant Name** box (prefixed by the panel **Constant Name** FIRST_PNL). Thus the complete name for this command button is FIRST_PLN_RUN. When

Chapter 2 • Basics of Creating the Graphic User Interface

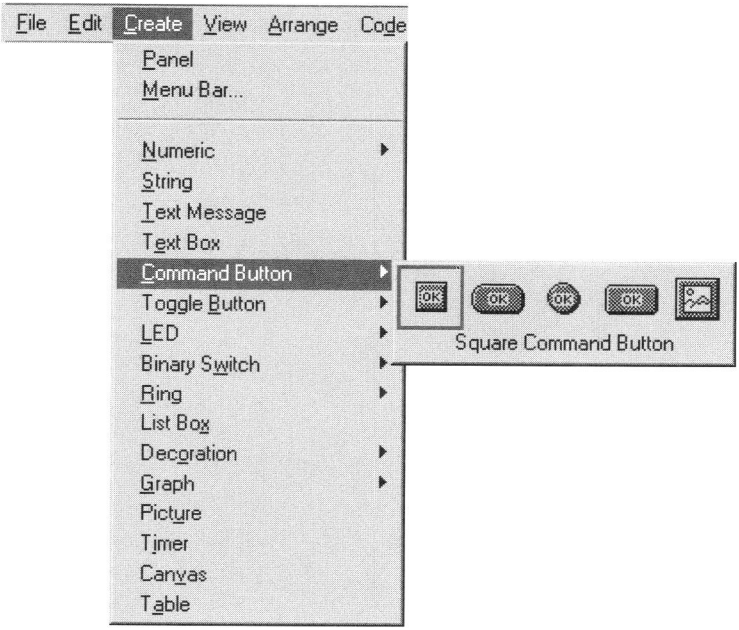

Figure 2–10
Creating the Command Button

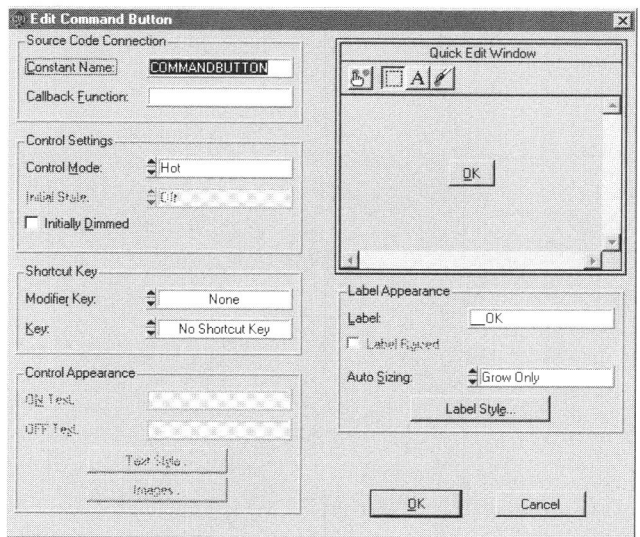

Figure 2–11
Command Button Dialog Box

Chapter 2 • Basics of Creating the Graphical User Interface

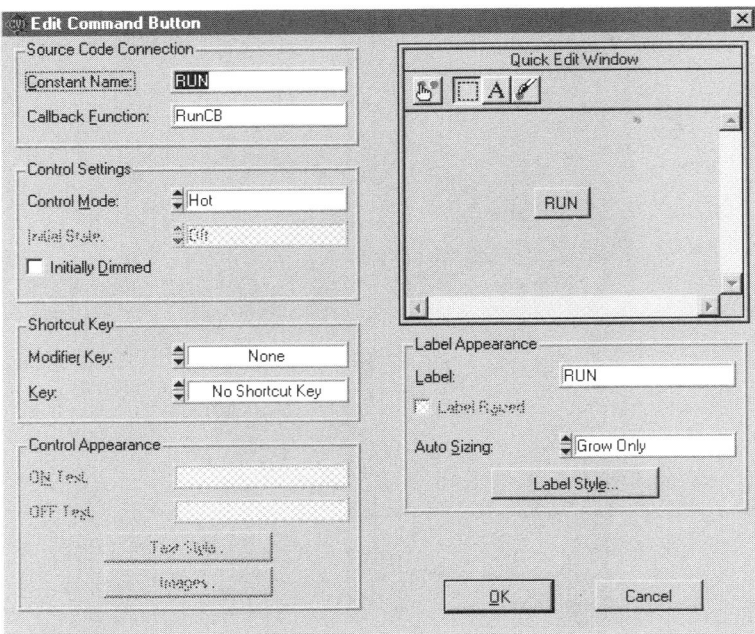

Figure 2–12
Attributes for RUN Command Button

this command button is selected while running the application, the code inside the *RunCB* callback function will be executed. The **Label** entered in the **Label Appearance** box (RUN) is what you will see as text on the command button on the GUI.

Similarly, create another command button on the panel and enter the following attributes.

For **Constant Name**,	enter EXIT.
For **Callback Function**,	enter ExitCB.
For **Label**,	enter EXIT.

Your GUI will now look like Figure 2–13. Move the mouse over one of the command buttons and while holding down the left mouse button, drag the command button to any location on the panel. Any object on the panel can be moved in this manner. This is referred to as drag-and-drop. Your GUI is now complete. As you can see, it is easy to create and move the objects on the GUI.

Figure 2–13
First Project GUI

From the **File** menu, select **Save As** and assign the project name of project2-1. *CVI* will create a user resource file and append the .uir extension to the file name. Select **File>>Add File to Project** to add the file to the **Project** window list.

After the GUI is completed, the next step is to create the skeleton code for the GUI by using the *CVI CodeBuilder* engine. The *CodeBuilder* creates syntactically and programmatically correct code. It takes all the callback function names defined in the user interface and creates the skeleton code for those functions. All you have to do is to add the C code to the callback functions to perform the functions you wish.

Although the *CodeBuilder* is a very helpful tool in generating the skeleton code, all the code can be entered manually using the *CVI* **Source Code Editor** just like any other C compiler.

When you use the *CodeBuilder,* all the header files are included in the C program. A *main* function is created with code to start the user interface. *CodeBuilder* also creates a function to terminate the program based on the callback function in the command button that you created in the user interface resource.

The *CodeBuilder* automatically generates code based on the layout of the user interface and the user-defined preferences set in the **Code>>Preferences** menu of the **User Interface Editor.** You can select the different default events that you want monitored in your program by selecting **Code>>Preferences>>Default Control Events** and **Code>>Preferences>>Default Panel Events** as shown in Figures 2–14 and 2–15.

CVI gives you the option to select the events from these lists and to create the skeleton code based on these selections. You are shown some of these events in Figures 2–14 and 2–15, and you will see other events later in the

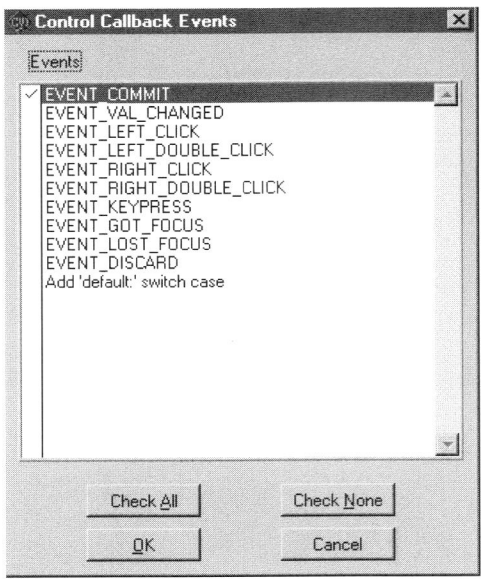

Figure 2–14
Default Control Callback Events

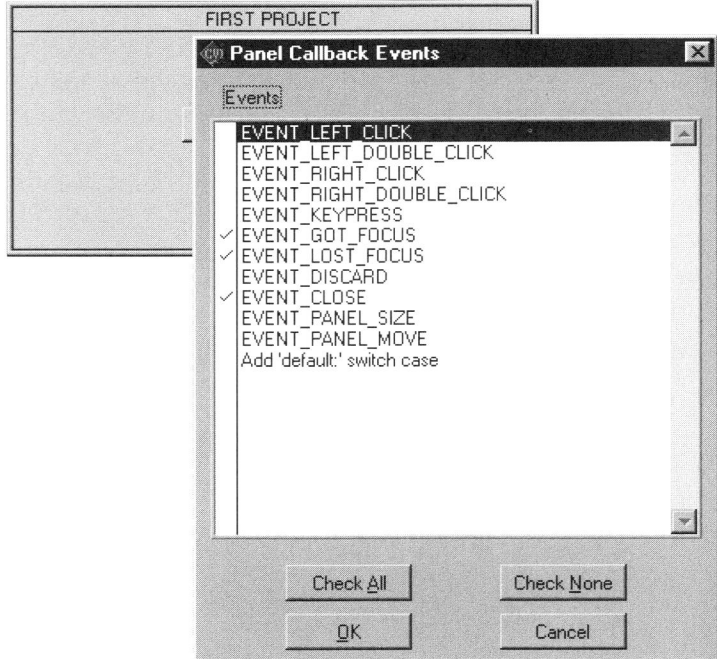

Figure 2–15
Default Panel Callback Events

chapter. For now, place the check marks for the events as shown in Figures 2–14 and 2–15 so your code will respond to these selected events only.

Create the code by selecting the **Code>>Set Target File>> <New Window>** from the **User Interface Editor.** The above step is usually not required if you are creating a new source file. This will create a new source code file containing the *CodeBuilder* generated code. You will see later that you need to open the source code file to add a callback function created using the control.

From the **Code** menu, select **Generate>>AllCode...** as shown in Figure 2–16.

A dialog box listing all the callback functions that you created for the project is displayed as shown in Figure 2–17. The bottom panel of this dialog box expects you to place a check mark next to the callback function that will be used to exit the project. This will enable *CodeBuilder* to assign the *QuitUserInterface* function to the selected callback function. The callback function *ExitCB* is invoked when the **EXIT** command button is selected since you assigned that in the **EXIT** command attributes box when you created this command button. Place a check mark next to the *ExitCB* function and click **OK.**

The skeleton code created by the *CVI CodeBuilder* includes the *main* function and the structure of the callback functions. The *RunUserInterface* and

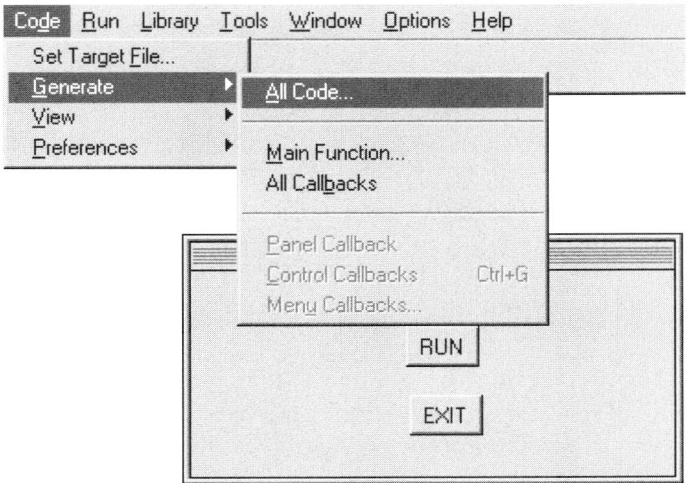

Figure 2–16
Automatic Code Generation

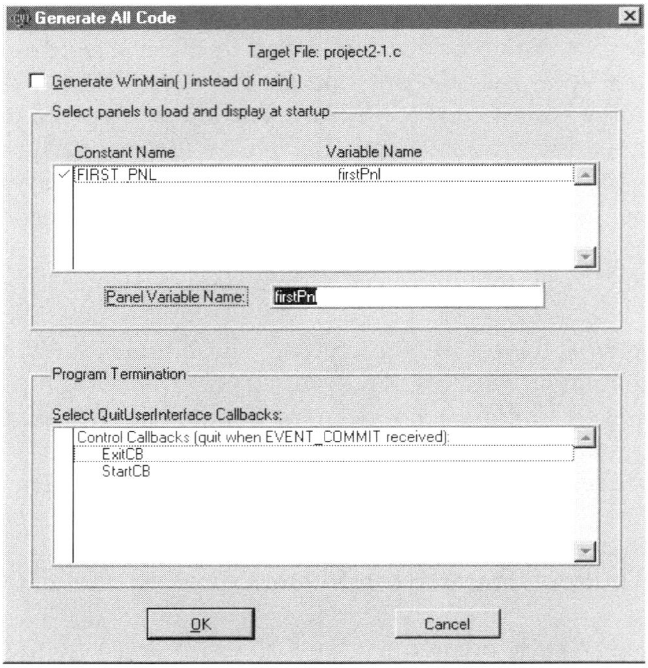

Figure 2–17
Code Generation Dialog Box

QuitUserInterface are included in the appropriate functions in the source code. In addition, the *CodeBuilder* creates the appropriate header files.

The *CodeBuilder*-generated code is shown in Figure 2–18.

Notice that the *CodeBuilder* generates the code for EVENT_COMMIT only as shown on lines 25 and 37, which respond to the mouse click events when the mouse button is released as per the selection made in Figure 2–14. All other mouse button clicks will be ignored by the code.

The *CodeBuilder* opens and places the code in the **Source Code Editor** window. Type in the highlighted line in the *StartCB* function as shown in Figure 2–19.

Save the source code file as `project2-1.c` using **File>>Save As.** Add this file to the **Project** window by selecting **File>>Add File to Project**. This will add the file name to the **Project** window file list as shown in Figure 2–20.

You have the option of setting the configuration of your project to **Debug** or to **Release.** For an explanation of the configurations, see the "Build Menu" section of Appendix B, "**Project** Window Environment." To create

```
1       #include <cvirte.h>
2       #include <userint.h>
3       #include "project2-1.h"
4
5       static int firstPnl;
6
7
8       int main (int argc, char *argv[])
9       {
10              if (InitCVIRTE (0, argv, 0) == 0)
11                  return -1;  /* out of memory */
12      if ((firstPnl = LoadPanel (0, "project2-1.uir", FIRST_PNL)) < 0)
13                  return -1;
14              DisplayPanel (firstPnl);
15              RunUserInterface ;
16              DiscardPanel (firstPnl);
17              return 0;
18      }
19
20      int CVICALLBACK StartCB (int panel, int control, int event
21                  void *callbackData, int eventData1, int eventData2)
22      {
23              switch (event)
24                  {
25                  case EVENT_COMMIT:
26
27                      break;
28                  }
29              return 0;
30      }
31
32      int CVICALLBACK ExitCB (int panel, int control, int event,
33                  void *callbackData, int eventData1, int eventData2)
34      {
35              switch (event)
36                  {
37                  case EVENT_COMMIT
38                      QuitUserInterface (0);
39                      break;
40                  }
41              return 0;
42      }
```

Figure 2–18
Project2–1 *CodeBuilder* Skeleton Code

the project in the **Debug** configuration, select **Build>>Configuration>> Debug** as shown in Figure 2–21.

To create the executable code in this configuration, select **Build>>Create Debuggable Executable** (short cut is <Ctrl-M>) from Figure 2–21. This will link all the required files and create an executable file, project2-1.exe. When the executable is created, *CVI* displays the message that the file is

Chapter 2 • Basics of Creating the Graphical User Interface

```
//The project is executed when the START command button is selected
int CVICALLBACK StartCB (int panel, int control, int event,
         void *callbackData, int eventData1, int eventData2)
{
     switch (event)
        {
        case EVENT_COMMIT:
                printf("CONGRATULATIONS!\nYou have just created and run your first project in LabWindows/CVI\n");
                break;
        }
     return 0;
}
```

Figure 2–19
START Command Button Callback Function

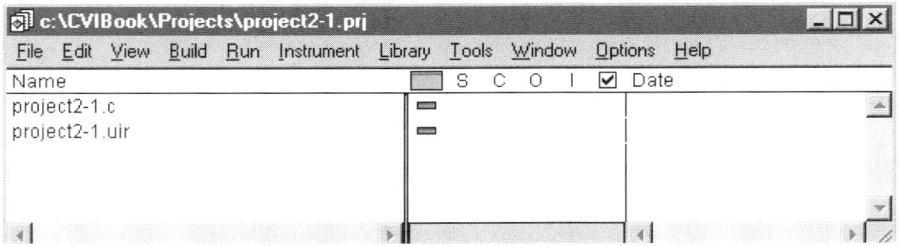

Figure 2–20
Project Window List

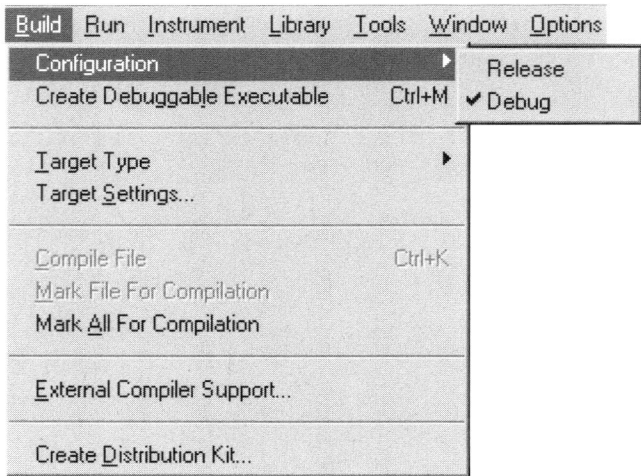

Figure 2–21
Build Configuration Sub-Menu

Chapter 2 • Basics of Creating the Graphic User Interface

being created, and when the file creation is complete, *CVI* displays a completion message as shown Figure 2–22.

Notice that the executable file is created with the project name appended by_dbg.exe.

If you had selected the **Release** configuration, the **Build** menu would update the sub-menu dynamically to display **Create Release Executable.** The executable file is similarly created with the project name appended by **.exe** extension.

In *CVI* versions prior to 5.5, the project had to be rebuilt each time it was loaded into *CVI*. In *CVI* 5.5 and later versions, the compiled state of the project is saved in files on the disk. After you have built your project, *CVI* compiles only files that have changed since the last time you build the project. These files are saved in folders that *CVI* creates when the project is built, and these folders contain cvibuild in the folder name.

If there are no build errors, you can run the project. You can run the project in the **Debug** or in the **Release** configuration. If you have the **Debug** configuration active, you can select the **Debug** menu item to run the project with debug options (short cut is <Shift –F5>) as shown in Figure 2–23.

If you want to run the project without attaching the debugger to the executable code, select **Execute** (or <Ctrl-F5> for shortcut).

Select **Execute** to run the project. The project will run and display the GUI that you have just created, as shown in Figure 2–24.

To start the application, click with the left mouse button on the **START** command button. The string you entered inside *printf* in Figure 2–19 will be displayed as shown in Figure 2–25. To execute again, click with the left mouse button on the **START** button. To exit the project, click with the left mouse button on the **EXIT** command button. That's all there is to creating and running a *CVI* project!

Try clicking on these command buttons using the right mouse button. You will notice that the project will not run and you cannot exit the project.

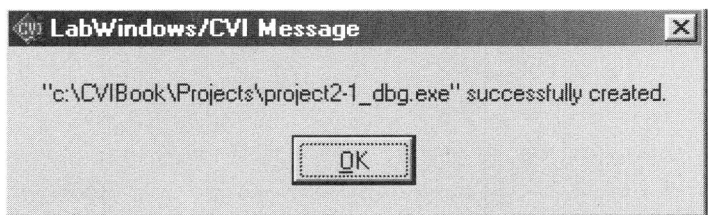

Figure 2–22
Executable File Creation Message

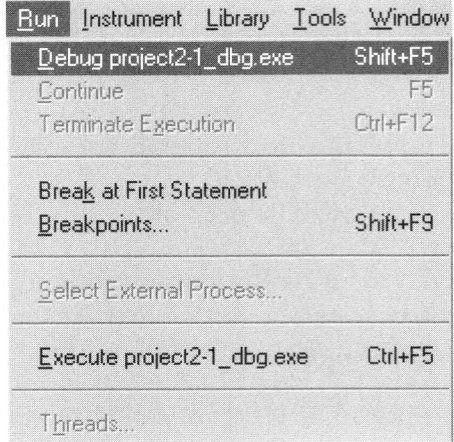

Figure 2–23
Project2–1 Run Menu

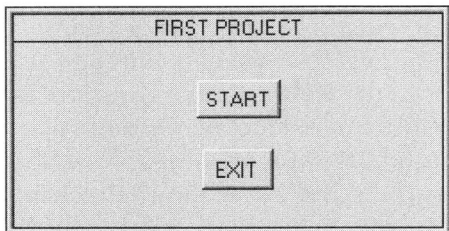

Figure 2–24
Project2–1 GUI

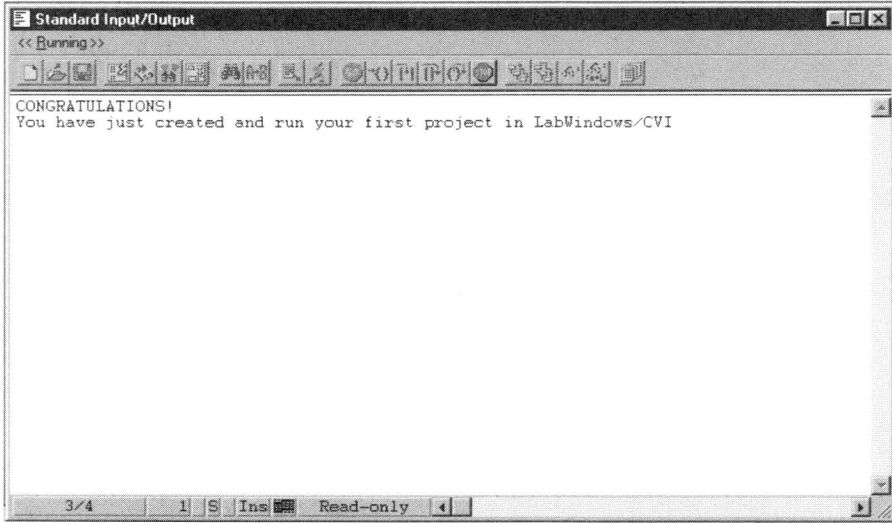

Figure 2–25
First Project Run

Chapter 2 • Basics of Creating the Graphic User Interface

All other mouse button clicks will not be recognized by the code since they were not enabled in the **Code>>Preferences** option above.

Remember the following shortcuts:

Command	Shortcut
Create Executable:	<Ctrl-M>,
Run Project with Debugger:	<Shift-F5>,
Run Project without Debugger:	<Ctrl-F5>.

Analyzing the *CodeBuilder*-Generated Code

Examine the *CodeBuilder*-generated source code in Figure 2–18. Notice that all the header files are included at the top of the code.

On line 1, the "include" file `cvirte.h` is used to support an external compiler if you are using one. The use of an external compiler will be discussed in Chapter 10, "External Compiler Support." The next "include" file `userint.h` supports the user interface library files. On line 3, the *CodeBuilder* includes the header file `project2-1.h` in the source code. This file is created by *CVI* to support the `project2-1.uir` file. The use of project header files is explained later in this chapter.

Line 5 shows the global variable `firstPnl` of type `static int`. This value is created on line 12 as a return value for the *LoadPanel* function that loads the GUI into memory and is referred to as *panel handle* by *CVI*. In *LoadPanel* library function, the file referenced is `project2-1.uir`. The *CodeBuilder* always assigns the panel handle(s) as static data variables because it does not export these variables to the linker, saving memory for the program.

In Windows programming terminology, a handle is a unique identifier used to call or to reference an object. A handle is an integer that remains associated with the object throughout the life span of the program.

Let us look at the *LoadPanel* library function. *LoadPanel* loads the GUI into memory. The *LoadPanel* function prototype is shown below.

```
int panelHandle= LoadPanel ( int parentPanelHandle,
                    char filename[ ], int panelResourceID);
```

The function arguments are explained in Table 2–1.

A *panelHandle* is a reference that is used by other objects and code to identify the named object. In this case, the *panelHandle* `firstPnl` refers to the

Table 2–1 LoadPanel *Function*

Input/ Output	Name	Type	Description
Input	parentPanelHandle	integer	panel handle where the panel is loaded as a child panel; zero is used for the parent panel, the top-level panel
	filename	string	name of the user interface file (.uir)
	panelResourceIID	integer	the Constant Name assigned to the panel in the **User Interface Editor**
Output	panelHandle	integer	value used to refer to this panel; negative values indicate error; Refer to Appendix A in the *User Interface Library Reference* manual or in Online Help

`project2-1.uir` interface file. You can change the panel handle to a more meaningful name, if you wish, by over-writing `firstPnl`.

DisplayPanel on line 14 displays the panel and all the controls associated with it on the screen. Recall this was loaded into memory by the *LoadPanel* function. The *DisplayPanel* has the following prototype.

```
int status= DisplayPanel (int panelHandle);
```

The function arguments are explained in Table 2–2.

Let us examine the *RunUserInterface* function called at line 15. This function was discussed above briefly in Figure 2–1. The *RunUserInterface* func-

Table 2–2 DisplayPanel *Function*

Input/ Output	Name	Type	Description
Input	panelHandle	integer	reference value of the panel currently in memory
Output	status	integer	negative value indicates error, refer to Appendix A in *User Interface Library Reference* manual or in Online Help

Chapter 2 • Basics of Creating the Graphic User Interface

Table 2–3 RunUserInterface *Function*

Input/ Output	Name	Type	Description
Output	returnCode	integer	value passed to *QuitUserInterface* function

tion responds to all the events for the callback functions. The *RunUserInterface* function creates a loop that processes all the callback functions continuously, and it returns the value that is passed to the *QuitUserInterface* function. When you want to quit the project by clicking on the **EXIT** command button, the *ExitCB* callback function is executed and the *QuitUserInterface* function is processed at line 38. The *QuitUserInterface* function terminates the event processing upon receiving a return value from the *RunUserInterface* function.

The *RunUserInterface* function has the following prototype:

```
int returnCode= RunUserInterface ( );
```

The function arguments are explained in Table 2–3.
The *QuitUserInterface* function has the following prototype.

```
int status= QuitUserInterface (int returnCode);
```

The function arguments are explained in Table 2–4.

Table 2–4 QuitUserInterface *Function*

nInput/ Output	Name	Type	Description
Input	returnCode	integer	value that the current call to *RunUserInterface* returns when it terminates
Output	status	integer	negative value indicates error, refer to Appendix A in the *User Interface Library Reference* manual or in Online Help

Overview of How to Develop a *CVI* Application

This section provides an overview of how to develop a *CVI* project. Some of these steps may seem unclear now but will become clearer as the section progresses.

The first step in developing a *CVI* project is to design your GUI by creating a user interface file using the **User Interface Editor.** This will create a file name with the extension `.uir`. *CVI* creates a corresponding header file with the same base name using a `.h` extension. The purpose and use of this header file will be explained later.

After creating the GUI, the second step is to create the skeleton code using the *CodeBuilder*. This opens and creates the source code in the **Source Code Editor** window. A source code file with a `.c` extension is created. You can then add the code to the skeleton code that is to be executed in the callback function. The third step is to compile the code and fix any errors. Lastly, build and run the project.

Creating the Second *CVI* Project

In your second *CVI* project, you will generate random voltages constantly and display these values on a voltmeter control one value at a time until the **EXIT** command button is selected.

The idea is to acquaint you with some of the countless capabilities of *CVI* user interface resources and at the same time get you started on creating your own projects as quickly as possible. Later you can modify this project by adding a General Purpose Interface Bus (GPIB) interface or a data acquisition card (DAQ) and monitoring the data from real instrument(s). Instead of monitoring voltages, you can monitor other stimuli: temperature sensors values, pressure, flow, position, frequency data, or any measurable entity.

To create this project, bring up a new panel as you did in the "Creating the First CVI Project" section.

Double-click on the panel, and the template shown in Figure 2–26 will be displayed. This attribute box contains the properties (attributes) of your panel. Some of the items on this template are explained below.

Every control that you create on the panel has a unique **Constant Name,** which you enter and which is used as the control's identifier for that object. The same is true for a panel on which you place the control(s). Recall that a panel is also a control. When the object (control) is first created, *CVI* assigns

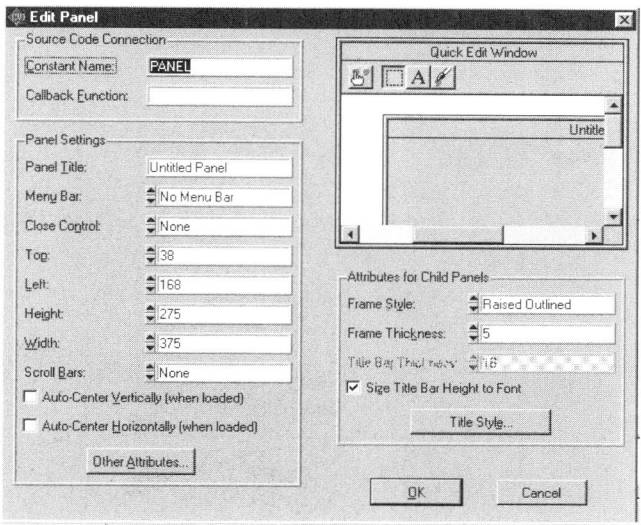

Figure 2–26
Edit Panel

a default name to it. You can see that in Figure 2–26, the default name assigned to the panel is **PANEL**. It is advisable to use meaningful names that will make sense to you because you will be using them quite frequently in your code. *CVI* uses the convention of always assigning uppercase letters for **Constant Names.** It is recommended that you do the same.

Change the **Constant Name** for the panel to MAIN_PNL by typing it into the **Constant Name** box. Type VOLTMETER READINGS in the **Panel Title** box. This title will be displayed in the Title Bar of the GUI when you run the project. Select **OK** to close the panel properties dialog box.

To place the controls on this panel, select **Create>>Numeric** from the menu on the **User Interface Editor.** Different numeric controls will be displayed consisting of knobs, gauges, meters, thermometers, and sliders as shown in Figure 2–27.

Numeric controls are used to input or display numeric values, depending on how you set their attributes in the associated control box. These attributes will be described later.

Move your mouse over all the numeric controls one by one to get familiar with the names that appear at the bottom of the numeric object's display.

Double-click the left mouse button on the **Meter** control and it will be placed on the panel. Move your mouse over the **Meter** and, by holding down the left mouse button, move it to the center of the panel.

Chapter 2 • Basics of Creating the Graphical User Interface

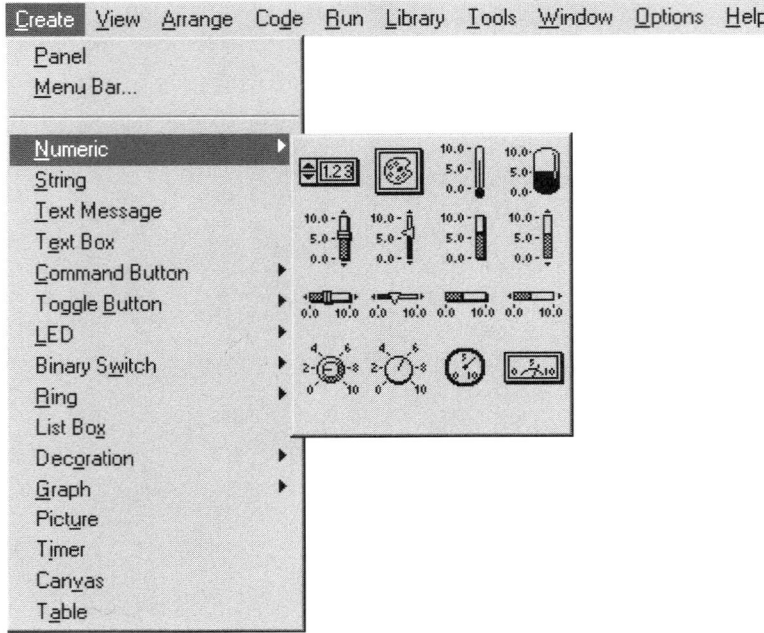

Figure 2–27
Numeric Controls

Instead of clicking on the **Create** menu command, you can right-click the mouse button on the panel to get the same menu (as shown in Figure 2–27). This shortcut is a convenient method for placing objects on the panel.

Double-click on the **Meter** control that you just placed on the panel. An **Edit Numeric Meter** dialog box appears in which you select the attributes for the **Meter** control shown in Figure 2–28. All controls have *almost* the same properties dialog box.

When the control is placed on the panel, *CVI* gives it a default name of **NUMERICMETER** for the **METER** control. Change the name to something that is more relevant to the control used. Call this object VOLT_METER (all uppercase letters). Although descriptive names for the **Constant Name** are recommended, be careful not to make these names too long or *CVI* will truncate these names. Leave the **Callback Function** blank, as this control will not use a callback function.

In the **Control Settings** group, leave the **Default Value** as 0.00. This is the value displayed on the control when *CVI* loads the GUI at run-time. You can change the value of the **Data Type** by clicking on the arrow boxes,

Chapter 2 • Basics of Creating the Graphic User Interface

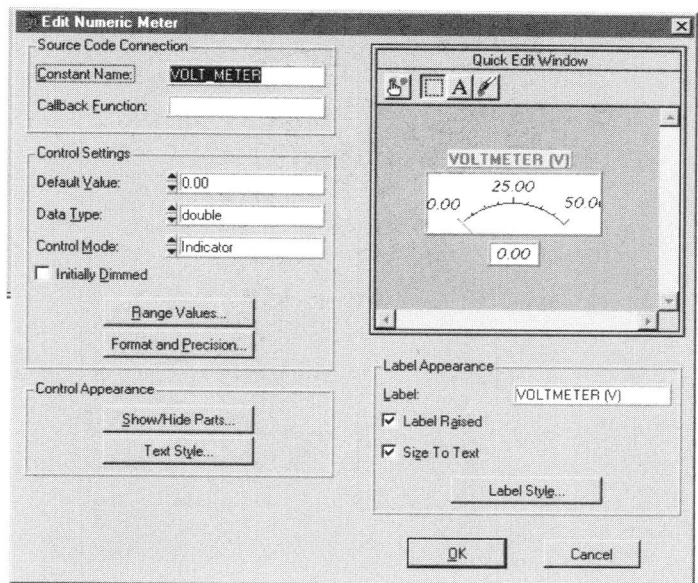

Figure 2–28
Numeric Meter Attributes

which gives you choices in a pull-down menu as shown in Figure 2–29. Leave the **Data Type** as **double** to read floating-point values.

The next control, **Control Mode,** needs some explanation. You can change the selection of this field by clicking on the arrow boxes next to **Control Mode** and selecting from the pull-down menu as shown in Figure 2–30. Set the **Control Mode** to **Indicator** for this control.

To understand the control mode, you need to be familiar with the way the events are generated when you modify the *GUI* settings on a control. *CVI* uses value-changed events that are implemented when the user modifies the settings on a control. Commit events are generated when the user performs an operation like changing the value on the control, selecting a command button or a menu item, selecting any object with the mouse, or releasing the mouse button over a control.

You can select the **Control Mode** option from the following list of choices.

Control Mode	Purpose
Normal	The **Normal** mode specifies that the user can operate the control to generate all controls except commit events.

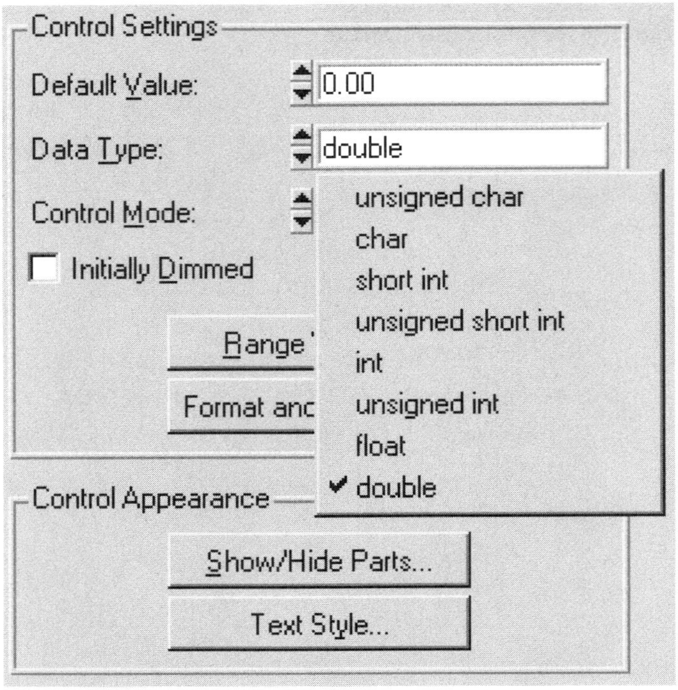

Figure 2–29
Data Type Pull-Down Menu

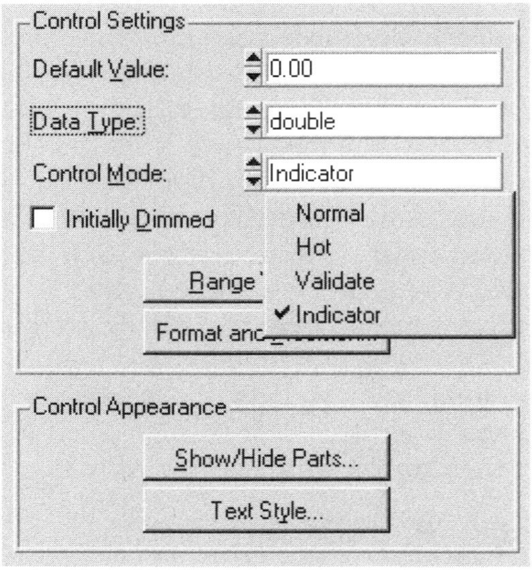

Figure 2–30
Control Mode Pull-Down Menu

Indicator	The **Indicator** mode specifies that only values can be displayed, that no events are generated, and that the user cannot enter values in the indicator box.

Hot	The **Hot** mode specifies that commit events are generated by this control and that the user can enter values in this box.

Validate	The **Validate** mode is similar to the **Hot** mode (and allows the user to enter values) but validates the range by checking the values before generating the commit event. When the value(s) is/are out of range, an out of range box pops up underneath the control box and displays the correct values, and the program is suspended until the correct value is entered.

Below the **Control Mode** dialog box in Figure 2–30 is the **Range Values** button used for numeric controls. Selecting the **Range Values** button opens a dialog box in which you can enter the minimum and maximum values for this control, and this dialog box is shown in Figure 2–31. Set the value `0.00` for **Minimum** and `50.00` for **Maximum** since you will be displaying values in the range 0 to 50 only. Select **OK** when done.

Select the **Format and Precision** button that is the next button below the **Range Values** in Figure 2–30, and enter `2` in the **Precision** box as shown in Figure 2–32. This will display two decimal places for the voltages. Leave the **Display Format** at **Floating Point** since you will display real numbers, and select **OK**.

The sub-panel on the bottom left side of the properties dialog box (Figure 2–28) and shown in Figure 2–30 sets the **Control Appearance.** Select

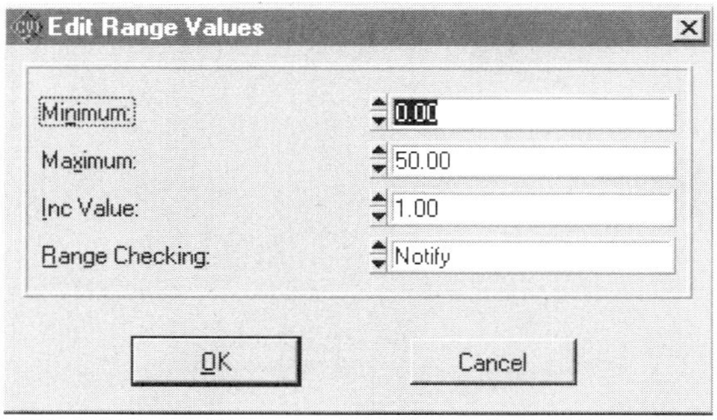

Figure 2–31
Range Values Dialog Box

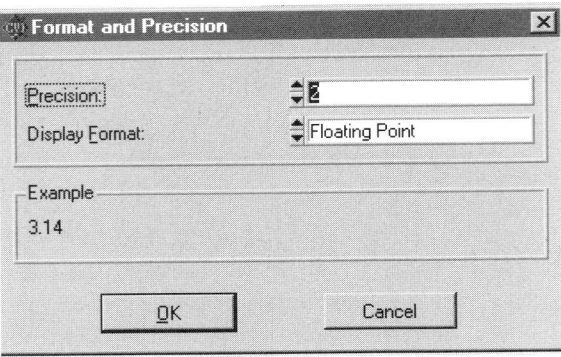

Figure 2–32
Format and Precision Dialog Box

Show/Hide Parts, and in the next dialog box, set **Marker Style** to **Full Markers** and **Tick Style** to **Full Ticks** to see the tick marks on the meter dial. Leave the **Show Inc/Dec Arrows** and **Show Digital Display** boxes checked as in Figure 2–33.

The **Show Inc/Dec Arrows** check box does not matter for this control box since the VOLT_METER control is used as an indicator and you will not be able to change the value on this control. The increment/decrement arrows are

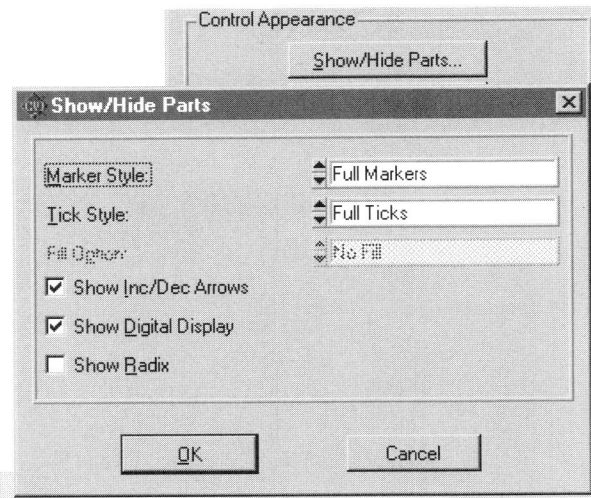

Figure 2–33
Show/Hide Parts Dialog Box

Chapter 2 • Basics of Creating the Graphic User Interface

the two arrows on the left side of the control box that allow you to change the values of that control. The increment/decrement arrows can be clicked when you want to change the control values for a particular control object.

Show Digital Display causes the numeric display control box below the meter to be visible. If you do not place a check mark on this box, the meter will appear without the numeric display control box. Place a check mark on this box. **Show Radix** box below **Show Digital Display** box when checked, displays the selected **Display Format** by means of an alphabetic character on the control. Leave it unchecked for now.

Below the **Show/Hide Parts** button in Figure 2–30 is the **Text Style** button. Selecting the **Text Style** box displays a dialog box, shown in Figure 2–34, in which you can change the **Font, Size, Justification,** and **Color.** You can select any color from the color display by clicking on the **Text Color** box. It is a good idea to experiment with a couple of different colors to help you decide. You can make the **Size** 12 points, check the **Bold** check box, and set the **Color** to black, or if you prefer, you can make your own selections. Select **OK** when finished.

The **Label Appearance** group is displayed in Figure 2–35 and is on the bottom right side of the control attribute box in Figure 2–28. For this project,

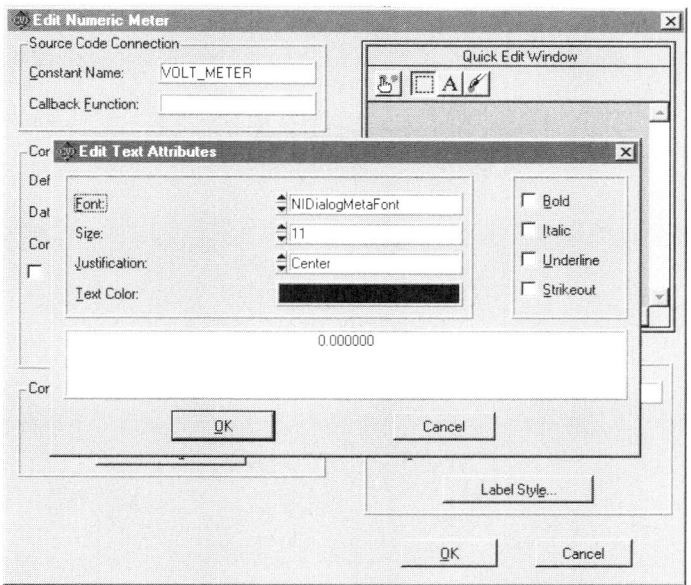

Figure 2–34
Text Attributes Dialog Box

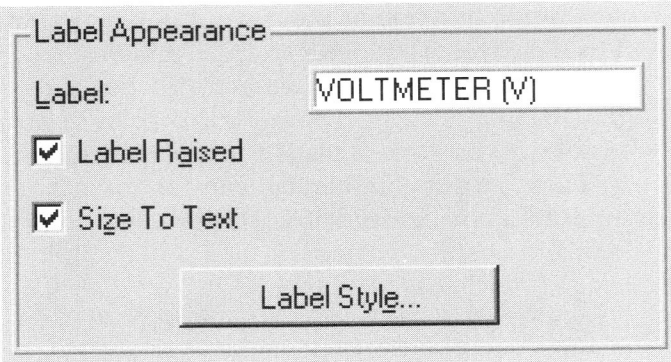

Figure 2–35
Label Appearance Dialog Box

enter VOLTMETER(V). Check the **Label Raised** box as it will make the label more distinct. Check the **Size To Text** box so the label box will re-size accordingly for the label text. Select **OK** on the **Edit Text Attributes** dialog box (Figure 2–34) after you have made your choices.

Select **OK** again in Figure 2–28 to accept all the changes you made to the meter. Your panel should now look like Figure 2–36.

Place two command buttons on the panel, one to start running your code and the other to exit. Right-click on the panel and select **Command Button**

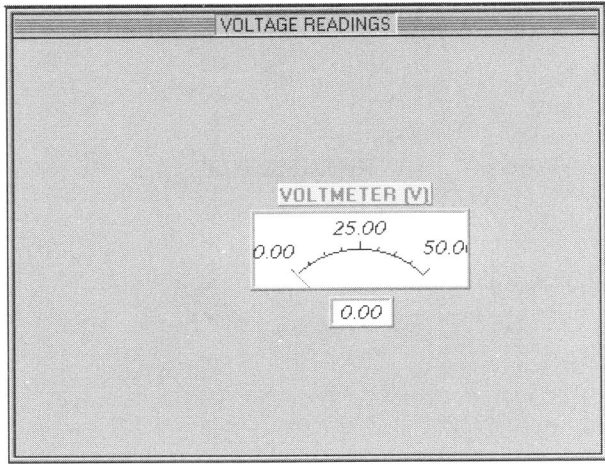

Figure 2–36
Voltmeter Control

Chapter 2 • Basics of Creating the Graphic User Interface   45

from the menu (shown in Figure 2–27). This will display various shapes of command buttons. Move your mouse over them and get familiar with their names, which reflect their shapes. Select the **Square Command Button** by clicking on it. This will put the selected command button on the panel. Drag-and-drop this command button to the left of the **Voltmeter.**

Next double-click on the command button to open the **Edit Command Button** dialog box to see its attributes. Enter the following information:

Constant Name:	START
Callback Function:	StartCB
Label:	START

The name you enter for the **Label** will appear on the command button. Finalize your selection by clicking **OK.**

You will need to create another command button to exit the project. A fast way to do it is to select the **START** command button that you just created, and from the **Edit** menu, make a **Copy** (using <Ctrl-C> for a shortcut) and **Paste** (or using <Ctrl-V> for a shortcut). It will make an exact copy of the command button with one exception: the **Constant Name** in the copied command button will be assigned the name START_1 instead of START because every control on the panel must have a unique identifier. *CVI* will not assign or accept the same name for two controls on the same panel but will change the name by adding a number to newly copied object to resolve the name conflict. Therefore, an underscore followed by a numeral is attached to the end of the original **Constant Name.** If you make another copy of the same command button, START_2 will be assigned as the new **Constant Name**.

Double-click on the command button to open the dialog for this command button and assign the following information:

Constant Name:	EXIT
Callback Function :	ExitCB
Label:	EXIT

Your panel will now look similar to Figure 2–37. Save this file as project2-2.uir from **File>>Save As** (from the **User Interface Editor**). To add this file to the project, select **File>>Add File to Project.**

The next step after creating the GUI is to use the *CodeBuilder* to generate the skeleton code. Leave the **Code>>Preferences** set to the values used in the previous project. Select **Code>>Generate>>All Code** and place a check

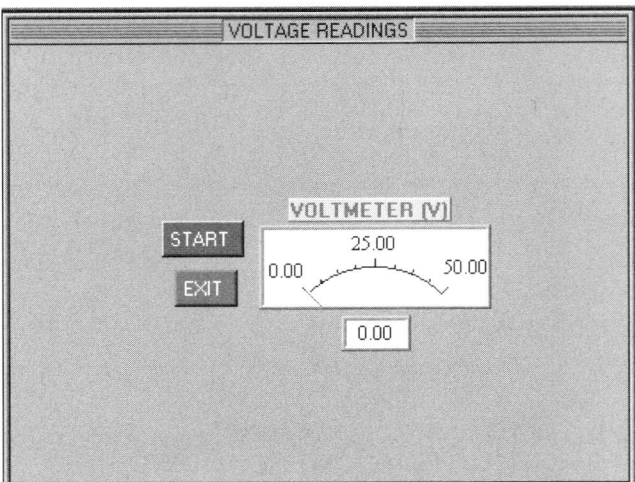

Figure 2–37
Panel with Voltmeter and Command Buttons

mark next to the *ExitCB* function as before to indicate that the **EXIT** command button will terminate the project.

Save the source code file as project2-2.c from **File>>Save As** and add it to the project. From the **Project** window, select **File>>Save As** to save the project as project2-2.prj.

In the *StartCB* callback function code, you need to include the highlighted code in Figure 2–38 to perform the designated function when the event is generated by selecting the **START** command button.

Using this code, the program will create random voltages and display them on the voltmeter one at a time with a one-quarter second delay between each display. The application starts to run when the **START** command button is selected. The time delay is necessary to enable you to read the values and can be varied to your liking. The *Delay* library function waits for the number of seconds specified in the function's argument (.25 seconds in this case). The purpose of the *ProcessSystemEvents* function was discussed previously. The application stops when the **EXIT** command button is selected, setting Start-Flag to 1 and thus suspending the program execution. If, for example, *ProcessSystemEvents* was not used in the *StartCB* callback function, the *while* loop will run indefinitely; selecting the **EXIT** command button to terminate the program will not invoke the *ExitCB* callback function code.

The completed source code for project2-2 is listed in Figure 2–39.

The code in the *main* function is the same as seen previously in project2-1. In the *ExitCB* callback function, the flag is set to break out of the *while* loop in

Chapter 2 • Basics of Creating the Graphic User Interface

```
int CVICALLBACK StartCB (int panel, int control, int event,
         void *callbackData, int eventData1, int eventData2)
{
    int count;
    switch (event)
    {
        case EVENT_COMMIT:
            while(StartFlag !=1) //Run as long as the flag is not set
            {
                volt_data=rand/(2*327.62);    // Generate random voltages
                //Write random values to the voltmeter
                SetCtrlVal (mainPnl, MAIN_PNL_VOLT_METER, volt_data);
                Delay(.25); //Add delay of .25 secs
                ProcessSystemEvents;
            }
            break;
    }
    return 0;
} //StartCB
```

Figure 2–38
StartCB Callback Function Listing

the *StartCB* callback function by checking the value of the `StartFlag`. The *StartCB* callback function needs to be examined closely.

The *StartCB* callback function, lines 32 through 51, generates the random voltage values and displays the results on the meter control. The *while* loop, lines 40 through 47, creates a random value at line 42 and assigns it to the variable *volt_data*. Using the *CVI* library function *SetCtrlVal*, the `volt_data` value is "written" to the **Meter** control. At line 44, the *Delay* library function pauses the execution for a time interval (.25 seconds in this case) specified in seconds in its argument.

The *SetCtrlVal* library function is used to set the value of a specified control on the indicated panel. This will become clear after looking at the prototype function. The *SetCtrlVal* function has the following prototype:

```
int status= SetCtrlVal ( int panelHandle, int controlID,
                                                      value);
```

The function arguments are explained in Table 2–5.

The first argument of *SetCtrlVal* is *mainPnl*. This is the *panelHandle* returned by *LoadPanel* function on line 24 of Figure 2–39. The next argument is the control identifier for which the value is to be modified. In this case, it is `VOLT_METER`, the **Constant Name** you assigned to the **Meter** control when you created the control on the panel. Recall that the complete ControlID consists of the **Constant Name** of the panel concatenated by _ and the **Constant Name** of the control. In this case, the **Constant Name** of the panel is

```
/***********************************************
Project Name: project2-2.prj

Purpose:            To demonstrate the use of command buttons
                    for Meter control


***********************************************/

#include <cvirte.h>     /* Needed if linking in external compiler; harmless otherwise */
#include <userint.h>
#include "project2-2.h"
#include <utility.h>
#include <ansi_c.h>

static int mainPnl;
double volt_data;
int StartFlag=0;

int main (int argc, char *argv[])
{
        if (InitCVIRTE (0, argv, 0) == 0)  /* Needed if linking in external compiler;
                                                            harmless otherwise */
                return -1;      /* out of memory */
if ((mainPnl = LoadPanel (0, "project2-2.uir", MAIN_PNL)) < 0)
                return -1;
        DisplayPanel (mainPnl);
        RunUserInterface ;
        return 0;
} //main


//Invoked with the START command button
int CVICALLBACK StartCB (int panel, int control, int event,
            void *callbackData, int eventData1, int eventData2)
{
        int count;
        switch (event)
        {
                case EVENT_COMMIT:
                        while(StartFlag !=1) //Run as long as the flag is not set
                        {
                                volt_data=rand/(2*327.62);   // Generate random voltages
                                //Write random values to the voltmeter
                                SetCtrlVal (mainPnl, MAIN_PNL_VOLT_METER, volt_data);
                                Delay(.25); //Add delay of .25 secs
                                ProcessSystemEvents;
                        }
                break;
        }
        return 0;
} //StartCB
```

Figure 2–39
Project2–2 Code Listing

Chapter 2 • Basics of Creating the Graphic User Interface

```
53      //Called when EXIT command is selected
54      int CVICALLBACK ExitCB (int panel, int control, int event,
55                      void *callbackData, int eventData1, int eventData2)
56      {
57          switch (event)
58              {
59              case EVENT_COMMIT:
60                      StartFlag=1;            //Set the flag to exit
61                      QuitUserInterface (0);
62                      break;
63              }
64          return 0;
65      } //ExitCB
```

Figure 2–39
Project2–2 Code Listing (*continued*)

MAIN_PNL, which is prefixed to VOLT_METER to make the complete ControlID for this control as MAIN_PNL_VOLT_METER. This means that the object belongs to the MAIN_PNL panel uniquely. This allows you to use the same **Constant Names** on different panels since they will be identified uniquely by the prefix of the parent panel name.

To view the contents of any file enclosed within quotes, move the mouse cursor to the file name in the code and right-click, and a pull-down menu is displayed. From this menu, select **Open Quoted Text** or (press <Ctrl-U>) and the listing for the "quoted file" is displayed. Move the cursor over project2-2.h included at the top of the source code and press <Ctrl-U> to open the header file.

Table 2–5   SetCtrlVal *Function*

Input/Output	Name	Type	Description
Input	panelHandle	integer	the value returned by *LoadPanel*
	controlID	integer	the resource ID of the control to modify
	value	same data type as control	the new value to set the control
Output	status	integer	negative value indicates error, refer to Appendix A in the *User Interface Library Reference* manual or in Online Help

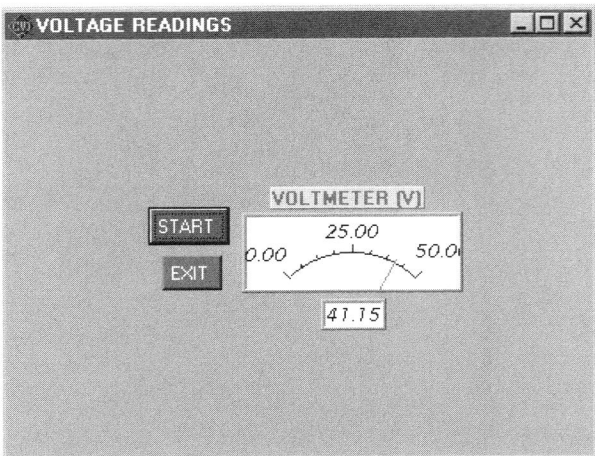

Figure 2–40
Project2–2 Run

Let us examine some of the features of the header file. Every user interface file has a corresponding header file that is created automatically when you save the user interface file. This header file has the same base name as the user interface file and is appended by .h (project2-2.h in this case). The header file contains all the **Constant Names** that you created in the user interface. The prototypes of the callback functions are also listed at the bottom of this header file.

Notice that all the **Constant Names** are assigned a unique number by *CVI*, and a warning about adding or removing anything from the header file appears at the top of the header file. You should take this warning seriously, and thus *CVI* creates the header file as read-only.

The **Constant Names** in the header are macros, which are assigned integer values. Whenever a particular **Constant Name** is referred to in the source code, the header file uses these macros to assign a number to the **Constant Name.** That number is substituted for that **Constant Name** in the source code.

Notice *CVI* adds comments in the header file, indicating the callback functions associated with the **Constant Names.**

The next step is to compile and build the project. From the **Project** window, select either **Build>>Configuration>>Debug** or **Build>>Configuration>>Release.** Select **Build>>Target>>Executable.** If there are no errors, execute the project by selecting **Run>>Project** or <Shift-F5>. You will see the random voltages generated one at a time and displayed in the **Meter** control, causing the **Meter** needle to display the random values. These values are also entered in the control box below the **Meter,** which is part of the **Meter** control. A snapshot of the run is shown in Figure 2–40. To exit the project, click on the **EXIT** command button at any time.

Remember the following shortcuts:

Command	Shortcut
Copy:	<Ctrl-C>,
Paste:	<Ctrl-V>,
Open Quoted Text:	<Ctrl-U>.

Callback Function and Generated Events

Let us look at the callback functions in more detail. When you click on any command button or control associated with a callback function, the selected events will be executed. The task that must be accomplished is specified in the callback function code under the selected events. For example, when you select the **START** command button to execute the application, *CVI* will jump to the function that you entered in the **Callback Function** for the **START** command button. After executing the code under the selected events, the program waits for the next event.

Control callback functions respond to events particular to that control. This is different from the panel callback though many events are common between them.

The prototype of a generic control callback is shown below.

```
int CVICALLBACK ControlCallback (int panel, int control,
                                 int event, void *callbackData,
                                 int eventData1, int eventData2);
```

Table 2–6 gives the events that *CVI* can generate for the control callback and the data that is passed via the callback function arguments. The arguments of the callback function are listed below.

- *panel*—This is the *handle* of the panel on which the control is located.
- *control*—This is the ConstantID of the control that responds to the event.
- *event*—This argument refers to the event that will trigger the callback function to execute.

- callback data—This argument is used to pass data between callback functions.
- *EventData1*—This is usually the x-coordinate of the mouse and sometimes the code of the key pressed. See Tables 2–6 and 2–7 for more information. Again for the Timer control, this argument has a different meaning, as shown in Chapter 3, "More Graphical User Interface."
- *EventData2*—This is usually the y-coordinate of the mouse and sometimes the pointer to the key code pressed. See Tables 2–6 and 2–7 for more information. For the **Timer** control, this argument has a different meaning, as you will see in Chapter 3, "More Graphical User Interface."

The prototype of a generic panel callback is shown below, and the events generated and the data passed via the callback arguments is given in Table

Table 2–6   *Control Callback Function Events*

panel	control	event	eventData1	eventData2
Panel Handle	Control ID	EVENT_COMMIT		
Panel Handle	Control ID	EVENT_VAL_CHANGED		
Panel Handle	Control ID	EVENT_LEFT_CLICK	X-coordinate of mouse	Y-coordinate of mouse
Panel Handle	Control ID	EVENT_LEFT_DOUBLE_CLICK	X-coordinate of mouse	Y-coordinate of mouse
Panel Handle	Control ID	EVENT_RIGHT_CLICK	X-coordinate of mouse	Y-coordinate of mouse
Panel Handle	Control ID	EVENT_RIGHT_DOUBLE_CLICK	X-coordinate of mouse	Y-coordinate of mouse
Panel Handle	Control ID	EVENT_KEYPRESS	Code of key pressed	Pointer to key code
Panel Handle	Control ID	EVENT_GOT_FOCUS		
Panel Handle	Control ID	EVENT_LOST_FOCUS		
Panel Handle	Control ID	EVENT_DISCARD		

Note: **callbackData* will not be discussed in this beginners' text.

Chapter 2 • Basics of Creating the Graphic User Interface

2–7. The blank columns in these tables indicate that the arguments are not applicable to the callback function.

```
int CVICALLBACK PanelCallback (int panel, int event,
                               void *callbackData,
                               int eventData1,
                               int eventData2);
```

The arguments in the *PanelCallback* have similar meanings as that of the *ControlCallback*. Notice however that there is no control argument for the *PanelCallback* because there is no control generating the event, so there is no need to specify the control identifier. You are generating the event by clicking on the panel only.

Table 2–7 *Panel Callback Function Events*

panel	event	eventData1	eventData2
Panel Handle	EVENT_LEFT_CLICK	X-coordinate of mouse	Y-coordinate of mouse
Panel Handle	EVENT_LEFT_DOUBLE_CLICK	X-coordinate of mouse	Y-coordinate of mouse
Panel Handle	EVENT_RIGHT_CLICK	X-coordinate of mouse	Y-coordinate of mouse
Panel Handle	EVENT_RIGHT_DOUBLE_CLICK	X-coordinate of mouse	Y-coordinate of mouse
Panel Handle	EVENT_KEYPRESS	Code of key pressed	Pointer to key code
Panel Handle	EVENT_GOT_FOCUS		
Panel Handle	EVENT_LOST_FOCUS		
Panel Handle	EVENT_CLOSE		
Panel Handle	EVENT_PANEL_SIZE		
Panel Handle	EVENT_PANEL_MOVE		
Panel Handle	EVENT_DISCARD		

Note: *callbackData* will not be discussed in this beginners' text.

Summary

This chapter introduced you to the fundamentals of creating user interfaces and taught you how to compile, build, and run *CVI* projects. The purpose of the first project was to show you how easy it is to create a *CVI* project. The second project introduced you to some of the numeric controls and command buttons. The concept of event-driven programming as used by *CVI* was explained in relationship to the callback functions. The ease of creating skeleton code from the GUI using the *CodeBuilder* was demonstrated, and the *CVI* library functions used in the code were explained.

The user interface consists of many controls, such as timers, graphs, stripcharts, knobs, and switches, and their use is discussed in the next chapter.

MORE GRAPHICAL USER INTERFACE

Chapter Highlights
- Using Timer Controls
- The Anatomy of the Panel
- Learning the Tool Bar
- Creating and Using the Strip Charts
- Project Window Icons
- Observations
- Summary

In this chapter, you will develop more *CVI* projects using different types of controls. You will systematically go through the process of building the project to understand the power and ease of adding controls and using their many features. You will be introduced to timers and the use of the function panels. The functionalities of the **User Interface Editor** toolbar are explained. Details are also given on creating and using strip charts.

Using Timer Controls

Using the timer control extends the event driven usage to include more advanced user operations. As you saw previously, an event can be generated by selecting an active control. Sometimes you may require certain events to occur automatically at regularly timed intervals without user intervention. To achieve this, **Timer Controls** generate the EVENT_TIMER_TICK events continuously at a periodic interval from the time the timer is enabled to the time it is disabled. The timer control generates only the EVENT_TIMER_TICK event continuously. The callback function arguments for the timer are the same as the other callback functions except for *eventData1* and *eventData2* arguments shown in Table 3–1.

During execution of the project, the timers are not displayed on the GUI. The user has no direct interaction with them through the GUI. There is a callback function associated with the timer that executes the code periodically at the specified time interval as long as the timer is enabled.

Here you will create another project that will build upon `project2-2` files that you created in the last chapter. By creating this project, you will learn how to add and use timer controls. Recall that in `project2-2`, random numbers were displayed on the meter at a time interval of one-quarter second. If you wanted to change the time interval, you had to go through the *Delay* function in the source code.

In this project, you will display the random numbers on the meter and control the time interval between data displays. You will have an option to let the data be displayed continuously until the **STOP** command button is selected.

Load `project2-2.prj` by double-clicking on this icon in the **Projects** folder. Double-click on `project2-2.uir` and select **File>>Save As** `project3-1.uir`. Figure 3–1 will be displayed, and `project3-1.uir` will replace `project2-2.uir` in the **Project** window.

The source code file for this project will be created using *CodeBuilder* and by adding code to the callback functions. You will create a new source code file, and therefore you need to remove `project2-2.c` from the **Project** win-

Table 3–1   *Continuous Events*

panel	event	eventData1	eventData2
	EVENT_TIMER_TICK	Pointer to time	Pointer to time since last tick

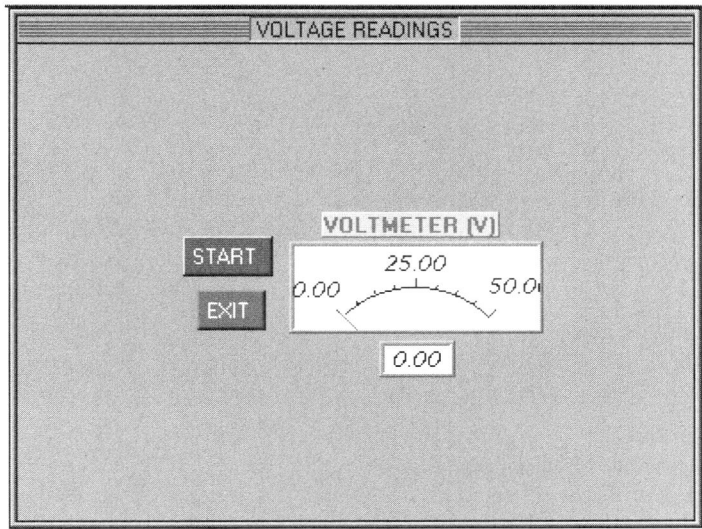

Figure 3–1
Project2–2 GUI

dow. Highlight the source code file and select **Edit>>Remove File** in the **Project** window. It is best to start with a new source file so the previous projects callback function code is not included. The file with the .c extension is removed from the **Project** window.

Save project2-2.prj as project3-1.prj using **File>>Save As** from the **Project** window.

You will now add a timer control, a knob, and a **STOP** command button to the GUI as in Figure 3–2. The knob will enable you to change the time interval between data display, and the **STOP** command button will stop the execution at your request.

To add the timer control on the GUI, right-click on the panel and select **Timer.** An object that looks like a stopwatch will be created on the panel. Double-click on the **Timer,** and the **Timer** attribute window, as displayed in Figure 3–3, is opened.

Enter the data listed below in the **Timer** attributes dialog box.

Constant Name	TIMER
Callback Function	ProcessTimerCB
Interval	0.250
Label	Variable Timer

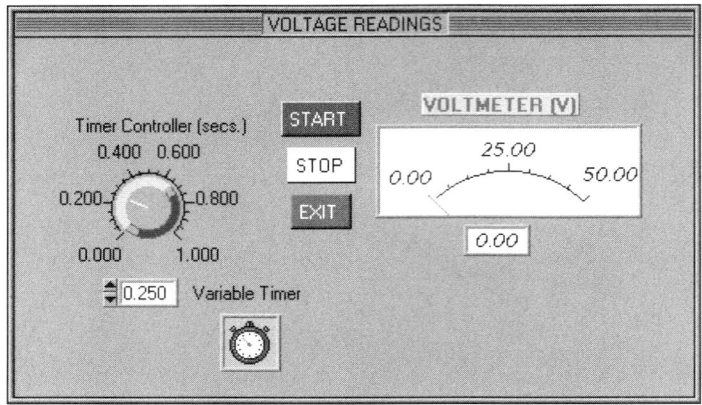

Figure 3–2
Project3–1 GUI with the Timer

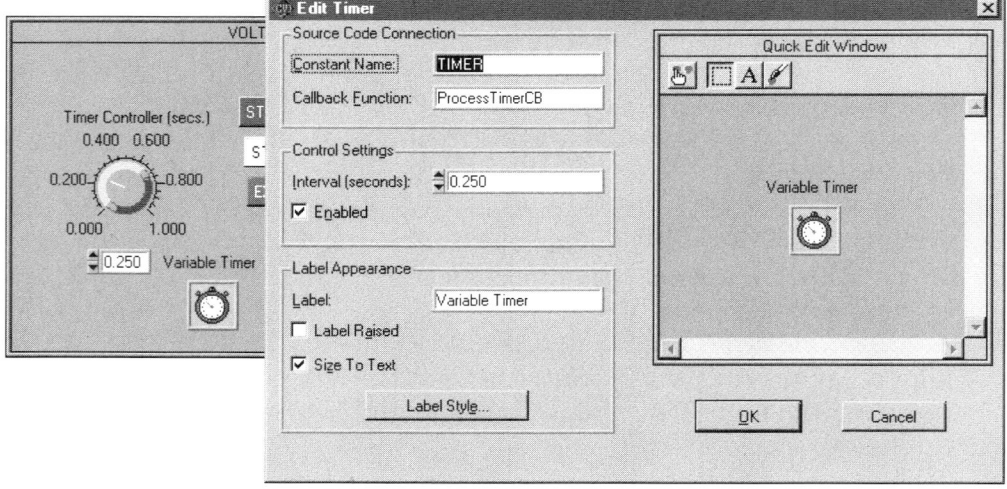

Figure 3–3
Timer Attributes Dialog box

Also, check the **Enabled** check box. This timer is initially set to process the code in the callback function *ProcessTimerCB* every 250 millisecond. You will vary the **Timer Controller** knob to change the rate of data display. The **Enabled** box is checked so the timer starts as soon as the *RunUserInterface* function is executed, though you can enable and disable the timer through the program.

Add the **STOP** command button to the panel. Remember that you can select the **START** command button and **Copy** (<Ctrl-C>) and **Paste** (<Ctrl-V>) on the panel. Double-click on the **START** command button to open the **Edit Command Button** dialog box and be sure it has the following attributes:

Constant Name	START
Callback Function	StartCB
Label	START

Also, do not check the **Initially Dimmed** check box.

The **STOP** and **EXIT** command buttons will have the attributes shown below, and the **Initially Dimmed** check boxes for these buttons will not be checked:

Constant Name	STOP	EXIT
Callback Function	StopCB	ExitCB
Label	STOP	EXIT

Now place the knob control on the panel. Right-click on the panel and select **Numeric>>Knob.** Double-click on the knob and enter the following information for the knob attributes:

Constant Name	TIMER_CNTRL
Callback Function	TimerControlCB
Default Value	0.250
Data Type	double
Control Mode	Hot
Label	Timer Controller (secs.)

This knob will be used to vary the interval at which you want to control the data displayed on the meter control. Select the **Range Values...** button for the knob control attributes and set the **Range Values** as shown in Figure 3–4.

Select the **Format and Precision...** button and set the attributes as shown in Figure 3–5.

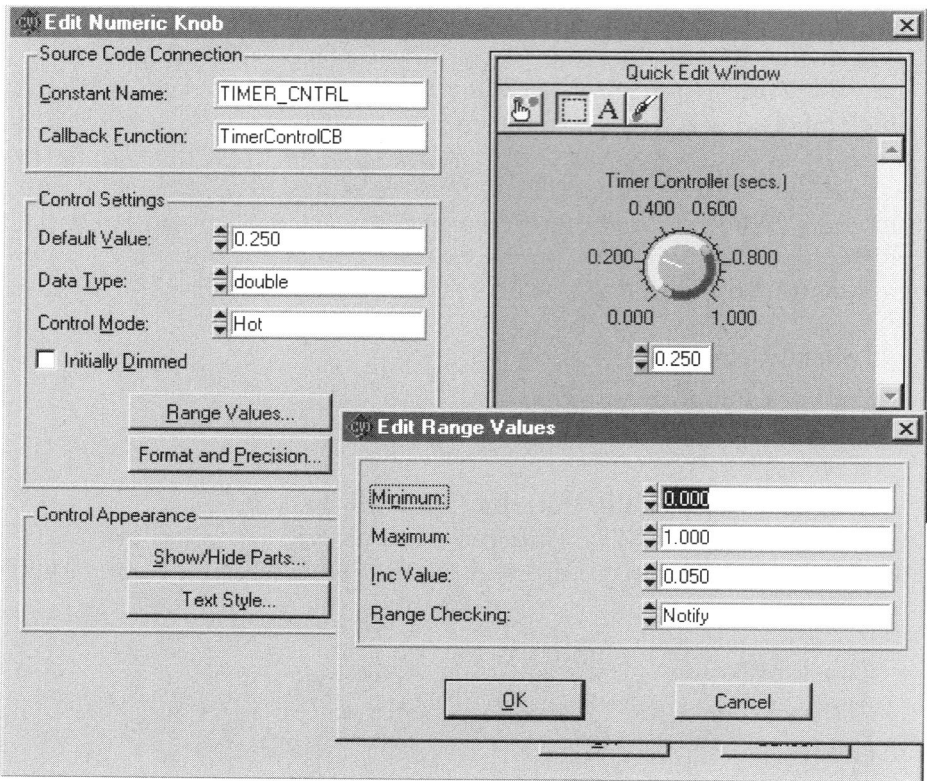

Figure 3–4
Timer Range Values Dialog Box

Select **Show/Hide Parts...** and modify the controls in the dialog box as shown in Figure 3–6.

The GUI is now complete and you can run the *CodeBuilder* to generate the skeleton code by selecting **Code>>Generate>>All Code** and marking *ExitCB* to contain the *QuitUserInterface* function as shown in the previous chapter. The *CodeBuilder*-created code is listed in Figure 3–7.

With the source code file added, the **Project** window will now look like Figure 3–8.

Most of the skeleton code is similar to that of the previous project(s), except for the *ProcessTimerCB* function at line 67. Any code that is placed in between EVENT_TIMER_TICK and *break* statements is executed once every timer tick.

Initially the timer uses the default **Interval** value of .250 seconds that you assigned when you created the **Timer** control. During run-time, as the

Figure 3–5
Format and Precision Dialog Box

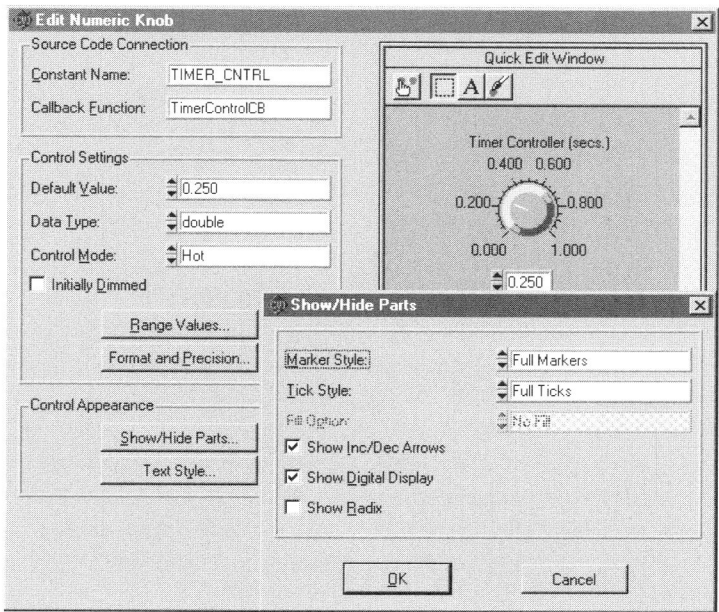

Figure 3–6
Show/Hide Parts Dialog Box

```
1    #include <cvirte.h>      /* Needed if linking in external compiler; harmless otherwise */
2    #include <userint.h>
3    #include "project3-1.h"
4
5    static int mainPnl;
6
7    int main (int argc, char *argv[])
8    {
9            if (InitCVIRTE (0, argv, 0) == 0)     /* Needed if linking in external compiler;
10   harmless otherwise */
11                   return -1;     /* out of memory */
12           if ((mainPnl = LoadPanel (0, "project3-1.uir", MAIN_PNL)) < 0)
13                   return -1;
14           DisplayPanel (mainPnl);
15           RunUserInterface ;
16           return 0;
17   }
18
19   int CVICALLBACK StartCB (int panel, int control, int event,
20                   void *callbackData, int eventData1, int eventData2)
21   {
22           switch (event)
23                   {
24                   case EVENT_LEFT_CLICK:
25
26                           break;
27                   }
28           return 0;
29   }
30
31   int CVICALLBACK ExitCB (int panel, int control, int event,
32                   void *callbackData, int eventData1, int eventData2)
33   {
34           switch (event)
35                   {
36                   case EVENT_COMMIT:
37                           QuitUserInterface (0);
38                           break;
39
40                   }
41           return 0;
42   }
43
44   int CVICALLBACK TimerControlCB (int panel, int control, int event,
45                   void *callbackData, int eventData1, int eventData2)
46   {
47           switch (event)
48                   {
49                   case EVENT_LEFT_CLICK:
50
51                           break;
52           }
53           return 0;
```

Figure 3–7
CodeBuilder-Generated Code for Project3–1

Chapter 3 • More Graphical User Interface

```
54      }
55
56      int CVICALLBACK StopCB (int panel, int control, int event,
57                      void *callbackData, int eventData1, int eventData2)
58      {
59            switch (event)
60                  {
61                  case EVENT_LEFT_CLICK:
62
63                        break;
64                  }
65            return 0;
66      }
67      int CVICALLBACK ProcessTimerCB (int panel, int control, int event,
68                      void *callbackData, int eventData1, int eventData2)
69      {
70            switch (event)
71                  {
72                  case EVENT_TIMER_TICK:
73
74                        break;
75                  }
76            return 0;
77      }
```

Figure 3–7
CodeBuilder-Generated Code for Project3–1 (*continued*)

Timer Controller knob's value is changed, the *TimerControlCB* function is activated and the new time interval passed to the timer. You will understand how this works when you enter the code in this callback function later.

Enter the code as listed in Figure 3–9. At the top of the code, at lines 15 and 16, the values for TRUE and FALSE are defined. The *main* function loads the project3-1 GUI and assigns it a handle: mainPnl at line 28. It then displays the GUI using the *DisplayPanel* function at line 30.

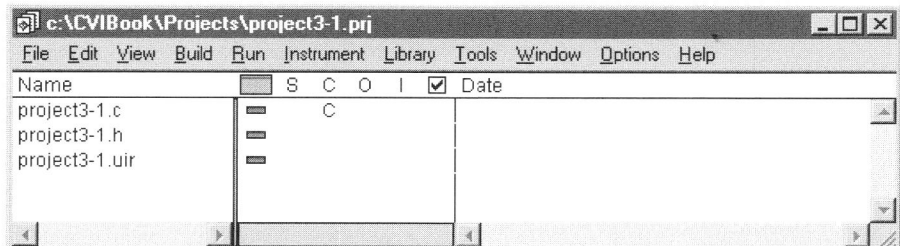

Figure 3–8
Project3–1 **Project** Window

```c
/***********************************************
Purpose:   To demonstrate the timer function
                  using random numbers

File Name: project3-1.c

***********************************************/

#include <cvirte.h>      /* Needed if linking in external compiler; harmless otherwise */
#include <userint.h>
#include <utility.h>
#include <ansi_c.h>
#include "project3-1.h"

#define TRUE 1
#define FALSE 0

static int mainPnl;
double volt_data, TimerInterval;

int   StartFlag=FALSE;

int main (int argc, char *argv[])
{
        if (InitCVIRTE (0, argv, 0) == 0)/* Needed if linking in external compiler; harmless
                                                                              otherwise */
                return -1; /* out of memory */
        if ((mainPnl = LoadPanel (0, "project3-1.uir", MAIN_PNL)) < 0)
                return -1;
        DisplayPanel (mainPnl);
        RunUserInterface ;
        return 0;
} //main

//Invoked from the command button "START" to run the project
int CVICALLBACK StartCB (int panel, int control, int event,
               void *callbackData, int eventData1, int eventData2)
{
        int count;
        switch (event)
        {
                case EVENT_COMMIT:

                        StartFlag = TRUE; //Flag to indicate the project is run
                        break;
        }
        return 0;
}  //StartCB

//Invoked from the command button "STOP" to Stop the project
int CVICALLBACK StopCB (int panel, int control, int event,
               void *callbackData, int eventData1, int eventData2)
{
        switch (event)
                {
                case EVENT_COMMIT:
                        StartFlag = FALSE;   //Flag to indicate the project is to stop running
                        break;
                }
        return 0;
} //  StopCB
```

Figure 3–9
Code Listing for Project3–1

Chapter 3 • More Graphical User Interface

```
63
64    //Function to obtain the new timer interval value and pass on to the timer
65    int CVICALLBACK TimerControlCB (int panel, int control, int event,
66                void *callbackData, int eventData1, int eventData2)
67    {
68        switch (event)
69                {
70                case EVENT_COMMIT:
71                    //Get the value from the knob
72                    GetCtrlVal (mainPnl, MAIN_PNL_TIMER_CNTRL, &TimerInterval);
73                    //Set the Timer interval to the new value obtained from the knob
74                    SetCtrlAttribute (mainPnl, MAIN_PNL_TIMER, ATTR_INTERVAL,
75                                                            TimerInterval);
76                    break;
77                }
78        return 0;
79    } //TimerControlCB
80
81    //This callback function is executed on every timer tick. The timer runs in the background
82    int CVICALLBACK ProcessTimerCB (int panel, int control, int event,
83                void *callbackData, int eventData1, int eventData2)
84    {
85        //Do only if the StartFlag is set.
86        if ((StartFlag == TRUE) && (event == EVENT_TIMER_TICK))
87        {
88                        volt_data=rand/(2*327.62);      // Generate random voltages
89                        //Write random values to the voltmeter
90                        SetCtrlVal (mainPnl, MAIN_PNL_VOLT_METER, volt_data);
91        }
92        return 0;
93    } //ProcessTimerCB
94
95
96    //Quit the project when EXIT command button is selected
97    int CVICALLBACK ExitCB (int panel, int control, int event,
98                void *callbackData, int eventData1, int eventData2)
99    {
100       switch (event)
101               {
102               case EVENT_COMMIT:
103                       QuitUserInterface (0);
104                       break;
105               }
106       return 0;
107   }//ExitCB
```

Figure 3–9
Code Listing for Project3–1 (*continued*)

When the **START** command button is selected to run the application, the *StartCB* callback function is called at line 36. This function sets the StartFlag to TRUE, and this value is used by the *ProcessTimerCB* function at line 86 to execute the code in *Process Timer CB*. Recall that you had enabled the **Timer** control by checking the **Enabled** box on the **Timer** attributes panel. The timer is running when the application starts, but because the `StartFlag` is initialized to `FALSE` on line 21, the timer is stopped and the code between lines 88 and 90 is not executed. It is important to understand that the timer is running in the background but is not executing any code, and so it appears that the application is stopped. The **START** command enables the code in the *ProcessTimerCB* function when `TRUE` is assigned to `StartFlag` in the *StartCB* function. The random number generator creates and assigns a value to `volt_data` at line 88 and writes that value to the **METER** control at every timer tick.

Select the **STOP** command button to stop the application. Doing so invokes the *StopCB* callback function that sets the `StartFlag` to `FALSE` at line 57, thus preventing the code in the *ProcessTimerCB* function from being executed.

When you change the value on the **Timer Controller,** the *TimerControlCB* callback function is invoked at line 65. The time interval is obtained from the **Timer Controller** at line 72 using the *GetCtrlVal* library function. The *GetCtrlVal* function obtains the current value of the selected control. The GetCtrlVal function has the following prototype.

```
int status = GetCtrlVal (int panel Handle, int control ID,
                                                    value);
```

Its function arguments are explained in Table 3–2.

This time interval is used to change the ATTR_INTERVAL of the timer control at line 74 using the library function *SetCtrlAttribute*.

Table 3–2 GetCtrlVal *Function*

Input/Output	Name	Type	Description
Input	panel Handle	integer	the value return by *LoadPanel*
	Control ID	integer	the resource ID of the control to obtain value
	value	same data type as control	the obtained value of control
Output	status	integer	negative value indicates error, refer to Appendix A in User Interface Library Manual or in Online Help.

The attributes of a control can be changed programmatically using this function. Different controls have different attributes that can be modified using this function. There are numerous attributes for the various controls and they are explained in Chapter 3 of *LabWindows/CVI User Interface Reference Manual*.

You can run `project3-1` as before and vary the **Timer Controller** to see the effect on the data display on the **Meter** control. Clicking the **STOP** command button will suspend the application. To re-start, select **START,** and the application will run again.

A snapshot of a `project3-1` run is shown in Figure 3–10. Note that the timer control is not visible on the GUI during execution.

If you were to move the **Timer Controller** to 0.0, the timer would run at the fastest speed possible for the system, which for Windows and Sun workstations is 1.0 millisecond.

The best way to become acquainted with the control objects attributes is to study them from the function panels, which are a graphical method of calling a sub-routine in a program. Function panels are screen-oriented interfaces to the *CVI* libraries that provide a template for selecting and setting the attribute parameters. The number and order of the function arguments is selected for you automatically by the function panel. All you have to do is to

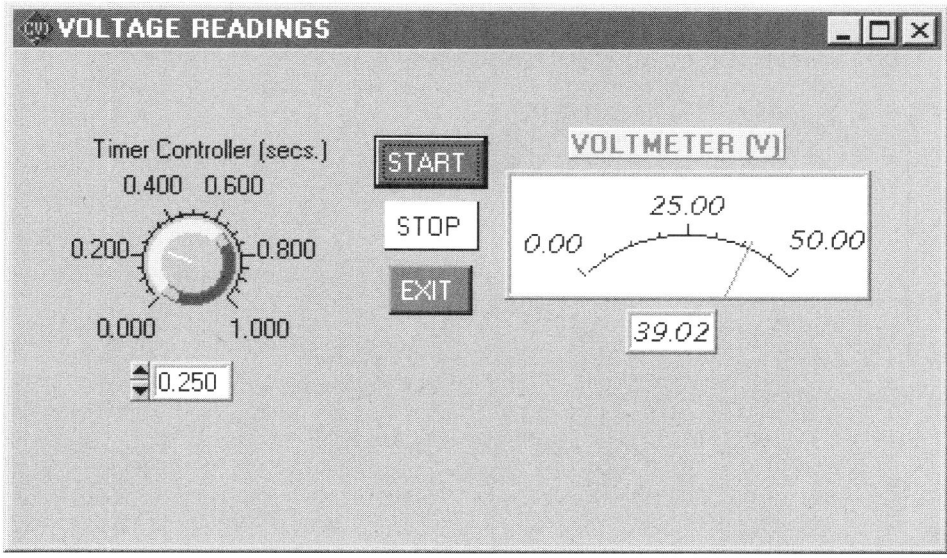

Figure 3–10
Project3–1 Run

select or enter the desired attributes. Right-clicking on the argument dialog boxes will give you a choice of all the available parameters for this function. As you insert the parameters, *CVI* fills in the function arguments in the text display at the bottom of this panel. A help box at the bottom of this function panel explains the purpose of the attributes as shown in Figure 3–11 for the *SetCtrlAttribute* function. The prototype for *SetCtrlAttribute* function is described in Chapter 4, "Enhancing the User Interface" in Table 4–4.

After you have selected the appropriate parameters, you can insert a function panel in your code at the cursor location by selecting **Code>>Insert Function Call** from the function panel menu (<Ctrl-I> is the shortcut).

To bring up the function panel for a library function in the code, move your cursor over the function and select **View>>Recall Function Panel** (<Ctrl-P> is the shortcut). This will display the function panel for the selected function as shown in Figure 3–11.

Remember the following shortcuts:

Command	Shortcut
View>>Recall Function Panel	<Ctrl-P>
Code>>Insert Function Call	<Ctrl-I>

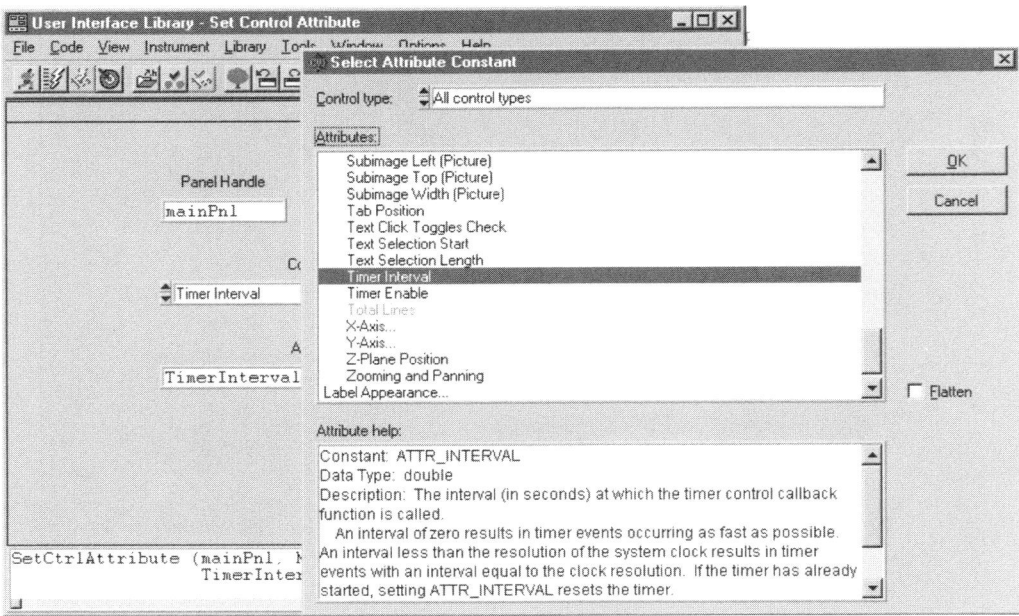

Figure 3–11
SetCtrlAttribute Function Panel

Chapter 3 • More Graphical User Interface

We have delved into creating the GUI and running the projects. Before we go further, let us look at some of the prominent features of the **User Interface Editor** that will help you create your own projects.

The Anatomy of the Panel

You have seen some of the attributes of the panel, but you need to look at some more which were not discussed previously. The panel properties template is displayed in Figure 3–12.

Note the **Panel Settings** group in this figure. The **Panel Title** is the text that appears in the panel's title bar.

Menu Bar refers to the list of command names across the top of the panel that may or may not have pull-down submenus, and which, when selected, can perform certain tasks, e.g., commands like **File, Edit,** and **View** in most Windows programs. If you are not going to add any menu bars to this panel,

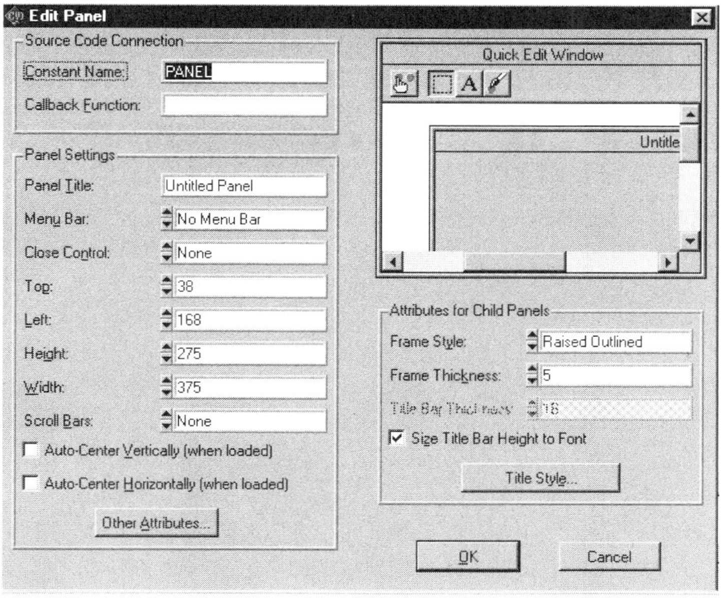

Figure 3–12
Panel Attributes

you can leave **MenuBar** set to No Menu Bar. In the next chapter, you will learn how to add menu bars.

The numbers in **Top, Left, Height,** and **Width** boxes set the dimensions of the panel size in pixels and you can change these settings to re-size the panel. The numbers in these boxes in Figure 3–12 are the location and size (coordinates and dimensions) of the panel as it exists presently. The values for **Top** and **Left** are referenced from the top left corner of the screen as shown in Figure 3–13.

When the panel properties template is closed, you can re-size the panel by holding down the left mouse button on the edges and corner of the panel and dragging the mouse when the cursor becomes a double arrow. You will notice that next time you open this template, the dimensions are changed to the new panel dimensions in the panel properties template.

You can have the option of adding horizontal, vertical, or both horizontal and vertical scroll bars to the panel by selecting **Scroll Bars.** Scroll bars are usually not required because you should create your user interface resource so that every object is visible in the project window without scrolling.

If you want your panel to display in the center of the screen when the project is run, you can select the **Auto Center Vertically (when loaded)** and **Auto Center Horizontally (when loaded).**

Selecting **Other Attributes** brings up the menu shown in Figure 3–14. From this dialog panel, you can set more panel attributes, like **Sizable, Mov-**

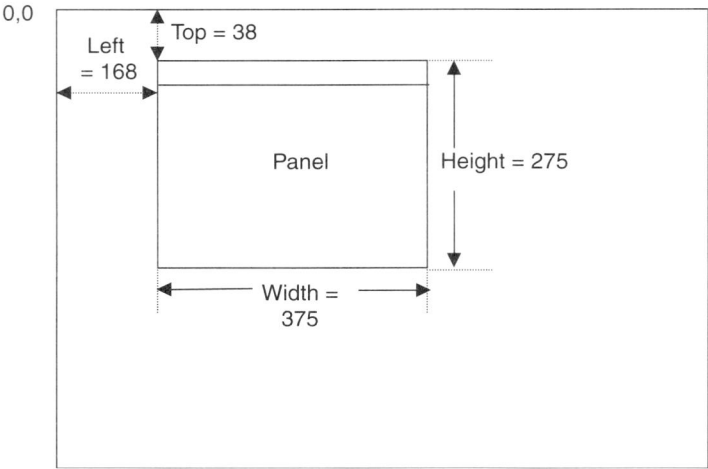

Figure 3–13
Panel Coordinates and Dimensions

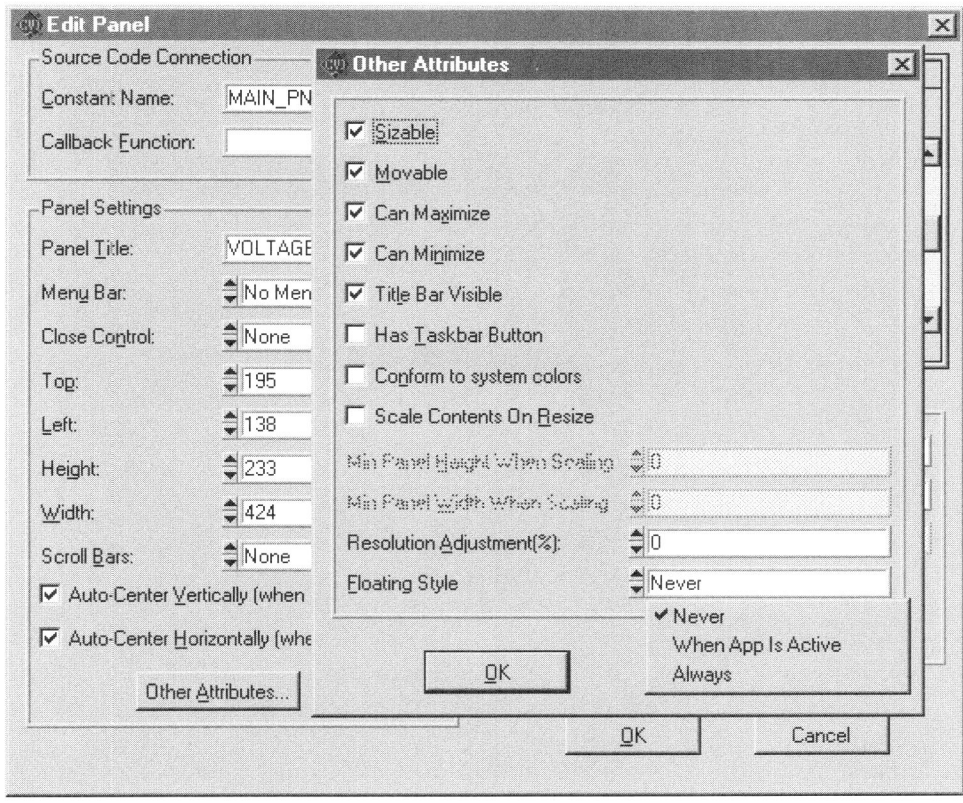

Figure 3–14
Other Attributes Panel

able, Can Maximize or Can Minimize, Title Bar Visible, Has Task Bar Button, Conform to System Colors,** and **Scale Contents On Resize.**

On the bottom right side of the Panel Attributes dialog box (Figure 3–12), there is an **Attributes for Child Panels** box from which you can select the **Frame Style, Frame Thickness,** and **Title Bar Thickness.** When selecting the **Frame Style,** you have a choice of **Raised, Beveled, Outline, Hidden, Step,** and **Raised Outline.** These are frame styles of the panel that you are creating. You can uncheck the **Size Title Bar Height to Font** box to choose your own font size for the title. Experimenting with these items will let you decide what looks good on the display. Recall that the relationship between parent and child panels was explained in Chapter 2, "Basics of Creating the Graphical User Interface" and shown in Figure 2–8.

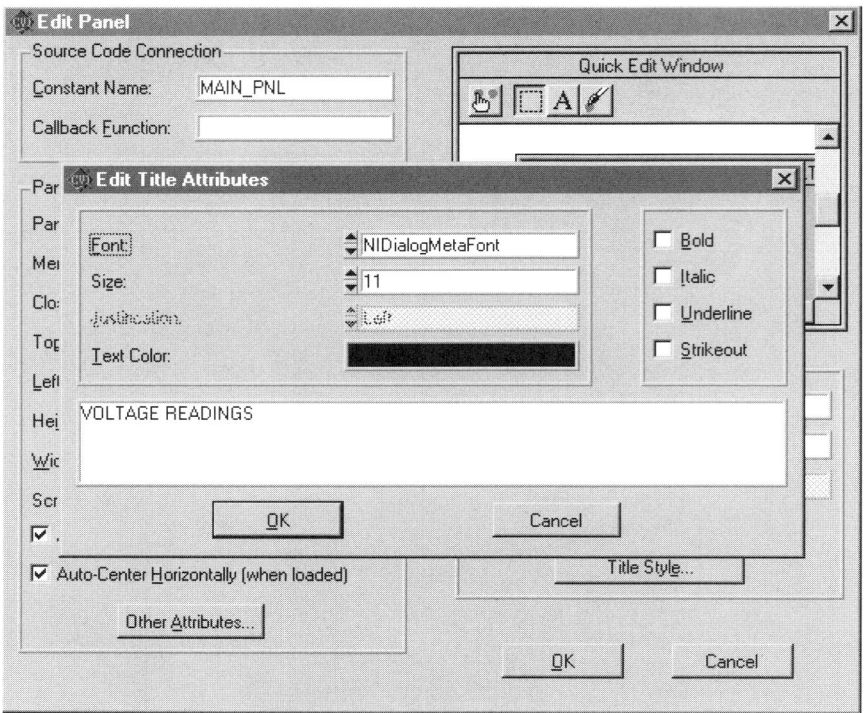

Figure 3–15
Edit Title Attributes

The button below the **Size Title Bar Height to Font** box (Figure 3–12) is **Title Style,** which, when selected, brings up the box shown in Figure 3–15. You select the **Font** and **Size** and the attributes of the fonts: **Bold, Italic, Underline,** or **Strikeout.** Clicking on the color bar to the right of **Text Color** displays a color panel, which you can use to change the color of the label. When you have made your selections, select **OK.**

Learning the Tool Bar

The tool bar in the **User Interface Editor** consists of the icons shown just below the menu bar as shown in Figure 3–16. This tool bar is known as the **Editing Tool Bar** or the **Shortcut Tool Bar.** It has many useful features, which you can access quickly when creating the GUI. The features of the **Tool Bar** icons are explained below.

Chapter 3 • More Graphical User Interface

Figure 3–16
Editing Tool Bar

The **Operating Tool** icon is used to operate on control objects. You saw the use of the **Operating Tool** previously for simulating event generation. When this object is clicked with the left mouse button, the CVI events are displayed on the top right side of the **User Interface Editor** in slow motion, enabling you to see the sequence of events.

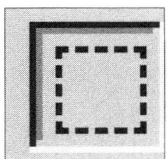

The **Editing Tool** is used to position or size objects and is selected when the **User Interface Editor** is loaded.

The **Text Editing Tool** tool can be used to modify the text on the objects.

The **Coloring Tool** is used for coloring the object on the GUI. Clicking on this icon with the right mouse button displays a color palette from which you can select the color to add to the object. After clicking on the **Coloring Tool,** move the cursor to the object, and right-click to display the color panel to select the color. Hold down <Ctrl> and notice that the brush takes the form of an eyedropper. Clicking this dropper over various objects enables you to color the objects to the previously selected color.

Creating and Using the Strip Charts

In this section, the creation and usage of the **Strip Chart** will be explained via a project. The **Strip Chart** is a graph control that displays data continuously while the data is being acquired. A **Strip Chart** simulates the same functionality as a strip chart recorder instrument connected to your measuring equipment. You can think of a **Strip Chart** as a data recorder that is recording data in real-time in a graphical form. A **Strip Chart** can have one or more traces on the same chart and all these traces can be updated simultaneously, unlike a graph control, which gives you a static view of the data.

You will now add a **Strip Chart** to project3-1 to display the random data on the **Meter** as well as on the **Strip Chart** as the data is being generated.

Load project3-1 in the **Project** window if not already loaded. Before you make any changes to the project, rename the files in the **Project** window to project3-2, keeping their corresponding extensions. Rename the project file to project3-2.prj and remove the source code file from the **Project** window.

Right-click on the panel and select **Graph.** A sub-menu opens showing two items: the first is the **Graph** and the second is the **Strip Chart.** Move your mouse over the object that says **Strip Chart** and click on it to place it on the GUI. Drag the **Strip Chart** object onto the panel to make the GUI look like Figure 3–17.

Double-click on the **Strip Chart** and set the attributes. Figure 3–18 shows the **Strip Chart** attributes dialog box. In the **Constant Name,** enter STRIP-CHART. In the **Points per Screen** enter 100 since we plan to plot 100 points per screen full, but you can use a different number if you desire.

The next attribute is the **Scroll Mode.** When you click on the box next to **Scroll Mode,** a pull-down menu appears giving you the choice of the following scrolling modes for the strip chart.

- **Continuous**—When **Continuous** scroll mode is selected, the strip chart adds the data to right edge of the strip chart and moves the data on the left side of the chart off the left edge as new data is added. The x-axis value is incremented continuously and gives the impression of a strip chart recorder with the "paper" continuously moving.

- **Block**—When **Block** scroll mode is selected, all the data is erased from the strip chart when the plot reaches the right edge of the chart. New data is added starting from the left again, with the x-axis value updated a block at a time.

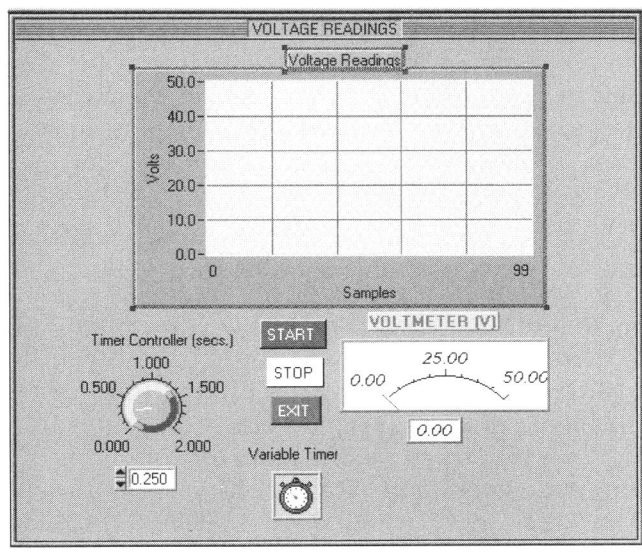

Figure 3–17
Project3–1 GUI with the Strip Chart

Figure 3–18
Strip Chart Attributes

- **Sweep**—When **Sweep** scroll mode is selected, the displayed strip chart data is not erased, rather the new data is written over the old data moving from left to right. When the data reaches the right-most edge of the strip chart, it starts to over-write from the left side again.

You will start this project by first running it in **Continuous Scroll Mode.** Later you can try running it in different scroll modes to see how the data display is affected.

Select **X-axis...** and enter `Samples` for the **Axis Name.** Click on **Y-axis...** and enter `Volts` as the **Axis Name.** Enter the **Minimum Value** as `0.0` and the **Maximum Value** as `50.0`. This will specify the range limits to be displayed on the strip chart. Set the **Grid Color** to gray and **Edge Style** to **Raised.** These last two items are used to improve the appearance of the Strip Chart. Enter `Voltage Readings` as the label for this chart in the **Label Appearance** section, and check **Label Raised.** If you want to change the color and the font of the label, you can change them from the **Label Style.**

The GUI for `project3-2` is now complete and should look like Figure 3–19. The *CodeBuilder* utility is used to create the skeleton code. Add it to the **Project** window as before. The complete source code for `project3-2` is listed in Figure 3–20.

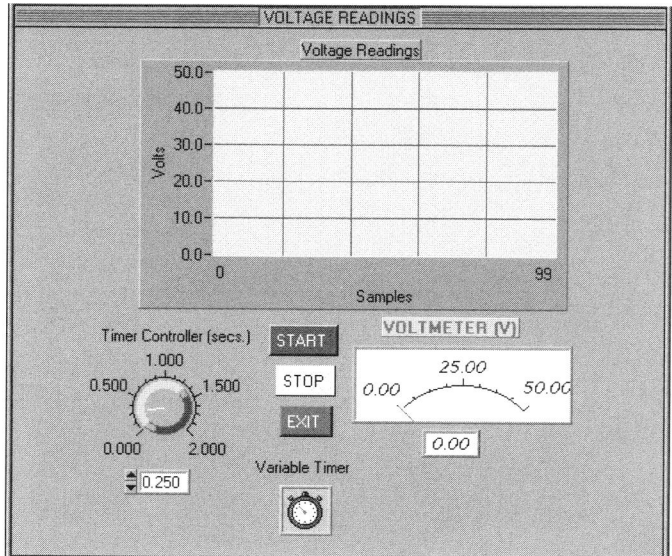

Figure 3–19
Project3–2 GUI

```c
/***********************************************************
Purpose: Project to plot data on the Strip Chart using the
timer control.

Project Name: project3-2.prj

***********************************************************/

#include <cvirte.h>     /* Needed if linking in external compiler; harmless otherwise */
#include <userint.h>
#include <utility.h>
#include <ansi_c.h>
#include "project3-2.h"


#define TRUE 1
#define FALSE 0

static int mainPnl;
double volt_data[1], TimerInterval;

int  StartFlag;

int main (int argc, char *argv[])
{
    if (InitCVIRTE (0, argv, 0) == 0)   /* Needed if linking in external compiler;
                                                                harmless otherwise */
            return -1;      /* out of memory */
    if ((mainPnl = LoadPanel (0, "project3-2.uir", MAIN_PNL)) < 0)   //Load the project3-2
                                                                                GUI
            return -1;
    DisplayPanel (mainPnl); //Display the GUI
    RunUserInterface ;
    return 0;
}//main

//Start the program
int CVICALLBACK StartCB (int panel, int control, int event,
            void *callbackData, int eventData1, int eventData2)
{
    int count;
    switch (event)
        {
            case EVENT_COMMIT:

                    StartFlag = TRUE;   //Flag control to start the timer
                    break;
        }
        return 0;
} //StartCB

//Invoked when the STOP button is selected
int CVICALLBACK StopCB (int panel, int control, int event,
            void *callbackData, int eventData1, int eventData2)
{
    switch (event)
            {
                case EVENT_COMMIT:
                    StartFlag = FALSE;    //Flag to stop the timer
```

Figure 3–20
Project3–2 Source Code Listing

```
61                          break;
62                  }
63          return 0;
64  } //StopCB
65
66  //Get the timer interval from the knob and pass it to the timer
67  int CVICALLBACK TimerControlCB (int panel, int control, int event,
68                  void *callbackData, int eventData1, int eventData2)
69  {
70          switch (event)
71                  {
72                  case EVENT_COMMIT:
73                          GetCtrlVal (mainPnl, MAIN_PNL_TIMER_CNTRL, &TimerInterval);
74                          SetCtrlAttribute (mainPnl, MAIN_PNL_TIMER, ATTR_INTERVAL,
75                                                                      TimerInterval);
76                          break;
77                  }
78          return 0;
79  }//TimerControlCB
80
81  //Invoked every timer tick by the Timer. Displays the data on the meter and the Strip Chart
82  int CVICALLBACK ProcessTimerCB (int panel, int control, int event,
83                  void *callbackData, int eventData1, int eventData2)
84  {
85
86          if ((StartFlag == TRUE) && (event == EVENT_TIMER_TICK))
87     {
88                          volt_data[0]=rand/(2*327.62);        // Generate random voltages
89                          //Write random values to the voltmeter
90                          SetCtrlVal (mainPnl, MAIN_PNL_VOLT_METER, volt_data[0]);
91                          //Plot random values to the the strip chart
92                          PlotStripChart (mainPnl, MAIN_PNL_VOLT_CHART, volt_data, 1, 0, 0,
93                                                                      VAL_DOUBLE);
94
95          }
96          return 0;
97  } //ProcessTimerCB
98
99  //Called by the EXIT command button
100 int CVICALLBACK ExitCB (int panel, int control, int event,
101                 void *callbackData, int eventData1, int eventData2)
102 {
103         switch (event)
104                 {
105                 case EVENT_COMMIT:
106                         QuitUserInterface (0);
107                         break;
108                 }
109         return 0;
110 } // ExitCB
```

Figure 3–20
Project3–2 Source Code Listing (*continued*)

Most of the callback functions are the same as in `project3-1` except for *ProcessTimerCB* function, which now has the *PlotStripChart* library function added at line 92. This function is inserted in the code by a function panel.

As mentioned above, function panels are used to facilitate the entering of the function arguments in *CVI* library functions with minimal typing and minimal chance of errors. You can insert the *PlotStripChart* library function using the function panel by moving your cursor just below the line where you entered *SetCtrlVal* function. Select **Library>>User Interface>>Controls/Graphs/Strip Charts…>>Graphs and Strip Charts>>Strip Chart Traces>> Plot Strip Chart** (see Figure 3–21).

This brings up the *PlotStripChart* library function template, shown in Figure 3–22, into which you can enter the arguments for this function. Notice that the window at the bottom of this function panel is initially devoid of any arguments. As you enter the values in the function panel, the bottom part of this function panel fills with the data you entered in the template.

If you know the name of the library function, you can type it where you want to insert it in the code, place the cursor on the function name, press <Ctrl-P> and the function panel will pop up. Enter the data in the appropriate fields and insert the created function into the source file by pressing <Ctrl-I>.

Before you start to enter data into the function panel, let us look at some features of this template. If you move your cursor over any data field and right click on it, a *Help* menu is displayed, explaining the data required in

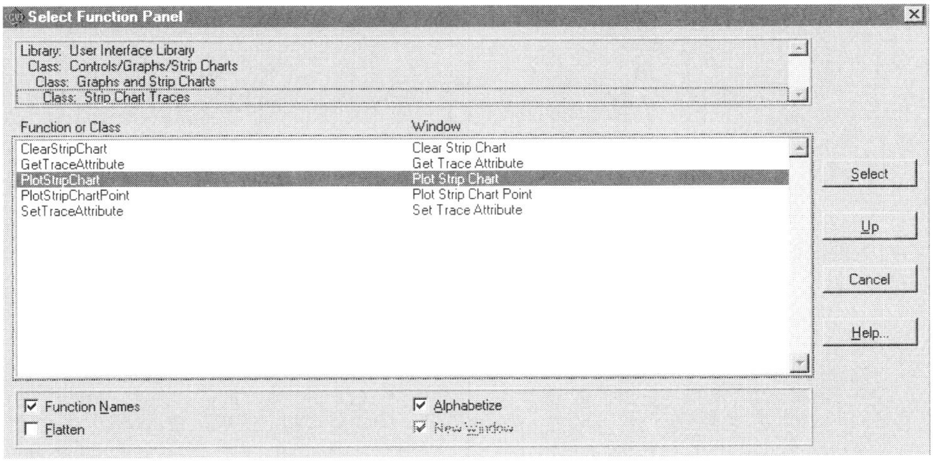

Figure 3–21
Select Function Panel

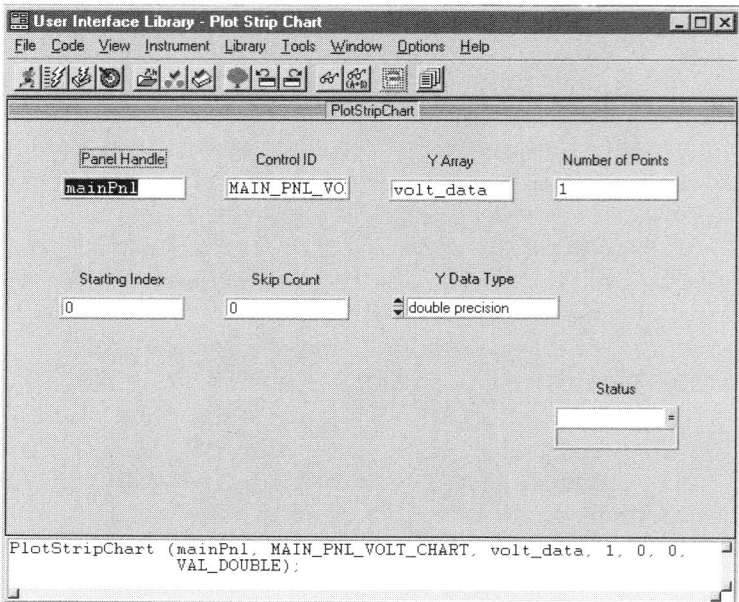

Figure 3–22
PlotStripChart Function Panel

that field. If you right click on the panel anywhere, you can see a description of the function. This *Help* feature saves you much time since you do not have to look for the functions in the *CVI* manuals and enter the function structure and the data manually.

Let us now see how we can easily enter data into the argument fields in this function panel. Place your cursor in the **PanelHandle** data field and select **Code>>Select Variable** (<Ctrl-A> is a shortcut for this command). A list of variables is displayed. Select the variable mainPnl since this is the *panelHandle* for this panel. Move the cursor to the **Control ID** field and select the **Constant Name** of the **Strip Chart** from **Code>>Select UIR Constant,** or you can just press <Enter> and select the **Constant Name** MAIN_PNL_ VOLT_CHART. Go to the next data field and similarly select the variable volt_data. This is the data you want to plot on the Y-axis of the **Strip Chart.** Leave 1 for the **Number of Plots** and leave the rest of the data unchanged. Notice the bottom window of the function panel is being filled simultaneously as you select the arguments of the function. You may type the data directly into the template, if you know the variables, instead of selecting them from the list.

Entering 1 for the **Number of Points** in the **Strip Chart** template means that there is only one variable plotted on this strip chart, which will turn out to be `volt_data[0]`. The *PlotStripChart* function considers the data plotted on the strip chart to be an array. If you want to plot more variables on the same chart, you can give another data value to the variable, e.g., `volt_data[1]`, `volt_data[2]`, `volt_data[3]`, and so on, and change the value in the **Number of Points** box accordingly on the **Strip Chart** attributes panel.

To enter the function panel in the source code, select **Code>>Insert Function Call** (<Ctrl-I>) from the function panel.

*Entering the function panel in the source code can also be achieved by clicking on the **Insert Function Call** icon on the toolbar.*

Close the function template and notice that the *PlotStripChart* function that you created was inserted in the code at the cursor location. The *PlotStripChart* library function plots the specified data on the strip chart with the specified arguments.

The prototype of *PlotStripChart* function is shown below.

```
int status = PlotStripChart( int panelHandle, int controlID,
        void *yArray, int numberOfPoints, int starting-
        Index, int skipCount, int YdataType);
```

The function arguments are explained in Table 3–3.

After adding the code to the *CodeBuilder*-generated code, select **Build>>Compile File** to compile the program. If there are no errors and the file is not saved, you will see a pop-up menu asking you to save the file. It is always a good idea to save the file before execution by selecting **File>>Save.** After that, select **Build>>Configuration>>Release,** then **Build>>Create Release Executable.** The executable file for the **Release** configuration will be created. The different characteristics of **Debug** and **Release** configurations are explained in Appendix B, "Project Window Environment." To run the project, select **Run>>Execute** or <Ctrl-F5>. You will notice on the user interface that the project GUI is loaded but is not executing yet. It is waiting for you to start the program by clicking on the **START** command button.

Note:

*You can run the project by clicking on the "Run" icon (shown here) in the **Source Code Editor** tool bar.*

Table 3–3 PlotStripChart *Function*

Input/ Output	Name	Type	Description
Input	PanelHandle	integer	panel handle where the panel is loaded
	controlID	integer	constant name assigned to the strip chart
	yArray	void*	the array containing the value(s) to plot along the Y-axis
	numberOfPoints	integer	the number of **yArray** points to add to the strip chart, e.g., in the above example, numberOfPoints is 1.
	startingIndex	integer	the zero-based index into the **yArray** where the first block of data begins (the default is 0)
	skipCount	integer	number of **yArray** elements to skip over after plotting each set of points (default is 0)
	YdataType	integer	specifies the data type of the **yArray** from one of the following data types: • VAL_CHAR • VAL_INTEGER • VAL_SHORT_INTEGER • VAL_FLOAT • VAL_DOUBLE • VAL_STRING • VAL_UNSIGNED_SHORT_INTEGER • VAL_UNSIGNED_INTEGER • VAL_UNSIGNED_CHAR
Output	status	integer	refer to Appendix A in *CVI User Interface Library Reference Manual* for error codes or in On-line Help

The project runs continuously until stopped by the user. The time interval of the data displayed on the strip chart and the **Voltmeter** is controlled by the **Timer Controller.** The **STOP** and **EXIT** command buttons work the same as they did in the previous project.

An instance of the project run is shown in Figure 3–23.

Remember the following shortcuts:

Command	Shortcut
Code>>Select Variable	<Ctrl-A>
Code>>Select UIR Constant	<Enter>
Recall Function Panel	<Ctrl-P>
Code>>Insert Function Call	<Ctrl-I>

Project Window Icons

Look at the **Project** window with the project files listed. You will notice certain icons and letters are marked in front of these files. An example is shown in Figure 3–24, and the following explanation refers to this example.

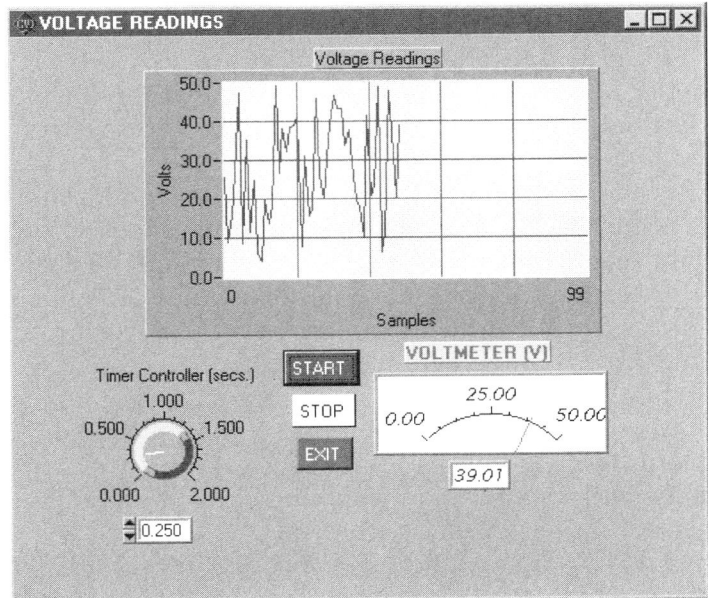

Figure 3–23
Project3–2 Run

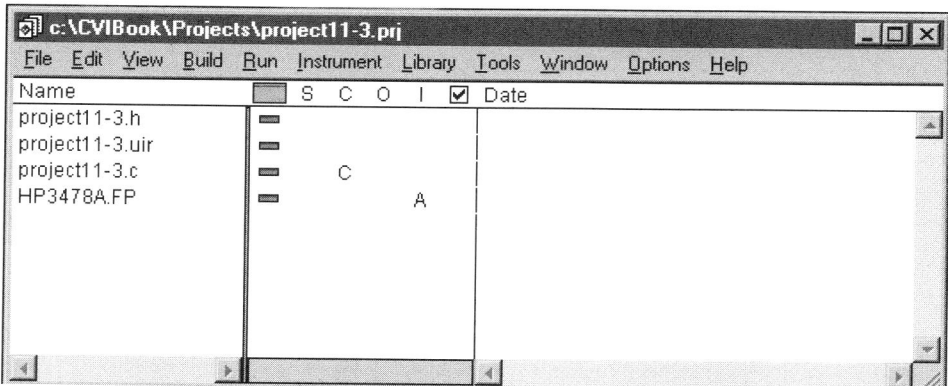

Figure 3–24
Project Window Icons

Letter codes and icon identifiers are associated with every project file. The heading row consists of the letters S, C, O, and I, and each has a certain meaning.

If the file is marked with a c under the header letter **C,** the file needs to be compiled before running. The compilation can be performed from the **Project** window without opening the .c file by double-clicking on the icon. When the file is compiled, this icon disappears.

An s letter under the **S** header represents that the file must be saved because it has been changed. You can do that by clicking on the s, which will disappear after the file is saved.

In *CVI* versions prior to 5.5, an o icon under the **O** column indicates that the file is compiled into an object file and applies only to source files. If you enable the **Create Object File** option from the **Source Editor,** *CVI* creates an object (.obj) file that contains the non-debuggable code in memory. Double-clicking on this icon toggles the option. In *CVI* 5.5 version, an o under the **O** column means that you should compile the file without debugging it since object files are always generated in *CVI* 5.5. If the o option is on (o is visible in the window list), the file is compiled without debugging; if it is off (o is not visible in the window list), the file is compiled with debugging when using the **Debug** configuration. Enabling this option reduces the amount of memory needed when building a project.

An I in the **I** column indicates that the file is an instrument driver program. We will look at this feature when we talk about instrument drivers in Chapter 11, "GPIB Communications."

Chapter 3 • More Graphical User Interface

An A under the **I** column indicates the (.fp) file is loaded into the **Instrument** menu and is attached to a program file. In Figure 3–24 the instrument driver for the digital multimeter (DMM) HP3478A is shown loaded in the instrument menu.

A U under **I** column indicates that the (.fp) file is loaded into the **Instrument** menu and is unattached to any program file. *CVI* attaches a program file when you double click on U.

If there is no letter in the **I** column next to the .fp file, no (.fp) file is loaded in memory.

This icon means that the file is currently closed. CVI opens this file when you double-click on this icon. If the file is a .fp file, the Function Tree Editor is opened.

This symbol next to the file name indicates that the file is open.

Every file has a date and time stamp that tells you when the file was last modified. (The date and time stamp has been removed from Figure 3–24.)

Observations

Let us look at some of the software metrics for this project. The total number of source lines for this project is 110 including all the comment lines. This number is small compared with other software tools doing a similar job.

The amount of hard-disk space required for each of the files in this project is shown in Table 3–4. The requirements are diminutive compared to other graphical languages' requirements for the same project.

When you **Build** the project, a folder with the name cvibuild.project3-2 is created in the project directory. If you selected the **Configuration** to **Debug,** an object file project3-2.nidobj with file size 39 Kbytes is created in the cvibuild.project3-2 folder. This is the object file compiled with the debugging option. At the same time, project3-2_dbg.cdb, a file of size 476 Kbytes containing the debugging information, is created and stored on the disk. This allows you to store the debugging information for your project when you leave *CVI* and re-start the project. The executable file project-3-2_dbg.exe, a 165 Kbyte file, is also created in the project directory.

Table 3–4 Memory Usage

File Name	Memory Usage (KB)
project3-2.c	3
project3-2.h	2
project3-2.uir	5
project3-2.prj	5

When you **Build** the project with the **Configuration** set to **Release,** the object file `project3-2.niobj` (2 Kbytes) is created is the `cvibuild.project3-2` folder. The executable `project3-2_dbg.exe` (144 Kbytes) is also created in the project directory. Notice that the **Release** files are considerably smaller than the **Debug** files.

Summary

You learned how to set the attributes of the panel and were introduced to the use of tool bars in the **User Interface Editor.** You learned some additional features of creating user interfaces and how to add the library functions using the function panel. The concept and use of timer controls and strip charts were discussed and demonstrated through projects. Lastly, the **Project** window icons were explained.

This chapter showed you how to create more user interface controls. The creation and usage of the timer control is a very important concept as it adds another dimension to the GUI—a dimension of time to execute various events at regular timer intervals. The projects demonstrated in this chapter used only one timer, though you can add as many timers as you like in the same project. For example, you could have added one timer to control the meter display and another timer to control the strip chart display, and both timers could be running independently using different time intervals.

Enhancing the User Interface

Chapter Highlights
- Project Overview
- Starting the Project
- Creating and Using the Menu Bar
- Generating the Source Code
- Creating and Using Tool Bars
- Adding a Text Box to the Project
- Creating the Tab Order
- Summary

"Enhancing the User Interface" introduces you to some of the more advanced features of the **User Interface Editor**. You will learn how to create menu bars to control the execution of the project using the menus and submenus. Creating and using tool bars with icons is introduced to show you another aspect of running and controlling your program. Many features that can enhance the GUI will be explained, including aligning objects on the panel and creating decoration boxes. Use of binary switches and LEDs (Light Emitting Diodes) will be explained and their operations shown. You will learn how to order the controls on the GUI to step through them in the user-specified order during execution using the <Tab> key.

Project Overview

Before you get started, let us discuss the first project in this chapter (named project4-1.prj). The project will use a random number generator to activate five input channels at a regular time interval of 250 milliseconds. You can selectively turn the channel monitoring on and off by means of the binary switches. The application will execute when you select the **START** command button and will terminate when you click on the **STOP** command button. You will have the choice of setting the LEDs to their default states by selecting **RESET** and running the application again.

Starting the Project

To get started, load the *CVI* program. The **Project** window will show the last project called (if any). Unload this project by selecting **File>>New>>Project(.prj)** or pressing (<Ctrl-N>). This will bring up a pop-up dialog box as shown in Figure 4–1.

Select **Yes.** The old project (if any) will be unloaded, and the **Compiler Options** dialog box will be displayed. After selecting the default compiler options, you will see the untitled1.prj window. Select **File>>New>>User Interface (*.uir)**. A blank **User Interface Editor** window is displayed as in Figure 4–2.

Select **Create>>Panel** and a new panel will be displayed where you will create the GUI. Double-click on the panel and enter CHANNEL for the **Constant Name**, Channel_CB for the **Callback Function,** and CHANNEL READINGS for the **Panel Title** in the **Edit Panel** box. Instead of trying to enhance

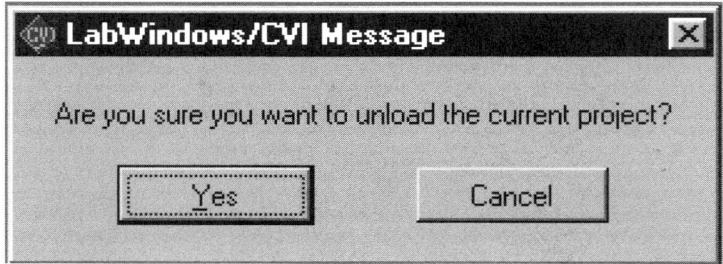

Figure 4–1
New Project Creation

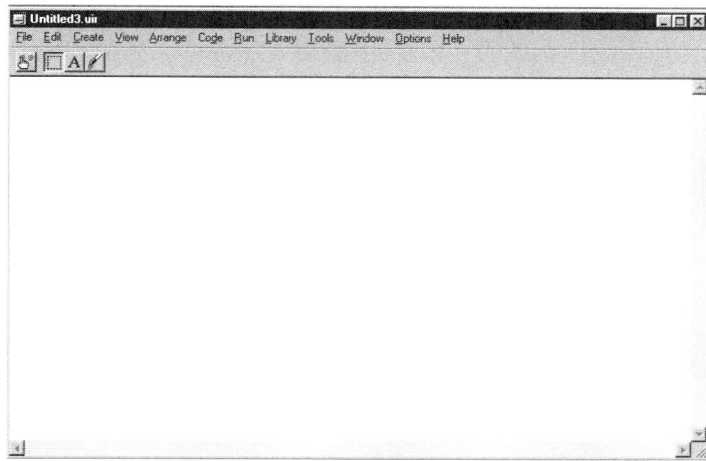

Figure 4–2
User Interface Editor Window

the GUI after it is completed, you will add the enhancements as you progressively build it. Let us begin by adding **LEDs** and **Binary Switches** to the **Decoration** box.

Decoration boxes consist of a variety of frames (or boxes) that you can place around the objects (or place objects on them) to make them prominent and to group controls according to their functions. A **Decoration** box is like a picture frame that enhances the picture. **Decoration** boxes can be accessed from the **Create** menu.

You will see different types of **Decoration** boxes. Move your mouse over them and get familiar with their names as they appear at the bottom of the box. Select the **Raised Box,** the very first item on the top row (see Figure 4–3).

A small box is formed on the panel. You can move it by dragging it with the mouse, and if you hold one of the corners, the mouse cursor becomes a double-arrow with which you can re-size the panel. Move the **Decoration** box to the left side of the panel. Re-size the box to approximately three inches wide by five inches high. On this **Decoration** box, you will create five additional decoration boxes in which you will place **LEDs** and **Binary Switches.** Each channel and the binary switch controlling this channel will be in one decoration box. Place five **Raised Frames** one below the other. Add the **LEDs** to the first decoration box by selecting **Create>>LED** and select the **Round LED.** You will notice that you cannot select the **Create** menu

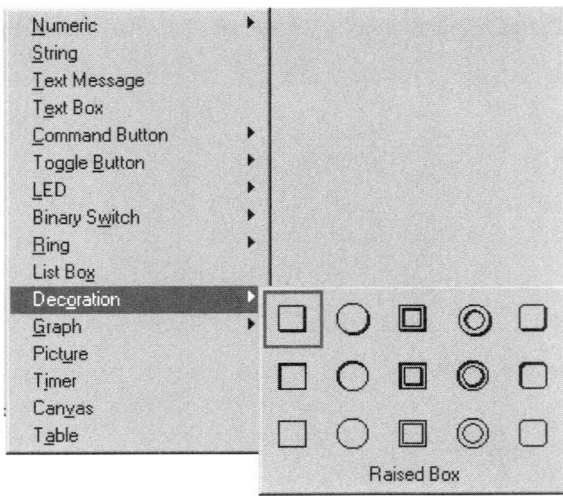

Figure 4–3
Decoration Boxes

if you right-click on certain **Decoration** boxes. In such situations, the right-click feature to display the **Create** panel works only when clicked on the panel.

Notice that by default the **Label** is placed above the **LED** for all the objects. Move the **Label** to the left side of the **LED** by clicking anywhere in the panel to de-select the control. Click on the **LED** label so only the label is selected, not the **LED.** Drag-and-drop it to the left side of the **LED.**

Move your mouse anywhere on the panel and right-click. The **Create** menu appears and you can create new objects. Similarly, add four more **LEDs** to each of the **Decoration** boxes, one below the other. The best way to do this is to copy the first **LED** and paste it onto the **Decoration** boxes one at a time. The **Decoration** boxes and the **LEDs** are shown in Figure 4–4.

When you place the **LEDs** on the panel, they are not aligned vertically and are unevenly spaced. Let *CVI* perform the alignment for you. Select the five **LEDs** by holding down the <Shift> key and clicking on each of them. If the controls are not on (a) **Decoration** box(es), you can select a group of controls by drawing a frame around them with the mouse. Unfortunately, when the objects are on some **Decoration** boxes or placed over another control, you cannot select the objects by drawing a frame around them without selecting the **Decoration** box along with the control.

Chapter 4 • Enhancing the User Interface

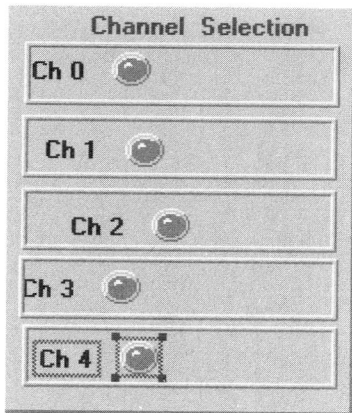

Figure 4–4
LEDs in **Decoration** Boxes

To align the objects in a straight column, select **Arrange>>Alignment** and move your cursor to the **Horizontal Centers** box and click on it. This is shown in Figure 4–5. You will notice that the **LEDs** are now aligned vertically.

With the LEDs still highlighted, select **Arrange>Distribution** and move your cursor to the **Vertical Centers** box in the top row and click on it. You will notice the vertical space between the **LEDs'** centers is now equidistant, and you will have perfect horizontal and vertical alignment of the **LEDs.** Your GUI should now look like Figure 4–6. You can align any objects on the panel using these alignment tools.

Double-click on the **LED** to bring up the attributes (properties) box, and for the first **LED,** enter the **Constant Name** as LED_0 and Label it Ch 0. You

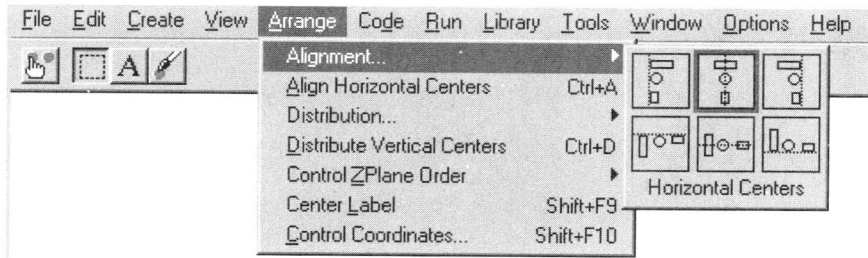

Figure 4–5
Objects Aligning Menu

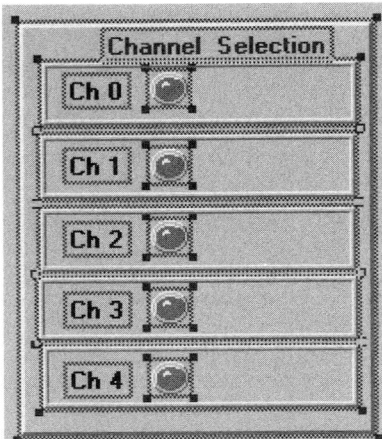

Figure 4–6
LEDs Aligned

will notice that the default value of the **LED** is set to `Off`. Change the color of the **Off** attribute color to green using the *Coloring Tool*. Recall that to do so, you select the *Coloring Tool* (shown in Figure 4–7), move the mouse over the **LED,** and right-click on it. A color template appears, as shown in Figure 4–8, from which you can select the color(s) of your choice.

Assign the following attributes to the **LEDs:**

LED	Constant Name	Label
0	LED_0	Ch 0
1	LED_1	Ch 1
2	LED_2	Ch 2
3	LED_3	Ch 3
4	LED_4	Ch 4

Figure 4–7
Coloring Tool

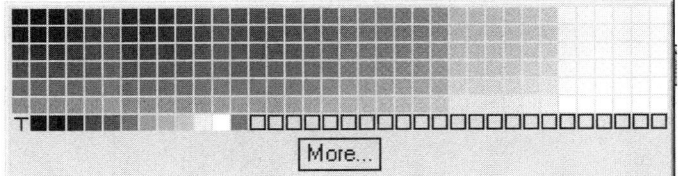

Figure 4–8
Color Template

Leave the rest of the attributes to their default values.

Now let us add the **Binary Switches** to the right side of the **LEDs** inside the decoration boxes. Right-click on the panel and select **Binary Switch,** and from the second row in the middle column, click on **Horizontal Grooved Switch.** Drag-and-drop it on the **Decoration** box. Double-click on this switch, and for **Constant Name,** enter SW_0. Change the default value to **ON** and remove the label by leaving the label attribute blank.

You can add the switches individually to each **Decoration** box, but it is quicker to copy and paste the first switch box four times in each of the remaining **Decoration** boxes. Similarly, align the **Binary Switches.** Your GUI with the **LEDs** and the **Binary Switches** should now look like Figure 4–9 without the Channel Selection title.

Change the **Constant Name** of these **Binary Switches** to the following:

Switch	Constant Name
0	SW_0
1	SW_1
2	SW_2
3	SW_3
4	SW_4

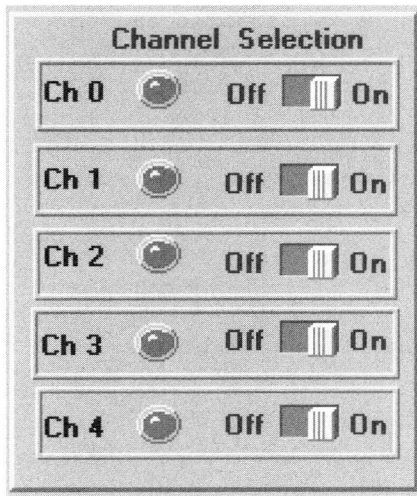

Figure 4–9
LEDs and **Binary Switches** after Alignment

In addition, you want to be sure that the switches are in the **On** state when the program starts executing (if not already so). Double-click on the switch control to bring up the properties dialog box and observe that the **Initial State** is set to **On** for all the switches.

To add a title to the **Decoration** box, select **Create>>Text Message,** double-click on this object, and enter `Channel Selection` in the **Default Value** box. Drag-and-drop to center this title on top of the **Decoration** box above the LEDs and switches.

Create a timer to process the channels every 250 milliseconds (msec) and activate the corresponding channel LED. Right-click on the panel and select **Timer**. Double-click on the **Timer** and enter `Timer_250MS` for the **Constant Name**. For the **Callback Function,** enter `Generate250LoopCB`. In the **Enabled** box, remove the check mark. This will stop the timer from starting as soon as the project starts to execute. You will want to programmatically start and stop the **Timer**. In the **Interval (seconds):** box, set the time interval to `0.250` to execute the code in the timer callback function every 250 milliseconds, as you will see later in this chapter. In the **Label** box, enter `Timer 250 msecs`.

Add four command buttons with the following attributes:

Label	Constant Name	Callback Function	Control Mode
START	START	StartCB	Hot
STOP	STOP	StopCB	Hot
RESET	RESET	ResetCB	Hot
EXIT	EXIT	ExitCB	Hot

These command buttons will perform the functionalities as indicated by their names, that is, **START** will begin the application execution, **STOP** will stop the application, etc. Align these command buttons vertically and equidistant from each other. Place a **Decoration** box of your liking around these controls. On some **Decoration** boxes, you will need to copy and paste these command buttons. Your *GUI* should now look like Figure 4–10. Name this GUI `project4-1.uir` and add it to the **Project** window.

Use the *CodeBuilder* to generate the skeleton code using *ExitCB* for exiting the project. Rename the source code file to `project4-1.c` and add it to the **Project** window.

Chapter 4 • Enhancing the User Interface

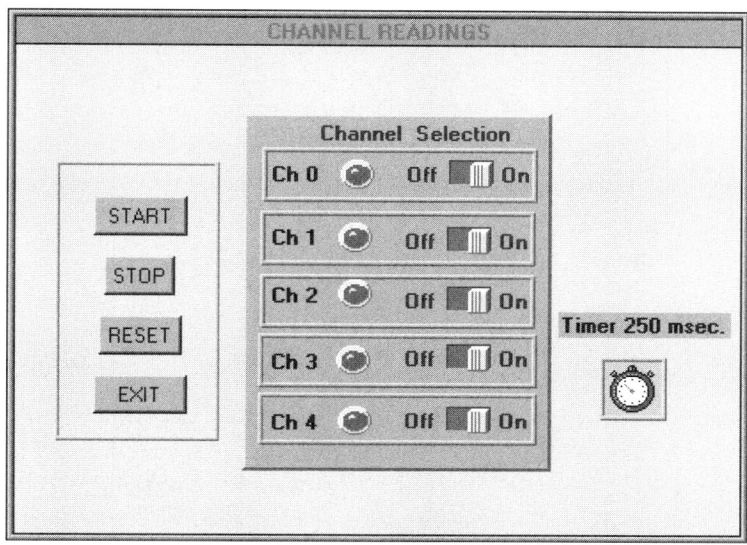

Figure 4–10
Project4–1 GUI

Examine the source code in Figure 4–11. As before, the header files containing the library function prototypes are listed at the top of the source code, including the header file project4-1.h.

The function prototype of the user-created function *TurnON_LED* is included and will be explained.

In the main function, the project4-1 panel is loaded and displayed and *RunUserInterface* processes the events.

The *StartCB* callback function is invoked by selecting the **START** command button. This function enables the 250-millisecond timer using the library function *SetCtrlAttribute*, as shown in Figure 4–12.

The *StopCB* callback function shown in Figure 4–13 is invoked by selecting the **STOP** command button. This function disables the timer using the *SetCtrlAttribute* function as shown in Figure 4–12 with the last argument set to 0.

The *ExitCB* callback function shown in Figure 4–14 will call the *QuitUserInterface* function and stop the processing of events. In this case, the timer is also disabled (in case the execution was not stopped from the **STOP** command button before exiting the program).

```
#include <ansi_c.h>
#include <cvirte.h>     /* Needed if linking in external compiler; harmless otherwise */
#include "project4-1.h"
#define NUM_LEDS    5
#define SW_CTRL_INDEX    0
#define LED_CTRL_INDEX   1
static int channel;
int attr[NUM_LEDS][2] = {
            {CHANNEL_SW_0, CHANNEL_LED_0},
            {CHANNEL_SW_1, CHANNEL_LED_1},
            {CHANNEL_SW_2, CHANNEL_LED_2},
            {CHANNEL_SW_3, CHANNEL_LED_3},
            {CHANNEL_SW_4, CHANNEL_LED_4}
            };
void TurnON_LED (int LED_Number); //to turn on LEDs
int main (int argc, char *argv[])
{
    if (InitCVIRTE (0, argv, 0) == 0)   /* Needed if linking in external compiler;
                                                         harmless otherwise */
        return -1;   /* out of memory */
    if ((channel = LoadPanel (0, "project4-1.uir", CHANNEL)) < 0)
        return -1;
    DisplayPanel (channel);
    RunUserInterface ;
    return 0;
}//main
```

Figure 4–11
Project 4–1 Main Function

```
//Start the application when START command button is selected
int CVICALLBACK StartCB (int panel, int control, int event,
            void *callbackData, int eventData1, int eventData2)
{
    switch (event)
        {
        case EVENT_COMMIT:
            // Start the 250 msec Timer
            SetCtrlAttribute(channel, CHANNEL_TIMER_250MS,
                                    ATTR_ENABLED, 1); //start timer
            break;
        }
    return 0;
} //StartCB
```

Figure 4–12
StartCB Callback Function

Chapter 4 • Enhancing the User Interface

```
//Stop the application, invoked from the STOP command button
int CVICALLBACK StopCB (int panel, int control, int event,
          void *callbackData, int eventData1, int eventData2)
{
    switch (event)
        {
        case EVENT_COMMIT:
            SetCtrlAttribute(channel, CHANNEL_TIMER_250MS,
                                        ATTR_ENABLED,0); //stop timer
            break;
        }
    return 0;
} //StopCB
```

Figure 4–13
StopCB Callback Function

The *ResetCB* function is shown in Figure 4–15 and, as the name implies, sets the LEDs and the binary switches to their default state. This callback function is invoked from the **RESET** command button and uses the *SetCtrlAttribute* library function.

The panel was assigned a callback function, *Channel_CB*, whose source code listing is shown in Figure 4–16.

This callback is invoked by clicking on the "X" on the top right corner of the panel, generating an EVENT_CLOSE event that causes the panel to close and exit the project. The timer is disabled and the *QuitUserInterface* function is called.

```
//Stops the Timer and Exit the program
int CVICALLBACK ExitCB (int panel, int control, int event,
          void *callbackData, int eventData1, int eventData2)
{
    switch (event)
        {
        case EVENT_LEFT_CLICK:
            //Stop the 250 msec. Timer
            SetCtrlAttribute(channel , CHANNEL_TIMER_250MS,
                                        ATTR_ENABLED, 0);  //Stop Timer
            QuitUserInterface (0);
            break;
        }
    return 0;
} //ExitCB
```

Figure 4–14
ExitCB Callback Function

```
//All LEDs and switches reset to their startup default states
int CVICALLBACK ResetCB (int panel, int control, int event,
            void *callbackData, int eventData1, int eventData2)
{
    int i;
    switch (event)
        {
        case EVENT_COMMIT:
            //Set all switches to On State
            //Set all LEDs to OFF
            for (i=0;i<NUM_LEDS;i++){
                SetCtrlAttribute(channel,
                        attr[i][SW_CTRL_INDEX],ATTR_CTRL_VAL,1);
                SetCtrlAttribute(channel,
                        attr[i][LED_CTRL_INDEX],ATTR_CTRL_VAL,0);
            }

            break;
        }
    return 0;
} //ResetCB
```

Figure 4–15
ResetCB Callback Function

In *CVI*, all user interface objects have attributes (properties) associated with them, like color, position, size, visibility, and many more. Some objects have specific attributes. Many attributes can be set when creating the GUI. You can change the value of many of the attributes programmatically using the *SetCtrlAttribute* function. You used this function in Chapter 3, "More Graphical User Interface" to set the timer attributes. Here this function is

```
//Close the panel when the 'X' on the panel window is clicked
int CVICALLBACK Channel_CB (int panel, int event, void *callbackData,
            int eventData1, int eventData2)
{
    switch (event)
        {
        case EVENT_CLOSE:
            SetCtrlAttribute(channel, CHANNEL_TIMER_250MS,
                                    ATTR_ENABLED,0); //stop timer
            QuitUserInterface (0);
            break;
        }
    return 0;
} //Channel_CB
```

Figure 4–16
Channel_CB Callback Function

used to set the LED and the binary switch attribute values. Using *SetCtrlAttribute* in *ResetCB* function, the LED values are set to 0, which is the Off state, and the binary switch set to 1, which is the On state.

The prototype for *SetCtrlAttribute* is shown below, and the function arguments are explained in Table 4–1.

```
int status =SetCtrlAttribute (int panelHandle, int ControlID, int
                    controlAttribute, controlValue¹.... );
```

Different control attributes have different data types and different valid ranges. A list of attributes, their data types, and valid values are provided in Tables 3–9 to 3–44 in Chapter 3, "Programming with the User Interface Library," of the *LabWindows/CVI User Interface Reference Manual* or refer to Online Help. Although you can obtain the value of all control attributes using *GetCtrlAttribute*, there are some control attributes that you cannot modify. The attribute list in Tables 3–9 to 3–44 of this manual indicates which attributes cannot be modified.

The *Generate250LoopCB* callback function listed in Figure 4–17 is used by the timer to execute the code every 250 milliseconds. A new random number is generated on every timer tick and assigned a LED_Number. This number is

Table 4–1   SetCtrlAttribute *Function*

Input/ Output	Name	Type	Description
Input	panelHandle	integer	panel handle where the panel is loaded
	controlID	integer	constant name assigned to the control
	controlAttribute	integer	a particular control attribute
	controlValue	¹depends on the attribute	new value of the control attribute
Output	status	integer	refer to Appendix A in *LabWindows/CVI User Interface Library Reference Manual* for error codes or in Online Help

```
//Timer controlled loop. Code executed every 250 msec.
int CVICALLBACK Generate250LoopCB (int panel, int control, int event,
            void *callbackData, int eventData1, int eventData2)
{
    int LED_Number;
    int SW_Val;
    switch (event)
        {
            case EVENT_TIMER_TICK:
                //invokes routine at every timer tick
                //Generate random LED number
                LED_Number=(rand/(1500)% NUM_LEDS);

                GetCtrlAttribute(channel, attr[LED_Number][SW_CTRL_INDEX],
                                            ATTR_CTRL_VAL, &SW_Val);

                if (SW_Val) TurnON_LED(attr[LED_Number][LED_CTRL_INDEX]);
                break;
        }
    return 0;
} //Generate250LoopCB
```

Figure 4–17
Generate250LoopCB Callback Function

passed to the appropriate section in the code and gets the control attribute value of the binary switch. The setting of the switch, On or Off, turns the corresponding LED On or Off.

If the binary switch is enabled, the user-defined function *TurnOn_LED* is called to turn On the LED whose Control ID is passed as an argument to this function.

The *TurnOn_LED* function is listed in Figure 4–18.

```
//Turn on the selected LED
void TurnON_LED(int LEDOn_Number)
{    // First turn all LEDs off,then turn the selected LED ON
    int i;
    for (i=0;i<NUM_LEDS;i++){
            SetCtrlAttribute(channel, attr[i][LED_CTRL_INDEX], ATTR_CTRL_VAL,
                                            0);//Turn off LED
    }
    SetCtrlAttribute(channel, LEDOn_Number, ATTR_CTRL_VAL, 1);//Turn on
selected LED
}//TurnON_LED
```

Figure 4–18
TurnOnLED Callback Function

Chapter 4 • Enhancing the User Interface

The *GetCtrlAttribute* function retrieves the various control attributes programmatically.

The function prototype for *GetCtrlAttribute* is shown below, and its arguments are listed in Table 4–2.

```
int status = GetCtrlAttribute (int panelHandle, int controlID,
                int controlAttribute, void *attributeValue);
```

If you want to view the complete listing of the source code, double-click on `project4-1.c`. The **Source Code** window will display the complete code.

Build and run the project. Click the **START** command button on the GUI to execute the application. The random number generator will turn the LEDs on and off for various channels. To disable the monitoring of a certain channel, turn off the binary switch for that channel. Notice that the LED for that channel does not light up as long as the corresponding binary switch is off. You can turn off/on as many switches as you desire and observe that the LEDs are being controlled by the switches. Selecting the **STOP** command button temporarily halts the execution. Selecting **RESET** will set the LEDs and the binary switches to their default values. To terminate the execution, select **EXIT**. A snapshot of the program execution is given in Figure 4–19.

Remember that <Ctrl-N> is the shortcut for the **New Project** command.

Table 4–2 GetCtrlAttribute *Function*

Input/Output	Name	Type	Description
Input	panelHandle	integer	panel handle loaded in memory
	controlID	integer	constant name assigned to the control
	controlAttribute	integer	a particular control attribute
	attributeValue	depends on the attribute	current value of control attribute
Output	status	integer	refer to Appendix A in *LabWindows/CVI User Interface Library Reference Manual* for error codes or in Online Help

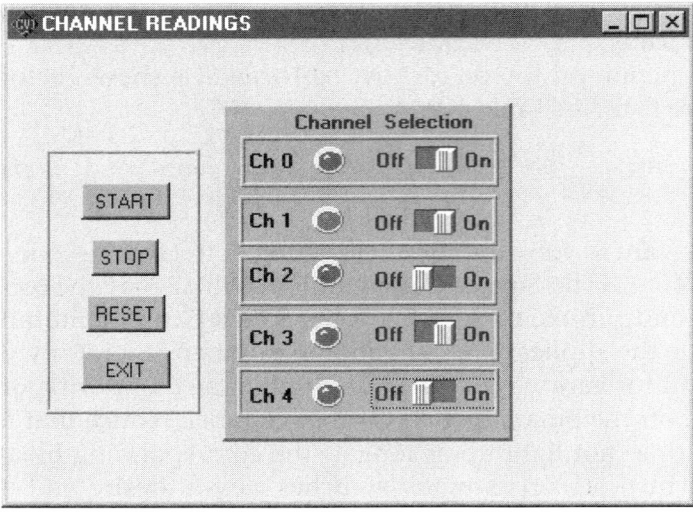

Figure 4–19
Project4–1 Run

Creating and Using the Menu Bar

Menu Bars are the pull-down menus that reside on the top part of the panel. When using Windows applications, you deal with the **Menu Bars** almost unconsciously. Look at the **User Interface Editor** and notice the line consisting of names like **File, Edit, Create, View, Arange,** and **Code.** These make up a menu bar associated with the panel. In this text, a menu or a menu command will be used synonymously to refer to one of the commands that you see on a typical Windows menu bar. In this case, **File** is one menu or menu command, **Edit** is another, and so on. Normally when you select a menu command from the menu bar, you see a list of sub-commands or menu items on the pull-down menu. In this text, the menu items, sub-menu commands, and sub-commands will be used synonymously.

Select **Edit>>Menu Bars** from the **User Interface Editor.** This brings up the **Menu Bar List** as shown in Figure 4–20.

If there are no previous menu bars created, a blank dialog box will appear. Select **Create...** from the **Menu Bar List,** and it opens an **Edit Menu Bar** dialog box as shown in Figure 4–21.

Previously created menu bars would appear in the list, and the **Edit, Cut, Copy,** and **Paste** buttons would be enabled. When no menu bars have yet been created, the buttons will be dimmed (unusable) as shown in Figure 4–20.

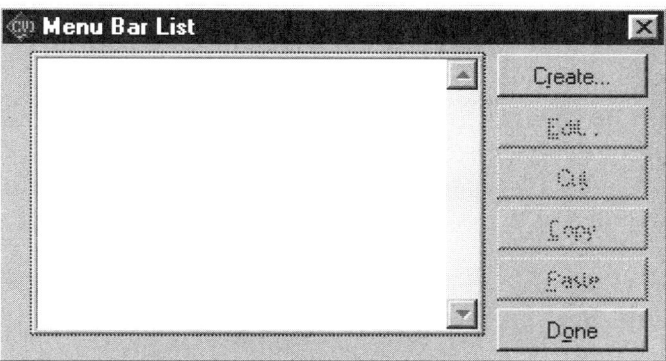

Figure 4–20
Menu Bar List

Figure 4–21
Edit Menu Bar Dialog Box

Create opens a new menu bar, where the menu and menu items are created. **Edit** opens the **Edit Menu Bar** from which the control boxes are modified if needed. **Cut** deletes the highlighted item in the menu bar and places it on the clipboard. **Copy** leaves the highlighted item in the menu bar and places it on the clipboard. **Paste** inserts the items **Cut** or **Copied** from the clipboard to the current cursor location. **Done** closes the **Menu Bar List** dialog box.

The attributes of the menu and sub-menu commands are set from the **Edit Menu Bar** dialog box shown in Figure 4–21. Let us look at these attributes and see what they mean.

The **Menu Bar Constant Prefix** is the resource identification of the menu bar, similar to the **Constant Name** assigned to the *panel* or *controls* for creating menu bars. The menu bar is identified by this name. All the items of the menu bar will have this identification prefixed to the item name. The normal convention is to use all uppercase letters in this field.

Item is the command name that is visible on the menu bar. It could be the name of the menu bar, sub-menu or menu command. Inserting two underscore characters before a letter in the item name underscores that letter in the item name on the menu bar. The <Alt> key plus the underscored letter can then be used to select this item from the keyboard.

Constant Name defines the name of the current item. For objects in the panel, this is a unique name. The **Constant Name** paired with the **Menu Bar Constant Prefix** identifies this menu item uniquely.

In the **Callback Function** box, enter the callback function name if you want certain code to be executed when this menu or menu item is selected.

Modifier Key refers to keyboard keys that are associated with the menu item to execute the function. The **Modifier Key** choices are shown in Figure 4–22.

A useful feature of the menu bar is the ability to assign shortcut keys for the menu items by means of the **Modifier Key.** Click on control next to the **Modifier Key** and select the **MenuKey (Ctrl)** from the list. Click on the control next to the **Shortcut Key.** You will get a choice of using any of the shortcut keys shown in Figure 4–23. For example, if you selected **S** from the list, whenever <Ctrl-S> is pressed during the execution of this program, this menu item will be selected.

Referring to Figure 4–22, you can check the **Dimmed** box if for some reason you do not want the user to access this menu item when the menu is loaded initially. This menu item can be un-dimmed programmatically.

In Figure 4–23, **Checked** indicates if this menu item initially has a check mark next to it. **Insert New Item,** shown on the top right side of this dialog box, gives you the choice of inserting the next item above or below the current item. **Insert Separator** inserts a line above or below the current item.

Chapter 4 • Enhancing the User Interface

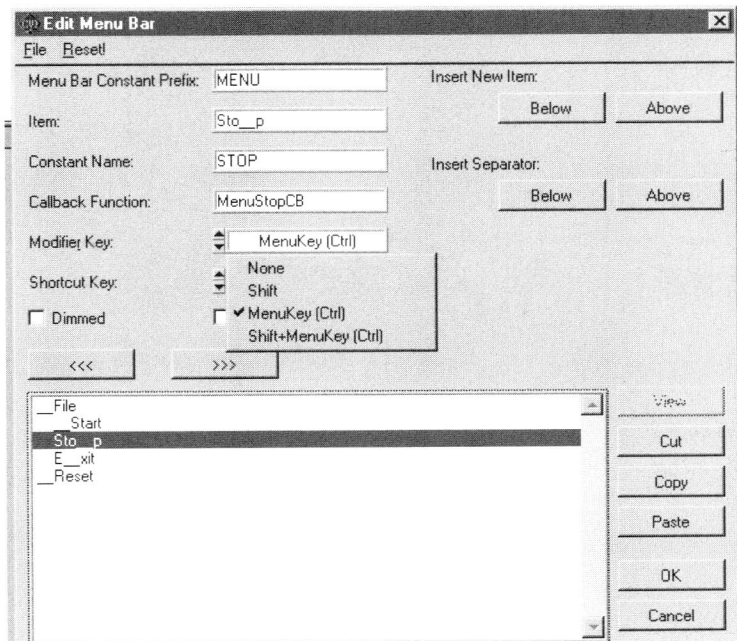

Figure 4–22
Edit Menu Bar Modifier Key Choices

   *The left hierarchy button moves the selected item one level up in the sub-menu.*

    The right hierarchy button moves the selected item one level down in the sub-menu.

View displays the current state of the menu bar and all the pull-down items that you have created so far. This is how your menu bar will look when the program executes. If necessary, you can change them now before going any further.

You will build upon `project4-1.prj` and add a menu bar that will perform the same functionalities as the command buttons. The menu bar will consist of the *Start, Stop,* and *Exit* callback functions in the first menu. In the next menu, using the same menu bar, create a menu to *Reset* the LEDs and binary switches.

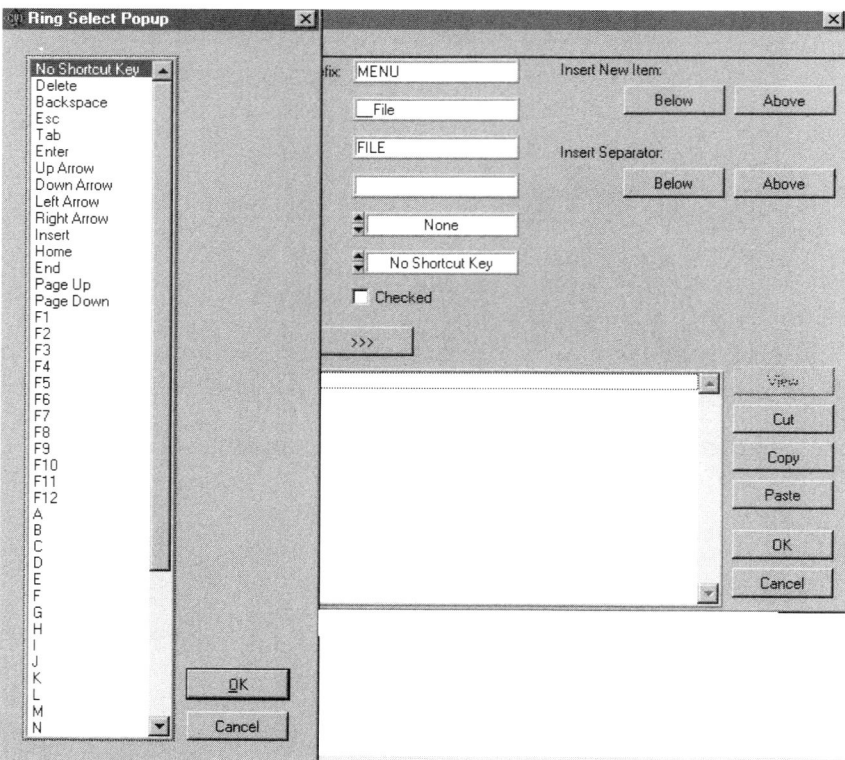

Figure 4–23
Edit Menu Bar Shortcut Key Choices

Load `project4-1.prj`. Rename the project and all `project4-1` files as `project4-2`. Delete `project4-1.c` from the **Project** window by highlighting the file and selecting **Edit>>Remove File** as displayed in Figure 4–24.

You will be creating a new source code file using the *CVI CodeBuilder*. Double-click on `project4-2.uir` to open the user interface file. Delete all the command buttons (including the decoration box) by drawing a frame around them, holding down the left mouse button mouse and pressing <Delete> on the keyboard. If you wish, you could leave the command buttons and use them as well to operate the application, but the emphasis in this project is to show you how to control the application using the menu commands.

Center the remaining objects on the panel. Your GUI for `project4-2` should now look like Figure 4–25.

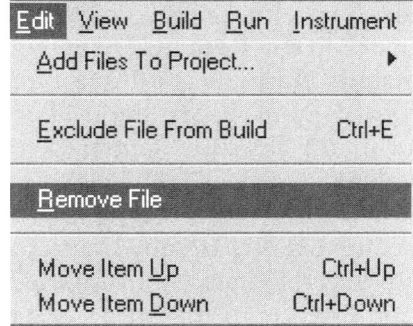

Figure 4–24
Edit Menu Commands (**Project** Window)

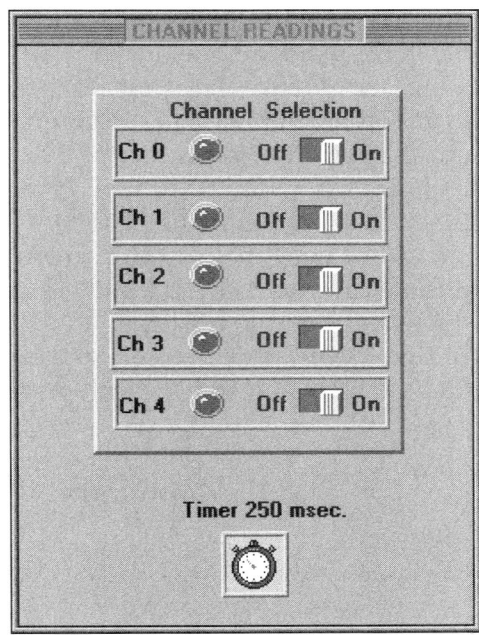

Figure 4–25
Project4–2 GUI

Select **Edit>>Menu Bars** and **Create**. For the **Menu Bar Constant Prefix,** enter `MENU`. Enter `__File` in the **Item** field. This will appear as the first menu command. The **Constant Name** is a unique name that is assigned to this menu. Enter `FILE` for this item. Leave the callback function blank since the **File** menu does not call any function. Leave the other attributes on this box set to their default values.

Click on the **Below** button under **Insert New Item:**. This will open another dialog box with the **MENU** already entered in the **Menu Bar Constant Prefix** box to indicate that this item is part of the same menu bar. Add **Start** as a menu item to the **File** menu that you just created. For the **Item,** enter `__Start`, and for the **Constant Name,** enter `START`. Enter `MenuStartCB` for the **Callback Function** for this menu item.

Place **Start** a level below the **File** menu by highlighting **Start** in the list box and clicking on the right hierarchy button. This will shift **Start** one level down to the right.

CVI keeps track of the menu **Constant Name** by following a hierarchy of concatenating the **Menu Bar Constant Prefix** and the **Item** name(s). In the previous menu item, the **Start** menu item will be addressed as `MENU_FILE_START`. Keeping the **Menu Bar Constant Prefix** and the **Constant Name** to less than ten characters each is recommended. Otherwise, when you select **OK** from the **Edit Menu Bar** dialog box, you will get a message indicating that the names have been truncated. Some **Constant Names** could be truncated to the same shortened name, resulting in an error.

Similarly, add the list of items below to the menu bar list, clicking on the right hierarchy button each time to keep these items at the same level as the **Start** command. To add a new item to the list, click on the **Below** button under the heading **Insert New Item.** Note that you can add an item above the previously added item by selecting the **Above** button under the heading **Insert New Item:**

	STOP Menu Item	**EXIT Menu Item**
Menu Bar Constant Prefix	`MENU`	`MENU`
Item	`St__op`	`E__xit`
Constant Name	`STOP`	`EXIT`
Callback Function	`MenuStopCB`	`MenuExitCB`

To create another menu, enter the same **Menu Bar Constant Prefix** (`MENU`), and for **Item,** enter `Reset` to create a **Reset** menu command. For the **Constant**

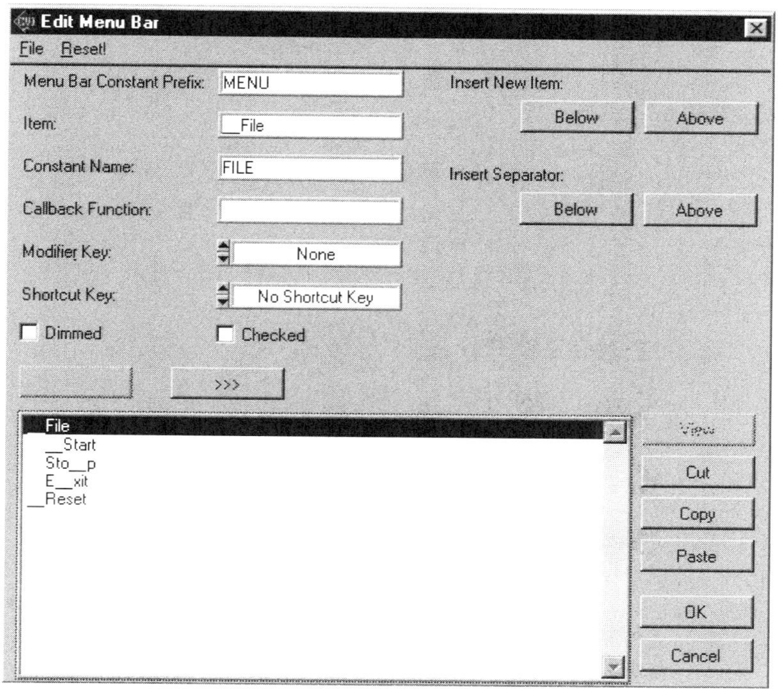

Figure 4–26
Menu Bar Items

Name, type RESET, and for the **Callback Function,** type MenuResetCB. In *CVI,* a menu command with a "!" means that the menu has no menu item(s) below it. The order of the menu items in the list can always be changed if you decide to do so later by simply using the **Cut** and **Paste** commands.

All the menu bar items needed for this project are now complete. The completed menu bar list should now look like the one shown in Figure 4–26.

It is now time to create the skeleton code using the *CVI CodeBuilder.*

Generating the Source Code

Previously, you saw how *CVI CodeBuilder* creates the source code for command button callback functions. You will see the same here, but you will notice some subtle differences in the way the callback functions are created when you use the menu bars instead of the command buttons or other control objects.

Let us begin by saving everything we have done so far by selecting **File>>Save All.** Open the user interface file (if not already open) by clicking on project4-2.uir in the **Project** Window.

Select **Code>>Set Target File>><New Window>** to create a new source code file. Select **Code>>Generate>>All Code.** Check the *MenuExitCB* callback function from the bottom sub-panel since this function will be used to exit the program. Select **OK,** and the *CodeBuilder* will display the skeleton code file. It will be named **Untitled1.c** by *CVI*. Save this file as project4-2.c using the **File>>Save As** command.

Include this file into the project by selecting **File>>Add Files to Project.** The file name will now appear in the **Project** Window. Let us look at some of the salient features of this skeleton code created by the *CodeBuilder* and shown in Figure 4–27.

The *main* function loads and displays the user interface resource and creates its panel handle. The *RunUserInterface* library function is called to process the user interface events as they are generated. You have seen this in the previous project. Nothing new here.

Look at the menu item callback function's code. One of the most obvious differences you will notice between the control callback functions and the menu callback functions is the lack of an event variable to check in menu callbacks. This event variable does not appear because when the user selects a menu item, that menu item callback function is invoked regardless of whether a GOT_FOCUS or DOUBLE_CLICK event was generated. Therefore, there is no switch (event) in the menu callback function's source code. Observe that the menu callback function is always of type void and that a *return* to the calling routine does not appear because the menu and menu item callbacks are not passing any *events* to the *RunUserInterface* function. There is no need to inform the *RunUserInterface* function that the callback function has been completed and that the *RunUserInterface* should process the next *event*.

The prototype for the menu callback function is different from the control callback function. There are no arguments for *events* or *eventData1* and *eventData2* in the menu callback functions. The generic menu callback function prototype is shown below.

```
void CVICALLBACK MenuStartCB (int menuBar,int menuItem,
                     void *callbackData,int panel);
```

Chapter 4 • Enhancing the User Interface

```c
#include <cvirte.h>
#include <userint.h>
#include "project4-2.h"
static int channel;
int main (int argc, char *argv[])
{
        if (InitCVIRTE (0, argv, 0) == 0)
                return -1; /* out of memory */
        if ((channel = LoadPanel (0, "project4-2.uir", CHANNEL)) < 0)
                return -1;
        DisplayPanel (channel);
        RunUserInterface ;
        DiscardPanel (channel);
        return 0;
}
int CVICALLBACK Generate250LoopCB (int panel, int control, int event,
                void *callbackData, int eventData1, int eventData2)
{
        switch (event)
                {
                case EVENT_TIMER_TICK:
                        break;
                }
        return 0;
}
void CVICALLBACK MenuStartCB (int menuBar, int menuItem, void *callbackData,
                int panel)
{
}
void CVICALLBACK MenuStopCB (int menuBar, int menuItem, void *callbackData,
                int panel)
{
}
void CVICALLBACK MenuExitCB (int menuBar, int menuItem, void *callbackData,
                int panel)
{
        QuitUserInterface (0);
}
void CVICALLBACK MenuResetCB (int menuBar, int menuItem, void *callbackData,
                int panel)
{
}
int CVICALLBACK Channel_CB (int panel, int event, void *callbackData,
                int eventData1, int eventData2)
{
        switch (event)
                {
                case EVENT_CLOSE:
                        break;
                }
        return 0;
}
```

Figure 4–27
Skeleton Code Generated by *CodeBuilder*

The `menuBar` is the handle of the menu bar on which the menu item resides. The handle is generated by the *LoadMenuBar* function as shown in Figure 4–28.

The `menuItem` is the **Control ID** of the menu item. This is the concatenation of the **Menu Bar Constant Prefix** and the **Constant Name** that was entered in the menu item list box. This name uniquely defines the menu item(s). The names have to be unique, or *CVI* will generate an error when you select **Done**.

The `callbackData` is a void pointer to user defined data.

The `panel` is the panel handle of the panel on which the menu bar is loaded.

Add the highlighted lines of Figure 4–28 to `project4-2.c` in the `main` function.

The *LoadMenuBar* function loads the menu bar. The prototype of *LoadMenuBar* and the definition of the arguments of this function are given below.

```
int  menuBarHandle= LoadMenuBar( int destinationPanelHandle,
           char filename[ ], int menuBarResourceID);
```

The arguments are explained in Table 4–3.

Add the highlighted lines to the skeleton code of the *MenuStartCB* callback function. This function is called by the **Start** menu item and enables the 250-millisecond timer. The code for this function is shown in Figure 4–29.

Add the highlighted line to the skeleton code of the *MenuStopCB* function as shown in Figure 4–30 listed below. The **Stop** menu item stops the 250-milliseconds timer enabled by *MenuStartCB*.

```
static int channel, menuHandle;
//Main function
int main (int argc, char *argv[])
{
    if (InitCVIRTE (0, argv, 0) == 0)    /* Needed if linking in external compiler;
                       harmless otherwise */
         return -1;    /* out of memory */
    if ((channel = LoadPanel (0, "project4-2.uir", CHANNEL)) < 0)
         return -1;
    //Load the menubar
    menuHandle = LoadMenuBar (channel, "project4-2.uir", MENU);

    DisplayPanel (channel);

    RunUserInterface ;
    return 0;
} //main
```

Figure 4–28
Project 4–2 Main Function

Chapter 4 • Enhancing the User Interface

Table 4–3 LoadMenuBar *Function*

Input/ Output	Name	Type	Description
Input	destinationPanel-Handle	integer	handle of the panel where the menu bar resides
	filename	string	name of the user interface file (.uir) that contains the menu bar
	menuBar ResourceID	integer	the menu bar constant prefix assigned to the menu bar in the User Interface Editor
Output	menuBarHandle	integer	value used to refer to this menu bar (negative values indicate error), refer to Appendix A in *LabWindows/CVI User Interface Library Reference Manual* or in Online Help

Add the highlighted lines to the skeleton code for *MenuResetCB* function as shown in Figure 4–31. This function is similar to the control callback function *ResetCB* explained above in Figure 4–15, except for the structure of the menu callback function. This function is invoked when the **Reset!** button is clicked on the menu bar.

```
/*************************************************
This callback function starts the program execution
when the "Start" menu item is selected. This function enables
the timer.
*************************************************/
void CVICALLBACK MenuStartCB (int menuBar, int menuItem, void *callbackData,
        int panel)
{
        //Start the 250 msec Timer,
        SetCtrlAttribute(channel , CHANNEL_TIMER_250MS, ATTR_ENABLED, 1);

} //MenuStartCB
```

Figure 4–29
MenuStartCB Callback Function

```
/***********************************************
This callback function stops the program execution when
the Stop menuitem is selected.
***********************************************/
void CVICALLBACK MenuStopCB (int menuBar, int menuItem, void *callbackData,
          int panel)
{
          SetCtrlAttribute(channel , CHANNEL_TIMER_250MS,
                                    ATTR_ENABLED, 0);   //Stop Timer

} //MenuStopCB
```

Figure 4–31
MenuStopCB Callback Function

Copy the code from `project4-1` for *Generate250LoopCB* and the *Channel_CB* callback function since it is the same as in `project4-1.prj`.

After you have entered the code in Figures 4–28 through 4–31, save your work, compile the program, and check for any compiler errors. If you find some errors, fix them now before proceeding.

Run the project. It should run similar to the previous project except the command buttons are replaced by the menu commands. Clicking on the menu and menu item commands will execute similar code as the command buttons shown previously. An instance of the run is shown in Figure 4–32.

```
/***********************************************
This routine resets the LEDs and the Switches to the default
state, to before when the program started executing.
***********************************************/
void CVICALLBACK MenuResetCB(int menubar, int menuItem, void *callbackData, int panel)
{
        int i;
        //Set all switches to On State
            //Set all LEDs to OFF
            for (i=0;i<NUM_LEDS;i++){
                    SetCtrlAttribute(channel,
                            attr[i][SW_CTRL_INDEX],ATTR_CTRL_VAL,1);
                    SetCtrlAttribute(channel,
                            attr[i][LED_CTRL_INDEX],ATTR_CTRL_VAL,0);
            }

} //ResetLEDCB
```

Figure 4–31
MenuResetCB Callback Function

Chapter 4 • Enhancing the User Interface

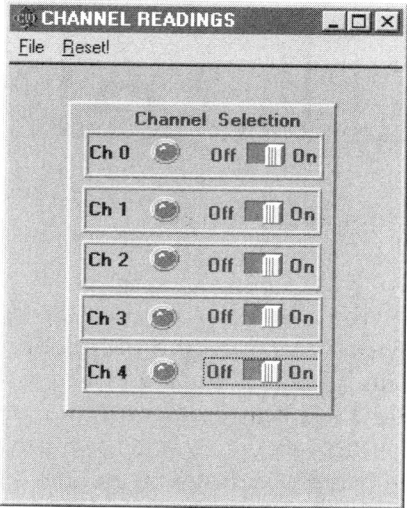

Figure 4–32
Project4–2 Run

You will notice that the menu bar on the user interface is visible only during run-time and not when you create the user interface; the menus are created and loaded programmatically and are not created on the user interface as objects.

Note: If the **Channel Selection** decoration box and the controls are created too close to the top of the panel, the menu bar writes over the top over the decoration box. To fix this problem, open the user interface file, select the decoration box with all the controls on it and drag it down sufficiently.

To exit the program, select **File>>Exit.**

Creating and Using Tool Bars

In this section, you will learn how to create a **Tool Bar,** which consists of a series of picture buttons shown on the panel just below the menu bar. If you look at the **User Interface Editor,** you will see the icons, shown in Figure 4–33, that make up the tool bar. Each icon is an eighteen by eighteen pixel image and, for Windows, could be in any of the .wmf, .bmp, .pcx, .ico, .dib and .rle file formats. The tool bars are not part of the GUI creation. They are not created using the **User Interface Editor** but programmatically by calling the **Tool Bar** functions from the tool bar function library and inserting them in the code at the appropriate location.

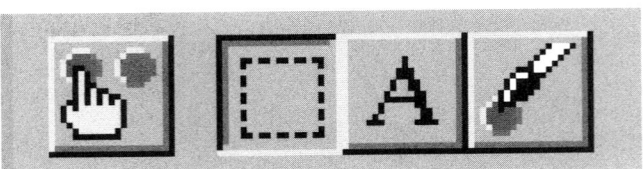

Figure 4–33
User Interface Editor Tool Bar

The discussion of the **Tool Bar** belongs with the menu bar discussion since they both perform the same kind of functions. The menu bar invokes a selection via the menu item and the **Tool Bar** via the icon.

To understand the **Tool Bar,** create another project with **Tool Bar** icons. Rename `project4-2.prj` to `project4-3.prj` and all the other files to `project4-3` files with the corresponding extensions.

To use the **Tool Bar,** load the tool bar instrument driver `toolbar.fp`, which is located in `toolslib\custctrl` folder (directory) under the CVI directory. To do so, go to the **Project** window and select **Instrument>>Load>>toolbar.fp** from the path. This will load the instrument drivers, **Tool Bar** and **Programmer's Toolbox,** which are visible when you click on the **Instrument** menu as shown in Figure 4–34.

The **Tool Bar** icons will perform the same function tasks as those of the menu bar items in `project4-2`: *Start, Stop, Exit,* and *Reset.* Open `project4-3.c` and add the two lines shown in Figure 4–35 above the *main* function.

`ToolbarType` is a *typedef* declared in `"toolbar.h"`. The `ToolbarHandle` is the name assigned to the tool bar handle when created.

Figure 4–34
Toolbar and Programmer's Toolbox Shown on the Instrument Menu

```
#include "toolbar.h"   //toolbar header
static ToolbarType  ToolbarHandle;
```

Figure 4–35
Project 4–3 Header

In the *main()* function, add the highlighted lines shown in Figure 4–36.
Examine the *main* function as it has many new features.

The project folder (directory) at line 13 is obtained by using the function *GetProjectDir(prj_dir)*. This function obtains the pathname of the current project directory in the variable `prj_dir`. As you will see later, this pathname is required to access the icon files. At line 19, *Toolbar_New* creates a new tool bar on the specified panel (`channel` in this case). The prototype for this function is shown below.

```
int Toolbar_New(int Parent_Panel, int Menu_Bar, char Title[ ],
    int Top, int Left, int Conform_Bitmap, int Conform_System,
    ToolbarType *New_toolbar);
```

The function arguments are listed in Table 4–4.

The *CVI* function *Toolbar_InsertItem* inserts a new **Tool Bar** item on the panel. The prototype for this function is given below.

```
int Toolbar_InsertItem(ToolbarType Toolbar, int Position,
    int Item_Type,  int Active, char Description, int Callback_Type,
    int Menu_Item,   CtrlCallbackPtr  Control_Callback_Function,
             void *Callback_Data, char Image_File[ ]);
```

An explanation of the function arguments is given in the Table 4–5.

The `icon_name` created on line 28 is a concatenation of the project folder (directory) name and the icon file name as shown.

```
sprintf(icon_name, "%s\\go.ico", prj_dir);
```

Here the icon is in the project directory.

Most of the icons used here are part of the *CVI* program located in the `cvi5\samples\custctrl\toolbar\buttons` directory. There are numerous icons included in the folder (directory) that you can try out in your applications. You can create your own icons by using National Instrument's **Icon Editor** by selecting **Start>>Programs>>National Instruments CVI>>Utility Programs>>Icon Editor** as shown in Figure 4–37.

To give the tool bar a dressy look, add separators, small vertical lines, between each of the **tool bar** icons. The separators are added to the **tool bar** at lines 26, 33, 41 and 50 of Figure 4–36.

```c
//Main function
int main (int argc, char *argv[])
{

    if (InitCVIRTE (0, argv, 0) == 0)    /* Needed if linking in external compiler;
                        harmless otherwise */
        return -1;       /* out of memory */
    if ((channel = LoadPanel (0, "project4-3.uir", CHANNEL)) < 0)
        return -1;

    //Get the project directory
    GetProjectDir (prj_dir);

    //Load the menubar
    menuHandle = LoadMenuBar (channel, "project4-3.uir", MENU);

    // Create the Toolbar
    Toolbar_New(channel,menuHandle,"Channel Readings Toolbar",
                                          0,0,1, 1, &ToolbarHandle);
    sprintf(icon_name, "%s\\go.ico", prj_dir);
    //Insert the Start icon
    Toolbar_InsertItem (ToolbarHandle, 1, kCommandButton, 1, "Start",
                        kMenuCallback, MENU_FILE_START, 0, 0, icon_name);
    //Insert the separator
    Toolbar_InsertItem (ToolbarHandle, 2, kSeparator, 1, "", kNoCallback,
                                          0, 0, 0, " ")
    sprintf(icon_name, "%s\\stop.ico", prj_dir);
    //Insert the Stop icon
    Toolbar_InsertItem (ToolbarHandle, 3, kCommandButton, 1, "Stop",
                        kMenuCallback, MENU_FILE_STOP, 0, 0, icon_name);
    //Insert the separator
    Toolbar_InsertItem (ToolbarHandle, 4, kSeparator, 1, "",
                                          kNoCallback, 0, 0, 0, " ");

    sprintf(icon_name, "%s\\reset.ico", prj_dir);
    //Insert Reset icon
    Toolbar_InsertItem (ToolbarHandle, 5, kCommandButton, 1, "Reset",
                        kMenuCallback, MENU_RESET, 0, 0, icon_name);
    //Insert the separator
    Toolbar_InsertItem (ToolbarHandle, 6, kSeparator, 1, "",
                                          kNoCallback, 0, 0, 0, " ");

    sprintf(icon_name, "%s\\quit.ico", prj_dir);
    //Insert the Exit icon
    Toolbar_InsertItem (ToolbarHandle, 7, kCommandButton, 1, "Exit",
                        kMenuCallback, MENU_FILE_EXIT, 0, 0, icon_name);

    //Insert the separator
    Toolbar_InsertItem (ToolbarHandle,8, kSeparator, 1, "", kNoCallback,
                                          0, 0, 0, " 52 ");
    DisplayPanel (channel);

    //Display the Toolbar on the panel
    Toolbar_Display(ToolbarHandle);
    RunUserInterface ;
    return 0;
} //main
```

Figure 4–36
Project 4–3 *Main* Function

Table 4–4 Toolbar_New *Function Arguments*

Argument	Purpose
Parent_Panel	the name of the panel where the **Tool Bar** will be loaded
Menu_Bar	the menu_bar handle associated with the **Tool Bar**
Title	the title of the **Tool Bar**
Top	the vertical coordinate where the upper left corner of the **Tool Bar** is placed
Left	the horizontal coordinate where the upper left corner of the **Tool Bar** is placed
Conform_Bitmap	can have either a '1' or a '0': '1' will use the color scheme of the toolbar for the icons, '0' will use the icons without any modifications
Conform_System	can have either a '1' or a '0': '1' will use the color scheme of the panel for the icons, '0' will use the icons without any modifications
*New_toolbar	the name of the new **Tool Bar** that is passed by reference and that is used in other **Tool Bar** functions to refer to this **Tool Bar** (similar in functionality of the *panelHandle*, used for referring to the panel and the objects placed on the panel)

After all the **Tool Bar** items have been inserted in the **Tool Bar,** the function *Toolbar_Display* at line 55 of Figure 4–36 displays the **Tool Bar** on the panel. The rest of the code is similar to `project4-2.c`.

Save all the changes and run `project4-3.prj`. To start the program, click on the **GO** icon. To suspend execution, select **STOP** icon, and click on **RST** to reset all the controls to their default values. Clicking on the **X** icon will terminate the project. A snapshot of the run is displayed in Figure 4–38.

The **Tool Bar** icons are visible on the line beneath the menu names. Instead of using the menu bar items that you created in the previous project, click on the **Tool Bar** icons to perform the same program functions. These icons perform the same functions as the *Start, Stop, Exit,* and *Reset* commands in the menu bar. If you want to know the function of a particular icon, move the mouse cursor over the icon and the text that you entered for

Table 4–5 Toolbar_InsertItem *Function Arguments*

Argument	Purpose
Toolbar	this is the handle of the Tool Bar that is returned from the *Toolbar_New* function; in this project you will use *ToolbarHandle* which is returned by *Toolbar_New*
Position	this argument specifies where in the Tool Bar this item is to be placed, starting at '1' as the left most item, and moving to the right in numerical order
Item_Type	in this argument, a choice of five types of items is available: KCommandButton: Command button; kToggleButton: Toggle button; kExclusiveToggleButton: Exclusive toggle button; kSeparator: Separator (no control); kRing: Ring Control
Active	this argument determines if the item will be visible or hidden; use '1' for visible and '0' for hidden
Description	this is the text the user sees when the cursor is moved over the tool bar icon, and it serves as a help feature
Callback_Type	there is a choice of three callback functions for the button, control, menu, or none, as shown below: kControlCallback : Control callback function; kMenuCallback: Menu callback function; kNoCallback: no callback function
Menu_Item	this argument takes the Control ID for the menu item to be associated with it if *Callback_Type* is set to menu callback; if no such association exists, this field can be ignored by entering a '0'; at lines 23–24 of Figure 4–36, the **MENU_FILE_START** item is associated with this **Tool Bar** item as both perform the "Start" function
Control_ Callback_ Function	this field is only applicable if the *Callback_Type* is set to control callback; this contains the name of the callback function that will be executed when an event occurs; in this project, it is set to zero because the Callback_type is menu callback
Callback_Data	this argument is the callback data to be passed to the callback function when the item receives an event; this parameter is ignored if the item is a separator or if the callback type is not *kControlCallback*
Image_File	this is the full pathname of the icon file

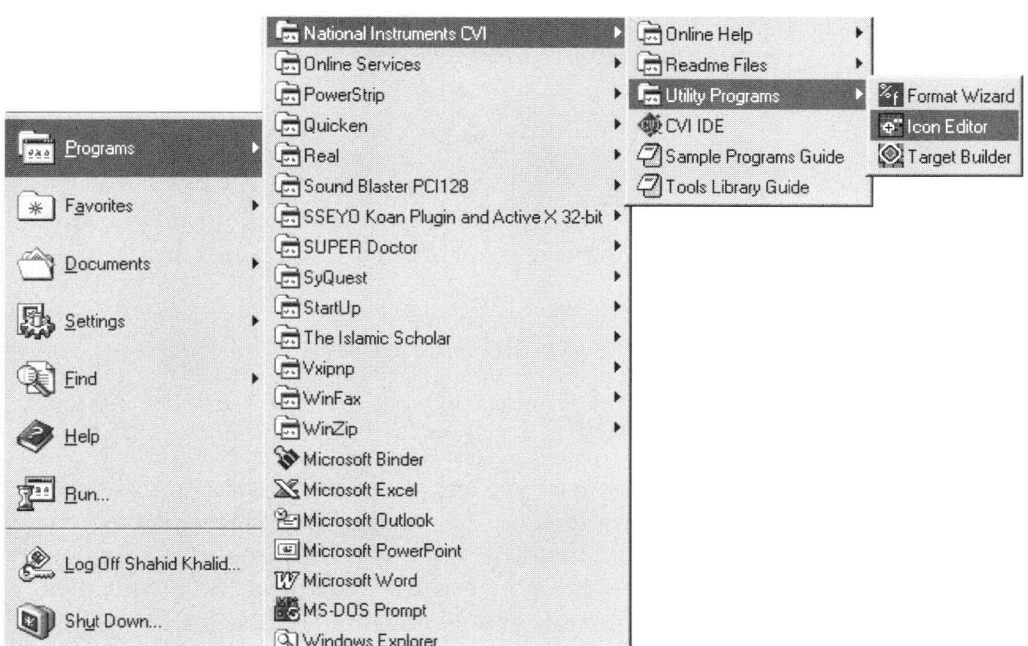

Figure 4–37
Accessing Icon Editor

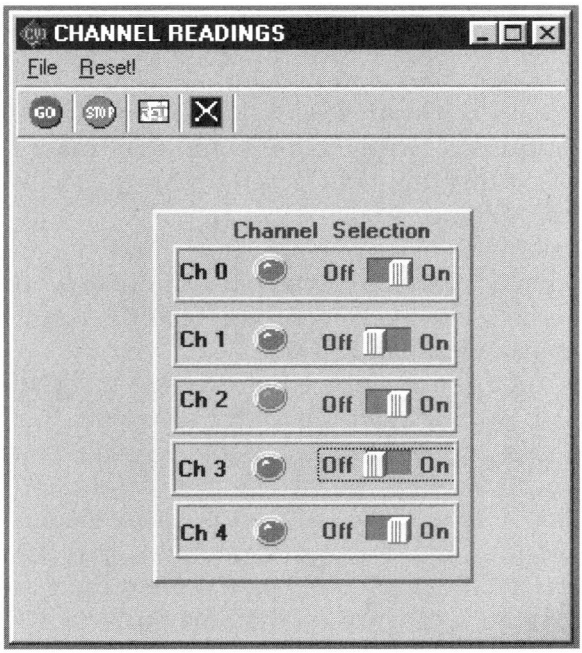

Figure 4–38
Project4–3 Run

this icon in the *Toolbar_InsertItem* function will be displayed. Note that the tool bar, like the menu bar, is only visible during the program execution because it is loaded programmatically.

Adding a Text Box to the Project

This project will use the same menu bar and tool bar commands as created in the previous project. In addition, this project will show you how to display, on the text box, the channel number and the time of its activation as well as the total execution time between the **Start** and the **Stop** times. You will also learn to display the system date and time on the GUI.

Load `project4-3.prj`, rename the project and all the files in the **Project** window to `project4-4`. This time rename the source code file also as you will be adding code to the same source code file. We will see how easy it is to add callback function skeleton code to the already existing source code.

You will create a **Text Box** in which you will display the data that the program generates. A **Text Box** is an object on the user interface on which text can be displayed by using the appropriate text insert command.

Place a **Decoration** frame on which to place the **Text Box.** Right-click on the panel and select **Decorations>>Raised Frame.** Size the frame to be larger than the **Decoration** box with the **LEDs** and the **Binary Switches.**

To create a **Text Box**, right-click on the panel and select **Text Box** from the menu as shown in Figure 4–39.

A small blank box is placed on the panel. Place it over the **Decoration** box that you just created. Re-size the text box to make it larger and center it on the **Decoration** box. Double-click on the **Text Box,** and its properties dialog is displayed as shown in Figure 4–40.

Enter `DISP` for the **Constant Name.** Leave the **Callback Function** box blank as you will not use any callback function with this **Text Box.**

In the **Controls Appearance** group the **Visible Lines** are automatically set by the size of the **Text Box** when you create and re-size it. You can change the value here to make the **Text Box** become larger or smaller by entering the number of lines you want displayed. For **Scroll Bars,** select **Both** to show the vertical and horizontal scroll bars so you can view the data that has scrolled past the visible area of the **Text Box.**

Enter the **Label** as `DATA DISPLAY` in the **Label Appearance** group. The remaining controls can be left to their default values. Select **OK** when done.

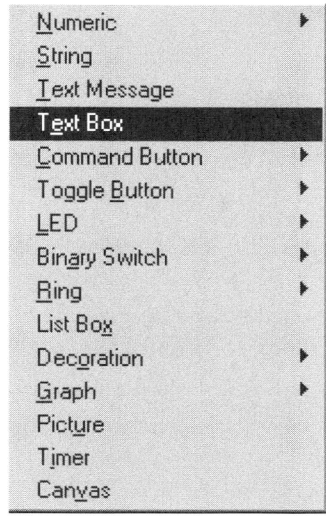

Figure 4–39
Creating a **Text Box** Control

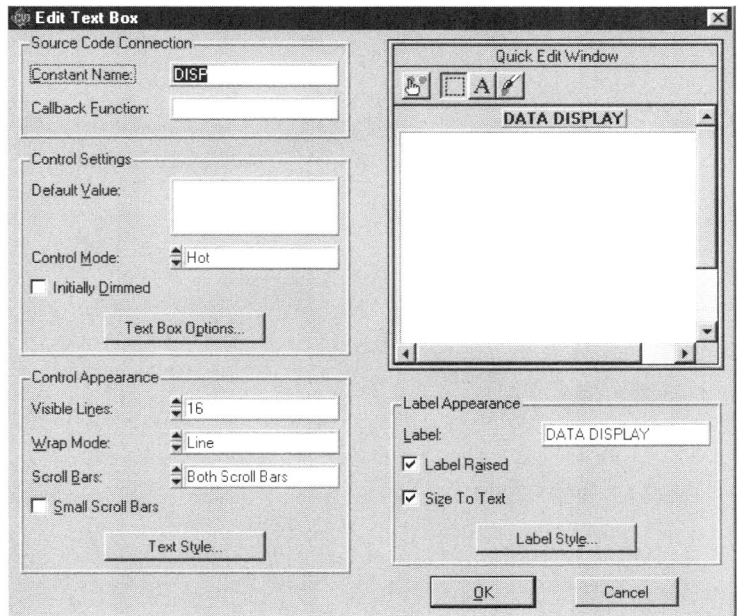

Figure 4–40
Text Box Properties Dialog Box

You will want to see the elapsed time from the time the **START** command button is selected to the time the **STOP** command button is clicked. For this, you need to add a **String** box below the **Text Box,** in which the elapsed time will be displayed. You want to use a **String** box rather than a **Numeric** box because the data you want to place in this box will consist of a floating-point value plus the characters "secs." You will first convert the floating-point number to a string before placing in this box, as listed in the *MenuStopCB* function. Right-click on the panel, select **String,** and move it below the **Text Box.** The **String** properties dialog box is shown in Figure 4–41.

Set the attributes for this string indicator as listed below:

Constant Name	TIME
Control Mode	Indicator
Label	Run Duration

Leave the other attributes for these string boxes at their default values.

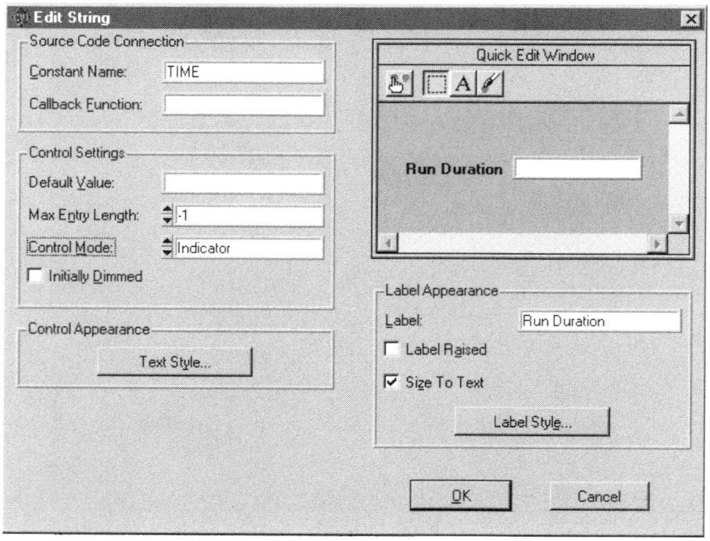

Figure 4–41
Creating a String Control

Chapter 4 • Enhancing the User Interface

To add date and time on the panel and update the time every second, add two **String** boxes, as before, and place them on the panel above the **DATA DISPLAY** text box. Set their attributes as shown below:

	First Box	**Second Box**
Constant Name	DATE_STR	TIME_STR
Control Mode	Indicator	Indicator
Label	Date	Time

To update and display the time every second on the GUI, you need to create a one-second timer. Right-click on the panel and select **Create>> Timer.** Double-click on the timer control and set the properties as shown in Figure 4–42, with the **Interval (seconds):** set to 1.000.

Your GUI should look similar to Figure 4–43.

Recall that to create project4-4, the only additional control you added with a callback function was the one-second timer (**Timer 1 sec.**) to project4-3. Instead of creating the skeleton code for all the callback functions and then adding your source code as in the previous project(s), *CVI* enables

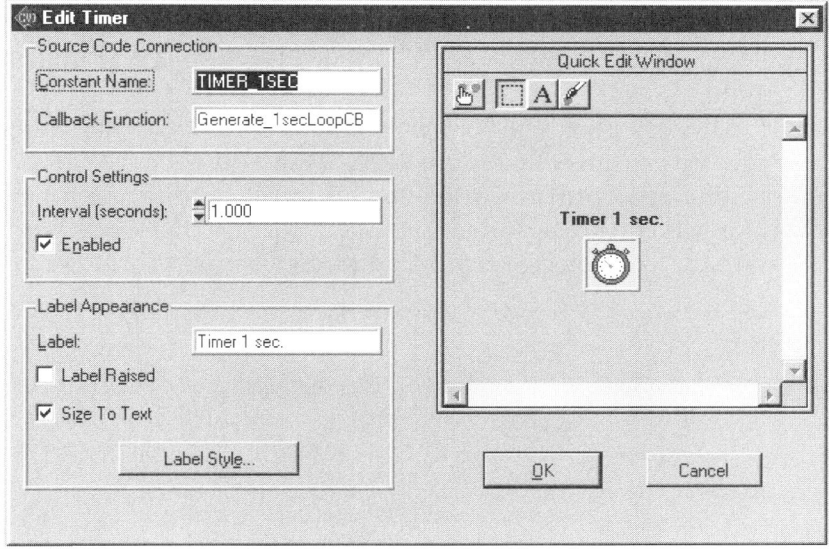

Figure 4–42
Timer 1 sec. Properties Box

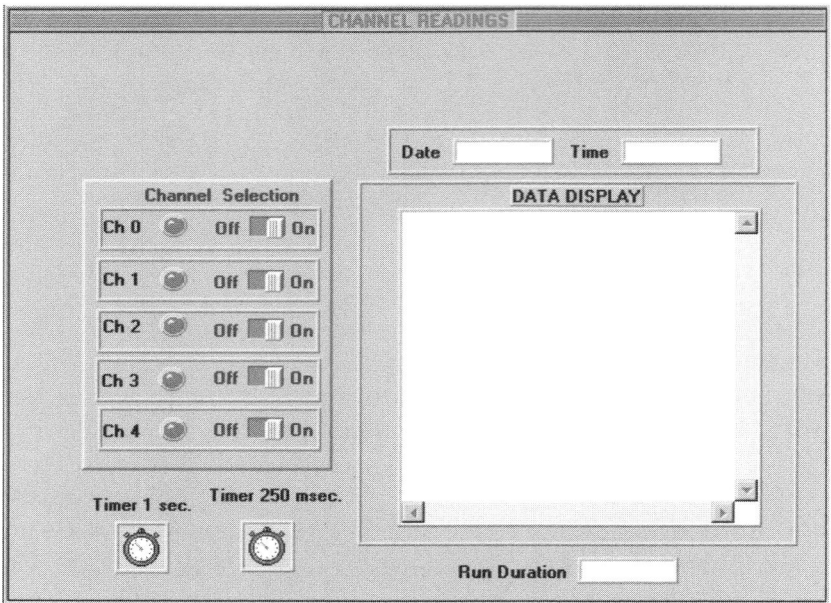

Figure 4–43
Project4–4 GUI

you to create and add the skeleton code for the individual callbacks associated with the controls.

To do so, right-click the **Timer 1 sec.** control, and the following two choices pop up as shown in Figure 4–44.

Select **Generate Control Callback,** and the skeleton code for the callback function for this control is added to the source code. This code is added to the source code file you selected from **Code>>Set Target File** at the location that

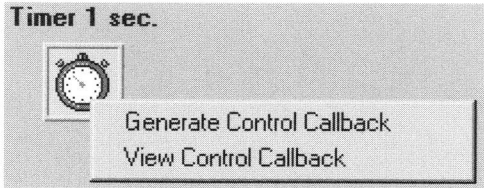

Figure 4–44
Control Callback Shortcuts

you specified in **Code>>Preferences** as explained in Chapter 2, "Basics of Creating the Graphical User Interface." The skeleton code generated for this control callback function is shown in Figure 4–45.

After the callback function is created, you can always go to it by right-clicking on the timer and selecting **View Control Callback** as shown in Figure 4–44.

Add the following line to this function between the *case* and the *break* statements in Figure 4–45. This will display the system time string in the text box.

```
SetCtrlVal(channel, CHANNEL_TIME_STR, TimeStr);
```

The **Timer 1 sec.** control has a callback function *Generate_1secLoopCB*, which is called every second. This function displays the present system time in the **Time** control box on the GUI. The *TimeStr* function returns an eight-character string in the form HH:MM:SS, where HH is the hours, MM is the minutes, and SS is the seconds.

The source code for this project is now complete. To examine the source code, double-click on project4-4.c from the **Project** window. The *main* function and the header files in project4-4 are very similar to project4-3, except for the highlighted lines shown in Figure 4–46.

The function

```
SetCtrlVal(channel, CHANNEL_DATE_STR, DateStr);
```

inserts the date string in the **Date** control indicator on the GUI using the library function *DateStr*. The *DateStr* function returns the system clock time in the format of a ten-character string in the form MM-DD-YYYY, where MM is the month, DD is the day of the month, and YYYY is the year. This function takes the date information from the computer system clock.

```
int CVICALLBACK Generate_1secLoopCB (int panel, int control, int event,
        void *callbackData, int eventData1, int eventData2)
{
    switch (event)
        {
        case EVENT_TIMER_TICK:
            break;
        }
    return 0;
}
```

Figure 4–45
Generate_1secLoopCB Callback Function Skeleton Code

```
#include <ansi_c.h>
#include <formatio.h>
#include <utility.h>
#include <cvirte.h>     /* Needed if linking in external compiler; harmless otherwise */
#include <userint.h>
#include "project4-4.h"
#include "toolbar.h"    //toolbar header
void TurnON_LED( int LED_Number);   //Turn on selected LED
static ToolbarType  ToolbarHandle;
static int channel, menuHandle;
double StartTime, StopTime, Duration;
char buf[1000], prj_dir[260], icon_name[260];
//Main function
int main (int argc, char *argv[])
{
     if (InitCVIRTE (0, argv, 0) == 0)  /* Needed if linking in external compiler;
                                 harmless otherwise */
              return -1;      /* out of memory */
     if ((channel = LoadPanel (0, "project4-4.uir", CHANNEL)) < 0)
              return -1;
     //Load the menubar
     menuHandle = LoadMenuBar (channel, "project4-4.uir", MENU);

     DisplayPanel (channel);

     //Display the date on the GUI
     SetCtrlVal( channel, CHANNEL_DATE_STR, DateStr);
     GetProjectDir (prj_dir);
```

Figure 4–46
Project 4–4 *Main* Function Code Differences

You have seen the *Generate250LoopCB* function previously in Figure 4–17, but here some extra functionalities have been introduced. The code for this function is listed in Figure 4–47, and the highlighted statements indicate the new lines that will be explained.

Every time a particular `LED_Number` is selected, the string consisting of the time string and the LED number is created at line 20 and displayed on the **Text Box** using line 21.

The new statements added to *MenuStartCB* callback function are highlighted and shown in Figure 4–48.

The highlighted lines in *MenuStartCB* need some explanation. The *ResetTextBox* function on line 5 of Figure 4–48 replaces all the text in the text box `CHANNEL_DISP` by the string in the third argument of the function, which is a null string. This function basically clears the text box and is done before you start the application in order to display the data on a clean display box. The arguments for this function are

Chapter 4 • Enhancing the User Interface

```
1     //Timer controlled loop. Code executed every 250 msec.
2     int CVICALLBACK Generate250LoopCB (int panel, int control, int event,
3                    void *callbackData, int eventData1, int eventData2)
4     {
5         int LED_Number;
6         int SW_Val;
7         switch (event)
8             {
9             case EVENT_TIMER_TICK:
10                     //invokes routine at every timer tick
11                     //Generate random LED number
12                     LED_Number=(rand/(1500)% NUM_LEDS);
13
14                     GetCtrlAttribute(channel, attr[LED_Number][SW_CTRL_INDEX],
15                                                 ATTR_CTRL_VAL, &SW_Val);
16
17                     if (SW_Val)
18                     {
19                         TurnON_LED(attr[LED_Number][LED_CTRL_INDEX]);
20                         sprintf(buf, "%s         %d",TimeStr, LED_Number);
21                         InsertTextBoxLine (channel, CHANNEL_DISP, -1, buf);
22                     }
23                     break;
24             }
25         return 0;
26     } //Generate250LoopCB
```

Figure 4–47
Additions to *Generate250LoopCB* Callback Function

```
    ResetTextBox(channel, CHANNEL_DISP, "");
```

where the `channel` is the panel handle of the user interface resource currently in memory and the `CHANNEL_DISP` is the **Control ID** of the text box.

The *DefaultCtrl* function at line 7 of Figure 4–48 resets the value of the control box to its original value before the program starts to execute. The arguments for this function are

```
    DefaultCtrl (channel, CHANNEL_TIME);
```

where the `channel` is the panel handle as just explained. `CHANNEL_TIME` is the Control ID of the box being reset. Notice that instead of using *ResetTextBox* you could have used *DefaultCtlr* provided there is nothing in the **Default Value:** box of the **CHANNEL_DISP** attributes box.

At line 9 of Figure 4–48 `buf` is assigned the string " Time Active Channels" to display as the header on the **Text Box.** Line 11 inserts this string using the *InsertTextBoxLine* function, which displays a line of text in the text box. The prototype of *InsertTextBoxLine* is

```
1    void CVICALLBACK MenuStartCB (int menuBar, int menuItem, void *callbackData,
2                   int panel)
3    {
4            //Clear the Display box
5            ResetTextBox (channel, CHANNEL_DISP, " ");
6            //Clear the Time Duration box
7            DefaultCtrl (channel, CHANNEL_TIME);
8            //Create a string for header
9            sprintf(buf, "            Time            Active Channels");
10           //Insert the header on the display
11           InsertTextBoxLine (channel, CHANNEL_DISP, -1, buf);
12           //Start the 250 msec Timer,
13           SetCtrlAttribute(channel , CHANNEL_TIMER_250MS, ATTR_ENABLED, 1);
14           //Record the start time
15           StartTime=Timer;
16
17   } //MenuStartC
```

Figure 4–48
Additions to *MenuStartCB* Callback Function

```
           int status= InsertTextBoxLine (int panelhandle,
                       int controlID, int lineIndex, char text[ ]);
```

and these arguments are explained in Table 4–6.

At line 15 of Figure 4–48, the library function *Timer* is called, and its value is assigned to the variable `StartTime`. The *Timer* function call captures the

Table 4–6 InsertTextBoxLine *Function*

Input/ Output	Name	Type	Description
Input	panelHandle	integer	handle of the panel that is curently in memory
	controID	integer	ID number of the control, where the data is written
	lineIndex	integer	zero-based index of the text box line above which to insert the new line, pass –1 to insert the new line at the end of the text box line or in Online
	text	char	text string inserted in the Text Box
Output	Status	integer	refer to Appendix A in *User Interface Library Reference Manual* or in Online Help

StartTime when the timer is started. Another call to *Timer* stops the timer. These times are obtained from the system clock.

You can imagine the *Timer* function to be similar to a stop watch. When the first call is made to the *Timer* function, the stop watch starts and the next call to the *Timer* function stops the stop watch. This function is explained in *LabWindows/CVI Standard Libraries Reference Manual*.

In the *MenuStopCB* callback function listed in Figure 4–49, the highlighted lines are new to project4-4. The StopTime variable at line 5 captures the time when the execution is stopped by calling *Timer*. The execution elapsed time is calculated at line 7 by taking the difference between the *StopTime* and the *StartTime*. At line 11, the string is created and displayed by line 13 on the **Run Duration** box on the GUI.

The other callback functions are the same as in project4-3. Save all the files as project4-4, compile, build, and run the project as before. The program will start executing and the system time along with the active channel number generated by the random number generator will be displayed in the text box. Turn off a few **Binary Switches** on the *GUI* by clicking the **Off** side of the switch. You will notice that the channels that are switched **Off** will not be activated and the data for only the channels for which the **Binary Switches** are in the **On** state will be displayed in the text box. When you turn the switches set to **Off** back to **On,** those channels will be activated (depending on the value of random number generator). If the data scrolls off the display area, you can see that data after you stop the program and scroll down using the scroll bar on the display. Also, notice that when the program is stopped, the elapsed time is displayed in the **Duration** box. A snapshot of the run is shown in Figure 4–50.

```
1    void CVICALLBACK MenuStopCB (int menuBar, int menuItem, void *callbackData,
2                int panel)
3    {
4        //Stop the Timer
5        StopTime=Timer;
6        //Calculate the elapsed time
7        Duration= StopTime-StartTime;
8        //Stop the 250 msec. Timer
9        SetCtrlAttribute(channel , CHANNEL_TIMER_250MS, ATTR_ENABLED, 0);  //Stop
10                                                                               Timer
11       sprintf(buf, "%f secs.",Duration);
12       //Display the time duration on the GUI
13       SetCtrlVal (channel, CHANNEL_TIME, buf);
14
15   } //MenuStopCB
```

Figure 4–49
Additions to *MenuStopCB* Callback Function

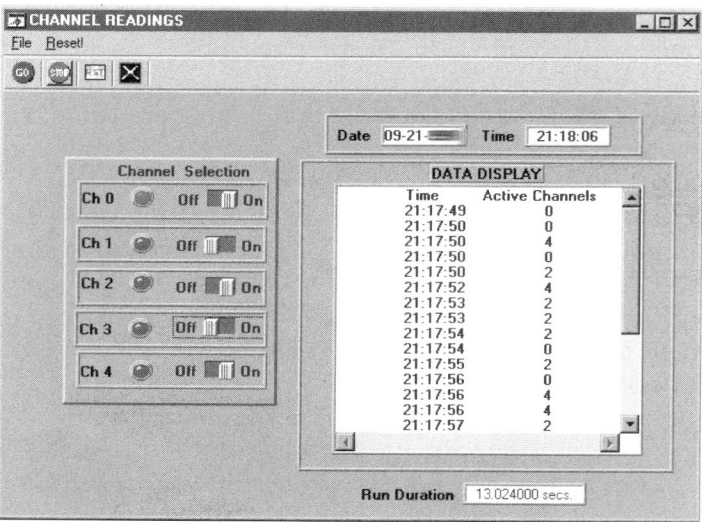

Figure 4–50
Project4–4 Run

The purpose of this project was to introduce more features of the **User Interface Editor,** but it is by no means comprehensive. More **User Interface Editor** controls are explained later in the text.

Before leaving the discussion of the user interface in this chapter, let us look at **Tab Order,** a user interface enhancement that facilitates the use of the controls on the GUI during run-time.

Creating the Tab Order

When you place the control objects on the panel, *CVI* assigns them a number in which they are accessed when you press the <Tab> key. The order in which you place these objects on the panel is the order in which they are accessed. Pressing the <Tab> key activates the next object on the panel in the ordered sequence. Usually the order in which you place the objects on the panel is not the order in which you want to access them during execution. You can re-arrange the access order by using the **Tab Order** key after the *GUI* has been completed.

To set the **Tab Order,** load the user interface file and select **Edit>>Tab Order.** The GUI will have the controls with the tab order numbers displayed, similar to the example shown in Figure 4–51, which shows some controls placed randomly on the panel. On this panel, the objective is to set the **Tab Order** from 0 to 4 along the top row moving across from left to right and down to the **Exit** button.

A special pointer cursor that looks like a hand with a # sign over the index finger appears, and you can use it to move with the mouse and click on the number on the control to go to the next number in the **Tab Order.**

If you hold down the <Ctrl> key, the special pointer cursor changes to the special eyedropper cursor with which you change the number in the **Click to set to** box. This box is shown on the top of the panel. When you move the special eyedropper cursor to any of the controls and click with the mouse, the **Click to set to** box is assigned the tab order number associated with that control. This feature is useful in incrementing the tab ordering from that number onwards.

Selecting the **OK** button accepts the new **Tab Order** and saves the GUI with the new tab order. Clicking on the **Close** icon (the **X** in the top right corner of the panel) restores the original tab order without saving the changes.

In this example, you created the **Tab Order** from top left to right and to the bottom. The new **Tab Order** is shown in Figure 4–52, with the numbers on the black background. When you are running a program and this GUI is called, your cursor will appear at the first **Tab Order** number, which in this case is the **OFF** command button at **Tab Order** number 0.

When you press the <Tab> key, the cursor jumps to the next item in the **Tab Order.** In this example, it is the **Untitled Control** box (on the right side

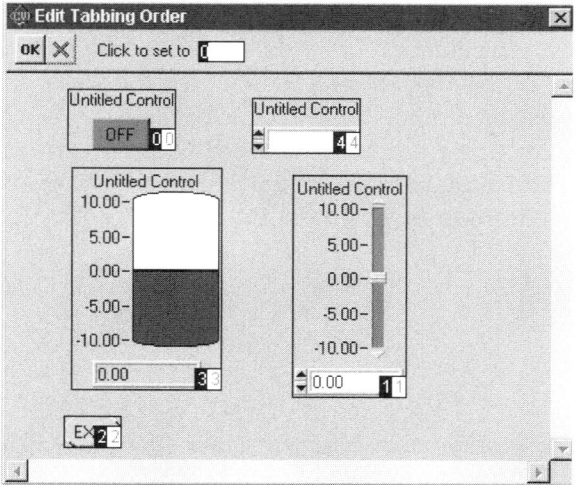

Figure 4–51
Tab Order Dialog Box

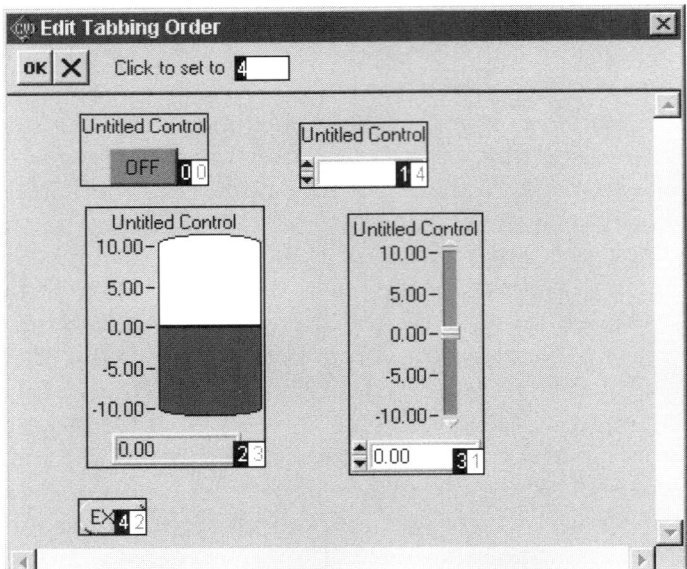

Figure 4–52
New Tab Order

of the same line) with a **Tab Order** number 1. After that, it moves to the first item in the next row, and so on.

The purpose of setting the tab order is to make the GUI more user-operable by moving the cursor to the next control automatically and making it convenient for the user to select or enter the data. In short, it makes the GUI easier to use and more "user-friendly."

In this chapter, the **User Interface Editor** menu commands used to create the project were explained. A complete description of the **User Interface Editor** menu commands is given in Appendix C, "User Interface Editor Environment."

Summary

This chapter explained more of the **User Interface Editor** features. An attempt was made to familiarize you with creating the menu bars and their callback functions. A **Tool Bar** was created, and you learned how to link it with the menu items. **Tab Order** and alignment of the objects was shown by means of examples, and the **Text Box** introduced.

You were shown how to create the **Decoration** boxes and how to align objects. You were shown how to create and add an individual callback function's skeleton code to the source code.

You have taken a first step into the huge realm of what you can create with the **User Interface Editor,** which is by no means complete at this time. There are many objects in the **User Interface Editor** with which you can experiment and discover as you create your own projects. More controls and library functions will be discussed in the coming chapters as needed.

5

SOURCE EDITOR AND DEBUGGING TECHNIQUES

Chapter Highlights
- Source Editor Environment Overview
- Exploring the Source Editor
- Manipulating the Code
- Using the Debugger
- Summary

The use of the **Source Editor** and the **Debugger** are introduced in this chapter. Understanding these tools will help make your coding and debugging tasks go faster and easier.

You will learn how you can enable and disable the line numbers in the source code, set breakpoints, and use the tags as bookmarks to conveniently move around the code. You will learn to exclude and un-exclude sections of code, perform conditional compilation, find and replace text, and many more functions, which will facilitate the coding and debugging tasks. The method by which you can add the C constructs to the code will be discussed, enabling you to add code structure templates with very minimal typing. You will learn how to compare two source code files. The necessary **Source Editor** commands to compile, build, and run the project will be shown here. Learning the many details of using the **Debugger** will enable you to find execution and logic errors in your code.

Source Editor Environment Overview

The **Source Editor** window is the area where you enter the code to create your program. You have used this in previous chapters when the *Code-Builder* created the skeleton code for you and where you added the code for the sample projects. **Source Editor** is basically an American Standard Code for Information Interchange (ASCII) text editor with many enhancements. Any ASCII text file can be loaded into this editor, and as you saw in the previous projects, the *CVI* function panels code can be inserted with the click of a button, which is a great time saver. The **Source Editor** has the capability of accepting lines as long as 254 characters and up to one million lines of text in a file.

When you double click on a source file (with a .c extension) in the **Project** window, this file opens in the **Source Editor** window as shown in Figure 5–1.

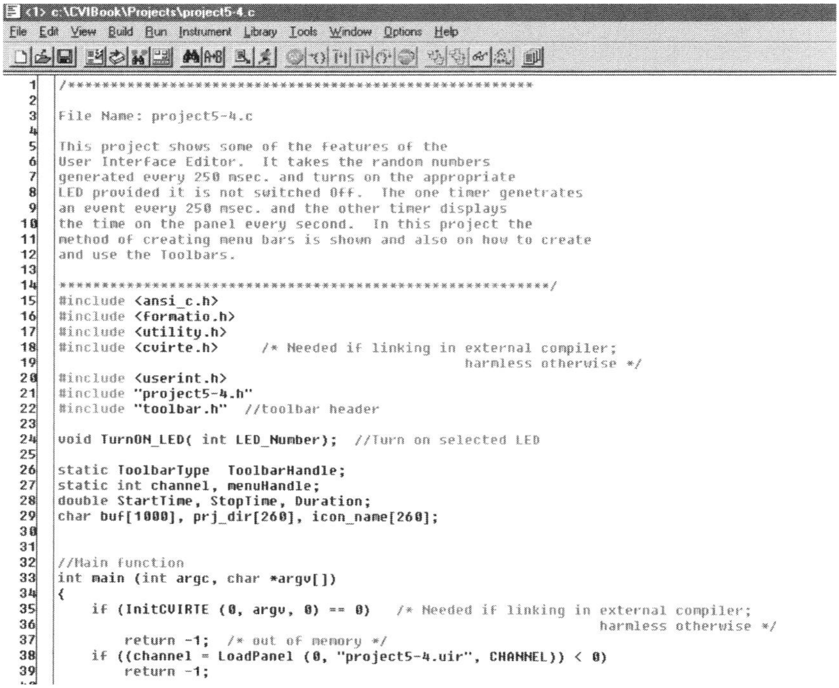

Figure 5–1
Source Code Window

Chapter 5 • Source Editor and Debugging Techniques

Recall that you can recognize the kind of editor window by noting the file name's extension in the editor's title bar. If the file name extension is .prj, it is in the **Project** window, if it is .uir, it is the **User Interface Editor** window, and if .c, it is the **Source Editor** window.

The context sensitive menu capability of this editor enables you to obtain different menus based upon where in the **Source Editor** window you right-click with the mouse. If you right-click in the text area of the code (where code is entered), the menu shown in Figure 5–2 is displayed.

A different menu, as shown in Figure 5–3, is displayed if you right-click with the mouse near the line numbers area of the **Source Editor.** This menu gives you the commands that are relevant to insert **Breakpoints** and **Tags** in that area of the source code. The marks that are placed in the column between the line numbers and the code are called the **Line Icons.** The column is called the **Line Icon** column. All these features will be explained to you later in this chapter.

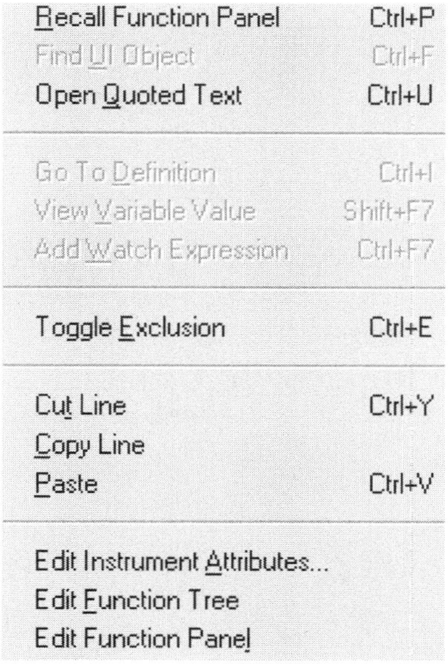

Figure 5–2
Menu Commands from Text Area

```
                Toggle Breakpoint         F9
                Breakpoints...         Shift+F9

                Toggle Tag             Shift+F2
                Next Tag                    F2
                Previous Tag            Ctrl+F2
                Clear Tags...
```

Figure 5–3
Menu Commands for Line Icons

Exploring the Source Editor

Let us look at the various **Source Editor** window features. A handy way to load the project is by double-clicking on the file name in the bottom part of the **File** menu where the last few project files that you opened previously are stored. Select any .c file from the **File** menu which will open the **Source Editor** window.

Source Editor Tool Bar

Look at the **Tool Bar** of the **Source Editor.** This **Tool Bar** is displayed just below the **Source Editor** menu bar and is shown in Figure 5–4. The icons on the **Tool Bar** are shortcuts for menu items. Each icon on the **Tool Bar,** proceeding from left to right, are explained and described.

Figure 5–4
Source Editor Tool Bar

Chapter 5 • Source Editor and Debugging Techniques

New File opens a new source code file, just like selecting **File>>New.**

Open File opens a **File Select** menu from which you can open a file and is the same as selecting **File>>Open.**

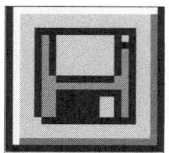

File Save saves the current opened file. It is the same as selecting **File>>Save.**

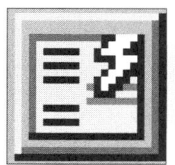

Insert Construct brings up a menu to insert a C construct in the code, just like selecting **Edit>>Insert Construct.**

Go To Definition is used to obtain a definition of the C identifier when you place the cursor on a C identifier and click on this icon. This is the same as **Edit>>Go To Definition.** This command works to find the function and the variable declarations. If the function you selected is defined in a header file, the header file is opened and the selected function definition displayed. Similarly, the variable definitions are displayed. Starting with CVI version 5.5, this command also works on pre-processor macros.

Find UI objects finds the user interface object that you highlighted in the code. This brings up the user interface resource file with the object highlighted. This is the same as **View>>Find UI Object.**

 **Recall Panel** *recalls the panel according to where the cursor is placed in the code. It is similar to* **View>>Recall Function Panel** *from the menu item.*

 *Clicking* **Find** *opens a dialog box in which you can type a string that you want to find. It is similar to* **Edit>>Find.**

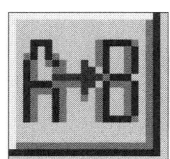 **Replace** *replaces the string that you enter into the pop-up dialog box with another. It is similar to* **Edit>>Replace** *from the menu item.*

 **Compile File** *compiles the opened source file, just like* **Build>>Compile File.**

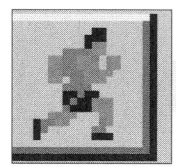 **Run Project** *is similar to* **Run>>Run Project.**

The icons shown on the tool bar in Figure 5–5 are related to debugging the code and will be defined in the debugging section of this chapter. The extreme right icon brings up the **Project** window.

Figure 5–5
Debugging Tool Bar Icons

Working with Function Panels

As discussed in the previous chapter, to recall a library function panel in your code, place the cursor on or near the function name and select **View>>Recall Function Panel** or <Ctrl-P>.

After you have made changes to the function panel, you can re-insert it in the code by selecting **Code>> Insert Function Call** or <Ctrl-I>. A *CVI* message appears as shown in Figure 5–6, giving the user the options of **Insert, Replace,** or **Cancel.** If you want to insert the function without replacing the previous one, select **Insert;** otherwise, select **Replace,** which will over-write the function.

If you find a function panel with more than one function panel window, *CVI* will display a list of panels from which you can select by highlighting the panel name and double-clicking on it or pressing <Enter>.

If you want to find the function panel, select **View>>Find Function Panel...** (or <Ctrl-Shift-P>) and a dialog box will appear in which you can enter the name of the function panel. If you only remember a part of the name, you can enter part of the function panel name(s) and matching strings will be displayed. Figure 5–7 shows the display after searching on `Ctrl`. All func-

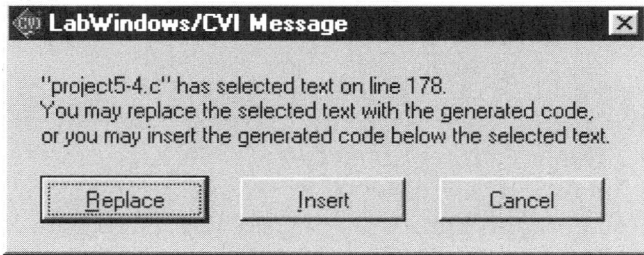

Figure 5–6
Function Panel Insert Dialog Box

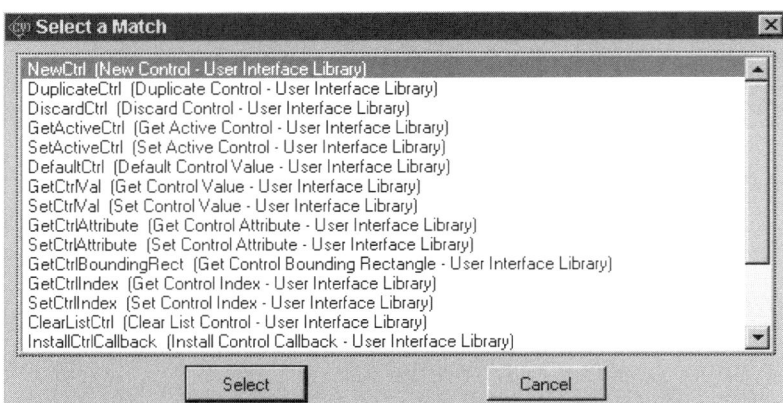

Figure 5–7
Matching Function Panel Names

tion panels with `Ctrl` anywhere in the function panel name are found and displayed.

Highlight the function name you want to use by double-clicking with the mouse and click on the **Select** button.

Remember the following shortcuts:

Command	Shortcut
View>>Recall Function Panel	<Ctrl-P>
View>>Find Function Panel	<Ctrl-Shift-P>
Code>> Insert Function Call	<Ctrl-I>

Interactive Execution Window

CVI allows you to execute a selected portion of code in the **Interactive Execution Window** (IEW) without using the complete C code. This allows you to test portions of your code without running the whole program. For example, you can execute code without a *main* function.

To use this feature, select **Window>>Interactive Execution** from the **Project** window. This will bring up an **Interactive Execution** window that looks similar to a blank **Source Editor** Window. For a demonstration of IEW, enter the test code shown in Figure 5–8. When you execute code from IEW, *CVI* will prompt you to include the header file (if not already included) and place it on the top of your test code. During the IEW execution, *CVI* excludes all declarations by marking them in a different color (usually gray). For this

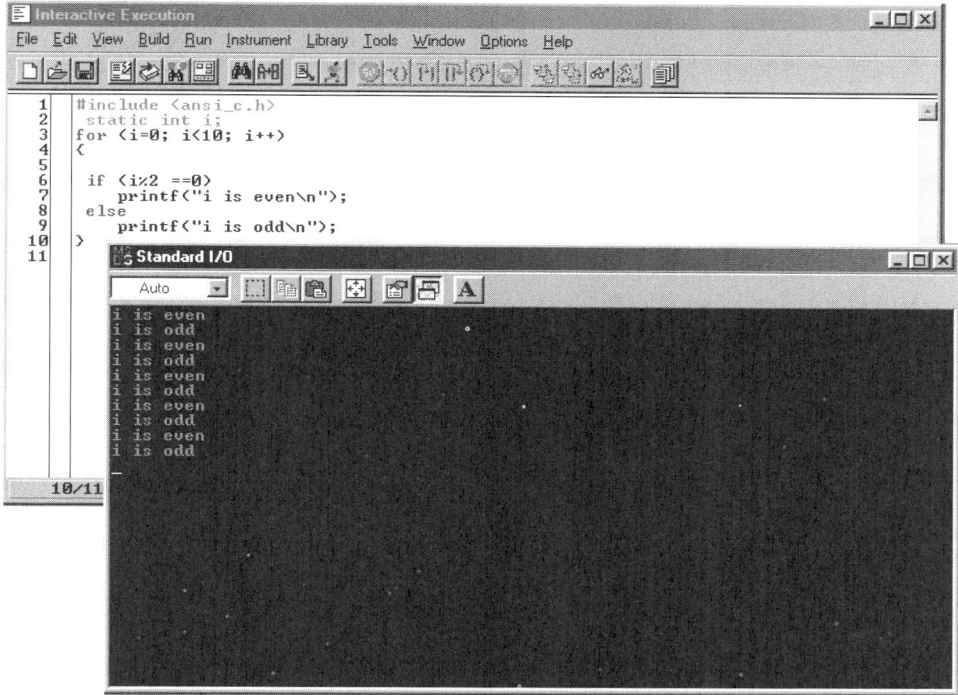

Figure 5–8
Interactive Execution Window Run Demonstration

reason, you must not place the executable statements on the same line as the declarations. *CVI* displays the data in the **Standard Input/Output** or in the console window depending on your **Target Settings.** For an explanation of using this feature, see the "Memory Display Window" section of this chapter and "Target Settings" of "Using the Build Menu" section in Appendix B, "Project Window Environment." After you have tested a section of your code using the IEW, you can place the code in your project source file with a reasonable degree of confidence. To run this code fragment, click on the **Run** icon in the IEW, and the output of your code will be displayed in your selected output window (**Standard I/O** in this case).

You can use the IEW feature to test the function panel call(s) also. As an example, add the *MessagePopup* panel to the last line of the above code. Insert this popup panel by selecting **Library>>User Interface>>Popup Panels>> Message/Prompt Popups...>>Message Popup**. This is shown in Figure 5–9. The pop-up panels (also called modal panels) are discussed in Chapter 6, "File

Chapter 5 • Source Editor and Debugging Techniques

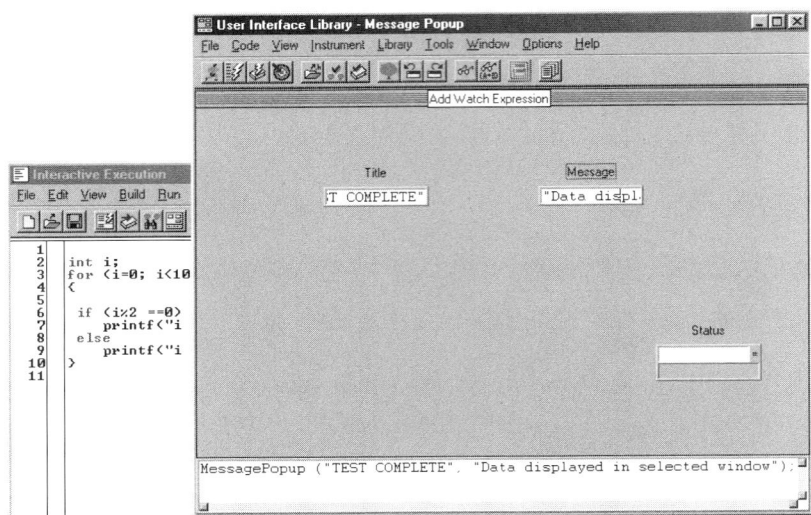

Figure 5–9
Inserting Popup Panels in IEW

Input/Output." Here you are only interested in observing how the function panel code is inserted and tested using the IEW.

Select **Run** from IEW, and as before, *CVI* will prompt you for the missing header file(s). The test code will execute as before and the **Message** pop-up panel that you inserted will be displayed as shown in Figure 5–10.

To test the **Message** pop-up call only without running the previous code, exclude the lines before the *MessagePopup* call by selecting **Edit>>Toggle Exclusion.** This will place the lines in a different color as indicated previously, and when you execute the IEW, only the **Message** pop-up function line will be executed. When you execute a function call from the function panel by selecting **Code>>Run Function Panel** (from the function panel window), *CVI* automatically excludes all previous lines in the IEW. There are certain rules for executing the code in the IEW, which are given in *LabWindows/CVI User Manual.*

Building Your Project

After you have completed the code and you want to run your project, you need to compile your code. Select **Build>>Compile File** to start the compilation. The compiler checks for syntax errors. You can add the file to

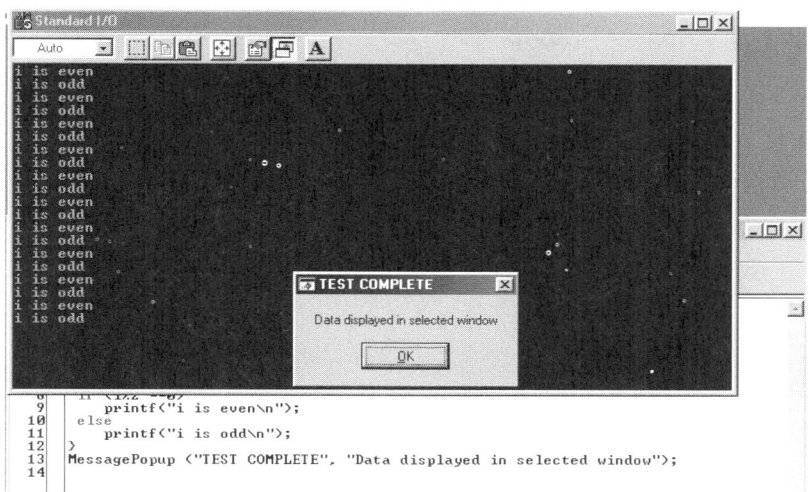

Figure 5–10
IEW Run Using the Pop-up Panel

the **Project** window by selecting **File>>Add File to Project**. If the file is already added to the **Project** window list, the **Add File to Project** will be disabled in the **File** menu. After the compilation, the **Build Errors** box appears and shows you any errors. If there are any compilation errors, you can step through them one at a time by selecting **Next Build Error** (shortcut is <F4>) or **Previous Build Error** (shortcut is <Ctrl-F4>) from the **Build** menu.

If you have multiple source files and *CVI* displays errors for multiple files, you can step to the next file with build errors by selecting **Build>>Build Errors in Next File** (shortcut is <Shift-F4>).

If there are missing include files, *CVI* displays a message indicating the missing header files and querying if you would like to add them at the top of the file. If you do, then it will add the header files. Otherwise, you will get compilation error(s) and the program will stop until you include the missing header files.

If there are some include statements that are necessary to add to the program, you can select **Build>>Insert Include Statements...** and the **Insert Include Statements Dialog Box** appears as shown in Figure 5–11. You can select the header file you want to include from this dialog box and click **OK**.

These files are added to the top of the source code file with the following preprocessor directive:

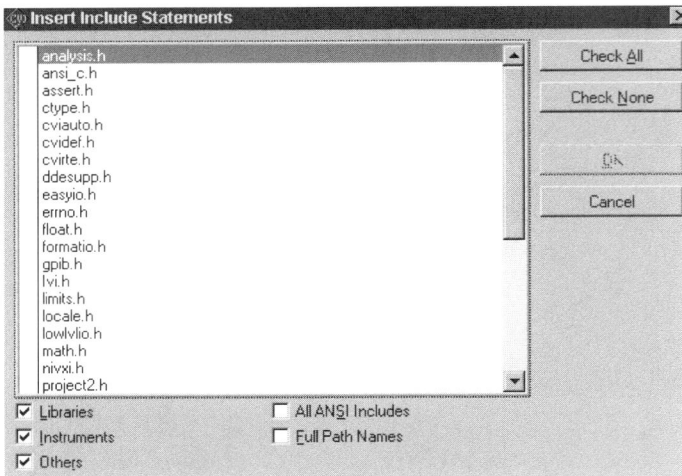

Figure 5–11
Insert Include Statements Dialog Box

```
#include <HeaderFileName.h>
```

To build the project, select **Build>>Create Debuggable Executable** if you set the **Configuration** as **Debug** from the **Project** window or **Build>>Create Release Executable** if you set **Release** from the **Configuration** menu. The executable for the **Debug** configuration is created with the project name concatenated by `_dbg.exe`. The executable for the **Release** configuration is created with the project name concatenated by `.exe`. This was explained in Chapter 2, "Basics of Creating the Graphical User Interface." Also for more information, see the "Using the Build Menu" section in Appendix B, "Project Window Environment" for details on setting active configuration and creating the executables.

After the compilation is completed, you can use the **Build>>Generate Prototype** command to generate a file containing the declarations for global and static functions and external declarations for the global variables. Figure 5–12 shows the prototype header. This is useful if want to use these declarations in the source and header file.

Remember the following shortcuts:

Command	Shortcut
Next Build Error	<F4>
Previous Build Error	<Ctrl-F4>

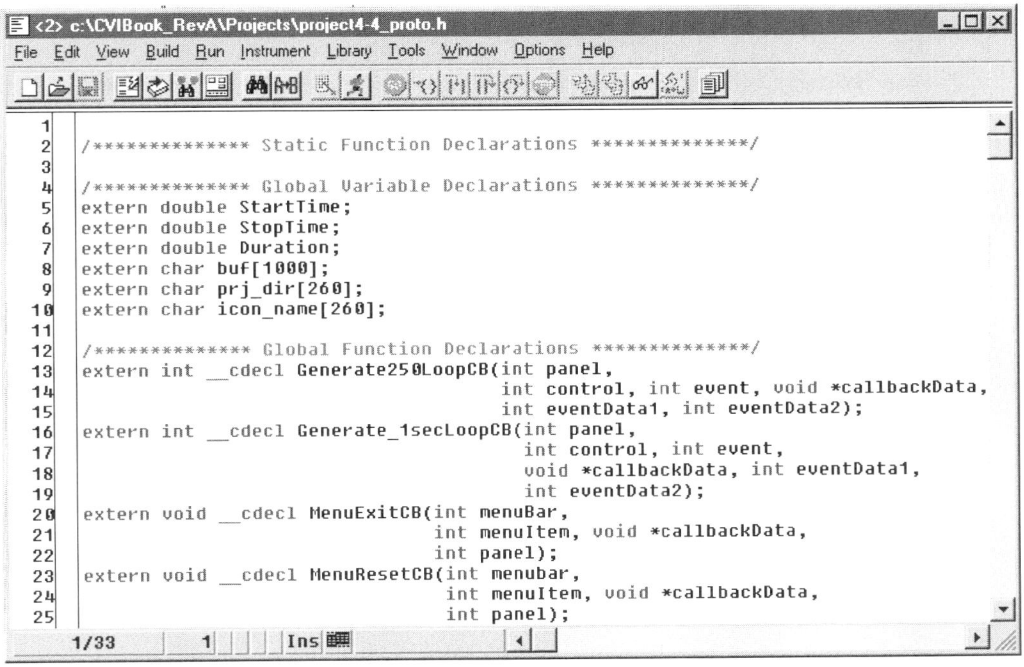

Figure 5–12
Prototype Generated Header File

Manipulating the Code

The **Source Editor** has many capabilities that make creating, editing, finding, replacing, and manipulating the text very convenient. You can set the **Source Editor** working environment according to your work habits.

Setting Line Numbers and Selecting Text

It is advisable to keep the line numbers displayed to keep track of where you are in the program and for referencing lines in the source code. To turn on the line numbers, select **View>>Line Numbers.** The line numbers are displayed in the extreme left column of the screen.

Load the source file by double-clicking on any source file (with .c extension) from the **Project** window list.

Using the Code Selection Modes

In the **Source Editor,** you can select the source code using three different modes. In the **Character Select Mode,** the text from the cursor to the the end of selection is highlighted in reverse video (white text on black background). You can select the text in the **Column Select Mode,** which highlights a rectangular block of text, or in **Line Select Mode** in which full lines of text are highlighted.

Let us start by selecting text using the three different modes. These text selection icons are on the bottom middle of the source code frame to the right of the **Ins** characters. They are displayed in sequence when you click on them with the mouse. You can select the text, just as in any word processing program, by holding the left mouse button and dragging to the end of your text selection.

The **Character Select Mode** button selects the text from the point at which you start selecting the text to the point at which you end the selection. The selected code using this mode is shown in Figure 5–13.

Figure 5–13
Character Select Mode Text Selection

Chapter 5 • Source Editor and Debugging Techniques   151

This icon is the **Line Select Mode,** which selects complete lines of text. The selected code using this mode is shown in Figure 5–14 in reverse video.

This icon is the **Column Select Mode,** and it selects a vertical block of text. It is useful when you have a table or a vertical block of data that you would like to select and copy to another part of the program. This is particularly useful when there is a column of comments at the end of the statements that needs to be removed all at once. The selected code using this mode is shown in Figure 5–15 in reverse video.

Besides these selection modes, remember you can always select a word by placing the mouse on the word and double-clicking it. You can select a line by clicking three times with the left mouse button. You can always undo the selection by clicking anywhere else in the screen or pressing <Esc>.

If you have highlighted text that is too big to fit inside your screen and you want to view it without scrolling the window, select **View>>Beginning/End of Selection** or <Ctrl-J>. This will move the cursor to the top of the block and to the bottom of the block each time this command is selected.

Figure 5–14
Line Select Mode Text Selection

```
277  //Turn ON the appropriate LED
278  void TurnON_LED( int LED_Number)
279  {
280        SetCtrlVal (channel, CHANNEL_LED_0, 0);   //Turn OFF LED0
281        SetCtrlVal (channel, CHANNEL_LED_1, 0);   //Turn OFF LED1
282        SetCtrlVal (channel, CHANNEL_LED_2, 0);   //Turn OFF LED2
283        SetCtrlVal (channel, CHANNEL_LED_3, 0);   //Turn OFF LED3
284        SetCtrlVal (channel, CHANNEL_LED_4, 0);   //Turn OFF LED4
285        SetCtrlVal (channel, LED_Number, 1);      //Turn ON the specified LED
286  } //TurnON_LED
287
```

Figure 5–15
Column Select Mode Text Selection

Balancing Parentheses

One of the most common mistakes in programming is forgetting to balance the opening and closing parentheses in the C code. This happens, for example, if you have a number of nested *if-else* statements and the curly braces ({ and }) are not balanced because you forgot to enter one of them. Finding where to place the missing brackets can be a confusing and time-consuming task, especially if you have multiple nested statements or nested loops.

CVI's **Source Editor** has a feature that enables you to check for balanced opening and closing parentheses. Conducting this check at various levels of parentheses can be immensely helpful in avoiding code errors. This same feature works if you want to check the (and) balancing parentheses, for example, in a complex *if* statement.

To find pairs of parentheses, curly braces, or brackets, just move the cursor inside the area of text that you want to display and enter **Edit>>Balance** or <Ctrl-B>. The text within the balancing parentheses will be highlighted. This check is useful when the code is not performing satisfactorily or when syntax errors occur due to missing opening or closing parentheses in a block of code. An example of using this check is shown in Figure 5–16 in which the cursor is placed anywhere between lines 159 and 162 and <Ctrl-B> is entered. The code between the two closest curly braces appears in reverse video. This is what you expected to see. If for some reason you had left out one of the curly braces or misplaced one of the braces, the code between the two misplaced parentheses would be highlighted and you would be able to discover your error quickly.

This is even true for an *if-else* construct in which you may have a complex logical expression with many opening and closing parentheses, (and) and

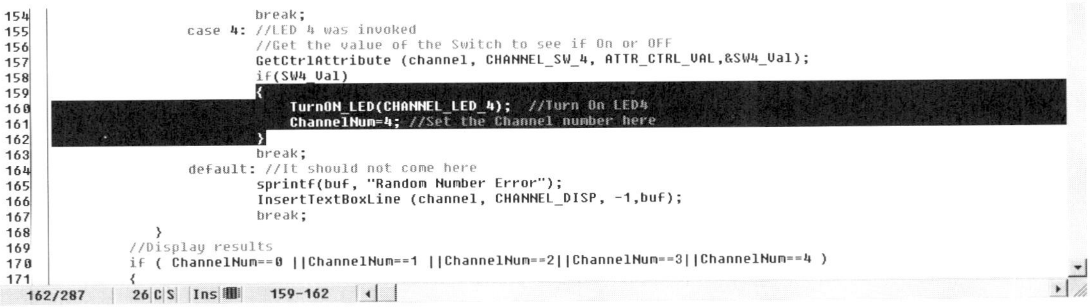

```
154             break;
155     case 4: //LED 4 was invoked
156             //Get the value of the Switch to see if On or OFF
157             GetCtrlAttribute (channel, CHANNEL_SW_4, ATTR_CTRL_VAL,&SW4_Val);
158             if(SW4_Val)
159             {
160                 TurnON_LED(CHANNEL_LED_4);   //Turn On LED4
161                 ChannelNum=4; //Set the Channel number here
162             }
163             break;
164     default: //It should not come here
165             sprintf(buf, "Random Number Error");
166             InsertTextBoxLine (channel, CHANNEL_DISP, -1,buf);
167             break;
168     }
169     //Display results
170     if ( ChannelNum==0 ||ChannelNum==1 ||ChannelNum==2||ChannelNum==3||ChannelNum==4 )
171     {
  162/287    26 C S  Ins    159-162
```

Figure 5–16
Highlighted Text Using the Edit>>Balance Command

you want to determine if the parentheses are balanced. Place the cursor inside the parentheses for which you want to find the corresponding parentheses and select <Ctrl –B>. The area between the two closest (balancing) parentheses will be highlighted in reverse video. This feature is also used to find the balancing brackets, [and], in your code.

Remember the following shortcuts:

Command	Shortcut
View>>Beginning/End of Selection	<Ctrl-J>
Edit>>Balance	<Ctrl-B>

Setting and Clearing Tags

Setting and clearing **Tags** (bookmarks) is a feature of *CVI* that allows you to move around your source code without getting lost. If you want to go to a certain line from another place in the code, you set the tagging feature at that line. If you want to return to the tagged line after working in another part of the code, you can select a function key (<F2> or <Ctrl-F2>) to return to the line where you set the tag. There is no limit on the number of **Tags** you can set in your source code.

To set the **Tag** in the text, go to the line you want to tag and enter **View>>Toggle Tag** or <Shift-F2>. This sets a green marker in the **Line Icon** column (the column between the line number column and the text). If there is a **Tag** on this line already, this command will clear the **Tag**. You can find your tags by pressing <Ctrl-F2> to find the previous tag or by entering <F2> to find the next **Tag**. The **Tags** are shown in Figure 5–17 at lines 146 and 161.

```
144             }
145             break;
146     case 3: //LED 3 was invoked
147             //Get the value of the Switch to see if On or OFF
148             GetCtrlAttribute (channel, CHANNEL_SW_3,ATTR_CTRL_VAL, &SW3_Val);
149             if(SW3_Val)
150             {
151                 TurnON_LED(CHANNEL_LED_3);   //Turn On LED3
152                 ChannelNum=3;   //Set the Channel number here
153             }
154             break;
155     case 4: //LED 4 was invoked
156             //Get the value of the Switch to see if On or OFF
157             GetCtrlAttribute (channel, CHANNEL_SW_4, ATTR_CTRL_VAL,&SW4_Val);
158             if(SW4_Val)
159             {
160                 TurnON_LED(CHANNEL_LED_4);   //Turn On LED4
161                 ChannelNum=4; //Set the Channel number here
162             }
163             break;
164     default: //It should not come here
165             sprintf(buf, "Random Number Error");
166             InsertTextBoxLine (channel, CHANNEL_DISP, -1,buf);
167             break;
168     }
169     //Display results
```

Figure 5–17
Code with Tags

You can set the scope of the files you want to tag by selecting **View>>Tag Scope.** This brings up another dialog box in which you can set the scope to **Window, All Open Windows,** or **All Files.**

If you have set too many tags in the file and want to view or remove some of them, you can do so by selecting **View>>Clear Tags...** A dialog box with the **Tag** lines is displayed as in Figure 5–18. You can click on the line numbers of the tags that you want removed. To remove all the **Tags,** select **Check All** and select **OK.**

Remember the following shortcuts:

Command	Shortcut
View>>Toggle Tag	<Shift-F2>
View>>Next Tag	<F2>
View>>Previous Tag	<Ctrl-F2>

Finding User Interface Objects (Controls)

When you are writing the code, there are many times when you want to refer back to the user interface file (.uir) to find the **Constant Name,** callback function name, control, or menu object. The **Source Editor** provides you with a convenient way to find the object. Place the cursor on the

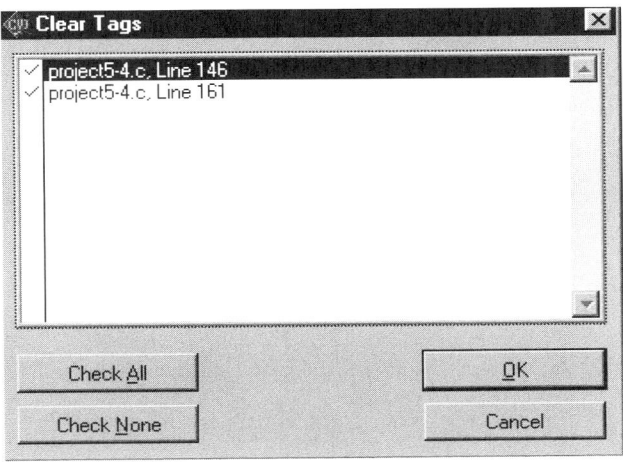

Figure 5–18
Clear Tags Dialog Box

object name and select **View>>Find UI Object** or press <Ctrl-F>. If the object is found, the panel with the object highlighted is displayed. If the object found is a menu object or a callback in the menu bar, a dialog box shows the list of matches. You can highlight the one you want to view by double-clicking on it.

Figure 5–19 shows the **User Interface** object selected when the *MenuStartCB* callback function in `project4-4.c` is highlighted and <Ctrl-F> is selected. The information for this object is shown with the **Constant Name, Label, Callback Name,** and **UIR File.**

From this dialog box, you can select **Edit** if you want to make changes to this menu item.

Remember that the shortcut for **View>>Find UI Object** is <Ctrl-F>.

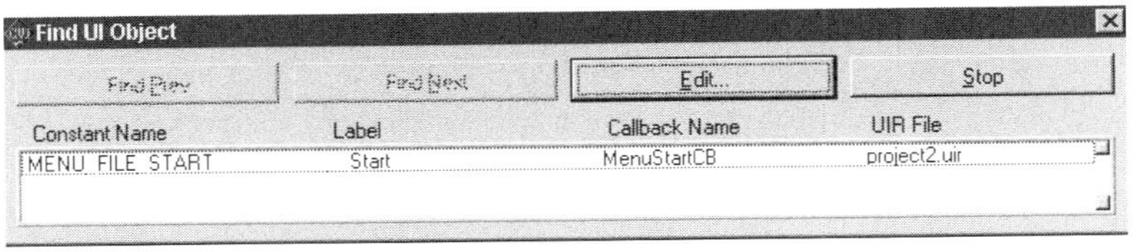

Figure 5–19
Find User Interface Object

Inserting C Constructs in Code

When you add code to your program, you need to use the C code constructs. In conventional compilers, you have to type the complete C construct manually. *CVI* inserts the structure of the construct for you, and you only have to add the variables and the logic within the constructs. The **Insert Construct** sub-menu is shown in Figure 5–20.

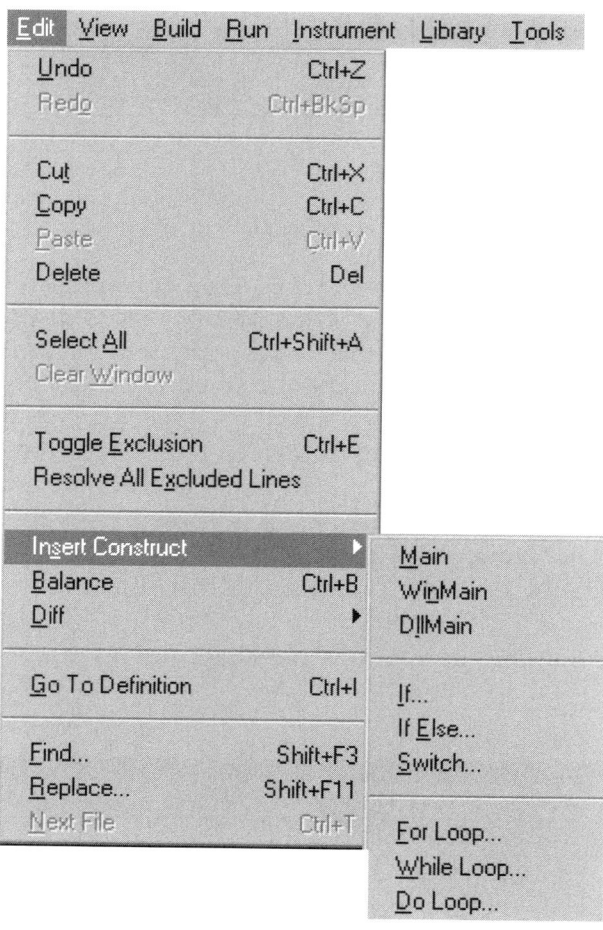

Figure 5–20
Insert Construct Sub-Menu

Let us look at an example and add a *switch* construct by selecting **Edit>>Insert Construct>>Switch** from the menu shown in Figure 5–20. As you enter the *case* numbers in the **Case:** box and select **Below,** the *case* structure is entered in the box below the previous *case* value, as illustrated in Figure 5–21. You have the option of entering the *case* value above the highlighted row by selecting the **Above** command button. The *switch* construct allows you to include the *default:* if you desire. To do so, select **Default Case:** and you have the following options: **At top, At bottom,** and **None.**

When you select **OK** from the dialog box, the complete *switch* construct is inserted at the location of your cursor, including the opening and closing curly braces and the break statements at the end of every *case*. In addition, the default statement is inserted at the end of this construct if you selected **At Bottom** from **Default Case:** as in this example shown in Figure 5–22. You can now add the appropriate code inside the *case* statements. Notice that if you had a very complex *case* statement, this construct would save you a lot of typing and give you syntactically correct structure.

If you prefer, you can always manually enter the C *constructs* without using the **Insert Construct** feature.

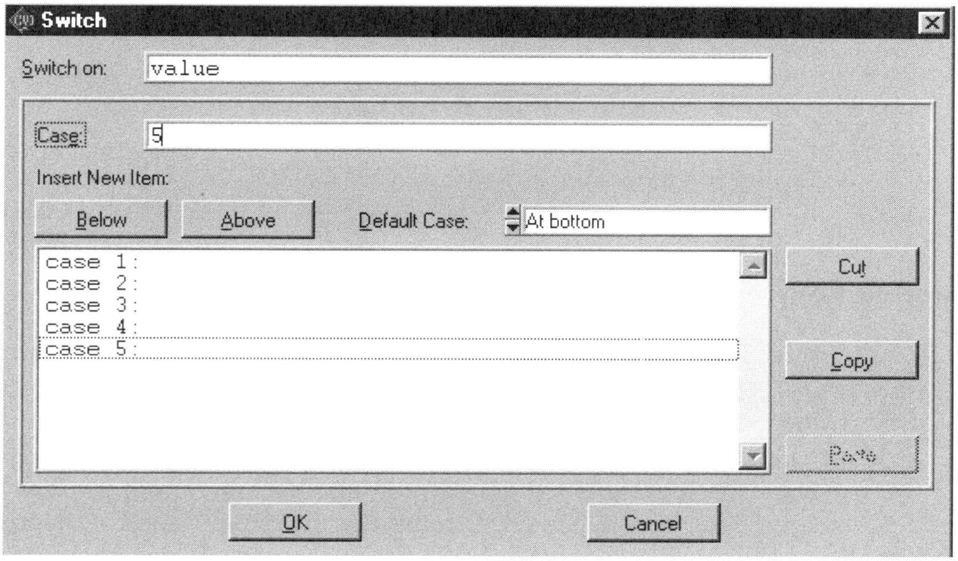

Figure 5–21
Switch Construct

```
62  switch (value)
63      {
64      case 1:
65
66          break;
67      case 2:
68
69          break;
70      case 3:
71
72          break;
73      case 4:
74
75          break;
76      case 5:
77
78          break;
79      default:
80
81          break;
82      }
83
```

Figure 5–22
Switch Construct Created Using the **Insert Construct**

Comparing Source Files

The **Source Editor** has a very useful feature for comparing two source files and finding the differences between the two. To use this feature, select **Edit>>Diff.** A **Diff** sub-menu, as shown in Figure 5–23, is displayed.

Before you start the comparisons, you should select **Ignore White Space** in the sub-menu because you do not care if the spaces, tabs, or other control characters match. This will place a check mark next to the sub-menu command. To start the comparison, open the source files you want to compare and select **Diff With...** or <Ctrl-Shift-D>. A split window with the two source files will be displayed. It is possible that the source files will not necessarily open to

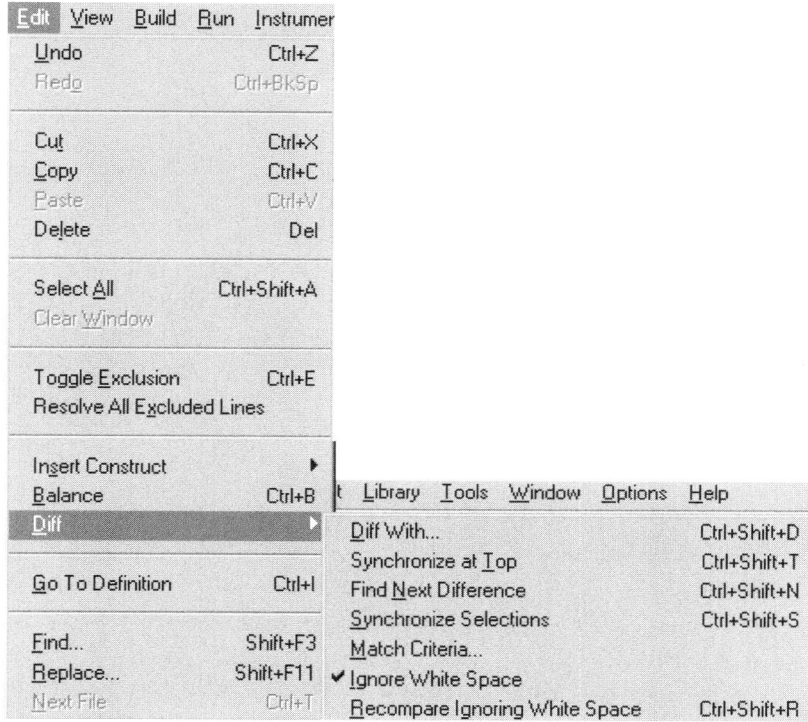

Figure 5–23
Diff Sub-Menu

their first lines. To resolve this, select **Synchronize at Top** or <Ctrl-Shift-T> so the comparison starts from the first lines of both the files. To start the comparison, select **Find Next Difference** or <Ctrl-Shift-N> to display the next lines with differences, which will appear in reverse video.

You can highlight a section from one of the files and select **Synchronize Selections** or <Ctrl-Shift-S> to find a matching section in the other file.

You can set **Match Criteria...** by selecting the number of lines from the pop-up dialog box shown in Figure 5–24. In this box, you enter the number of lines that must match to mark the end of differing sections in a file.

After you have found all the differences in the two files, you can determine if the only difference in the selections do not involve white space characters by selecting **Recompare Selections Ignoring White Space** or <Ctrl-Shift-R> to perform the difference check again.

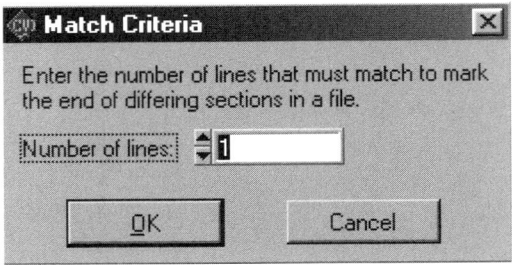

Figure 5–24
Match Criteria Dialog Box

Find, Search, and Replace Commands

The **Find** and **Replace** commands are the two most used commands in the **Source Editor**. If you want to find text in your code, select **Edit>>Find** or <Shift-F3>. A dialog box will pop up in which you enter the string to search. The **Find** dialog box is shown in Figure 5–25. This dialog box has a number of options, which you select by checking the boxes. Enter your search string in the **Find What:** box.

If you want your string search to match exactly as entered with upper and lower case letters, check the **Case Sensitive** box. If you want only the **Whole Word** searched and not the subset of matching strings, check this box. To select expressions with wild card characters in the string, check the **Regular Expression** box. You can use the **Multiple Files** option to include open source files and files from other projects.

The **Select Text Only** option searches only within the highlighted text region that you selected. If you want to search the text from the top of the file again after the first search through the file is completed, check the box next to **Wrap.**

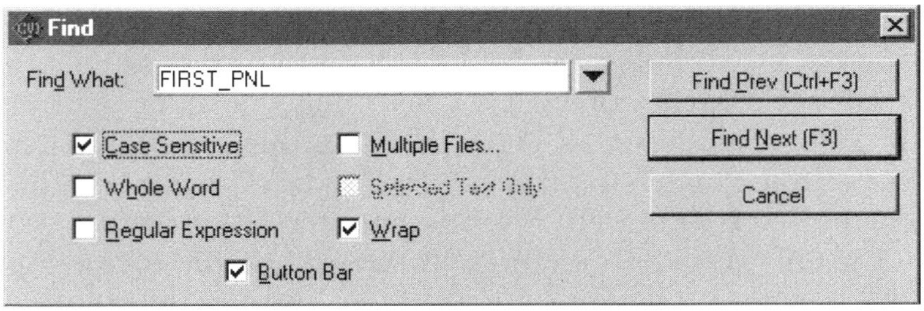

Figure 5–25
Find Dialog Box

Chapter 5 • Source Editor and Debugging Techniques 161

Figure 5–26
Find Button Bar

Check the **Button Bar** if you want the **Find Button Bar** to be displayed when you search for the text. This is shown in Figure 5–26.

Find Next and **Find Previous** on both the **Find Dialog Box** and the **Find Button Bar** searches for your selection in the specified direction. The *CVI Source Editor* finds the command, highlights it in the text, and stops with the **Find Button Bar** displayed (if enabled in the dialog box), waiting for your input. You have the option to **Find Prev, Find Next,** or **Stop** at the current searched line or to **Return** to the line from which you started the search. The shortcut for performing the search in the forward direction is <F3> and in the reverse direction is <Ctrl-F3>. To continue search in the next file, select **Edit>>Next File** provided that the **Multiple Files** option was checked in the **Find** dialog box.

One of the features of the **Find What:** box is that it saves your previous search text in a pull-down list box from which you can select the string instead of re-typing it.

To **Replace** one string by another, select **Edit>>Replace** or press <Shift-F11>. A dialog box, as shown in Figure 5–27, appears which is very similar to the **Find** dialog box shown in Figure 5–26. All the functionalities of the

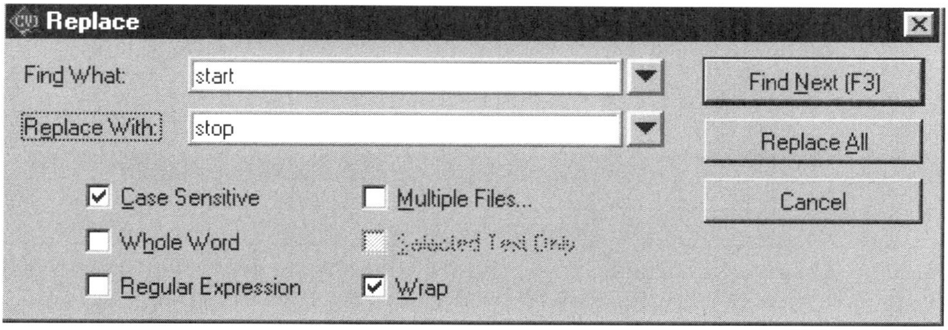

Figure 5–27
Replace Dialog Box

options in this box are the same as those of the **Find** dialog box except that in the **Replace With:** box, you enter both the replaced and the replacing string.

When you click on **Find Next** button in this dialog box, a **Replace Button Bar,** as shown in Figure 5–28, appears.

If you select **Find Next,** the next search string is found without making the current replacement, and if **Replace** is selected, the current find is replaced by the **Replace With** string. Click on the **Replace All** button to find all the string matches and to replace them with the new string in one click. This is a very fast and convenient way to replace strings, but it can be very dangerous if the wrong strings are replaced or if sub-strings are replaced. Saving the file before using this feature is recommended so you can recover the file if necessary due to incorrect string replacement. The **Stop** and **Return** commands are similar to the **Find Button Bar** commands.

Remember the following shortcuts:

Command	Shortcut
Edit>>Find	<Shift-F3>
Find Forward	<F3>
Find Previous	<Ctrl-F3>
Edit>>Replace	<Shift-F11>
Edit>>Diff>>Diff With…	<Ctrl-Shift-D>
Edit>>Diff>>Synchronize at Top	<Ctrl-Shift-T>
Edit>>Diff>>Find Next Difference	<Ctrl-Shift-N>
Edit>>Diff>>Synchronize Selections	<Ctrl-Shift-S>
Edit>>Diff>>Recompare Selections Ignoring White Space	<Ctrl-Shift-R>

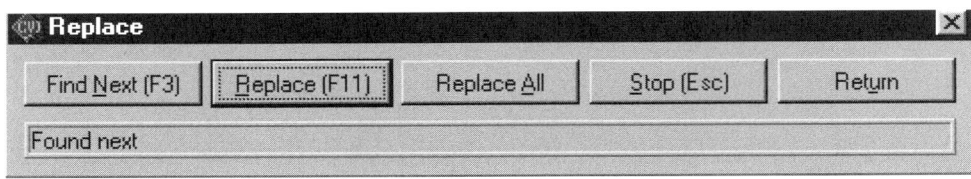

Figure 5–28
Replace Button Bar

Using the Debugger

It is more than likely that every source code you write may not work the first time you run it. The more complex the software, the more the latent bugs, and the more time it takes to find the problems. Learning the **Debugger** will help you find the bugs with relatively little effort. The **Debugger** is not a tool that will guarantee that all the errors have been found, rather it will assist you in stepping through the program and make you aware of the possible errors in the code, guiding you through the program's logical path(s).

By learning the intricacies of **Debugger,** you can save a great deal of effort. Before using the **Debugger,** you need to set the **Debugging Level** from the **Project** window. From the **Project** window, select **Options>>Build Options** and set **Debugging Level** to **Standard** or **Extended.** Without setting this option, you cannot enable **Breakpoints** in the code. The **Build Options** dialog box is shown in Figure 5–29.

You must enable **Line Icons** and **Line Numbers.** Enabling the **Line Icons** allows you to insert the **Breakpoints** and the **Tags** in the column adjacent to the line numbers. To enable this option, select **View>>Line Icons** from the **Source Editor.** This will put a check mark next to this menu item. Similarly, enable the line numbers from **View>>Line Numbers.**

Before proceeding to learn the **Debugger,** it will be necessary to understand the tool bar icons associated with the debugging environment. The tool bar shown in Figure 5–30 is enabled when you run the **Debugger**.

Continue *is used to run the project to the end of the file or the next* ***Breakpoint.*** *It is similar to* ***Run>>Continue.***

Go To Cursor *runs the debugging program to the cursor location and is similar to* ***Run>>Go To Cursor.***

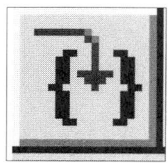

Step Into command *is used during debugging. With this command, you can use the* ***Debugger*** *to step into the function, just as with* ***Run>>Step Into.***

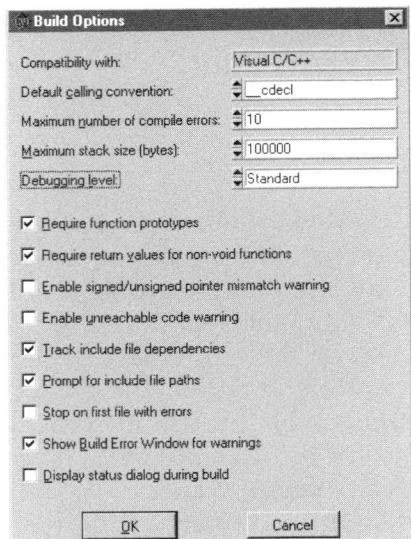

Figure 5–29
Build Options Dialog Box

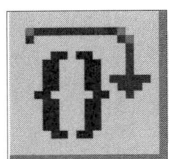

Step Over *steps over the function, though it doesn't go step by step into the function code, and returns the function results to the calling program. It is similar to* **Run>>Step Over.**

Finish Function *finishes the function that it was debugging and returns to the calling program. It is similar to* **Run>>Finish Function.**

Terminate Execution *stops the project from running, just as with* **Run>> Terminate Execution.**

Figure 5–30
Debugging Tool Bar Icons

Up Call Stack *moves you up one level in the function call stack and is similar to* ***Run>>Up Call Stack.***

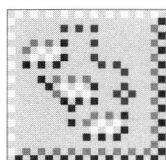

Down Call Stack *moves you down one level in the function call stack. Note that* ***Up Call Stack*** *and* ***Down Call Stack*** *are not enabled at the same time; you can only select one or the other. It performs the same function as selecting* ***Run>>Down Call Stack.***

View Variable Value *is used to view the value of the variable during debugging and is similar to* ***Run>>View Variable Value.***

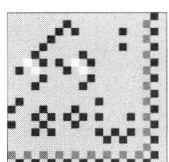

Add Watch Expression *is used in the debugging stage when you want to add a watch expression, just as with* ***Window>>Watch.***

ProjectWindow *is not part of the debugging commands but can be used any time to access the* ***Project*** *window.*

Using Breakpoints

Before you start with the actual procedure of using the breakpoints, let us talk about the purpose of the breakpoints. **Breakpoints** enable you to *suspend* the execution of the program without exiting the program, allowing you to view variable values at that instant. You can also set conditional breakpoints at certain lines that are maintained outside the code.

Setting a breakpoint is accomplished by moving the cursor to the **Line Icons** column at the line in which you want to set the breakpoint and clicking with the left mouse button. Move the cursor to line 48 in the **Line Icons** column and click with the mouse button. A red diamond, as shown in Figure 5–31, appears on line 48. The program will execute up to line 48 and stop before it executes this line. Line 48 is also highlighted when the execution stops.

When setting a breakpoint, a few items must be considered. You can set a breakpoint only at an executable line. It cannot be set at a comment line or a blank line. If you do so, the program will remind you that the **Breakpoint** cannot be set at this line. In addition, when the program is stopped at the breakpoint, you cannot change the code anywhere in the program until the program has been terminated.

If you would like to step through the complete code, you can set the breakpoint at the first line manually, or better still, from the **Run** menu, select **Break at First Statement.** By doing this through the **Run** menu, you can be absolutely sure that the program will find the first statement and stop there.

To clear the breakpoint from the line, you can move the cursor to the breakpoint and click on the red diamond, select **Run>>Toggle Breakpoint,** or press <Ctrl-F9>. You can also set the breakpoints using these methods. You can also set breakpoints by programmatically inserting the function *Breakpoint* at the point in the code at which you want to stop the program.

To step a line at a time from the breakpoint, press <F10>. It will step you to the next line and stop there before executing it. Pressing <F10> again will take you to the next line. This way you can examine the complete code execution one line at a time and follow the path of the program flow. You can use the debugging icons explained earlier instead of the menu items or the short cuts if you wish.

```
41      //Load the menubar
42      menuHandle = LoadMenuBar (channel, "project5-4.uir", MENU);
43
44      DisplayPanel (channel);
45
46      //Display the date on the GUI
47      SetCtrlVal( channel, CHANNEL_DATE_STR, DateStr());
48 ◆    GetProjectDir (prj_dir);
49
50      // Create the Toolbar
51      Toolbar_New(channel,menuHandle,"Channel Readings Toolbar",
52                                       0,0,1, 1, &ToolbarHandle);
53      sprintf(icon_name, "%s\\go.ico", prj_dir);
54      //Insert the Start icon
55      Toolbar_InsertItem (ToolbarHandle, 1, kCommandButton, 1, "Start",
56              kMenuCallback, MENU_FILE_START, 0, 0, icon_name);
```

Figure 5–31
Breakpoint Set

If you feel that you have no need to step over all the lines in the code, you can easily bypass a section of code. Move your cursor down to the line at which you want the program execution stopped and select **Run>>Go To Cursor** or press <Ctrl-F7>. The program will execute all the remaining lines from the breakpoint and stop at the cursor.

If there is a function in your code that needs a performance check, select **Run>>Step Into** or press <Ctrl-F8>. *CVI* takes you into the function and stops at the function name. From there, you can either step through it line by line or bypass a section of code as described earlier. If you want to run this function to completion, enter **Run>>Finish Function** or press <Ctrl-F10>, and you will return to the line after the function call.

Suppose you did not want to go into the function code. Select **Run>>Step Over** or press <F10> as before. This will not go into the function but continue with the next line in the code. In other words, you will bypass stepping through the function code.

To run to the next breakpoint in the code, select **Run>>Continue** or press <F5>. This will execute the code between the two breakpoints and stop at the next breakpoint. If there are no other breakpoints in the execution sequence, the program will run to completion. To terminate the program execution, select **Run>>Terminate Execution** or press <Ctrl-F12>. Figure 5–32 shows the program execution suspended at the breakpoint at line 48.

You may have set many breakpoints in your code and forgotten where they are. It becomes difficult to step through the code and find all the breakpoints if the code is long. *CVI* solves this problem by displaying all the **Breakpoints** and their line numbers in the program by opening a **Breakpoints** dialog box, as shown in Figure 5–33, from **Run>>Breakpoints...** or press <Shift-F9>.

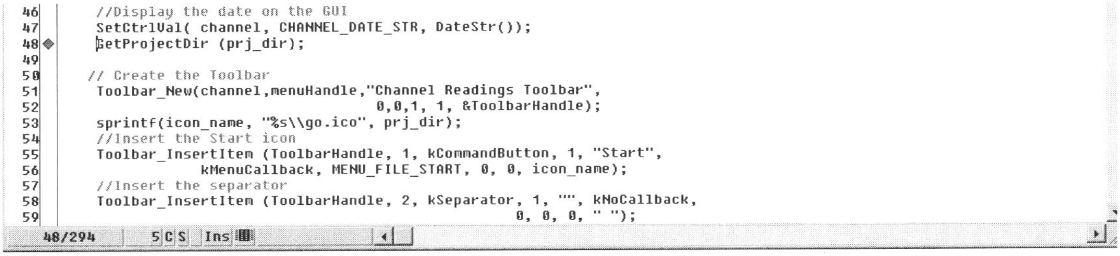

Figure 5–32
Execution Stopped at Breakpoint

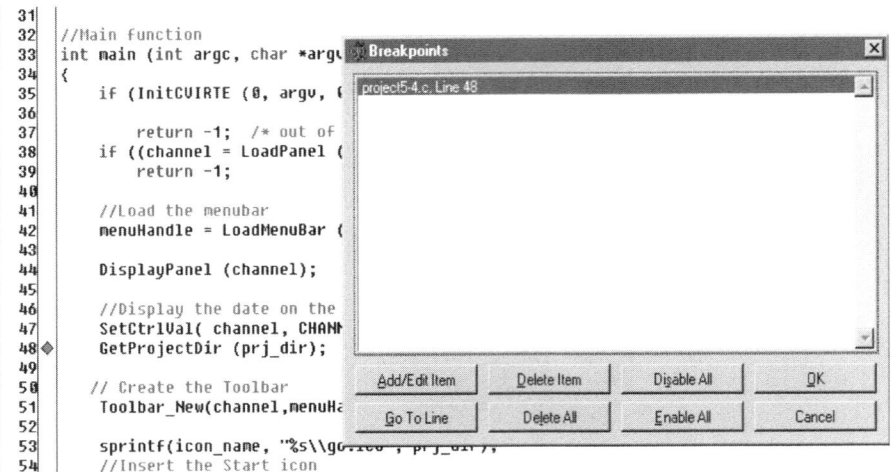

Figure 5–33
Breakpoint Dialog Box

To edit a breakpoint, select the **Add/Edit Item** button, which brings up another dialog box, shown in Figure 5–34. From this box, you can select the source file you want to edit by entering the name in the **File:** box. Using the **Line:** box, you can change the line number of the breakpoint.

If you want a certain line to be executed more than once in the code before the breakpoint is enabled, set the number in the **Pass Count**: field.

In the **Condition** box, you set an expression that *CVI* evaluates before executing the source code. A *True* condition sets the breakpoint and the program pauses, while a *False* condition continues with the execution. You can check the **Disabled** box to disable the breakpoint for this line number. The breakpoint will still be visible on the line but in a different color (possibly light gray) and the word Disabled will be added to the breakpoint line in the **Breakpoints** dialog box.

After you have made the appropriate selections on this box, you will return to the window shown in Figure 5–26, and you can make further selections. If you are finished pausing the program at the breakpoints but you want to keep the reference to them, you can **Disable All.** You can enable all the breakpoints from **Enable All.** To remove all the breakpoints, select **Delete All.**

To go to the line where the breakpoints are set, highlight the line in the **Breakpoint** dialog box, and select **Go to Line.** After you have made all your selections, select **OK**.

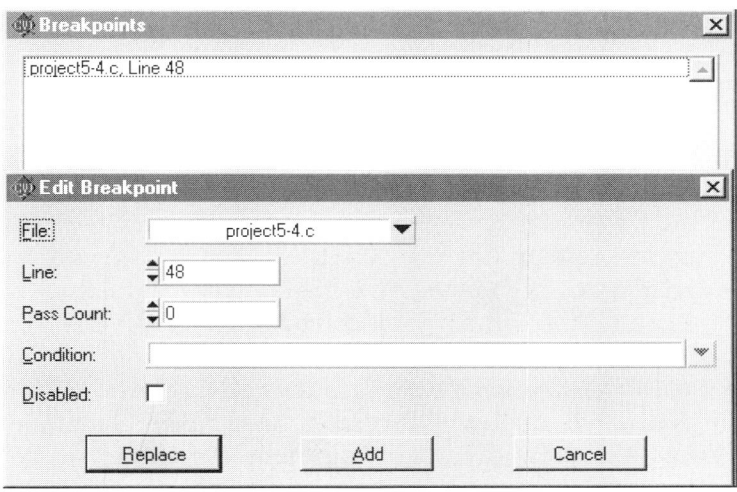

Figure 5–34
Edit Breakpoint Dialog Box

Remember the following shortcuts:

Command	Shortcut
Run>>Toggle Breakpoint	<Ctrl-F9>
Run>>Step Over	<F10>
Run>>Go To Cursor	<Ctrl-F7>
Run>>Step Into	<Ctrl-F8>
Run>>Finish Function	<Ctrl-F10>
Run>>Continue	<F5>
Run>>Terminate Execution	<Ctrl-F12>
Run>>Breakpoints	<Shift-F9>

Exclude/Include Lines

Many times while testing your software you may want to bypass a section of code so it does not compile or execute. You can accomplish this by commenting out those lines of code, and when everything is working to your satisfaction, you can un-comment or remove those lines. This is quite time-consuming if you have to do a number of lines to comment out, i.e., to add /* and */ around or to include // on every line.

CVI has a handy feature for commenting lines of code without the trouble of putting comment marks around the code. You can exclude the lines of code by highlighting them and selecting **Edit>>Toggle Exclusion** or pressing <Ctrl-E>. You will notice that the color of the lines of code will change (to light gray on some systems), and they will not be compiled or executed when you run the program.

To include the lines into the code again, highlight the lines and select **Edit>>Toggle Exclusion** or press <Ctrl-E> again. This will include the previously excluded code in the source file and return the lines to their original color. You can do the same by selecting **Edit>>Resolve All Excluded Lines.** A dialog box, as shown in Figure 5–35, is displayed.

To put the excluded lines back into the code, select the **Unexclude** button from the dialog box. If you wish to comment out the section of the excluded code, select **Comment.** It will add comment markers to all the selected lines. To perform conditional compilation around a block of lines, click on the **#if 0** button and the conditional compilation block is created around the selected lines as shown in Figure 5–36. In this case, the excluded lines are 176 through 181. To include these lines again into the compilation, either remove the `#if 0` and `#endif` statements or replace `#if 0` with `#if 1`.

Selecting **Delete** will remove these lines altogether, and **Skip** will cause the program to go over these lines without changing their present state.

When you close or exit the program with the lines excluded, *CVI* pops up a dialog box as shown in Figure 5–37, giving you a choice of **Options, Unexclude All,** or **Cancel.** When you select **Options,** the box shown in Figure 5–35 is displayed and you can set your preferences as before.

When you are running the program with breakpoints and you have a panel that has been loaded and displayed by *CVI*, every time you step through the code the panel flashes on the screen momentarily. The code executes faster if you select the **Run>>Activate Panels when Resuming** option by placing a check mark on this menu item. If you disable this option, the panels are activated when the code causes certain events to be processed or

Figure 5–35
Resolve All Excluded Lines Dialog Box

```
174         //Display results
175  #if 0  /* formerly excluded lines */
176         if ( ChannelNum==0 ||ChannelNum==1 ||  ChannelNum==2 ||ChannelNum==3 || ChannelNum==4 )
177         {
178             sprintf(buf, "%s               %d",TimeStr() , ChannelNum);
179             InsertTextBoxLine (channel, CHANNEL_DISP, -1, buf);
180         }
181         break;
182  #endif   /* formerly excluded lines */
183       }
184       return 0;
185  } // TimerLoop250CB
186
```

Figure 5–36
Conditional Compilation

displayed on the panel. This option is saved from one session to another and does not have to be saved every time.

Working with Variables, Arrays, and Strings

You have learned how to work with breakpoints in the code and how to suspend and resume program execution. The next step is to learn what to do when you have paused the execution to debug the code.

Select **Windows>>Variables.** A **Variables** window will be displayed, as in Figure 5–38, with the breakpoint set at line 249.

There are two portions of this window: the top sub-window lists the global variables and the bottom sub-window the function variables. The sub-window on the top displays project global variables that are not declared *static*, Interactive Execution window variables, and the program global variables declared as *static*.

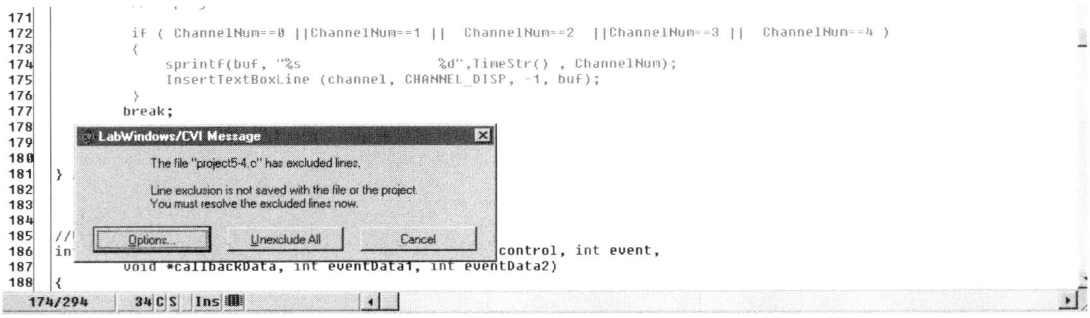

Figure 5–37
Excluded Lines Exit Dialog Box

172   Chapter 5 • Source Editor and Debugging Techniques

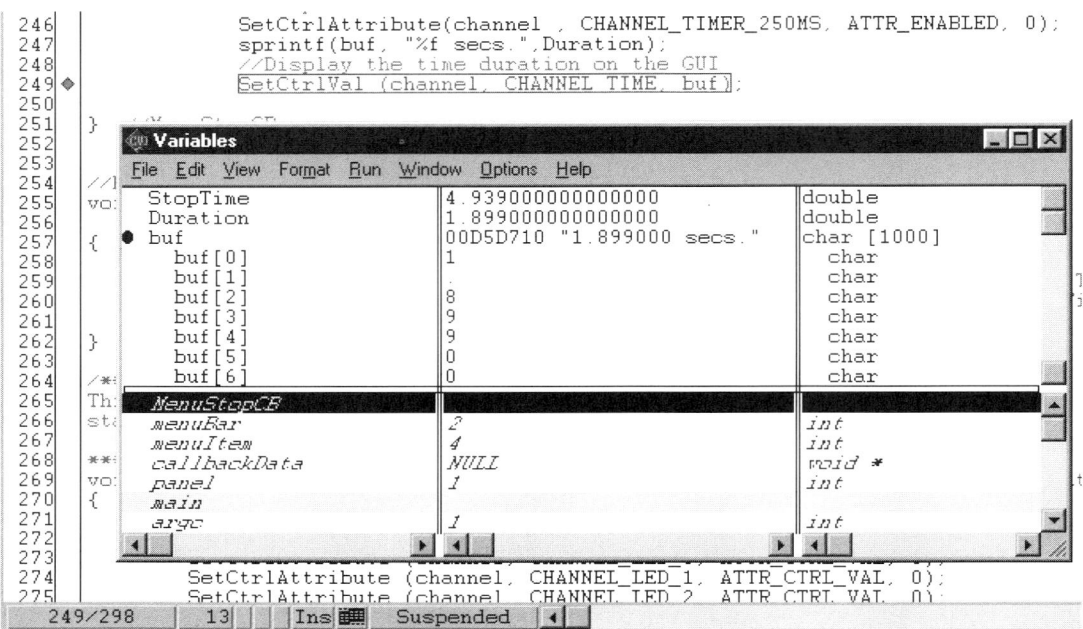

Figure 5–38
Viewing Variables at the Breakpoint

The sub-window on the bottom is for the current, active function. Each function has a different section, and it contains the name of the function at the top, followed by the parameter list (arguments) of the function in *italics*, followed by the local variables of that function.

Let us examine the **Variables** window. The first column contains the variable names, the second column the value of the variable or the address of the variable if it is an array, and the last column the data type of the variable.

Besides viewing the values of the variable during execution, you can also change their values at run-time in this window.

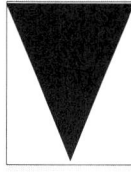

*The inverted triangle means that the variable next to it is the starting pointer to a block of data, such as an array, string, or structure. Clicking on this icon or selecting **View>>Expand Variable** expands the variable so the values of all the elements of this variable are seen. When you click on the variable, the inverted triangle changes into a solid circle.*

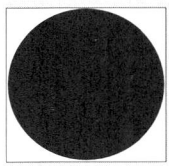

The solid circle means that the variable on this line is the starting pointer to a block of data that is currently being shown in the expanded form. To close the expanded form, you can click on this icon or select **View>>Close Variable,** *which will display only the starting pointer, and the solid circle will change to an inverted triangle.*

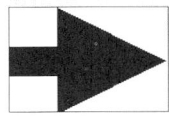

The right arrow indicates that the variable on this line is a member of a structure that is a pointer to another structure of the same type. Select **View>>Follow Pointer Chain** *or click on this arrow to replace it with the child structure that the pointer references.*

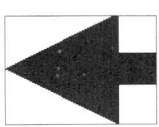

The left arrow indicates that the variable on this line is the child structure in a chain. Select **View>>Retrace Pointer Chain** *or click on this arrow to replace the current structure with its parent.*

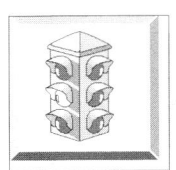

Caution: The View menu used to control these variables is from the Variables menu and not the View menu in the Source Editor menu.

A debugging feature lacking from *CVI* versions prior to 5.5 allows you to view the value of a variable or an expression by moving the cursor over them while the program is suspended at a break point. For this feature to be active, select **Options>>Environment** from the **Project** window and place a check mark on the **Enable Data Tool Tips** box.

Suppose you have a segment of a code with a *for loop* in which the counter is a local variable `loop`. You can change its value by highlighting this variable; select **Run>>View Variable Value,** press <Shift-F7>, or just double-click on the variable. A dialog box, as shown in Figure 5–39, is displayed.

Chapter 5 • Source Editor and Debugging Techniques

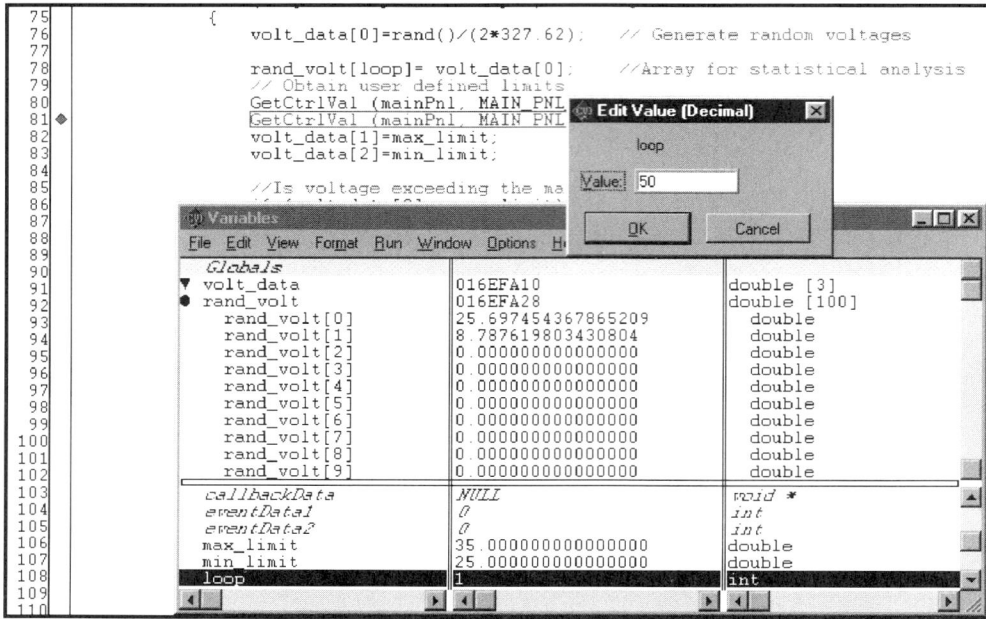

Figure 5–39
Changing Values of the Variable

Enter 50 in this box as the new value of loop. Selecting **OK** inserts the value of loop in the **Variables** window.

Now when you step through the program or enter <F5> to continue the program, it will run the complete *for* loop using 50 as the value for the loop variable and will stop again at the breakpoint. Select **Variable** from the **Window** menu, view the loop variable, and notice its value is now incremented to 51.

Every time you run the code from the breakpoints, the **Variable** window disappears and you have to open it again. You can re-size both the code and the **Variables** window so they both fit on the desktop without over-lapping each other, and both windows will remain visible when you run the **Debugger**.

Likewise, as shown in Figure 5–39, the array values can also be modified. Double-click on the rand_volt[2] and enter any value of the correct type (that is *double* for this variable). You will notice that the new value is entered in the **Variable** window for rand_volt[2]. When you run the program again, you will notice that the new value is used in the calculations.

Highlight the array and bring up **Array Display** window, as shown in Figure 5–40, by selecting **View>>Array Display,** which shows the rand_

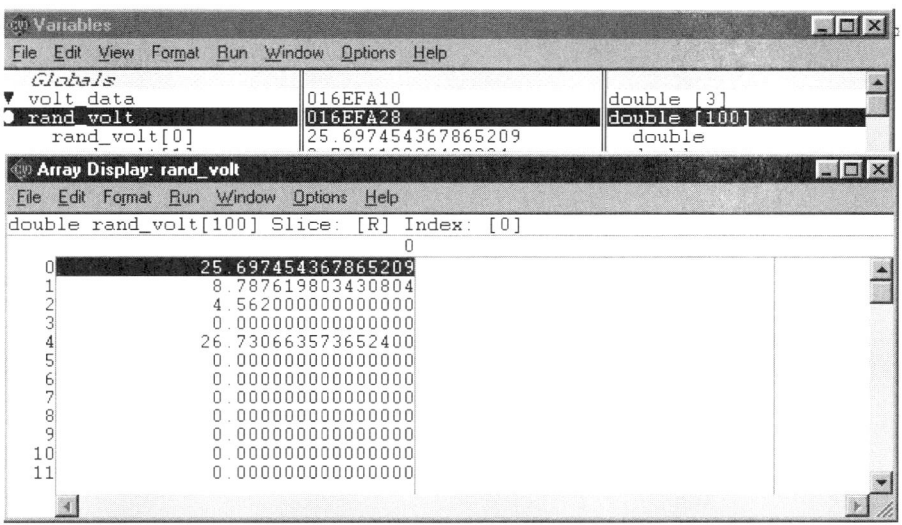

Figure 5–40
Array Display Window

volt array. The **Slice** indicator shows the array by row **[R]** or column **[C]** from the **Options>>Reset Indices** command in the **Variables** window. This is particularly useful when there are multi-dimension arrays. The **Index:** shows the highlighted element index. You can quickly view the other indices by using the scroll bar on the **Array Display** window.

For multi-dimensional arrays with two or more dimensions, you can specify the rows and columns for the dimensions using the **Options>>Reset Indices,** and you can specify the constant values to fix the other dimensions. For further information on Multi-Dimensional Arrays, see Chapter 7 of *LabWindows/CVI User Manual.*

The **String** variables can be viewed similarly. At the breakpoint, open the **Variables** window menu like you did before. Highlight the string variable and select **View>>String Display** from the **Variable** window, press <Shift-F4>, or simply double-click on the variable in the **Variables** window.

By using the **Edit** menu, you can edit the string variable provided it is in ASCII format. You can change the base of viewing the string and the array variables by selecting **Format** and clicking on **Decimal, Hexadecimal, Octal, Binary,** or **ASCII.** The **Format** menu is shown in Figure 5–41. The same is true for the array, but if the array consists of real numbers, the **Format** menu will show **Floating Point, Scientific,** and **Fix Precision.**

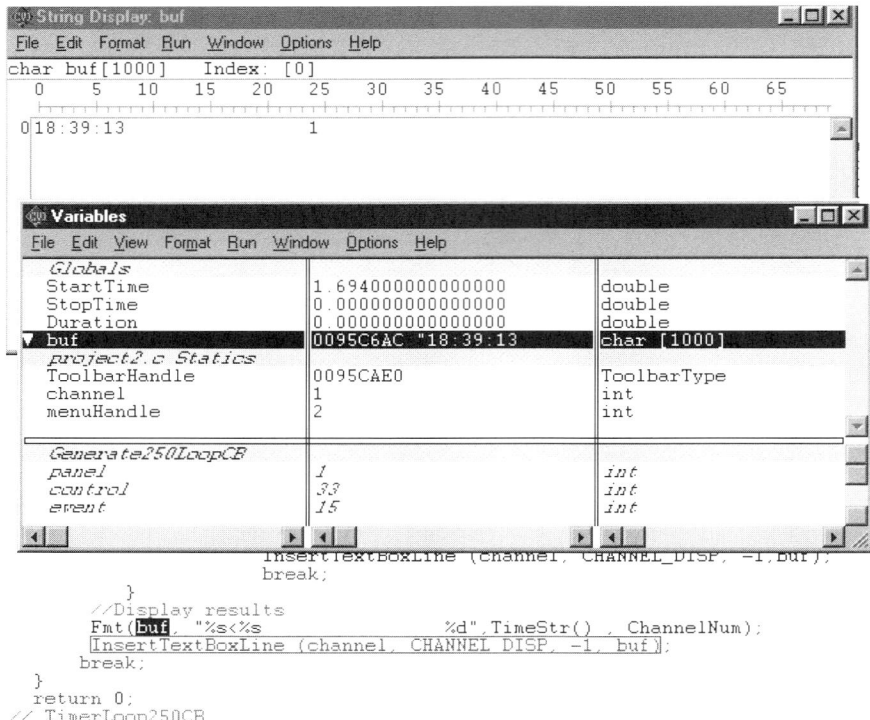

Figure 5–41
Format in String Display Window

Remember the following shortcuts:

Command	Shortcut
Run>>View Variable Value	<Shift-F7>
View>>String Display	<Shift-F4>

Watch Window

The **Watch** window is similar to the **Variables** window and is enabled from **Windows>>Watch.** In this window, you can decide what variables and expressions you want to monitor continuously and set a breakpoint when their value changes. To select the **Watch** expression, open the **Variables** window and select **Options>>Add Watch Expression.** A dialog box, as shown in Figure 5–42, is displayed.

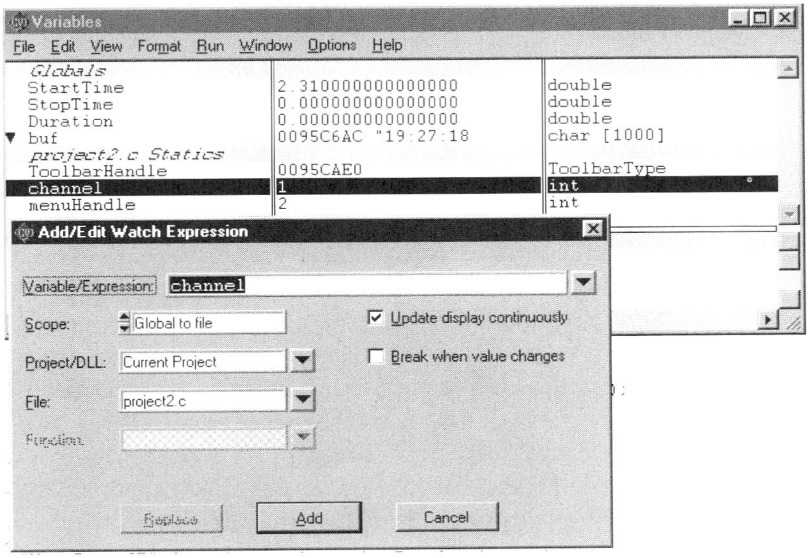

Figure 5–42
Add/Edit Watch Expression

In the **Variable/Expression** box, enter the expression you want to place in the **Watch** window to monitor. The **Scope** refers to whether the variable or expression is global to the project, global to the file, local to a function, or global to the **Interactive Execution** window. **Project/DLL** indicates whether the watch expression refers to the current project or the DLL file. **File** refers to the file name in which the variable or expression is located. If the variable is local, the **Function** box contains the function name. If it is global, the **Function** box is dimmed.

When checked, the **Update Display continuously** box updates and displays the variable and expression value at every statement while the program is running. The **Break when value changes** box, when checked, suspends the program at the change of the value of the variable or expression. This feature is useful when you are trying to find when the value of the variable has changed.

Select **Add** to include the variable or expression to the **Watch** window. Select **Replace** to change the previous attributes of the same variable or expression. Select **Cancel** if you want to exit the **Watch** dialog box without making any changes.

Memory Display Window

The **Memory Display** window did not exist in *CVI* versions prior to 5.5. You can use this window to view the contents of the memory location of your program while it is suspended. This window performs the same functions from the **Source Editor** window as from the **Project** window, which is discussed in the **Windows** menu section in Appendix B, "Project Window Environment." From the **Source Editor,** select **Windows>>Memory Display.** The **Memory Display** using the debugger is shown in Figure 5–43.

In *CVI* versions prior to 5.5, you could view the debugging data in the Standard I/O window. *CVI* 5.5 creates a separate process to execute your program even when it is debugging the program. **Standard I/O** window is not owned by your program and is inaccessible for your outputs even while the program is suspended during debugging. There are two options for bypassing this situation. First, you can use the Utility Library *ErrorPrintf* function to send the debug output to the **Run-time Errors** window. You can also check the box next to **Use Console Window for Standard I/O when Debugging** in **Options>>Environment** in the **Project** Window to use a Windows console window instead of the Standard I/O window while debugging. The differences between enabling the two displays are explained in Appendix B, "Project Window Environment."

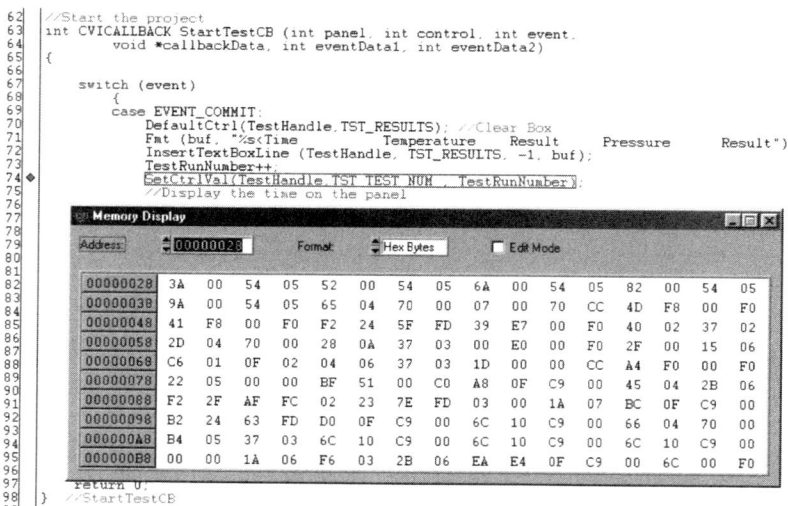

Figure 5–43
Memory Display Window

Summary

In this chapter, you learned how to work with the **Source Editor** and the **Debugger.** You learned how to enable and disable the line numbers and the line icons and to use the **Tags** as bookmarks to jump around the code without losing your place in a long code listing. You learned the three ways of selecting text. You were introduced to inserting the function panels into the code and finding the user interface objects from the source code. The convenience with which you can add C constructs in the code was discussed as was comparing of the two source code files. The **Find, Search,** and **Replace** commands were discussed in detail. Compiling, building, and running the project and fixing the compiler errors were discussed.

You learned how to set and disable the breakpoints and to step through the code and into functions. Exclude and include lines for debugging, viewing the values of the variables, arrays, and strings, and setting the variables and expressions in a **Watch** window to monitor their values were also demonstrated.

The purpose of this chapter was to help you become proficient in using the **Source Editor** and the **Debugger.** By mastering these two useful tools, you will be able to accomplish many of the tasks of creating and debugging the code. Understanding the **Source Editor** and the **Debugger** of any compiler for any language is a must; without these tools a programmer is left handicapped. Mastering these in *CVI* is no exception!

FILE INPUT/OUTPUT

Chapter Highlights
- What is File I/O?
- Using Files
- Overview of the Project
- Creating the User Interface
- Generating the Code
- Pop-Up Panels
- Adding Code to the Project
- Summary

In this chapter, you will be introduced to the concept of File Input and Output (File I/O). File I/O means to write and read information to/from a file. By means of examples, you will be shown how to open a file, read a file, and then close it. You will be introduced to the various pop-up panels (also called modal panels) that are used in *CVI*. The project in this chapter describes the use of the *CVI* library functions related to file manipulation through function prototypes and examples to facilitate understanding.

What is File I/O?

Files form an important part of programming. When you write a letter in your favorite word processing program or when you create a spread sheet, you save the data to a file. Without writing data to files, your work would be lost when you turn off the computer.

The same is true if you are running a test and need to save the data that is displayed on the screen. It will be lost as soon as the program is terminated. You should have a way to save the data on the hard disk or some other storage medium for later analysis and record-keeping.

File I/O simply means to write data to a file and to retrieve the data from the file when you need it. The first step is to open the file, the second is to write data to it, and the last is to close the file. You need to know when you open the file whether the file is to be opened for reading or for writing. You need to know whether the data is going to be appended to the existing data in the file or written to the beginning of a new file and whether the data will be stored in a binary or ASCII format. All of these decisions are dependent on your project requirements and on how you are going to use this data later.

Using Files

The *CVI* library function *OpenFile* defines whether to open the file for reading, writing, or both. *OpenFile* should know whether the file will be truncated or appended, how the file pointer will be positioned, and what type of data will be read or written to file, binary or ASCII. Recall that the file pointer is an index that references the data location in the file to read or to write data. The *OpenFile* function prototype is

```
int handle = OpenFile(char *filename, int read/writeMode,
                                      int action, int fileType);
```

and the function arguments are explained in Table 6–1.

Once the file is created, it is always referred to by its *handle* name that is created by the *OpenFile* function.

The next step is to write data to the file, and this is accomplished by the *CVI* library function *WriteFile*. The *WriteFile* function prototype is

```
int n = WriteFile(int fileHandle, char *buffer, unsigned int count);
```

and the function arguments are explained in Table 6–2.

Table 6–1 OpenFile *Function*

Input/ Output	Name	Type	Description
Input	filename	string	contains the pathname of the file to open
	read/writeMode	integer	specifies in which mode the file is opened: VAL_READ_WRITE opens the file for reading and writing, VAL_READ_ONLY opens the file for reading only, VAL_WRITE_ONLY opens the file for writing only
	action	integer	VAL_TRUNCATE truncates the file, deletes the old contents, and starts from the beginning of the file, VAL_APPEND doesn't truncate the file and appends all the data to the end of the old file, VAL_OPEN_AS_IS doesn't truncate the file and positions the file pointer to the beginning of the file
	fileType	integer	decides between creating a file as BINARY or ASCII: VAL_BINARY opens a file for binary data, VAL_ASCII opens a file for ASCII data
Output	handle	integer	creates the file handle to be used to refer to the file

The *fileHandle* used in the *WriteFile* function is obtained from the *OpenFile* function. This is the unique handle defining the file attributes. If you want to write to the **Standard Output** or console window, then use a 1 instead of the file handle. In ASCII mode, *WriteFile* function replaces each linefeed charac-

Table 6–2 WriteFile *Function*

Input/Output	Name	Type	Description
Input	fileHandle	integer	creates the file handle to be used to refer to the file
	buffer	string	data buffer to write to file
	count	Integer	number of bytes to write to the file, should be less than the buffer size
Output	n	integer	number of bytes written to the file, error is indicated by a return value of –1

ter with a carriage return (\r) and a linefeed (\n) in the output file. When you read the file, the carriage return (\r) is not returned; only the linefeed (\n) is sent back. The *WriteFile* function returns the number of bytes that are written to the file. The return value of –1 indicates an error.

After you have finished writing to the file, the next step is to close the file. The *CloseFile* function takes only one argument, the *fileHandle* of the file you want to close. A prototype of the *CloseFile* follows.

```
int status= CloseFile(int fileHandle);
```

The status returns the result of the function. A status of 0 means there are no problems in closing the file; a -1 means that the file handle is bad. There may be a number of reasons for getting a bad file handle. It could be that the handle you are using is not the correct one. The file may already be closed somewhere else in the program and you are trying to close it again, or the file handle may have never been created.

Before you go any further, let us discuss some of the key items of creating files. It is best to open the file for the read or the write mode. For example, if you know that you will only use the file for writing data and not for reading data at that time, then it is best to specify only the writing mode. Likewise, if the file is for reading only, give the file that privilege only; do not set it to read/write mode to reduce the chances of over-writing the data in the file by mistake. A good rule is to give a file the least privilege(s) necessary.

Another important point to remember is that you should close the file as soon as its job is done. Closing the file immediately after it is read or written will protect the file from being corrupted. If you think you will use the file again later on in the program, you can always re-open the file.

Once the file is written to the hard disk with a file name, you can always access it by the usual means. The whole purpose of creating a file is to access the data when you need it. You have a choice of looking at the file using a standard word processor or a text editor such as Notepad or WordPad if the file is written in ASCII. You can also transfer it to a spreadsheet format so it can be loaded into an application like Microsoft® Excel. This is what you will soon see when you create the next project.

Reading a data file that you have written is accomplished through a *ReadFile* function. First you have to open that file for reading if the file is not already set in that mode. You assign the file a handle, which is used to refer to this file and to create a large enough buffer to read the data. You also specify the number of bytes to read. This function reads the bytes from a file until a linefeed is encountered. The number of bytes read is placed in the buffer and the excess bytes over the specified maximum are discarded so that the buffer does not overflow. The ASCII NUL byte is added at the end of data in the buffer.

The prototype of *ReadFile* function is

```
int n= ReadFile(int fileHandle, char buffer[ ], int count);
```

and the function arguments are explained in Table 6–3.

Note in all these functions, if the *fileHandle* is specified as 0, then the data is read from **Standard Input.** (The standard input is the keyboard and the standard output is the computer screen.)

There is a way to read a line at a time from the file by using the *ReadLine* function. This function reads the bytes from a file until it encounters a line-

Table 6–3 ReadFile *Function*

Input/Output	Name	Type	Description
Input	fileHandle	integer	file handle to be used to refer to the file
	buffer	string	data buffer to store the read data
	count	integer	number of bytes to read from the file, should be less than the buffer size
Output	n	integer	number of bytes actually read from the file

feed or the maximum number of bytes specified in the argument, excluding the linefeed character. It then goes to the next line, and at the next *ReadLine* command, it reads the next line and so on. When this function places the data into the buffer, it places an ASCII NUL (\0) character to make it into a string. If the file is opened in an ASCII mode, then the carriage return and linefeed are treated as just a linefeed. *CVI* uses -2 as the End Of File (EOF) character. If the EOF is reached, then the *ReadLine* returns a -2. This is a very useful fact to use if you want to read all the lines to the end of file and don't how many lines are in the file. An example of how to use this feature is demonstrated through a *while* loop shown in Figure 6–1.

This will step through all the lines of the file associated with the fileHandle and read the data into the `lineBuffer` till the *End of File EOF* is encountered.

If n = –1, then *CVI* considers that there is an error reading from the file. You can check for the returned value on every call to *ReadLine* as part of safe programming practice. Let us look at the prototype of the *ReadLine* function.

```
int n= ReadLine ( int fileHandle, char lineBuffer [ ],
                               int MaximumNumofBytes);
```

The function arguments are explained in Table 6–4.

WriteLine is a similar function for writing the data into a file a line at a time. The *WriteLine* function is complementary to the *ReadLine* function. The prototype of this function follows, and the function arguments are explained in Table 6–5.

```
int n= WriteLine(int fileHandle, char *lineBuffer,
                               int NumberofBytes);
```

This function writes the number of bytes in the variable **NumberofBytes** in the **lineBuffer** to the file and then adds a linefeed to it. If the **Numberof-Bytes** is set to –1, *WriteLine* writes only the bytes in the **lineBuffer** before the ASCII NUL and then puts a linefeed at the end of the line. If there is an error

```
n=ReadLine( fileHandle, lineBuffer, MaxBytesToRead);
while (n != -2) //Read till end of file
{
        n=ReadLine( fileHandle, lineBuffer, MaxBytesToRead);
}
```

Figure 6–1
Sample to Demonstrate Reading All the Lines from a File

Chapter 6 • File Input/Output

Table 6–4 ReadLine *Function*

Input/Output	Name	Type	Description
Input	fileHandle	integer	file handle to be used to refer to the file
	lineBuffer	string	data buffer to store the complete read data
	MaximumNumofBytes	integer	maximum number of bytes to read into the lineBuffer excluding the ASCII Nul
Output	n	integer	number of bytes actually read from the file excluding linefeed

in writing to the file, then a –1 is returned. To write to **Standard Output** or console window, replace the fileHandle by 1.

There is a function *GetFileInfo* that gives us information on the existence of a file on the disk and its size. Let us look at the *GetFileInfo* function. The prototype of this function follows, and the function arguments are explained in Table 6–6.

```
int status = GetFileInfo ( char *filename, long *filesize);
```

Table 6–5 WriteLine *Function*

Input/Output	Name	Type	Description
Input	fileHandle	integer	file handle to be used to refer to the file
	lineBuffer	string	data buffer to write to file
	NumberofBytes	integer	number of bytes to write to the file
Output	n	integer	number of bytes actually written to the file including linefeed

Table 6–6 GetFileInfo *Function*

Input/Output	Name	Type	Description
Input	filename	string	pathname of the file to check
Output	status	integer	indicates whether the file exists: 1 – file exists, 0 – file does not exist, –1 – Maximum number of files already open
	filesize	long	file size (if the file exists)

The functions discussed above are some of the more commonly used file read and write functions. Most of these library functions will be used in the project that you will build in the next section.

Overview of the Project

`Project6` shows you how to use the file input/output and how to work with pop-up panels. This project consists of displaying random values for temperature and pressure at regular time intervals. This data is shown on the display using the time stamp showing when the data was generated, its value, and the Pass/Fail results. Two sets of random values are generated. One set of random data is generated every second and one set is generated every other second. This data will then be written to a file. After the program is stopped, the user will have a choice of reading the data on the display, looking at the ASCII data file using *Notepad* or any other text editor, or for a data file created in an *Excel* format, displaying the data in an *Excel* spreadsheet. During the project, pop-up panels will be used to prompt the user and to obtain information from the user.

Although this project uses random numbers for its input data, you can use instruments connected to your system to obtain and record the data in a file.

Most of the methods of creating a GUI will be familiar to you if you have followed the previous projects. Therefore, the details of creating a GUI will not be discussed, and you will be given only the attributes used to create the objects for the user interface.

Creating the User Interface

Create a new panel and set the attributes for this panel.

Name	Attribute
Constant Name	`TST`
Panel Title	`TEST PANEL`
Menu Bar	`No Menu Bars`
Close Control	`Exit`
Top	`59`
Left	`79`
Height	`617`
Width	`876`
Scroll Bars	`None`
Auto Center Vertically	Check box
Auto Center Horizontally	Check box

The other attributes of the panel like the color and label style are for you to choose (using *metafonts* for the **Font Types** is recommended since they are scalable fonts). Scalable fonts are useful if you have to use your GUI on a monitor with a different resolution than the monitor on which it was created. The size of the panel is given above so that all the objects will fit within the panel dimensions. For best results, leave attributes not mentioned above set to their default settings.

Add a decoration box to the top area of the panel and add four string boxes and one numeric box. Only the numeric box is the **Test Number** box; all others are all string boxes. In the numeric control, the **Test Number** for the test you are running is added by the program. The name of the person conducting the test will be entered in the **Operator Name** box. The **File Name** of the test conducted will be entered into this box. The data and time strings will be displayed in the other two boxes so marked. Set the following attributes of the objects.

Label					
	`Operator Name`	`Test Number`	`File Name`	`Date`	`Time`
Constant Name	`OPERATOR`	`TEST_NUM`	`TEST_NAME`	`DATE`	`TIME`
Control Mode	`Hot`	`Indicator`	`Indicator`	`Hot`	`Hot`
Data Type		`unsigned int`			

Your GUI should look like Figure 6–2.

Add a **Text Box** to the panel and make it sufficiently large as shown in Figure 6–3. Data will be displayed in this box. Click on the *Coloring Icon* and move it over the **Text Box** to color it black. Set the following attributes of the **Text Box**.

Label	RESULTS
Constant Name	RESULTS
Control Mode	Hot
Visible Lines	22
Wrap Mode	Line
Scroll Bars	Both Scroll Bars
Text Style	
Fonts	Times New Roman
Size	11
Justification	Left
Text Color	Yellow

Note: The right justification of the attributes under the **Text Style** denotes that the dialog items appear when the **Text Style** command button is selected.

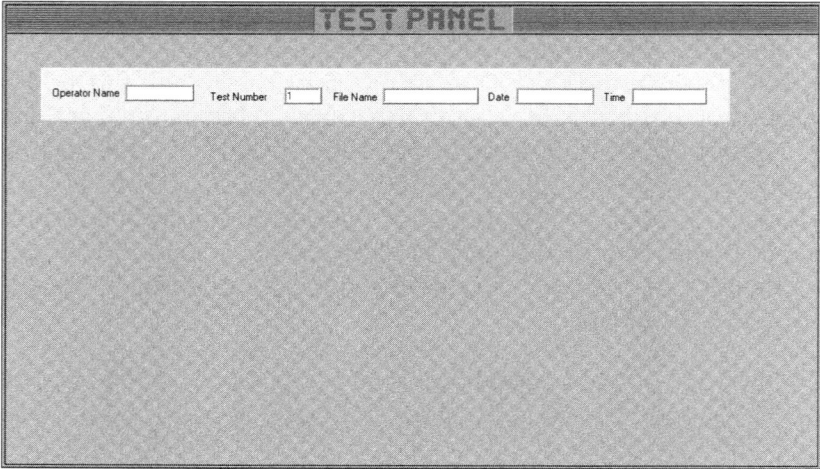

Figure 6–2
Test Panel GUI with Top Decoration Box Controls

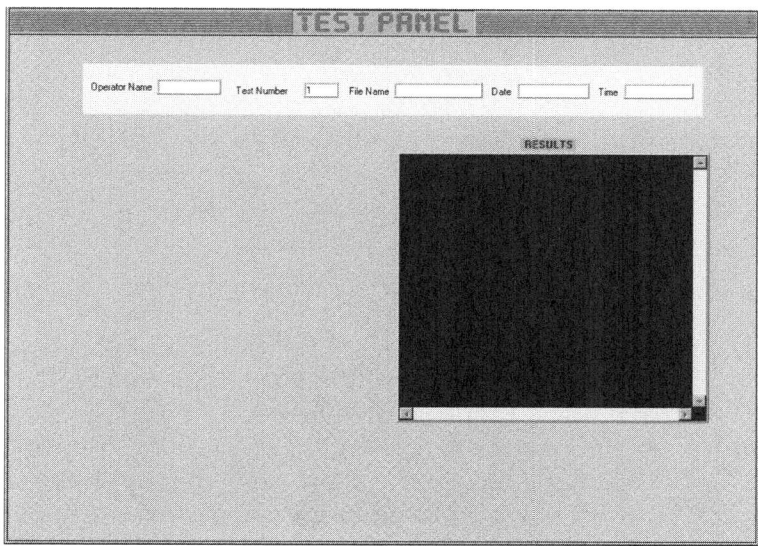

Figure 6–3
Test Panel GUI with Results Text Box

You will notice that here the use of **Text Style** is suggested because when displaying data on screen, it is better to have proportionally sized fonts as they align better in columns when displayed. Using yellow for the **Text Color** will contrast better on a black background of the **Text Box.**

Here you will create a decoration box with a **Thermometer** control and with two numeric control boxes that will contain the user-specified maximum and minimum limits for the temperature range.

Add a decoration box and select **Create>>Numeric>>Thermometer** and place it on the decoration box you just created. Create a frame on the decoration box and place two numeric boxes inside this frame. Add a **Text Message** and in the **Default Value** box, enter `Limits`. Center this box inside the frame. Add the following attributes to the thermometer and the two numeric controls.

Label	Temperature Sensor Gauge	Minimum	Maximum
Constant Name	THERM	TEMP_MIN	TEMP_MAX

Default Value	55.0	55.0	225.0
Data Type	double	double	double
Control Mode	Indicator	Hot	Hot
Label Raised	Check box	Do not check box	Do not check box
Size to Text	Check box	Check box	Check box

Your GUI should now look like the one shown in Figure 6–4.

Create a another decoration box about the same size as the **Temperature Sensor** box and add a frame to the decoration box. On this decorator box you will place the controls to measure the **Pressure Sensor** values. Select and place two numeric boxes, one below the other, within this frame; these boxes will be used to set the maximum and minimum limits. Add a **Text Message** and in the **Default Value** box, enter Limits. Center this box within the frame. Add the following attributes to the gauge and the two numeric objects.

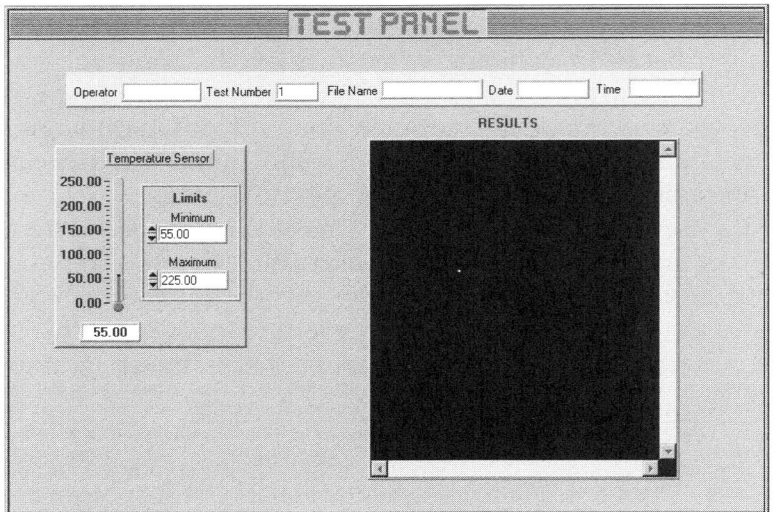

Figure 6–4
Test Panel GUI with Temperature Sensor

Chapter 6 • File Input/Output

Label	Pressure Sensor Gauge	Minimum	Maximum
Constant Name	PRESSURE	PRES_MIN	PRES_MAX
Default Value	20.0	20.0	85.0
Data Type	double	double	double
Control Mode	Indicator	Hot	Hot
Label Raised	Check box	Do not check box	Do not check box
Size to Text	Check box	Check box	Check box

Your GUI should now look like Figure 6–5.

Create two more decoration boxes and place the command buttons as shown in Figure 6–6. Add the following attributes to the command buttons in the top decoration box.

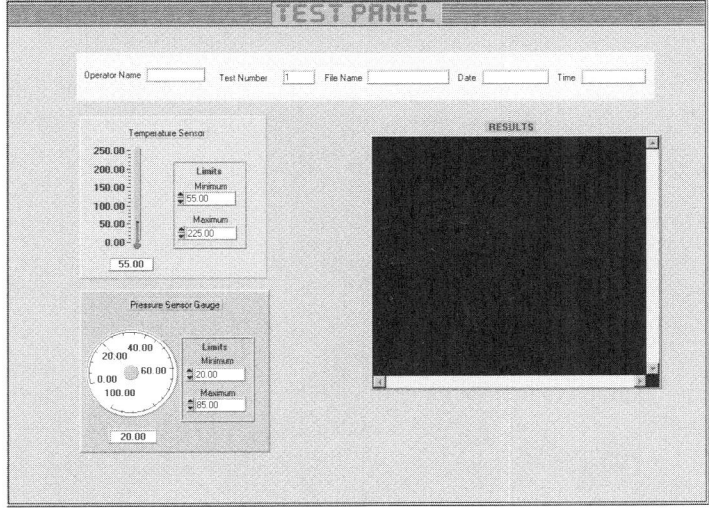

Figure 6–5
Test Panel GUI with Pressure Sensor Box

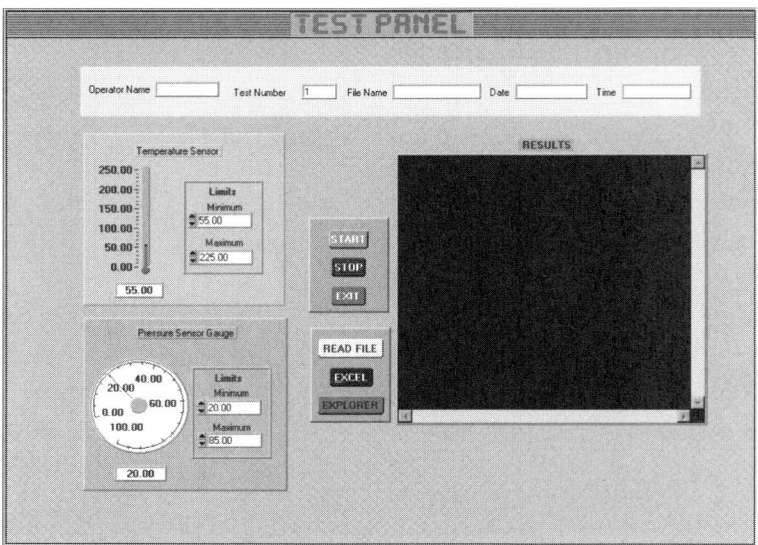

Figure 6–6
Test Panel GUI with Command Buttons

Label	START	STOP	EXIT
Constant Name	START	STOP	EXIT
Callback Function	StartTestCB	StopTestCB	ExitPanelCB
Control Mode	Hot	Hot	Hot

For the command buttons in the bottom decoration box, add the attributes as shown below:

Label	READ FILE	EXCEL	EXPLORER
Constant Name	READFILE	EXCEL	EXPLORER
Callback Function	ReadFileCB	RunExcelCB	ExplorerCB
Control Mode	Hot	Hot	Hot

Chapter 6 • File Input/Output

Add a command button to clear the **Text Box** as shown in Figure 6–7. Set the following attributes for the **CLEAR** button.

Label	CLEAR
Constant Name	CLEAR
Callback Function	ClearTextBoxCB
Control Mode	Hot

The function of all these command buttons is self-explanatory and will be seen when you run the project.

You will now add two timers. One will invoke the code within its callback function every second and the other after two seconds. The one-second timer will display the temperature value every second. The two-second timer will display the pressure value every two seconds.

Select **Create>>Timer** to place two timers on the panel. You can place them anywhere on the panel as they will not be displayed during run-time. Set the following attributes for the timers. The completed GUI is shown in Figure 6–7.

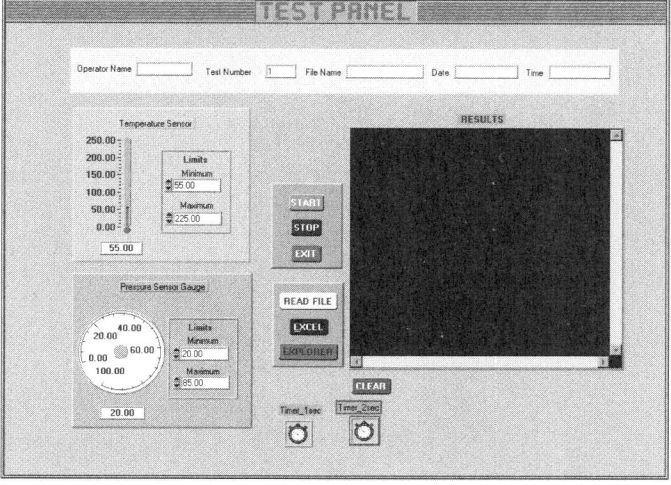

Figure 6–7
Completed Test Panel GUI

Label	Timer_1sec	Timer_2sec
Constant Name	TIMER_1SEC	TIMER_2SEC
Callback Function	Timer_1SecCB	Timer_2SecCB
Interval	1.00	2.00
Enabled	Do not check box	Do not check box

Generating the Code

The GUI is now complete with all the callback functions assigned to the command buttons. Now it is time to generate the code by using the *CVI CodeBuilder*.

From the user interface resource file, select **Code>>Generate>>All Code**. Check on the *ExitPanelCB* function for *CVI* to add the *QuitUserInterface* function, since this is the exit command for the project. A source file, Untitled1.c, will be created. **Select File>>Save As** and change its name to project6.c in the same folder (directory) as the .uir and .h files. Go to the **Project** Window and select **Edit>>Add Files to Project>>All Files (*.*)**. Add the .c, .h, and .uir files to project6. Save it as project6.prj.

All the callback function skeleton code has been created, and you need only add some more code to the functions. The code created by *CodeBuilder* should look like the listing in Figure 6–8.

Let us digress for a moment from building the code and discuss each of the pop-up panels in detail in the following section since you will be using these pop-up panels in the user interface. Pop-up panels are also called modal dialog boxes. You need to understand their functionality and become familiar with their names and properties.

Pop-Up Panels

The **User Interface Library** contains a number of pop-up panels which can be accessed from the **Library>>User Interface>>Popup Panels…** A pop-up panel, like the name suggests, appears on the display over the existing

```
include <cvirte.h>  /* Needed if linking in external compiler;
                                        harmless otherwise */
#include <userint.h>
#include "project6.h"

static int tst;

int main (int argc, char *argv[])
{
        if (InitCVIRTE (0, argv, 0) == 0)
/* Needed if linking in external compiler; harmless otherwise */
                return -1;       /* out of memory */
        if ((tst = LoadPanel (0, "project6.uir", TST)) < 0)
                return -1;
        DisplayPanel (tst);
        RunUserInterface ;
        return 0;
}

int CVICALLBACK StartTestCB (int panel, int control, int event,
           void *callbackData, int eventData1, int eventData2)
{
        switch (event)
                {
                case EVENT_COMMIT:

                break;
                }
        return 0;
}


int CVICALLBACK ExitPanelCB (int panel, int control, int event,
           void *callbackData, int eventData1, int eventData2)
{
        switch (event)
                {
                case EVENT_COMMIT:
                        QuitUserInterface (0);
                break;
                }
        return 0;
}

int CVICALLBACK StopTestCB (int panel, int control, int event,
           void *callbackData, int eventData1, int eventData2)
{
        switch (event)
                {
                case EVENT_COMMIT:

                        break;
                }
        return 0;
}

int CVICALLBACK ClearTextBoxCB (int panel, int control, int event,
           void *callbackData, int eventData1, int eventData2)
{
```

Figure 6–8
CodeBuilder Listing for Project6

```
        switch (event)
                {
                case EVENT_COMMIT:

                break;
                }
        return 0;
}

int CVICALLBACK ExplorerCB (int panel, int control, int event,
            void *callbackData, int eventData1, int eventData2)
{
        switch (event)
                {
                case EVENT_COMMIT:

                break;
                }
        return 0;
}

int CVICALLBACK ReadFileCB (int panel, int control, int event,
            void *callbackData, int eventData1, int eventData2)
{
            case EVENT_COMMIT:

            break;
            }
    return 0;
}
int CVICALLBACK RunExcelCB (int panel, int control, int event,
            void *callbackData, int eventData1, int eventData2)
{
    switch (event)
            {
            case EVENT_COMMIT:
            break;
            }
    return 0;
}
int CVICALLBACK Timer_1SecCB (int panel, int control, int event,
            void *callbackData, int eventData1, int eventData2)
{
    switch (event)
    {
            case EVENT_TIMER_TICK:
            break;
    }
     return 0;
}
int CVICALLBACK Timer_2SecCB (int panel, int control, int event,
            void *callbackData, int eventData1, int eventData2)
{
```

Figure 6–8
CodeBuilder Listing for Project6 (*continued*)

```
    switch (event)
    {
    case EVENT_TIMER_TICK:
    break;
    }
    return 0;
}
```

Figure 6–8
CodeBuilder Listing for Project6 (*continued*)

displayed panel(s), gets the user's response, and then disappears. The program will not continue until the user makes a selection from this dialog box. Some of the pop-up panels are explained here.

Confirm Pop-Up Panel

The *ConfirmPopup* panel enables the user to make a decision of some sort and select one of the available choices. This panel is shown in Figure 6–9.

The prototype function for *ConfirmPopup* is

```
int status= ConfirmPopup (char title[ ], char message[ ] );
```

and the function arguments are explained in Table 6–7, using Figure 6–9 as an example.

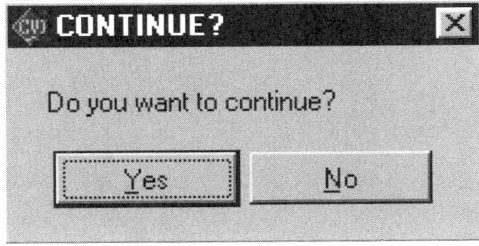

Figure 6–9
Confirm Pop-Up Panel

Table 6–7 ConfirmPopup *Function*

Input/Output	Name	Type	Description
Input	title	string	title of the dialog box: "CONTINUE?"
	message	string	message to show on the dialog box: "Do you want to continue?"
Output	status	integer	the status value: 1 – if **Yes** box is selected 0 – if **No** box is selected; for error messages, see Appendix A or the *LabWindows/CVI User Interface Reference Manual* or in Online Help

Message Pop-Up Panel

The *MessagePopup* panel displays a single or multi-line message and waits for the user to respond before continuing with the program. To get a multi-line message, include a \n (for new line) in the string making up the message. This panel is shown in Figure 6–10.

The prototype function for *MessagePopup* is

```
int status= MessagePopup (char title[ ], char message[ ] );
```

and the function arguments are explained in Table 6–8, using Figure 6–10 as an example.

Generic Message Pop-Up Panel

The *GenericMessagePopup* panel shows a message with an optional input box and gives the user the choice of selecting one of the three user-defined buttons. To add a multi-line message, include \n for new line. This panel is shown in Figure 6–11.

The prototype function for *GenericMessagePopup* follows.

Chapter 6 • File Input/Output

Figure 6–10
MessagePopup Panel

```
int button= GenericMesagePopup (char title[ ], char message[ ],
   char buttonLabel1[ ], char buttonLabel2[ ], char buttonLabel3[ ],
           char responseBuffer[ ], int maxResponseLength,
             int buttonAlignment, int activeControl,
                      int enterButton, int escapeButton);
```

The function arguments are explained in Table 6–9, using Figure 6–11 as an example.

Table 6–8   MessagePopup *Function*

Input/Output	Name	Type	Description
Input	title	string	title of the dialog box: "DATA FILE ERROR"
	message	string	message to show on the dialog box: "Data not saved!\nDo you want to continue?"
Output	status	integer	for error messages, see Appendix A of the *LabWindows/CVI User Interface Reference Manual* or in Online Help

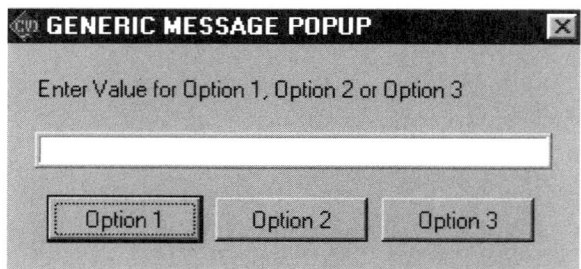

Figure 6–11
GenericMessagePopup Panel

Table 6–9 GenericMessagePopup *Function*

Input/ Output	Name	Type	Description
Input	title	string	title of the dialog box: "GENERIC MESSAGE POPUP"
	message	string	message to show on the dialog box: "Enter Value for Option 1, Option 2 or Option 3"
	buttonLabel1	string	user-defined label on button 1: "Option1"
	buttonLabel2	string	user-defined label on button 2: "Option2"
	buttonLabel3	string	user-defined label on button 3: "Option2"
	maxResponseLength	integer	maximum number of bytes the user is allowed to enter in the input box
	buttonAlignment	integer	selects the buttons location: Non-zero value—aligns buttons along the right hand side, Zero value—aligns buttons on the bottom of the panel

(continued)

Table 6–9 GenericMessagePopup *Function (continued)*

Input/Output	Name	Type	Description
	activeControl	integer	selects one of the buttons or the input string as the active control (the button that accepts the keystrokes); select the appropriate button from the function panel; the ring box from the function panel gives you a choice of button1, button2, button3, input string
	enterButton	integer	selects which button has <Enter> as its shortcut key; the ring box gives you a choice of none, button1, button2, button3
	escapeButton	integer	selects which button has <Esc> as its shortcut key; the ring box gives you a choice of none, button1, button2, button3
Output	responseBuffer	string	buffer to receive user input data from input box
	button	integer	for error messages, see Appendix A of the *LabWindows/CVI User Interface Reference Manual* or in Online Help

Prompt Pop-Up Panel

The *PromptPopup* panel asks the user for an input value in the form of a string. It allows the program to continue when **OK** is clicked. If you enter a <Return> only in the string box, then a NUL value is assigned to the string. This panel is shown in Figure 6–12.

The prototype function for *PromptPopup* is

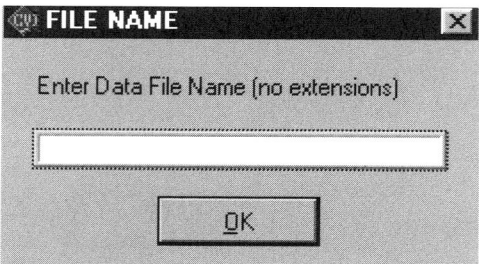

Figure 6–12
PromptPopup Panel

```
int status= PromptPopup (char title[ ], char message[ ],
          char responseBuffer[ ], int maxResponseLength );
```

and the function arguments are explained in Table 6–10, using Figure 6–12 as an example.

Table 6–10   PromptPopup *Function*

Input/ Output	Name	Type	Description
Input	title	string	title of the dialog box: "FILE NAME"
	message	string	message to show on the dialog box: "Enter Data File Name (no extensions)"
	maxResponseLength	integer	maximum number of bytes the user is allowed to enter in the input box
Output	responseBuffer	string	buffer to store the user's input from the pop-up panel
	status	integer	for error messages, see Appendix A of the *LabWindows/CVI User Interface Reference Manual* or in Online Help

File Select Pop-Up Panel

The *FileSelectPopup* panel displays a list of files and allows the user to select one of the files. The prototype function for *FileSelectPopup* is

```
int status= FileSelectPopup (char defaultDirectory [ ],
    char defaultFileSpec[ ], char fileTypeList[  ], char title [ ],
    int buttonLabel, int restrictDirectory, int restrict Extension,
    int allowCancel, int allowMakeDirectory, char pathName[ ]);
```

and function arguments are explained in Table 6–11, using Figure 6–13 as an example.

Install Pop-Up Function

The *InstallPopup* function displays and activates a user-created panel as a dialog box, which must be called from the thread in which the panel was created. Again this modal dialog box will not let the program continue while this panel is visible. The prototype function for *InstallPopup* is

```
int status= InstallPopup (int panelHandle);
```

and the function arguments are explained in Table 6–12.
 Caution: *InstallPopup* function does not work inside a *C loop construct*.

Remove Pop-Up Function

The *RemovePopup* function removes either the active pop-up panel or all the panels as specified by the user.
 The prototype function for *RemovePopup* is

```
int status= RemovePopup (int removePopup);
```

and the function arguments are explained in Table 6–13.

Multi-File Select Pop-Up Panel

The *MultiFileSelectPopup* panel displays a list of files from which the user can select one or more files or cancel. The prototype function for *MultiFileSelectPopup* is

Table 6–11 FileSelectPopup *Function*

Input/Output	Name	Type	Description
Input	defaultDirectory	string	the starting folder (directory) name; current working folder (directory) is used if " " is used
	defaultFileSpec	string	string specifying which files to display, e.g., `*.csv` will display only `.csv` files
	fileTypeList	string	list of file types to display separated by `;` when restrictExtension is set to 0
	title	string	title of the pop-up panel
	buttonLabel	integer	select OK, Save, Select, or Load from the ring menu for the label that will appear on the pop-up dialog box
	restrictDirectory	integer	if non-zero, the user is restricted from changing the drives or the directories; zero gives the user a complete choice
	restrictExtension	integer	if non-zero, user can only select files with the specified extension; if zero, the user can select any extension
	allowCancel	integer	if non-zero, the user has the choice to cancel out of the pop-up panel; if zero, the user must make a selection
	allowMakeDirectory	integer	if non-zero, the user can create a new folder (directory) from the FileSelect pop-up; if zero, the user is restricted from doing that
Output	status	integer	for error messages, see Appendix A of the *LabWindows/CVI User Interface Reference Manual* or in Online help
	pathName	string	buffer in which user's selection is returned

Chapter 6 • File Input/Output

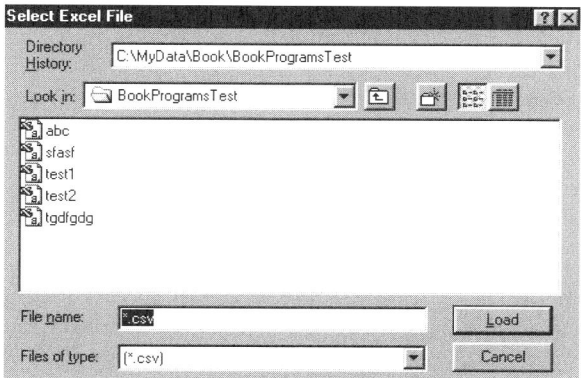

Figure 6–13
FileSelectPopup Panel

```
int status= MultiFileSelectPopup (char defaultDirectory [ ],
    char defaultFileSpec[ ], char fileTypeList[  ], char title [ ],
        int restrictDirectory, int restrictExtension, int allowCancel,
                        int numberOfSelectedFiles, char **fileList);
```

and the function arguments are explained in Table 6–14.

Directory Selection Pop-Up Panel

The *DirSelectPopup* panel displays a folder (directory) selection dialog box and waits for the user to select a directory. The prototype function for *DirSelectPopup* is

Table 6–12 InstallPopup *Function*

Input/ Output	Name	Type	Description
Input	panelHandle	integer	the panel handle number of the panel currently in memory
Output	status	integer	for error messages, see Appendix A of the *LabWindows/CVI User Interface Reference Manual* or in On-line Help

Table 6–13 RemovePopup *Function*

Input/Output	Name	Type	Description
Input	removePopup	integer	selects whether to remove all pop-up panels or just the active one: 1 refers to all, 0 refers to active only
Output	status	integer	for error messages, see Appendix A of the *LabWindows/CVI User Interface Reference Manual* or in On-line Help

```
int status= DirSelectPopup (char defaultDirectory [ ],
   char title [ ], int allowCancel, int allowMakeDirectory,
     char pathname[ ]);
```

and the function arguments are explained in Table 6–15.

Graph Pop-Up Panels

The *XGraphPopup, YGraphPopup, XYGraphPopup,* and *WaveformGraphPopup* plot data in various forms on the pop-up panels. These functions are available by selecting **Library>>User Interface>>Popup Panels...** Read Chapter 4 of the *LabWindows/CVI User Interface Reference Manual* for more details.

The *YgraphPopup* panel displays the plot of Y-values against the X-axis on a pop-up graph, which disappears when **OK** is selected. The *YGraphPopup* is shown in Figure 6–14.

In the *XGraphPopup,* the X-indices are marked along the Y-axis and the values are plotted along the X-axis, as shown in Figure 6–15. It is a rotation of the Y-graph plot shown in Figure 6–14.

There are a few other pop-up panels that will not be discussed since they are not used too often. The use of the majority of the pop-up panels has been explained here. For more information on these panels, see Chapters 1 and 4 of the *LabWindows/CVI User Interface Reference Manual.*

Let us now look at adding logic to the skeleton code created by *CodeBuilder* in Figure 6–8.

Table 6–14 MultiFileSelectPopup *Function*

Input/Output	Name	Type	Description
Input	defaultDirectory	string	the starting folder (directory) name; current working folder (directory) is used if " " is used
	defaultFileSpec	string	string specifying which files to display, e.g., *.c will display only .c files
	fileTypeList	string	list of file types to display separated by ; when **restrictExtension** is set to 0
	title	string	title of the pop-up panel
	restrictDirectory	integer	if non-zero, the user is restricted from changing the drives or the directories; zero gives the user a complete choice
	restrictExtension	integer	if non-zero, user can only select files with the specified extension; if zero, the user can select any extension
	allowCancel	integer	if non-zero, the user has the choice to cancel out of the pop-up panel; if zero, the user must make a selection
	numberOfSelectedFiles	integer	number of files selected by the user
	fileList	array of strings	buffer containing the filenames selected by the user
Return Value	status	integer	for error messages, see Appendix A of the *LabWindows/CVI User Interface Reference Manual* or in Online Help

Table 6–15 DirSelectPopup *Function*

Input/Output	Name	Type	Description
Input	defaultDirectory	string	the starting folder (directory) name; current working folder (directory) is used if " " is used
	title	String	title of the pop-up panel
	allowCancel	integer	if non-zero, the user has the choice to cancel out of the pop-up panel; if zero, the user must make a selection
	allowMakeDirectory	integer	if non-zero, the user is allowed to make a new directory
Output	pathname	string	buffer in which the user's selection is saved
	status	integer	for error messages, see Appendix A of the *LabWindows/CVI User Interface Reference Manual* or in Online Help

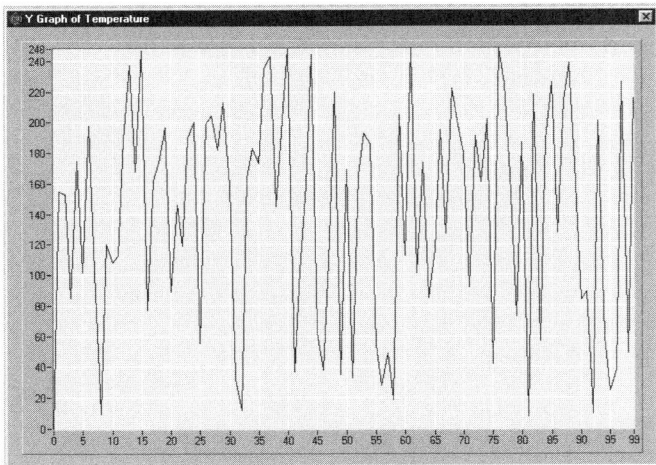

Figure 6–14
YGraphPopup Panel

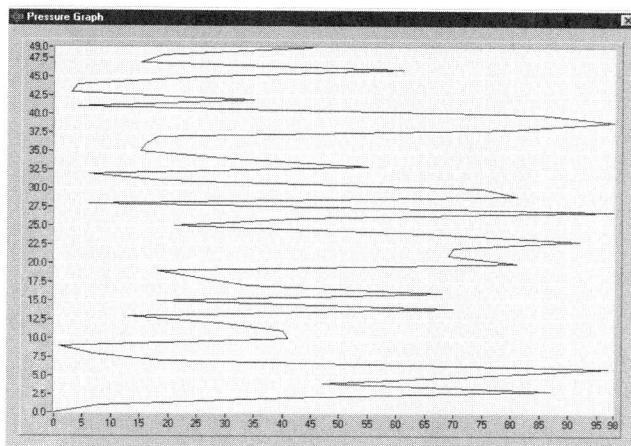

Figure 6–15
XGraphPopup Panel

Adding Code to the Project

In the code listing you saw in Figure 6–8, the code is created by the *CVI CodeBuilder*. This code gives you the skeleton code for the functions based on the callback functions. *CVI CodeBuilder* inserts the *QuitUserInterface* function in the *ExitCB* function. If you run the project and click on any of the command buttons except the **EXIT** button, nothing will happen; only the **EXIT** button will take you out of the project. This is a convenience feature for exiting the program, otherwise only <Ctrl-F12> would exit the program.

The *CVI CodeBuilder* adds the *panel handle* by using the panel's **Constant Name** that you assigned when you created the panel. In this case, it assigns tst for the panel handle. Let us use a more descriptive name and replace tst with TestHandle throughout the code. It is good practice to use more descriptive names to better understand the code. Using *handle* with a variable that describes the kind of handle; e.g., panel handle, file handle, or menu bar handle, is advisable.

To the code shown in Figure 6–8, add the lines shown in Figure 6–16 just below *DisplayPanel* in the *main* function.

The first line of code shown in Figure 6–16 is the *GetProjectDir* function, which gets the current project folder (directory) in the string *prj_dir*. The second statement obtains the value of today's date using the *DateStr* function

```
GetProjectDir (prj_dir); //Get the project directory
SetCtrlVal(TestHandle, TST_DATE, DateStr);
Fmt (buf, "%s<Time      Temperature      Result      Pressure      Result");
InsertTextBoxLine (TestHandle, TST_RESULTS, -1, buf);
```

Figure 6–16
Adding Code to *Main* Function

and inserts it in the **Date** box on the top decoration box. *Fmt* inserts the string of characters on the right side of < sign in the string variable *buf*. This is the heading that is displayed on the top of the **Text Box** every time the project is run. *Fmt* and *Scan* functions are explained in detail in Appendix D, "Formatting Functions."

Add the following variables as globals above the main program.

```
char prj_dir[260],buf[500];
```

You declare the variables as globals because you will be using these variables in several functions. When using filenames or folder (directory) structure variables, *CVI* requires a minimum length of 260 for the variable length. The variable *buf* is a string variable that will be used to store and display the data.

Defining the variables is part of C programming syntax and will not be mentioned again unless the variable must be declared in a particular way. The code listing is available, and if you have any concerns you can look at it and then enter the variable types appropriately. The appropriate headers and the function prototypes should be included above the *main* function. If there is a *CVI* function that requires a header file, then *CVI* will ask you to add that header (.h file) when you compile the code. You can let *CVI* insert it at the top of the program if you select **Yes**.

Main Function

The *main* function listing including the header and the library function/variable declarations will now look like Figure 6–17. *SetPanelPos* centers the panel on the screen.

StartTestCB Function

The StartTestCB callback function is invoked when the **START** command button is selected. This function starts to run the program. The listing for the *StartTestCB* function is shown in Figure 6–18.

```
//Header Files
#include <ansi_c.h>
#include <utility.h>
#include <formatio.h>
#include <cvirte.h>   /* Needed if linking in external compiler; harmless otherwise */
#include <userint.h>
#include "project6.h"

//Function Prototypes
void WriteToFile(void);
void CheckFileName (char *FileToCheck, char *extension);

//Static variables
static int WriteFileHandle, ReadFileHandle,ReadFileHandle, TestRunNumber=0, TestHandle, ExcelFileHandle;

//Other Global variables
double TempVal,PresVal, MinTempVal,MaxTempVal, MinPresVal,MaxPresVal, TempArray[500], PressureVal[500] ;
int Status, hours, minutes, seconds, TestNum, response, WriteButtonVal,WriteFileFlag=0, TempIndex=0, PressCounter=0;
char WriteFileName[260], PathName[260], Excelbuf[500], FileNameToWrite[260], prj_dir[260], buf[600],
OperatorID[260], Date[260], Time[260], ReadDataFile[260], ExcelFileName[260];

//Main program for project6
int main (int argc, char *argv[])
{
      if (InitCVIRTE (0, argv, 0) == 0)       /* Needed if linking in external compiler;
                                                                harmless otherwise */
               return -1;      /* out of memory */
      if ((TestHandle = LoadPanel (0, "project6.uir", TST)) < 0)
               return -1;

      DisplayPanel (TestHandle);
      GetProjectDir (prj_dir); //Get the project directory
      SetCtrlVal(TestHandle, TST_DATE, DateStr);
      Fmt (buf, "%s<Time       Temperature     Result     Pressure     Result");
      InsertTextBoxLine (TestHandle, TST_RESULTS, -1, buf);
      SetPanelPos (TestHandle, VAL_AUTO_CENTER, VAL_AUTO_CENTER);

      RunUserInterface ;

      return 0;
}// main
```

Figure 6–17
Project6 Main Function

```
1    //Start the project
2
3    int CVICALLBACK StartTestCB (int panel, int control, int event,
4                void *callbackData, int eventData1, int eventData2)
5    {
6    switch (event)
7    {
8    case EVENT_COMMIT:
9           TestRunNumber++;
10          SetCtrlVal(TestHandle,TST_TEST_NUM , TestRunNumber);
11          //Display the time on the panel
12          SetCtrlVal(TestHandle, TST_TIME, TimeStr);
13          response = ConfirmPopup ("WRITE TO FILE",
14                          "Do you want to write data to file?");
15
16          if (response) //Yes write to File
17                WriteToFile;
18
19
20          DefaultCtrl(TestHandle,TST_RESULTS); //Clear Box
21          Fmt (buf, "%s<Time      Temperature    Result    Pressure     Result");
22          InsertTextBoxLine (TestHandle, TST_RESULTS, -1, buf);
23
24          //Start the  Timers
25          SetCtrlAttribute(TestHandle ,TST_TIMER_1SEC , ATTR_ENABLED, 1);
26          SetCtrlAttribute(TestHandle ,TST_TIMER_2SEC , ATTR_ENABLED, 1);
27
28          SetInputMode (TestHandle, TST_START, 0);   //Disable the Start button
29          SetInputMode (TestHandle, TST_READFILE, 0);  //Disable the Read File
30                                                                   button
31          SetInputMode (TestHandle, TST_STOP, 1);   //Enable the Stop button break;
32
33        case EVENT_RIGHT_CLICK:
34          MessagePopup("", "Starts the project running");
35        break;
36    }
37    return 0;
38    } //StartTestCB
```

Figure 6–18
StartTestCB Function Listing

At line 9 of Figure 6–18, an integer variable, TestRunNumber, is incremented each time the project is started during the same run. This value is then displayed in the **Test Number** box by line 10. Line 12 sets the time string in the **Time** indicator box.

At line 13 of the *StartTestCB*, you call a *ConfirmPopup* panel like the one shown earlier. The *ConfirmPopup* panel displays the message **"Do you want to write data to file?"** You can select either **Yes** or **No** on the pop-up panel. If the response is **Yes**, the *WriteToFile* function is called and opens the file to receive data. If the response is **No**, a data file is not created and the program continues executing, displaying the data in the **RESULTS** box only.

Line 20 clears the text box by calling the *DefaultCtrl* function. Lines 21 through 22 display the header on top of the text box. At lines 25 and 26, the attributes of the timers are enabled to start the timers using the *CVI* function

```
SetCtrlAttribute(TestHandle ,TST_TIMER_1SEC ,
                                ATTR_ENABLED, 1);
```

The first argument is the *panelHandle*, the second is the **Constant Name** of the timer, and the third is a **Control Attribute,** which enables the timer here. The last argument in this function is set to 1 to start the timer. A 0 in this field stops the timer.

The statements at lines 28 through 31 use the library function *SetInputMode,* which enables/disables the buttons by un-dimming or dimming the specified control. Here you are disabling the **START** and **READFILE** buttons but enabling the **STOP** button. The only available button to use when the project is running is the **STOP** button. The first argument in this function is the *panelHandle,* and the second argument is the **Constant Name** of the button you are enabling/disabling. If the last argument is 1, the function is enabled, and if 0, it is disabled.

Lines 33 and 34 illustrate that the right mouse button can also be used to generate events. There is nothing that prevents us from using the right mouse button. The convention for Windows is that the left mouse button is the predominant button and is used mostly for primary functions. The right mouse is an additional button and is usually used for informative or secondary tasks. When you click the right mouse button over a command button, a message will pop up indicating the purpose of this button. In this example, the names are self-explanatory, but suppose icons were used instead of buttons; the right-click button would be very helpful in explaining the purpose of the icon.

WriteToFile Function

This function is called if the user selects **Yes** from the *ConfirmPopup* in line 13 of the *StartTestCB* function as shown in Figure 6–18. The purpose of each line in the listing will be explained. Be sure to include the function prototype above the *main* function as shown here.

```
void WriteToFile(void);
```

Figure 6–19 shows the listing of the *WriteToFile* function.

Line 5 of Figure 6–19 calls the *PromptPopup* panel, which prompts the user to

```c
//Create the Files to start writing data to file for ASCII and for Excel
void WriteToFile(void)
{
//Open the File for writing data
Status = PromptPopup ("FILE NAME", "Enter Data File Name (no
                                    extensions)",FileNameToWrite, 12);

//Add ".out" to the file name
Fmt(WriteFileName, "%s<%s.out",  FileNameToWrite);
SetCtrlVal(TestHandle,TST_TEST_NAME , WriteFileName);

//Create a comma separator file for Excel
 Fmt(ExcelFileName, "%s<%s.csv",  FileNameToWrite);

//Function to check the file name
CheckFileName (WriteFileName, ".out");
//Check .csv file
CheckFileName (ExcelFileName, ".csv");
//Create the file name path for .out
Status = MakePathname (prj_dir, WriteFileName, PathName);
WriteFileHandle = OpenFile (PathName, VAL_WRITE_ONLY, VAL_TRUNCATE,
                                      VAL_ASCII);

//Create the file name path for .csv file
Status = MakePathname (prj_dir, ExcelFileName, PathName);
//Open a file for Excel, comma separated format
ExcelFileHandle = OpenFile (PathName, VAL_WRITE_ONLY, VAL_TRUNCATE,
                                      VAL_ASCII);
//Get the Operator ID from the panel
GetCtrlVal(TestHandle,TST_OPERATOR ,OperatorID);
//Get the Test Number from the panel
GetCtrlVal(TestHandle,TST_TEST_NUM ,&TestNum);
//Get the date and time from the panel
GetCtrlVal(TestHandle,TST_DATE ,Date);
GetCtrlVal(TestHandle,TST_TIME ,Time);
//Create a file header with all the panel information
Fmt(buf, "%s<Operator: %s   Test #: %d   File Name: %s   Date: %s   Time: %s\n",
           OperatorID, TestNum, WriteFileName, Date, Time);
//Write header to file
WriteFile (WriteFileHandle, buf, StringLength(buf));
//Create the header to write to Excel spreadsheet
Fmt(Excelbuf,"%s<Operator: %s, Test #: %d, File Name: %s, Date: %s,   Time: %s\n",
                      OperatorID, TestNum, WriteFileName, Date, Time);
//Write to Excel spreadsheet
WriteFile (ExcelFileHandle, Excelbuf, StringLength(Excelbuf));
//Write the separator string to file
Fmt(buf,
"%s<*******************************************************************\n");
WriteFile (WriteFileHandle, buf, NumFmtdBytes);
//Write the results header to the file
Fmt (buf, "%s<Time    Temperature    Result    Pressure    Result\n");
WriteFile (WriteFileHandle, buf, NumFmtdBytes);

//Write the results header to the file for Excel
Fmt (Excelbuf, "%s<Time,    Temperature,    Result,    Pressure,    Result\n");

WriteFile (ExcelFileHandle, Excelbuf, StringLength(Excelbuf));

}//WriteToFile
```

Figure 6–19
WriteToFile Function Listing

```
Enter Data File Name (no extensions)
```

The user enters the file name (without the file extension) in the dialog box. The user-entered file name is stored in `FileNameToWrite`, a variable. The length of the file name is restricted to twelve characters as shown by the last argument in *PromptPopup* on line 5.

You are going to append our own extension, `.out` to the file name, which is done in line 9 using the *Fmt* function. Just remember that the string on the right is inserted in the string variable on the left. In this line, the `%s` on the right side of `<` gets the string contained in the variable `FileNameToWrite`, and then the `.out` is appended to the string and written to `WriteFileName`.

```
Fmt(WriteFileName, "%s<%s.out", FileNameToWrite);
```

Line 10 displays `WriteFileName` on the **File Name** indicator box on the panel.

At line 9, the *ASCII* file, with an extension of `.out`, is created and can be read with any text editor, including Notepad or WordPad. Excel can have different file extensions like `.xls` and `.csv`. The `.xls` extension is usually the default format of the Excel spreadsheet, and `.csv` is used for comma-separated files. At line 13, a `.csv` file is created to enable the file to be read through Excel.

To check if the user has entered a valid file name and that the file does not already exist, the function *CheckFileName* at lines 16 and 18 is invoked. This routine will be discussed after completing the discussion of this function.

At line 20, the complete pathname of the file is created by appending the file name to the project directory, which is obtained in the main program. The prototype of *MakePathname* is

```
MakePathname (char directoryName[ ], char FileName[ ],
                                     char PathName[ ]);
```

and the function arguments are explained in Table 6–16.

You create a *WriteFileHandle* at line 21 to open a file with the pathname you just created in write-only mode, and the file is truncated if the file is opened again. The *OpenFile* function is very easily created from the function panel displayed in Figure 6–20 by selecting **Library>>Formatting and I/O Library>>File I/O>>OpenFile.** When the function panel is displayed, enter the variables and select the attributes as shown in Figure 6–20. Of course, you can always enter the data manually if you wish. Lines 25 though 27 are similar to lines 20 and 21, but the pathname and the *OpenFile* is selected for *Excel* file.

Table 6–16 MakePathname *Function*

Input/Output	Name	Type	Description
Input	directoryName	string	folder (directory) path for file name
	FileName	string	file name and extension
Output	PathName	string	completed pathname with the folder (directory), file name, and file extension appended

In lines 30 through 35, the data from the panel header is obtained and assigned to the variables OperatorID, TestNum, Date, and Time. At line 37, the header string is placed in a default buf and then written to the file as a header with the information.

The *WriteFile* library function is used at line 40 to write data to the selected file. The first argument is the file handle you created in this function

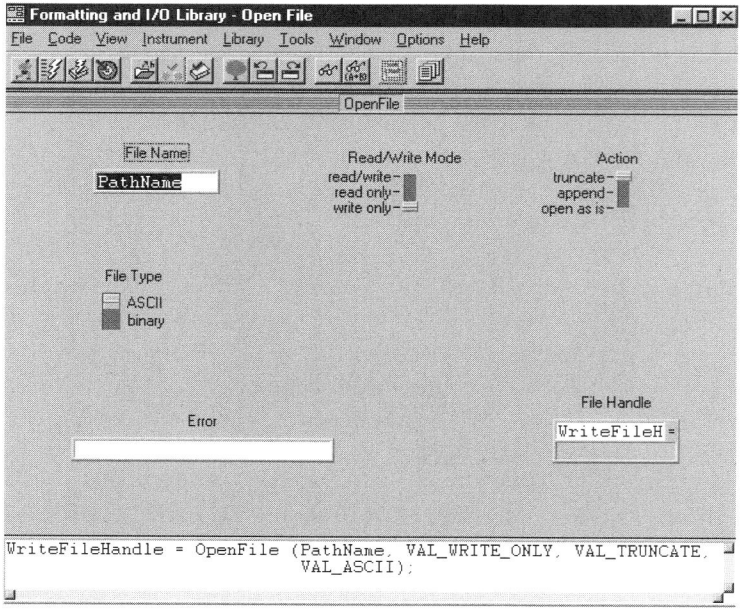

Figure 6–20
OpenFile Function Panel

at line 21 with the user-entered file name. The file handle is used whenever you want to refer to this file since it contains the file information. The second argument is the string containing the data in the variable *buf* that you want to write to the file. The function *StringLength(buf)* is similar to the C function, *strnlength* as it returns the length of the string of *buf*.

Similarly, the header data is written to the Excel file in a comma-separated format in lines 42 through 45.

The next few lines write the separator string of * to the file headers. A new function, *NumFmtdBytes*, is introduced at line 52. The *NumFmtdBytes* function counts the number of bytes in last *Fmt* or *Scan*. In this case, it counts the number of times * appears in the following code.

```
Fmt(buf,
"%s<**************************************************
*******\n");
```

The *NumFmtdBytes* function spares you from manually counting the occurrences of * in the *buf* string. You can insert *NumFmtdBytes* in the last argument of the *WriteFile* function, and the byte length is inserted automatically.

The other lines in this routine create the appropriated headers and write to the respective files. Note that the header for the Excel file must have commas between all the variables or values (since it is a comma-separated file) so they will be inserted into the appropriate columns.

CheckFileName Function

The *CheckFileName* function is a user-created function and is not part of the *CVI* library function. Its code is listed in Figure 6–21. This function determines if the user has entered a valid file name and not just hit <Enter> for a blank file name. It also checks for a previously existing file with that name, and if such a file is found, it asks the user to either enter a new file name or to over-write the existing file. This is an extra protection in case the user wants to keep the old file name and data in that file. The *WriteFile* function does not check for illegal file characters; in its present state, it will accept any characters the user inputs for the file name. It will be an interesting exercise for you to write a modification to this routine to remove all invalid characters, like control characters, escape characters, numerals occurring at the beginning of the file name, or any other characters that ANSI C disallows in file names.

At line 13 of Figure 6–21, the library function *GetFileInfo* is called. This function was explained in Table 6–6. Besides checking for previously exist-

```
1    /*This function checks to see if a valid filename is entered, and if the
2     file does not already exist. If it does then the user is asked if
3    he/she would like to over-write the file name.
4    */
5
6    void CheckFileName (char *FileToCheck, char *extension)
7    {
8    int n, response, count;
9    long size;
10   char save_file[260];
11
12      //Check if file name already exists
13      n=GetFileInfo( FileToCheck, &size);
14      while ( StringLength(FileToCheck) <= 4 ) // Loop while size is not valid, 4 is for the
15                                                                              extension
16      {
17          MessagePopup ("FILE ERROR", "Please enter a valid File Name");
18          PromptPopup ("FILE NAME","Enter File Name (no extension)",save_file, 30);
19          strcat(FileToCheck,extension);      //Append the extension
20          n=GetFileInfo( FileToCheck, &size);
21      }
22          if (n==1) //File exists
23          {
24           response = ConfirmPopup ("FILE CREATION ERROR","File already exists.
25                                           Do you want to over-write it?");
26
27          if (response ==0) //"NO"
28          { // Do not over-write the file
29            while (n) // Loop while n=1, i.e. the file exists
30            {
31
32              PromptPopup ("FILE NAME", "Enter File Name (no extension)",
33                                              FileToCheck, 30);
34              strcat(FileToCheck,extension);    //Append the extension
35              n=GetFileInfo( FileToCheck, &size);
36              if (n==1)
37                  MessagePopup ("SELECT FILE ERROR", "File already exists, select
38                                              another name");
39          } //while (n)
40     } //}/* if (n==1)*/
41   } //CheckFileName
```

Figure 6–21
CheckFileName Function

ing files with that name on the disk, you can also determine if the file is non-empty by creating the structure shown in Figure 6–22.

Three different pop-up panels are being used in this function: *MessagePopup, PromptPopup,* and *ConfirmPopup* at lines 17, 18, and 24 of Figure 6–21. Their use has been described in the "Pop-Up Panels" section above. The *CheckFileName* function is well-commented, and there is no need to go into any more detail.

```
if (filesize != 0)
{
        //File is not empty

}
else
{
//File is empty, take appropriate action here. Warn the user!
}
```

Figure 6–22
Code Structure to Check File

Timer_2SecCB Function

Let us now look at some of the features of *Timer_2SecCB*. The two-second timer executes the code within this function every two seconds, and this function updates the **Pressure Sensor** data every two seconds using the random numbers. The source code listing for this function is shown in Figure 6–23.

The random value of the pressure is obtained using *modulus* 99 to keep it less than 100, the upper limit of the pressure gauge. A counter, Press-Counter, is incremented every two seconds. This counter will serve as an index into the array, PressureVal, that will store the pressure values. The

```
/*Get the random values for the pressure gauge every 2 seconds and
 show it on the indicator box of the gauge.
*/
int CVICALLBACK Timer_2SecCB (int panel, int control, int event,
      void *callbackData, int eventData1, int eventData2)
{
   switch (event)
   {
    case EVENT_TIMER_TICK:
        PresVal= ( (rand/2)% 99); //Random Values for the pressure
        PressureVal[PressCounter]= PresVal; //Enter value in an array, starting index
                                                                                =0
        SetCtrlVal(TestHandle, TST_PRESSURE,PresVal);
        PressCounter++; //increment counter
        break;
   }
    return 0;
} // Timer_2SecCB
```

Figure 6–23
Timer_2SecCB Function

current pressure value is displayed on the pressure gauge indicator box and the gauge needle moves accordingly. This variable will be used later in the program to illustrate the *GraphPopup* panel.

Timer_1SecCB Function

Now let us look at a more detailed *Timer_1SecCB* callback function, as listed in Figure 6–24. This routine is executed every second and updates the temperature values every second and the pressure values every other second.

GetSystemTime obtains the time in hours, minutes, and seconds on line 17 of Figure 6–24. The `seconds` value is used later on line 50 to obtain the temperature value only or both the temperature and pressure values depending on whether the seconds value is odd or even. Line 19 displays the time string on the panel in the **Time** indicator box.

Random values for the temperature are generated at line 20 using *modulus* 249 to keep the value within the upper limit of 250. Line 23 displays the value in the **Temperature** indicator box and, simultaneously, the temperature in the thermometer indicator box. Again, an array of the random numbers is created using an incremental counter as an index at line 21.

Lines 25 through 29 obtain the minimum and maximum limits of the temperature and pressure from the numeric controls on the panel. If the temperature is outside the minimum and maximum limits then at line 32 the string "FAIL" is copied into `TempResult`, otherwise "PASS" is copied into `TempResult` at line 38. The library function *CopyString* is used for copying strings. Let us look at the CopyString prototype.

```
void CopyString( char targetString[ ], int targetIndex,
        char *sourceString, int sourceIndex, int maximumNumBytes);
```

The purpose of this function is to copy the *sourceString,* starting at *sourceIndex*, to the *targetString,* starting at the *targetIndex*. The string is copied until the value of *maximumNumBytes* is reached or an ASCII NUL character encountered. This function also appends an ASCII NUL character to the copied string. The function arguments are explained in Table 6–17.

Similarly, on line 41, the pressure values are compared against the limits and a "PASS" or a "FAIL" is written to the `PressureResult`. At line 50, the `seconds` is checked to determine even or odd value. If the `seconds` is even then both the temperature and pressure values are displayed and lines 52 through 63 are executed. If the `seconds` is odd then only temperature is displayed at lines 68 through 77. In either case, temperature data is generated

```c
/*
Get the random value every second for the thermometer, compare the results
 for pass/fail for both the temperature and the pressure.
 The temperature results are compared every second for the temperature,
 and every two seconds the pressure. The results are displayed in the Text Box
 and also written to both the files created for text output and Excel spreadsheet.

*/

int CVICALLBACK Timer_1SecCB (int panel, int control, int event,
            void *callbackData, int eventData1, int eventData2)
{
char TempResult[50], PressureResult[50];
  switch (event)
  {
   case EVENT_TIMER_TICK:
        GetSystemTime (&hours, &minutes, &seconds);
        //Display the time on the panel
        SetCtrlVal(TestHandle, TST_TIME, TimeStr);
        TempVal= ( rand% 249); //. Random Value for temperature
        TempArray[TempIndex]=  TempVal;
        TempIndex++;
        SetCtrlVal(TestHandle, TST_THERM,TempVal);
        //Get the limits for the temperature
        GetCtrlVal(TestHandle,TST_TEMP_MIN  ,&MinTempVal);
        GetCtrlVal(TestHandle,TST_TEMP_MAX  ,&MaxTempVal);
        //Get limits for pressure
        GetCtrlVal(TestHandle,TST_PRES_MIN  ,&MinPresVal);
        GetCtrlVal(TestHandle,TST_PRES_MAX  ,&MaxPresVal);

        //Check if the Temperature is within limits
        if ((TempVal <MinTempVal) ||(TempVal >MaxTempVal))   //Temperature Fail
        {
                CopyString(TempResult,0,"FAIL",0, StringLength("FAIL"));
        }
        else  //Temperature Pass
        {
                CopyString(TempResult,0,"PASS",0, StringLength("PASS"));
        }
        //Check if the Pressure is within limits
        if ((PresVal <MinPresVal) ||(PresVal >MaxPresVal))   //Temperature Fail
        { //Pressure Fail
                CopyString(PressureResult,0,"FAIL",0, StringLength("FAIL"));     }
        else  //Pressure Pass
        {
                CopyString(PressureResult,0,"PASS",0, StringLength("PASS"));
        }


        if ((seconds % 2)==0)  //If seconds are even
        { //Display both the  temperature and pressure
                sprintf(buf, "%s%20.2f%10s%12.2f%12s",
                            TimeStr,TempVal,TempResult, PresVal, PressureResult);
                InsertTextBoxLine (TestHandle, TST_RESULTS, -1, buf);
                if (response)
                { //Create the string to write to file
                   sprintf(buf, "%s%12.2f%15s%12.2f%12s\n",
                            TimeStr,TempVal,TempResult, PresVal, PressureResult);
```

Figure 6–24
Timer_1SecCB Function

```
59                          //Write to file
60                              WriteFile (WriteFileHandle, buf, StringLength(buf));
61                              sprintf(Excelbuf, "%s,%12.2f,%15s,%12.2f,%12s\n",
62                                      TimeStr,TempVal,TempResult, PresVal, PressureResult);
63                          WriteFile (ExcelFileHandle, Excelbuf, StringLength(Excelbuf));
64                          }
65              }
66              else    //If seconds are odd, display only temperature
67              {   //Create the string to write to file
68                      sprintf(buf, "%s%20.2f%10s",TimeStr,TempVal,TempResult);
69                      InsertTextBoxLine (TestHandle, TST_RESULTS, -1, buf);
70                      if (response)
71                      {
72                      //Create the string to write to file
73                      sprintf(buf, "%s%12.2f%15s\n",TimeStr,TempVal,TempResult);
74                      //Write to file
75                      WriteFile (WriteFileHandle, buf, StringLength(buf));
76                      sprintf(Excelbuf, "%s,%12.2f,%15s\n",TimeStr,TempVal,TempResult);
77                      WriteFile (ExcelFileHandle, Excelbuf, StringLength(Excelbuf));
78                      } //if (response)
79              } //else
80              break;
81              }
82          return 0;
83  } // Timer_1SecCB
```

Figure 6–24
Timer_1SecCB Function (*continued*)

Table 6–17 CopyString *Function*

Input/Output	Name	Type	Description
Input	targetIndex	integer	starting position in target string
	sourceString	string	starting position in the source string
	sourceIndex	integer	starting position in source string
	maximumNumBytes	integer	number of bytes to be copied excluding the ASCII NUL character
Output	targetString	string	buffer where the string is copied

every second and is displayed for both even and odd `seconds` value. The value of `response` obtained from *StartTestCB* callback function in Figure 6–18 is checked to determine if the user wants to write the data to a file or display it on the screen only. If the data is to be written to a file, then lines 55 through 63 are executed to write the temperature and pressure data and lines 72 through 77 to write the temperature data. Note that in any case the data is always displayed to the screen whether it is written to files or not.

Line 57 sets up the string in `buf`, using the format specified in `sprintf` to write the data to the file. Similarly the same data is written in a comma-separated format to `Excelbuf` at line 61. This is written to a file with `.csv` extension for reading through Excel. Note that the different file handles are used to write to the respective files.

StopTestCB Function

The *StopTestCB* callback function is called when you want to stop the program by selecting the **STOP** command button. Figure 6–25 lists the code for the *StopTestCB* function.

At lines 14 and 15 of Figure 6–25, the one-second and the two-second timers are disabled, which stops the random number generation and the processing of the code in the *Timer_1SecCB* and the *Timer_2SecCB* functions (Figures 6–24 and 6–23, respectively). At lines 16 and 17, the value of `response` is checked to verify if the user has chosen to write the data to the file. If so, the files are closed using the *CloseFile* function with the appropriate file handles. If you try to close a file that has not been opened, you will get an error message.

The **START** and **READFILE** command buttons are enabled, and the **STOP** button is disabled at lines 18 through 20. This is opposite to the settings for the *StartTestCB* function; the **START** and **READFILE** buttons were disabled and the **STOP** button was enabled.

At line 22, the *YGraphPopup* is created for 100 points using the temperature data in array `TempArray`. Recall that you had created an array in *Timer_1SecCB* function for this purpose.

Line 25 shows the *XGraphPopup* to display the pressure data for 50 points in the variable `PressureVal`, which is plotted along the X-axis with respect to the array index along the Y-axis. These are shown in Figures 6–14 and 6–15.

Line 29 displays the message `Stops the program from running` when you right-click the mouse on the **STOP** button.

```
1    /* This function stops the timers and closes all the files,
2    enables the Start and ReadFile buttons and disables the Stop
3    button. A Y-Graph is plotted on a popup panelfor the Temperature
4    and a X-Graph is plotted for the Pressure values. Right click
5    on the button shows the function of the Stop button.
6    */
7    int CVICALLBACK StopTestCB (int panel, int control, int event,
8            void *callbackData, int eventData1, int eventData2)
9    {
10          switch (event)
11          {
12          case EVENT_COMMIT:
13              //Stop the timers
14              SetCtrlAttribute(TestHandle ,TST_TIMER_1SEC , ATTR_ENABLED, 0);
15              SetCtrlAttribute(TestHandle ,TST_TIMER_2SEC , ATTR_ENABLED, 0);
16              if (response) CloseFile(WriteFileHandle);    //Close the files
17              if (response) CloseFile(ExcelFileHandle);
18              SetInputMode (TestHandle, TST_START, 1);   //Enable the Start button
19              SetInputMode (TestHandle, TST_READFILE, 1); //Enable the Read File button
20              SetInputMode (TestHandle, TST_STOP, 0);   //Disable the Stop button
21              //Plot Y Graph for Temperature
22              YGraphPopup ("Y Graph of Temperature", &TempArray, 100,
23                                                      VAL_DOUBLE);
24              //Plot X Graph for Pressure
25              XGraphPopup ("Pressure Graph", PressureVal, 50, VAL_DOUBLE);
26              break;
27
28          case EVENT_RIGHT_CLICK:
29              MessagePopup("", "Stops the program from running");
30              break;
31      }
32          return 0;
33   }// StopTestCB
```

Figure 6–25
StopTestCB Function

ClearTextBoxCB Function

The callback function *ClearTextBoxCB* is invoked when the **CLEAR** command button is selected. This function clears the **Text Box** and writes the header at the top of the box. The source code listing for this function is shown in Figure 6–26.

The *DefaultCtrl* library function is shown on line 9 of Figure 6–26. This function sets the value of the control box to its default value when this control is created.

```
1    //Clears the Text Box
2    int CVICALLBACK ClearTextBoxCB (int panel, int control, int event,
3            void *callbackData, int eventData1, int eventData2)
4    {
5        switch (event)
6        {
7        case EVENT_COMMIT:
8            //Set the box to its default value, i.e clear the box
9            DefaultCtrl( TestHandle, TST_RESULTS);
10           //create and display the header on top of the  text box
11           Fmt (buf, "%s<Time    Temperature    Result    Pressure    Result");
12           InsertTextBoxLine (TestHandle, TST_RESULTS, -1, buf);
13       break;
14
15           //Display the message for the Clear button
16       case EVENT_RIGHT_CLICK:
17           MessagePopup("", "Clears the display");
18       break;
19   }
20       return 0;
21   }// ClearTextBoxCB
```

Figure 6–26
ClearTextBoxCB Function

The function prototype for *DefaultCtrl* is as follows:

```
int status = DefaultCtrl( int panelHandle, int controlID);
```

The function arguments are explained in Table 6–18.
Lines 11 and 12 create and display the buffer to display the header on the **Text Box.** As before, line 17 displays the message when you right-click on the **CLEAR** button.

Table 6–18 DefaultCtrl *Function*

Input/Output	Name	Type	Description
Input	panelHandle	integer	the ID of the panel on which the control is located
	controlID	integer	the *Constant Name* of the control box
Output	status	integer	refer to Appendix A of *LabWindows/CVI User Interface Reference* for error codes or in Online Help

ReadFileCB Function

The *ReadFileCB* function is called when the **READ FILE** command button is selected. This callback function reads and displays the data written to the file during the program run to the **RESULTS** box. Its source code listing is shown in Figure 6–27.

Line 12 of Figure 6–27 sets the folder (directory) to the project directory, and line 14 clears the **RESULTS** box. In this function you have a choice of selecting any data file with a `.out` extension (by default) that is displayed in the folder (directory) listing by the *FileSelectPopup* function as shown on line 15.

At line 20, you create a file handle to read the selected file as a read-only file. This is important since you do not want to over-write the data file by mistake. You read the file a line at a time using the library function *ReadLine* shown on line 24. It reads the data into the `linebuffer` with a maximum of `BYTES_TO_READ`, which is set to eight hundred bytes in the `#define` statement on line 7.

Lines 27 through 33 read the lines in a *while* loop and display each line to the **RESULTS** box. The return value of *ReadLine* is checked at line 29 every time the data is read from the file. This value is assigned to a variable `count` whose value is checked for -2. As mentioned before, *CVI* uses -2 as the marker that indicates the end of the file. If `count` is not -2, the next line from the file is read. If the value of `count` is -2, no more data is read from the file as the end of the file is reached. The **End of File Encountered** message is displayed and the file is closed at lines 35–39.

Notice that at line 39, the file is closed as soon as you do not need it anymore, which is good practice for reducing the chance of file corruption.

Lines 43 and 44 show the message **Displays the Data written to the file** when the mouse is right-clicked on the **READ FILE** button.

RunExcelCB Function

This function is called when the **EXCEL** command button is selected. One of the many features of *CVI* is its ability to run executables from within the program. You can run Microsoft® Word, Powerpoint, CorelDraw, Notepad or any program that has an extension of `.exe`, `.com`, or `.bat`.

Here you are going to load the file with the `.csv` extension and then launch Microsoft® Excel using the *CVI LaunchExecutable* function on line 66 of Figure 6–28. The only argument that is required by *LaunchExecutable* is

```
/Read the Data File and display the results
int CVICALLBACK ReadFileCB (int panel, int control, int event,
            void *callbackData, int eventData1, int eventData2)
{
int count;
char linebuffer[801], FileToRead[260];
#define BYTES_TO_READ   800

        switch (event)
                {
                case EVENT_COMMIT:
                        SetDir (prj_dir); //Get the project directory
                        //Clear the display
                        DefaultCtrl(TestHandle,TST_RESULTS);
                        FileSelectPopup (prj_dir, "*.out", "*.*",
                                        "Select File to read on display",
                        VAL_LOAD_BUTTON, 0, 0, 1, 0, FileToRead);

                        //Open File to Read only
                        ReadFileHandle = OpenFile (FileToRead, VAL_READ_ONLY,
                                                    VAL_OPEN_AS_IS, VAL_ASCII);

                        //Read Line at a time
                        count=ReadLine (ReadFileHandle, linebuffer, BYTES_TO_READ);
                        InsertTextBoxLine (TestHandle, TST_RESULTS, -1, linebuffer);

                        while ( count !=-2)    //End of File (EOF)in LabWindows is "-2"
                        {
                                count=ReadLine (ReadFileHandle, linebuffer,
                                                        BYTES_TO_READ);
                                InsertTextBoxLine (TestHandle, TST_RESULTS, -1,
                                                                linebuffer);
                        }// while ( count !=-2)

                        Fmt (buf, "%s<End of File Encountered\n");
                        //Display data in the buf to the Text Box
                        InsertTextBoxLine (TestHandle, TST_RESULTS, -1, buf);

                        CloseFile(ReadFileHandle);   //Close the file after reading
                        break;

                //Display the message with the right-click of mouse
                case EVENT_RIGHT_CLICK:
                        MessagePopup("", "Displays the Data written to the file");
                        break;
                }
        return 0;
} //ReadFileCB
```

Figure 6–27
ReadFileCB Function

the complete pathname of the executable file. The default path name for Excel is

```
c:\Program Files\Microsoft
                    Office\Office\Excel.exe
```

The source code listing for the *RunExcelCB* function is shown in Figure 6–28.

Sometimes, the user may not have installed the Excel file in the default path on a different drive or directory. This function must be flexible enough to handle such a situation. The strategy used in the *RunExcelCB* function is to first look for the Excel in its default path at line 21. If the file is found, then it jumps to line 59 and executes lines 59 through 66. If Excel is not found in the default path, then the user must enter its path name in the *GenericMessagePopup* panel at line 44. If the path name is unknown, then clicking the **CANCEL** button on this pop-up panel will terminate this action. Otherwise, after entering the path name, the user selects **DONE** to run Excel. If the file is not found, the same pop-up panel is displayed and the user is asked to enter the correct path name. This is repeated in a *while* loop until the correct path is entered and the file is found or the user selects **CANCEL.** If the Excel executable is found, the user is asked to select the Excel file with the .csv extension using the *FileSelectPopup* at line 62.

The path name for Excel is copied into the variable NewExcelLocation along with the file name that you selected from *FileSelectPopup*. Line 70 displays the command button's purpose when right-clicked with the mouse.

Excel will load up the data file created in comma-separated value format (with .csv extension) into columns that you can manipulate using any of Excel's features. After you are done, you can close Excel, which will put you back into the project.

Note: In C, the \ is an escape sequence character, which means anything after \ is a special control character. To represent a \ in C, you therefore have to use \\. For this reason, \\ is used in defining the path name at line 21.

ExplorerCB Function

The *ExplorerCB* function is called when the **EXPLORER** command button is selected. This function puts you into Windows Explorer and lets you access the Explorer functions. You can open files for reading by using Notepad or WordPad, delete files if necessary, go to different directories, and when you close Explorer, return to the project. Figure 6–29 shows the listing of *ExplorerCB*. If the

```
/****************************************************************
    This function loads the .csv file and lauches the Excel program.
    This program first checks  the default Excel path name. If Excel
    is not found, prompts the user to enter the Excel pathname and
    confirms if the file exists in the user specified path. This is done
    till the Excel executable is found.

*****************************************************************/
int CVICALLBACK RunExcelCB (int panel, int control, int event,
                    void *callbackData, int eventData1, int eventData2)
{
        char ExcelPath[260], new_pathname[260], ExcelLocation[260],
                                                    NewExcelLocation[260];
        long size;
        int result;

        switch (event)
        {
         case EVENT_COMMIT:
         // Check if file exists in the default path
         result=GetFileInfo ("c:\\Program Files\\Microsoft Office\\Office\\Excel.exe",
                                                    &size);
/        /If Excel does not exist in the default location
         if (result==0)
        { // Inform user Excel is not found
            result = GenericMessagePopup ("FILE NOT FOUND",
                "Excel not found in default path.\nEnter new pathname (no file name)",
                "DONE", "CANCEL", 0, new_pathname,
                256, 1, VAL_GENERIC_POPUP_BTN1,
                VAL_GENERIC_POPUP_BTN1,
                VAL_GENERIC_POPUP_BTN3),

                if (result ==2) return 0;  //return if the CANCEL key is selected

                //create the user defined path name
                sprintf(ExcelLocation, "%s\\Excel.exe", new_pathname);
                //Check if Excel exists in this path
                result =GetFileInfo (ExcelLocation, &size);

                while(result !=1) //if result is 1, then file exists, otherwise prompt for
                                                            pathname till found
                {
                  //Enter new path name
                  result=GenericMessagePopup ("FILE NOT FOUND",
                  "Excel not found in specified path.\nEnter new pathname (no file
                  name)", "DONE",  "CANCEL", 0, new_pathname, 256, 1,
                  VAL_GENERIC_POPUP_BTN1, VAL_GENERIC_POPUP_BTN1,
                  VAL_GENERIC_POPUP_BTN3);

                  if (result ==2) return 0;      if the CANCEL key is selected

                  //create the user defined path name
                  sprintf(ExcelLocation, "%s\\Excel.exe", new_pathname);
                    //Check if Excel exists in this path
                    result =GetFileInfo (ExcelLocation, &size);
                } //while
        }//if (result==0)
        else
          sprintf(ExcelLocation, "c:\\Program Files\\Microsoft Office\\Office\\Excel.exe");
```

Figure 6–28
RunExcelCB Function

```
60
61              //Excel executable is found, now select the file to open in Excel
62              FileSelectPopup (prj_dir, "*.csv", "*.*", "Select Excel File",
63                               VAL_LOAD_BUTTON, 0, 0, 1, 0, ExcelPath);
64              //Launch Excel and open the selected file name
65              Fmt(NewExcelLocation, "%s<%s %s",ExcelLocation, ExcelPath);
66              LaunchExecutable(NewExcelLocation);
67
68          break;
69          case EVENT_RIGHT_CLICK:
70              MessagePopup("", "Launch the Excel executable");
71          break;
72          }
73      return 0;
74  } //RunExcelCB
```

Figure 6–28
RunExcelCB Function (*continued*)

path of Explorer is not in the default directory, you need to use the same code as used in *RunExcelCB* to enable the user to input the correct pathname.

ExitPanelCB Function

This function is called when you select the **EXIT** command button. You have seen the *ExitPanelCB* function many times before. It calls only the *QuitUserInterface* function to stop the *RunUserInterface* function and exit the project. The listing for this function is shown in Figure 6–30.

```
//Launch the Explorer
int CVICALLBACK ExplorerCB (int panel, int control, int event,
        void *callbackData, int eventData1, int eventData2)
{
        switch (event)
                {
                case EVENT_COMMIT:
                LaunchExecutable("explorer.exe");
                break;
            case EVENT_RIGHT_CLICK:
                MessagePopup("", "Launch the Explorer executable");
                break;
            }
    return 0;
} //ExplorerCB
```

Figure 6–29
ExplorerCB Function

```
//Exit the project
int CVICALLBACK ExitPanelCB (int panel, int control, int event,
         void *callbackData, int eventData1, int eventData2)
{
    switch (event)
    {
        case EVENT_COMMIT:
                QuitUserInterface (0);
                break;

        case EVENT_RIGHT_CLICK:
                MessagePopup("", "Exits the Project");
                break;
    }
    return 0;
} //ExitPanelCB
```

Figure 6–30
ExitPanelCB Function

Now you are ready to run the project. Everything is compiled and error-free (hopefully). If your code is different from this listing, please examine it and make the needed corrections.

Try all the features of this project by clicking on the command buttons. Display your data on the screen after the output data file is created, going through Explorer and double-clicking on the output file. Explorer will ask you what application you want to use to open the file. Select either Notepad or WordPad. If you use a text editor, you can see the data in ASCII format. Try to load the data file using Excel and see how easy it is to go from one application to another without leaving the *CVI* project.

Summary

This chapter showed you how to use the file I/O functions in *CVI*. Some new GUI objects were used, and the project run using two timers and different types of graphing pop-up panels to plot the data on the Y-Graph and the X-Graph panels. A section was devoted to explaining the pop-up panels, and different ones were used in the project. You were shown how to access Windows Explorer and Microsoft Excel executables and to create file that can be opened by Microsoft Excel.

LIST BOXES & RINGS

Chapter Highlights

- Introduction
- List Box Example 7–1
- List Box Example 7–2
- Using Ring Controls
- Designing the Test Executive Project
- Project7 Code Generation and Analysis
- Summary

This is a comprehensive chapter on understanding, creating, and using the *List Boxes*. You will be shown most of the *List Box* library functions and how to use them by means of examples. There are two examples here that will walk you through the different aspects of *List Boxes*. The concepts in these examples are used to develop a simple Test Executive.

One section will be devoted to the design and usage of *Rings*, which look very similar to *List Boxes* in some respects but have less functional capabilities.

Introduction

List Boxes in *CVI* have numerous applications and versatile features that add to the ease of programming. The many functions and capabilities of *List Boxes* can be used in a variety of ways. *Text Boxes*, which were discussed earlier, are somewhat similar to *List Boxes* but cannot perform many of the indexing capabilities of the *List Boxes* as you will see.

List Boxes may look similar to *Text Boxes*, but *List Boxes* are more versatile in their functionalities. *CVI* has a host of library functions to facilitate the use of list controls, including functions for inserting a list item in a certain location in the List Box, deleting a selected item from the list, replacing an item, and getting the name and the index number from the box. *CVI* functions can mark selected items with a check mark, get the number of list items and the number of checked items, get the label length from the list index, and display data in different colors on the same line as well as perform many other tasks. The *List Box* is created by selecting **Create>>List Box** as shown in Figure 7–1.

A *List Box*, as shown in Figure 7–2, is placed on the panel.

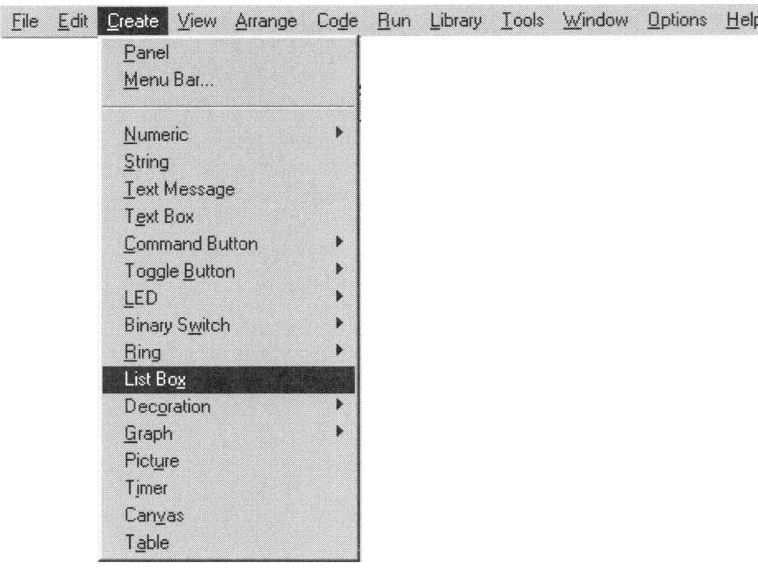

Figure 7–1
Creating a List Box

Chapter 7 • List Boxes & Rings

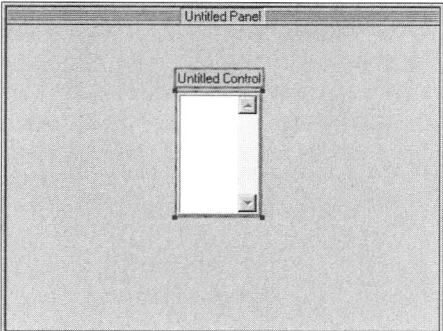

Figure 7–2
List Box

There is nothing that looks exciting about this box until you look at the attributes of the *List Box* by double-clicking on it and opening the attributes dialog box as shown in Figure 7–3. The top few attribute control boxes are the same as the other controls you have seen.

Some attributes are unique to the *List Box*, such as the **Label/Value Pairs...** and the **List Box Options...** as shown in the middle of the attribute box.

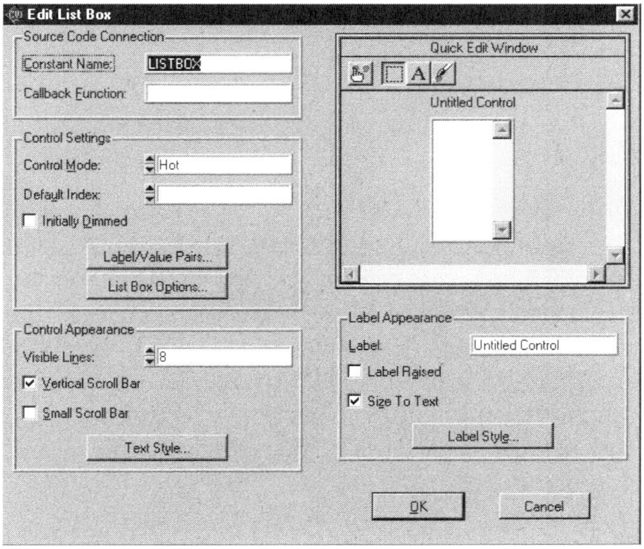

Figure 7–3
List Box Attributes

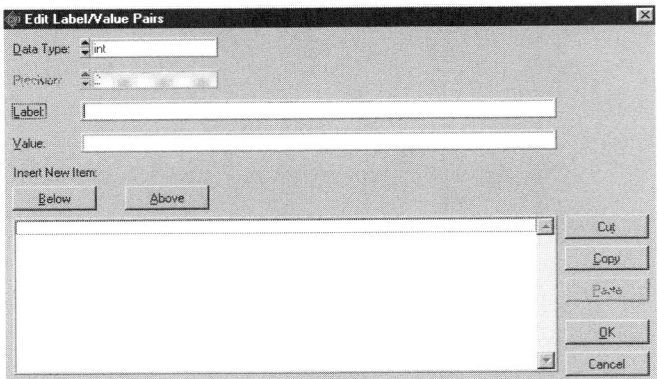

Figure 7-4
Label/Value Pair Control Box

Double-clicking on this control opens the **Label/Value Pairs** control box, shown in Figure 7-4. In this control box, the labels can be entered in the **Label** control and the associated value in the **Value** control. You can create as many items as you prefer in the list (within memory limits) and achieve various programming objectives, as you will see later in the examples and in the **Test Executive** project later in this chapter.

List Box Example 7-1

Let us start the topic of *List Boxes* by using a simple example to introduce a few basic programming concepts of the *List Boxes*. After that, we would like to build upon this example with more features, and then finally create a more elaborate project, using more of the list controls functions.

Note that the examples that are demonstrated here will be created as fully functional independent projects. To run *CVI*, you will have to create a user interface, a header, and a source file, and add them to the project.

Let us now start to create the project for Example 7-1. Create a panel with the following attributes

    Constant Name   PNL
    Callback Function   (None)
    Panel Title   EXAMPLE 7-1

Menu Bar	No Menu Bar
Close Control	None
Top	38
Left	142
Height	560
Width	438
Size Title Bar Height to Font	(Check Mark)

On this panel, create the *user interface resource* as shown in Figure 7–5. Place two *List Boxes* of the same size on the panel, one below the other. Label the upper *List Box* as **Initial List** and the lower box as **Selected List.** The purpose of this example is to move the selected item from the upper (**Initial List** box) to the lower (**Selected List** box) and display the label and value of the item(s) moved when the **MOVE DOWN** command button is selected. Then the highlighted item in the **Selected List** is moved back to the **Initial List** box when the **MOVE UP** command button is selected. This example will be helpful in understanding the underlying concepts of how the *List Boxes* are controlled.

Double-click on the **Initial List** box and for the **Constant Name,** enter TEST_BOX. Click on the **Label/Pair** Box and enter the following items in the **Label** and **Values** box.

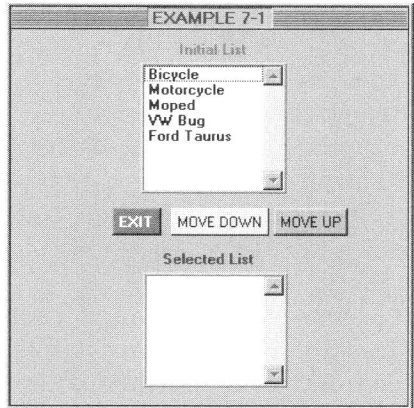

Figure 7–5
Example 7–1 User Interface Resource

Label	Value
Bicycle	200
Motorcycle	1500
Moped	500
VW Bug	15000
Ford Taurus	18300

Each label is assigned a value; it could be quantity, price, inventory number, etc. The purpose here is to show you that the item and its value are moved together from one **List Box** to another.

Double-click on the **Selected List** box and for the **Constant Name,** enter `SEL_BOX`. Do not enter anything in the **Label/Pair** box here.

Create three command buttons from the **Create** menu and label them as shown in Figure 7–5. Create the following attributes for the command buttons.

Label	Exit	Move Down	Move Up
Constant Name	Exit	MOV_DWN	MOV_UP
Callback Function	ExitCB	MoveItemDownCB	MoveItemUp
Control Mode	Hot	Hot	Hot

Run the CVI *CodeBuilder* and select `ExitCB` for the *QuitUserInterface* library function. The skeleton code for the three callback functions, *ExitCB, MoveItemDownCB,* and *MoveItemUp,* and for the *main* function will be created. Change the handle name created in the *main* function by the *CodeBuilder* to `ListHandle` to make it more meaningful. Rename it globally in the code by replacing this variable using <Shift-F11> or by typing it.

Now you can enter the code for each of the callback functions in the table above, and look at their purpose. The code for the complete project is listed in Figure 7–6. Let us examine the code.

The *main* function is straightforward and needs no explanation. Let us look at the callback function *MoveItemDownCB* listed between lines 38 and 80 of Figure 7–6. This function is activated when the **MOVE DOWN** command button is selected on the GUI. The *GetNumListItems* library function at line 50 checks the number of items(s) in the upper list box selected and returns the value in the variable `count`. The prototype for *GetNumListItems* is

```
1  /************************************************************************
2  File Name: Example7-1.c
3
4  Description: This code shows the preliminary functionalities of
5  the List Box. When the "Move Down" button is clicked, the selected item
6  in the upper box is inserted into the lower List Box. When the "Move Up" button
7  is selected then the selected item in the lower box is moved up. This project
8  also shows you how to align the items displayed using the escape sequence.
9
10
11 *************************************************************************/
12
13 #include <formatio.h>
14 #include <cvirte.h>     /* Needed if linking in external compiler; harmless
15 otherwise */
16 #include <userint.h>
17 #include "Example7-1.h"
18
19 static int ListHandle;
20
21 int main (int argc, char *argv[])
22 {
23      if (InitCVIRTE (0, argv, 0) == 0)     /* Needed if linking in external compiler;
24 harmless otherwise */
25           return -1;   /* out of memory */
26      //Load the panel
27      if ( (ListHandle = LoadPanel (0, "Example7-1.uir", LST)) < 0)
28           return -1;
29      //Display the panel
30      DisplayPanel (ListHandle);
31      RunUserInterface ;   //Start getting user events
32      return 0;
33 }
34
35
36
37 /*
38 This function moves the selected item to the lower List Box,
39 when the "Move Down" button is selected.
40 */
41 int CVICALLBACK MoveItemDownCB (int panel, int control, int event,
42           void *callbackData, int eventData1, int eventData2)
43 {
44      int item_index=0, item_value, count;
45      char item_label[260], buf[500];
46      switch (event)
47      {
48           case EVENT_COMMIT:
49             //Check if any items are selected
50             GetNumListItems(ListHandle, LST_INIT_BOX,&count);
51             if (count !=0)
52             {
53                  //Get the index of the item selected
```

Figure 7–6
Listing of Example 7–1

```
54                        GetCtrlIndex(ListHandle, LST_INIT_BOX, &item_index);
55                        //Get Value of item from index
56                        GetValueFromIndex(ListHandle, LST_INIT_BOX,
57                                            item_index, &item_value);
58                        //Get the Label from the index
59
60                        GetLabelFromIndex(ListHandle,LST_INIT_BOX,item_index,
61                                                        item_label);
62                        //Create a buffer to display the Label and the value
63                        Fmt(buf, "%s<%s\033p150r%d", item_label, item_value);
64                        //Insert the item into the "Selected" box
65                        InsertListItem(ListHandle, LST_SEL_BOX , -1, buf,
66                                                        item_value);
67                        //Delete the item from the upper box after moving to the
68                                                                    lower box
69                        DeleteListItem(ListHandle ,LST_INIT_BOX,item_index,1 );
70            }
71
72         else
73          //Message for the List Box being empty
74         MessagePopup ("INSERT SEQUENCE ERROR", "There is
75                                        nothing to MOVE
76                                                       DOWN");
77    break;
78    }
79    return 0;
80 }//MoveItemDownCB
81
82 /*This function moves the selected item to the upper List Box,when the "Move
83 Up" button is selected.
84 */
85 int CVICALLBACK MoveItemUp (int panel, int control, int event,
86         void *callbackData, int eventData1, int eventData2)
87 {
88      char item_label[260];
89      int item_index, count, item_value;
90
91      switch (event)
92      {
93      case EVENT_COMMIT:
94          //Check if there are any items in this box
95            GetNumListItems(ListHandle,LST_SEL_BOX  ,&count);
96          if (count !=0)
97          {
98                  //Get the index of the selected item
99                  GetCtrlIndex(ListHandle, LST_SEL_BOX ,
100                                                 &item_index);
101                 //Get the value from the index in the List Box
102                 GetValueFromIndex(ListHandle, LST_SEL_BOX ,
103                                            item_index,
104                                                        &item_value);
105                 //Get the label from the index
106                 GetLabelFromIndex(ListHandle,LST_SEL_BOX
```

Figure 7–6
Listing of Example 7–1 (*continued*)

```
107                                                      ,item_index,
108 item_label);
109                         //Insert the item in the List Box at the upper
110                         InsertListItem(ListHandle,LST_INIT_BOX, -1, item_label,
111                                                              item_value);
112                         //Delete the item from the lower box
113                         DeleteListItem(ListHandle ,LST_SEL_BOX  ,item_index,1 );
114                 }
115             else
116                         //Message for the List Box being empty
117                         MessagePopup ("REMOVE SEQUENCE ERROR",
118                                      "There is nothing to MOVE UP");
119
120      break;
121       }
122      return 0;
123 } //MoveItemUp
124
125 /* Exit the project */
126 int CVICALLBACK ExitCB (int panel, int control, int event,
127         void *callbackData, int eventData1, int eventData2)
128 {
129         switch (event)
130         {
131             case EVENT_COMMIT:
132                     QuitUserInterface (0);
133                     break;
134         }
135      return 0;
136
137 }// ExitCB
```

Figure 7–6
Listing of Example 7–1 (*continued*)

```
int status = GetNumListItems(int ListHandle, int LST_INIT_BOX, int * count);
```

and the function arguments are explained in Table 7–1.

If count is zero, a pop-up panel displays the message at line 74. If count is not zero, the highlighted item can be moved from the upper *List Box* to the lower *List Box* between lines 52 and 70.

At line 54, the library function *GetCtrlIndex is* invoked to obtain the index of the selected item. The prototype for this function is

```
int status= GetCtrlIndex(int ListHandle, int LST_INIT_BOX, int *
                                                      item_index);
```

and the function arguments are explained in Table 7–2.

Table 7–1   GetNumListItems *Function*

Input/ Output	Name	Type	Description
Input	ListHandle	integer	the panel handle of the user interface
	LST_INIT_BOX	integer	Constant Name of the List Box prefixed by the panel Constant ID
Output	count	integer	number of checked items in the box
	status	integer	refer to Appendix A of *LabWindows/CVI User Interface Library Reference Manual* for error codes or in Online Help

The index of the selected item is returned in the variable `item_index`. If there are no items in the List Box, the `item_index` returns -1. The *GetCtrlIndex* library function could have been used instead of *GetNumListItems* to verify the number of items in the List Box. The *GetNumListItems* library function is introduced here as it will be used quite often.

GetValueFromIndex is called on line 56 to obtain the value of the selected item from the index, `item_index`. The function prototype is

Table 7–2   GetCtrlIndex *Function*

Input/ Output	Name	Type	Description
Input	ListHandle	integer	the panel handle of the user interface
	LST_INIT_BOX	integer	Constant Name of the List Box prefixed by the panel Constant ID
Output	item_index	integer	the zero-based index of the selected item in the List Box; returns a –1 if no item is in the List Box
	status	integer	refer to Appendix A of *LabWindows/ CVI User Interface Library Reference Manual* for error codes or in Online Help

```
int status=GetValueFromIndex(int ListHandle,
      int LST_INIT_BOX, int  item_index, void * item_value);
```

and the function arguments are explained in Table 7–3.

At line 60, the library function *GetLabelFromIndex,* as the name suggests, obtains the label of the selected item in the variable `item_label` from the `index_item`. The prototype for this function is

```
int status = GetLabelFromIndex(int ListHandle,
      int LST_INIT_BOX, int item_index, char item_label);
```

and the function arguments are explained in Table 7–4.

You will notice that the `item_index` variable is the output from the *GetCtrlIndex* library function at line 54. The `item_index` variable is then input into *GetValueFromIndex* at line 56 to obtain its value and similarly used by *GetLabelFromIndex*; by obtaining the index from the *List Box*, all the list box information is available by calling other library functions.

The *Fmt* library function at line 63 needs some explanation. You can look up the details on the *Fmt* function in Appendix D, "Formatting Functions." Here you are adding formatting modifiers by embedding escape sequence code of \033, followed by p which stands for the number of pixels to move to the right before displaying the next item value in the *List Box*. After speci-

Table 7–3 GetValueFromIndex *Function*

Input/ Output	Name	Type	Description
Input	ListHandle	integer	the panel handle of the user interface
	LST_INIT_ BOX	integer	Constant Name of the List Box prefixed by the panel Constant ID
	item_index	integer	the zero-based index of the selected item in the List Box; returns a –1 if no item is in the List Box
Output	item_value	void*	returns the value of the item selected
	status	integer	refer to Appendix A in *LabWindows/ CVI User Interface Library Reference Manual* for error codes or in Online Help

Table 7–4 GetLabelFromIndex *Function*

Input/Output	Name	Type	Description
Input	ListHandle	integer	the panel handle of the user interface
	LST_INIT_BOX	integer	Constant Name of the List Box prefixed by the panel Constant ID
	item_index	integer	the zero-based index of the selected item in the List Box; returns a –1 if no item is in the List Box
Output	item_label	string	returns the label of the item selected
	status	integer	refer to Appendix A of *LabWindows/CVI User Interface Library Reference Manual* for error codes or in Online Help

fying the number of pixels to move, a code of 1 (letter *l*) is entered to specify left-justification, or c for centered, or r for right-justification.

This line is shown in Figure 7–7, and a string is created consisting of the `item_label`, moved 150 pixels to the right. The data is right-justified, and the integer value of `item_value` is displayed.

InsertListItem at line 65 is used to insert the string created in `buf` into the lower *List Box*. The prototype of *InsertListItem* is

```
int status= InsertListItem(int ListHandle,
            int LST_SEL_BOX ,int  item_index,
                             char buf, item_value¹);
```

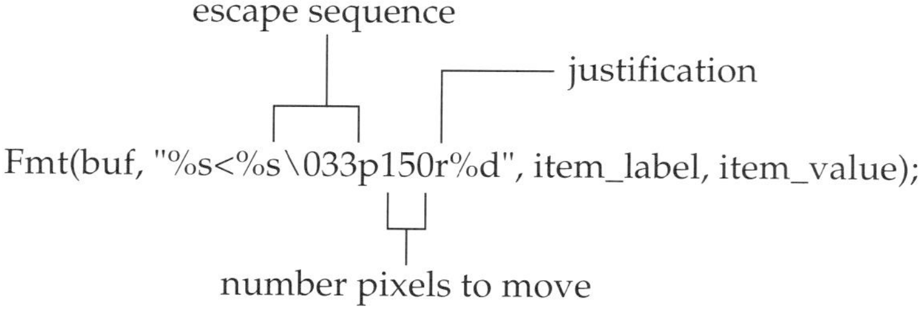

```
                escape sequence
                       |
                       |         ┌─── justification
                       |         |
       Fmt(buf, "%s<%s\033p150r%d", item_label, item_value);
                              |
                              |
                    number pixels to move
```

Figure 7–7
String Sequence Structure

and the function arguments are explained in Table 7–5.

The library function *DeleteListItem* at line 69 is used to delete items from the upper *List Box* after it inserts that item in the lower *List Box*. This function can delete one or more items from the *List Box* starting at the index specified. The prototype of this function is

```
int status = DeleteListItem( int ListHandle,
                    int LST_INIT_BOX, int item_index,
                           int numberOfItemsToDelete);
```

and the function arguments are explained in Table 7–6.

To re-cap, the *MoveItemDownCB* callback function moves the *List Box* item from the upper *List Box* to the lower box. This function checks for items to be moved. If there are no items selected, a message is displayed. Otherwise, the number of items in the *List Box* is obtained from the library function *GetNumListItems.* The *GetValueFromIndex* library function obtains the value of the selected list item. The label is obtained from the index of the selected item by using the *GetLabelFromIndex* function. The output string is created

Table 7–5 InsertListItem *Function*

Input/Output	Name	Type	Description
Input	ListHandle	integer	the panel handle of the user interface
	LST_SEL_BOX	integer	Constant Name of the List Box prefixed by the panel Constant ID
	item_index	integer	zero-based index of item placement in the List Box; if the item is to be placed at the end of the list, then pass a –1
	buf	string	returns the label of the item selected
	item_value	¹Same l type as the list control	the value of the item in the List Box; data type must be the same specified in the Label/Value Pair in the attribute box of the List Box
	status	integer	refer to Appendix A of *LabWindows/CVI User Interface Library Reference Manual* for error codes or in Online Help

Table 7–6 DeleteListItem *Function*

Input/Output	Name	Type	Description
Input	ListHandle	integer	the panel handle of the user interface
	LST_INIT_BOX	integer	Constant Name of the List Box prefixed by the panel Constant ID
	item_index	integer	zero-based index of the first item to delete from the List Box
	numberOfItemsToDelete	integer	number of items to delete; to delete all items starting from the *item_index* to the end, enter a –1
Output	status	integer	refer to Appendix A of *LabWindows/CVI User Interface Library Reference Manual* for error codes or in Online Help

and displayed in the lower *List Box*, right-justified and moved 150 pixels to the right from the first data field by the *InsertListItem* function. The item is then deleted from the upper *List Box* by the *DeleteListItem* function.

The *MoveItemUp* library function is very similar to the *MoveItemDown* library function, but it moves the list item from the lower box to the upper box. No further discussion is necessary as it uses the same functions.

Run this project and notice that by clicking on a different sequence of items, you can move the list items up and down to whatever order you prefer. This is a trivial example, but it serves the purpose of introducing the basic *List Box* functions.

List Box Example 7–2

The purpose of this example is to introduce some additional *List Box* functions. In particular, you will see how the *List Boxes* items can be checked and unchecked, displayed to another list box, and removed from the displayed list box in the order they are checked. You will also be shown how spacing and formatting of the items is accomplished using vertical lines between the items and how to add different colors to different data items.

Start Example 7–2 by creating the GUI shown in Figure 7–8. Some items in this GUI are new and will be explained. Create a panel and enter PNL for the **Constant ID** and `Example 7-2` as the **Panel Title**. Create two *List Boxes* as shown in Figure 7–8.

Define the attributes for the list box on the left. Double-click on this list box and enter `LISTBOX` as the **Constant Name**. Enter `ListBoxSelectCB` in the **Callback Function** box. Enter `TEST LIST` for the **Label**. Click on the **Label/Pair Box** and enter the following Label/Value Pairs.

Label	Value
Test1	1
Test2	2
Test3	3
Test4	4
Test5	5

Click on **List Box Options** and on the dialog box that appears, click the **Check Mode** and **Text Click Toggles Check** as shown in Figure 7–9. From the pull-down box, you have the choice of selecting either the **Check Box** or the **Check Mark. Check Box** places a box next to the text and **Check Mark**

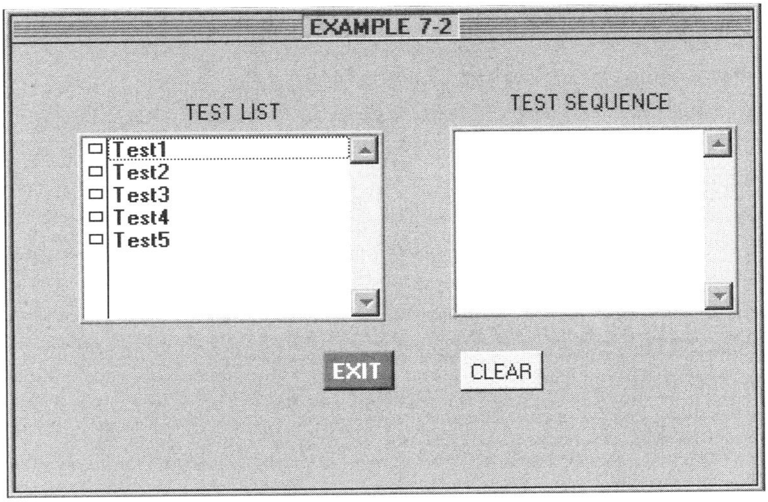

Figure 7–8
Example 7–2 GUI

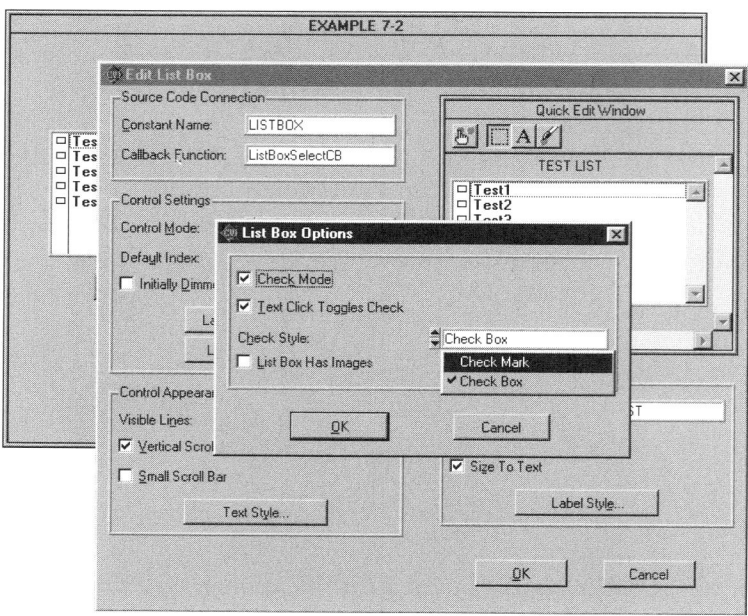

Figure 7–9
List Box Options

places a check mark next to the item. The **Check Box** is used in this example. The **Check Mark** option can be used similarly. The *CVI* library functions that control these are similar.

Now create the attributes for the list box on the right side of the **TEST LIST** box. Double-click on this list box and for the **Constant Name,** enter DISPLAY. For the **Label** of this box, enter TEST SEQUENCE. There are no **Label/Pair Box** settings for this list box as this will display the items copied from the **TEST LIST** box. Create two command buttons and set the following attributes.

Label	EXIT	CLEAR
Constant ID	EXIT	CLEAR
Callback Function	ExitCB	ClearCB
Control Mode	Hot	Hot

The GUI for this example is now complete. The next step is to create the code with the CVI *CodeBuilder* and to set the *QuitUserInterface* library function to the *ExitCB* callback function.

Let us add our code to these functions. The *ExitCB* is complete since it just calls the *QuitUserInterface* library function.

Examine the code listing for Example 7–2 shown in Figure 7–10. Let us start by examining the *ClearCB* function between line 113 and 122. The *ClearCB* function calls the *ClearListCtrl* library function only, which clears all items in the **TEST SEQUENCE** list box. The prototype of this function consists of only two arguments, the *PanelHandle* and the *ControlID* of the list box, as shown in line 118.

In the code created by the *CodeBuilder*, rename the panel handle *lst* to *PanelHandle,* a more descriptive name. The functions in the *main* function are ones you have seen in previous projects and will not be discussed here.

The bulk of code for this example is in the callback function *ListBoxSelectCB* beginning on line 38, and this function will be explained in detail.

The *MakeColor* library function at line 51 is a new library function, and it creates the color via the red, green, and blue (RGB) arguments of this function, as shown in the following prototype.

```
int color = MakeColor(int red, int green, int blue);
```

The color intensity levels are created using various values for red, green, and blue. The decimal numbers for each of these arguments can range from 0, the absence of that color completely, to 255, the full intensity of that shade. Any value in between gives a variation of that hue. The combination of all these shades creates a RGB color that is a four-byte integer value represented in hexadecimal format as 0X00RRGGBB. RR, GG, and BB are the respective red, green, and blue components of the color. By varying the values of these three arguments, you can create any color you desire.

Using 255 in the first argument creates a pure red color. At lines 52 and 53, using 255 in the second and third arguments respectively creates the green and blue colors. White is formed by including 255 in all the arguments and black by entering 0 in all the arguments of the *MakeColor* function.

At line 57, the *GetNumCheckedItems* library function is used. The purpose of this function is to determine the number of checked items in the list box. The prototype of this function is

```
int status= GetNumCheckedItems (int PanelHandle, int LST_LISTBOX,
                                            int * CheckedItems);
```

and the function arguments are given in Table 7–7.

GetCtrlIndex at line 59 obtains the zero-based index of the checked item.

```
1   /****************************************************
2   File Name: Example7-2.c
3
4   This Example demonstrates the use of check item
5   feature of the list box. This Example introduces
6   list box functions associated with the checking of
7   and displaying of list box items.
8
9   ****************************************************/
10
11  #include <cvirte.h>     /* Needed if linking in external compiler; harmless
12  otherwise */
13  #include <userint.h>
14  #include <ansi_c.h>
15  #include <formatio.h>
16  #include "Example7-2.h"
17
18  static int PanelHandle;
19  int inc=0, SelectAllFlag=0, item_val, item_index;
20  char buf[500], item_label[100];
21
22  int main (int argc, char *argv[])
23  {
24    if (InitCVIRTE (0, argv, 0) == 0)      /* Needed if linking in external compiler;
25                                              harmless otherwise */
26          return -1; /* out of memory */
27    if ( (PanelHandle = LoadPanel (0, "Example7-2.uir", LST) ) < 0)
28          return -1;
29    DisplayPanel (PanelHandle);
30    RunUserInterface ;
31   return 0;
32  }
33  /*****
34  This function displays the data in the Display List box as soon as the item is
35  selected. It also deletes the item from the display if the item is un-selected.
36  **********************************************************************/
37
38  int CVICALLBACK ListBoxSelectCB (int panel, int control, int event,
39         void *callbackData, int eventData1, int eventData2)
40  {
41         int itemIndex, CheckNotChecked,index_value,CheckedItems,i,TotalItems,
42                                        selectIndex=0, red, green, blue,
43  white,black;
44    switch (event)
45    {
46     case EVENT_COMMIT:
47       //Color is created as RGB. First argument of MakeColor is for red.
48       // Takes values from 0 to 255. 255 is the highest. Second argument is green,
49                                                        and third is
50                                                                  blue.
51       red = MakeColor (255, 0, 0);   //red color created
52       green= MakeColor (0,255, 0); //green color
53       blue= MakeColor(0,0,255);          //blue color
```

Figure 7–10
Example 7–2 Listing

Chapter 7 • List Boxes & Rings

```
54      white = MakeColor (255, 255, 255); //white color
55      black=MakeColor(0,0,0);      //black color
56      //Get the number of checked items from the list
57      GetNumCheckedItems (PanelHandle, LST_LISTBOX, &CheckedItems);
58      //Get index of selected item
59      GetCtrlIndex (PanelHandle, LST_LISTBOX, &itemIndex);
60      //If item is checked then CheckNotChecked=1
61      IsListItemChecked (PanelHandle, LST_LISTBOX, itemIndex,
62                                                  &CheckNotChecked);
63      if(CheckNotChecked) //If item is checked
64      {
65        inc++; // counter for number of items displayed in list box
66        GetLabelFromIndex (PanelHandle, LST_LISTBOX, itemIndex,
67                                                  item_label);
68        GetValueFromIndex (PanelHandle, LST_LISTBOX, itemIndex,
69                                                  &item_val);
70        //Make a format field for coloring displayed items and creating a vertical
71                                                  line
72                                                  after each item
73
74        sprintf(buf, "\033bg%06X\033fg%06X\033p50c%s\033fg%06X
75        \033vline\033fg%06X\033p100c%d\033fg%06X \033vline",
76        white,  red,  item_label,  black,  blue,  item_val, black);
77        InsertListItem (PanelHandle, LST_DISPLAY,-1, buf, item_val);
78      }//if
79      else //If Unchecked
80      {
81      if (inc>0)    //Check if the index is non-negative
82      {
83      //Get value of the item from the list box using the selected index
84      GetValueFromIndex (PanelHandle, LST_LISTBOX, itemIndex,
85                                                  &item_val);
86      //Get the index from the Display box based on the item value.
87      GetIndexFromValue (PanelHandle, LST_DISPLAY, &selectIndex,
88                                                  item_val);
89      //Delete item from display using the selected index
90      DeleteListItem (PanelHandle, LST_DISPLAY, selectIndex, 1);
91      inc—; //decrement counter
92      }// if (inc>0)
93      }//else
94     break;
95  }//case
96  return 0;
97  }//ListBoxSelectCB
98
99  //Exit the function
100 int CVICALLBACK ExitCB (int panel, int control, int event,
101      void *callbackData, int eventData1, int eventData2)
102 {
103  switch (event)
104  {
105   case EVENT_COMMIT:
106     QuitUserInterface (0);
```

Figure 7–10
Example 7–2 Listing (*continued*)

```
107    break;
108  }
109  return 0;
110 }//ExitCB
111
112 //Clear the display
113 int CVICALLBACK ClearCB (int panel, int control, int event,
114      void *callbackData, int eventData1, int eventData2)
115 {
116  switch (event) {
117    case EVENT_COMMIT:
118        ClearListCtrl(PanelHandle, LST_DISPLAY);   //Clear display
119    break;
120  }
121  return 0;
122 }// ClearCB
```

Figure 7–10
Example 7–2 Listing (*continued*)

The library function *IsListItemChecked* at line 61 determines whether the item in the list box is checked, and its prototype is shown here.

```
int status IsListItemChecked (int PanelHandle,
    int LST_LISTBOX, int itemIndex,  int * CheckNotChecked);
```

and the function arguments are explained in Table 7–8.

Table 7–7   GetNumCheckedItems *Function*

Input/ Output	Name	Type	Description
Input	PanelHandle	integer	the panel handle of the user interface
	LST_ LISTBOX	integer	Constant Name of the list box prefixed by the panel Constant ID
Output	Checked Items	integer	returns the number of checked items in the list box
	status	integer	refer to Appendix A of *LabWindows/ CVI User Interface Library Reference Manual* for error codes or in Online Help

Chapter 7 • List Boxes & Rings

Table 7–8 IsListItemChecked *Function*

Input/Output	Name	Type	Description
Input	PanelHandle	integer	the panel handle of the user interface
	LST_LISTBOX	integer	Constant Name of the list box prefixed by the panel Constant ID
	itemIndex	integer	returns the zero-based index of the selected item
Output	CheckNotChecked	integer	returns a 1 if the item is checked, 0 otherwise
	status	integer	refer to Appendix A of *LabWindows/CVI User Interface Library Reference Manual* for error codes or in Online Help

At line 61, the value returned from this function (`CheckNotChecked`) is used to determine whether to process the code between lines 65 and 77 or that between lines 80 and 93. Lines 65 to 77 are used to obtain all the information necessary to display the data in the list box. The first variable is *inc*, which is incremented to count the number of items displayed in the list box. This counter is decremented in line 91 to adjust for the number of items removed from the display list box and avoid "array out of bound" or "negative array index" errors.

The *GetLabelFromIndex* library function at line 66 is used to obtain the label of the selected item using the value of the index of the item in the list as obtained by the library function *GetCtrlIndex* in line 59. At line 68, *GetValueFromIndex* obtains the value of the selected item from the list box.

The *GetLabelFromIndex* and *GetValueFromIndex* library functions enable you to get both the label and the value of the items in the list box from the item index. Now you have all the information you need to display the results on the **TEST SEQUENCE** list box.

The label and value of the selected item is displayed in the list box using the *InsertListItem* library function in line 77. This function inserts the data in *buf* associated with the *item_val* at the end of the list in the display list box.

The `sprintf` at line 74 creates the string to be input into `buf`. This statement is shown again in Figure 7–11. It looks complicated, so let us analyze it carefully.

The *escape sequence* is indicated by `\033`. The required function code is appended to this escape sequence.

Escape sequences are control codes representing certain formatting characters that direct the display or printing of the data. In C, these are characters following the `\` character. For example, the escape sequence for a new line is `\n` and for a carriage return is `\r`.

The first part of this string is `\033bg`. The `bg` represents the background color. The next string is `%06X` and represents the hexadecimal value of the color specified. In this case, the variable for this format specifier is white, so the color *white* was created using the *MakeColor* function for the white background used here. The next part is the *escape sequence* followed by `fg%06X`. The `fg` is for the foreground color, and the hexadecimal number for red is used here, using the *MakeColor* function as before to create a red foreground on a white background.

The `\033p50c` means to move fifty pixels (`p50`) to the right from the present location and to center (`c`) the value in the column. Similarly, if the pixel value were followed by `l` (letter *l*), the value would be left-justified and by `r` right-justified.

The last escape sequence that needs to be explained is `\033vline`. The `vline` code is to make a black vertical line. This creates a dressy, columnar look and enhances the output data.

Using this method, various individual items can be displayed in different colors, which is particularly useful when displaying test data with Pass/Fail results. Setting the color of Pass to green and Fail to red will make the results more conspicuous.

Using the variable `item_val`, the index of the item is obtained in the variable `selectIndex` from the displayed list using the library function *GetIndexFromValue* at line 87. The prototype of this function is

```
int status=GetIndexFromValue (int PanelHandle,
    int LST_DISPLAY, int * selectIndex, item_val);
```

```
sprintf(buf,
"\033bg%06X\033fg%06X\033p50c%s\033fg%06X\033vline\033fg%06X\033p100c%d\033fg%06X\033vline",
    white, red, item_label, black, blue, item_val, black);
```

Figure 7–11
Formatted String Using Escape Sequences

Table 7–9 GetIndexFromValue *Function*

Input/Output	Name	Type	Description
Input	PanelHandle	integer	the panel handle of the user interface
	LST_DISPLAY	integer	Constant Name of the list box prefixed by the panel Constant ID
	item_val	[1]Depends on type of control	specifies the value to look for in the list box
Output	select_index	void *	returns the zero-based index of the first value specified; a –1 is returned if no matching value is found
	status	integer	refer to Appendix A of *LabWindows/CVI User Interface Library Reference Manual* for error codes or in Online Help

and the function arguments are explained in Table 7–9.

Once the index of the matching value is obtained, the *DeleteListItem* library function removes the item from the display list box on line 90. The prototype and the arguments of the library function *DeleteListItem* was shown above in Table 7–6.

Run this example a couple of times to get comfortable with the *CVI* list box functions. The best way to understand the capabilities of these functions is to insert **Breakpoints** at various points in the code, step through the program, and look at the index, values, and labels using the *Debugger*.

Using Ring Controls

Rings are very similar to *List Boxes*, and they utilize almost the same functions. A list box consists of only one type of box in which the items are shown in the form of a list, whereas *Rings* consist of a variety of different controls as shown in Figure 7–12. With *Rings,* some of the controls you use can be very similar to a list box; you can view all the items by clicking on the *Ring* controls, as you will see later in this chapter. *Rings* consist of sliders, switches, knobs, dials, gauges, meters, tanks, thermometers, and picture rings, from which distinct values can be accessed and used in your project. To create a *Ring,* select **Create>>Ring** and select one of the icons shown in Figure 7–12.

For demonstration of some of the *Ring* controls, a user interface is created as shown in Figure 7–13 with different types of *Ring* controls for possibly configuring instrumentation and their interfaces. This is a simple example designed to show you the various *Ring* controls.

The objects in the left-most decoration box show two different kinds of *Ring* controls. The *Ring* control titled **Equipment** is called the **Menu Ring.** This displays a pull-down list either when either clicking on the displayed menu item or the down arrow on the side of the box and is shown in Figure 7–14. This *Ring* can be used, for example, to select an instrument that you want to configure from the pull-down list.

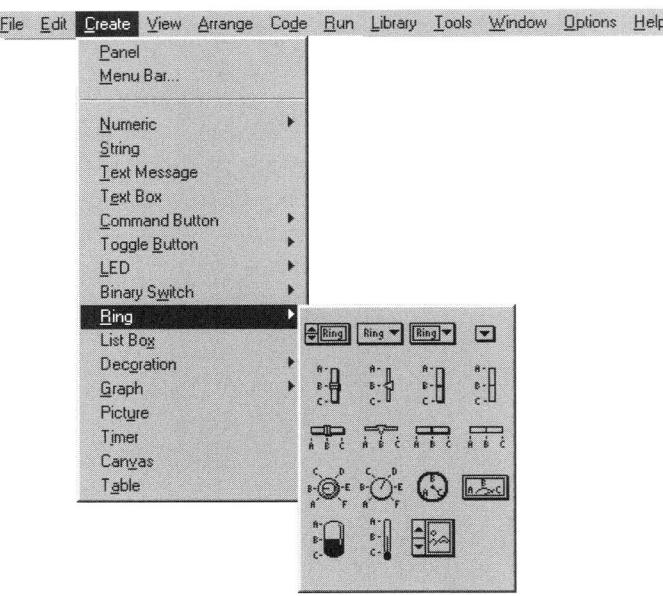

Figure 7–12
Creating a Ring Control

Chapter 7 • List Boxes & Rings

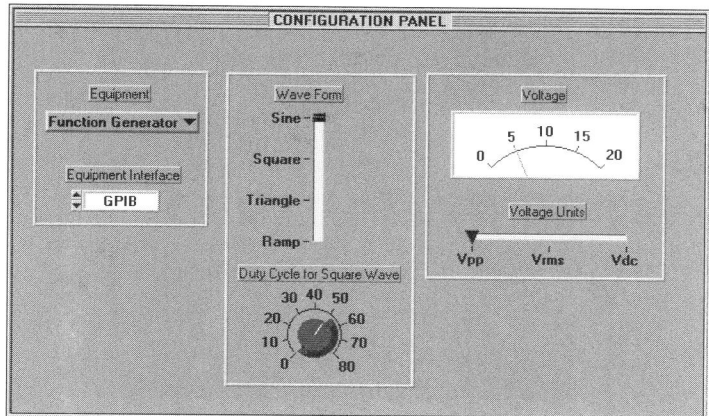

Figure 7–13
Ring Control Demonstration Panel

The *Ring* control titled **Equipment Interface** can be used, for example, to select the instrument interfaces. This is simply called a *Ring* and is shown in Figure 7–15. When the control box is clicked, it shows the complete selection list. When the increment/decrement arrows are selected, only one item at a time is shown in the *Ring* control window. In this example, you can select from one of the communication interfaces: GPIB, RS-232, or TCP/IP.

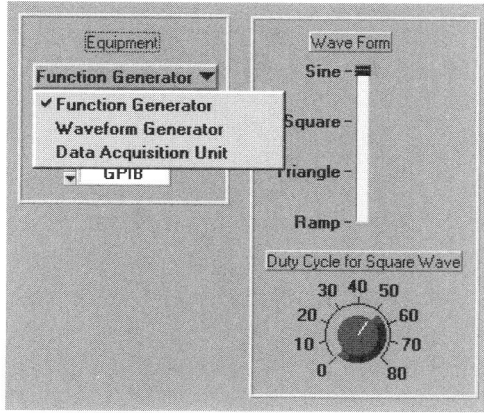

Figure 7–14
Menu Ring Control Box

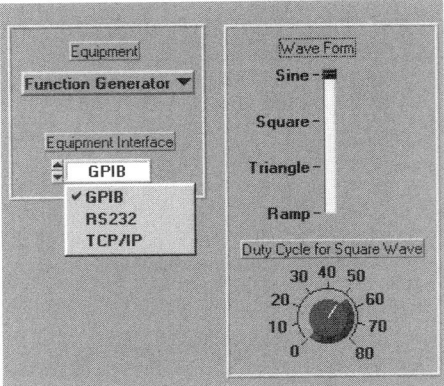

Figure 7–15
Ring Control Box

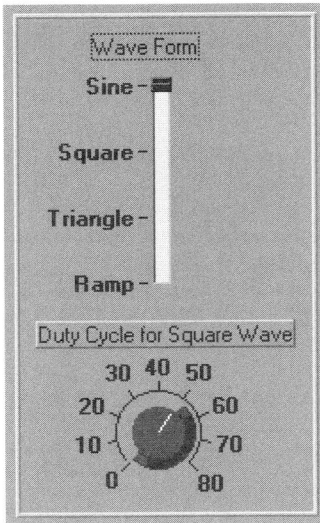

Figure 7–16
Vertical Level Slide and Knob

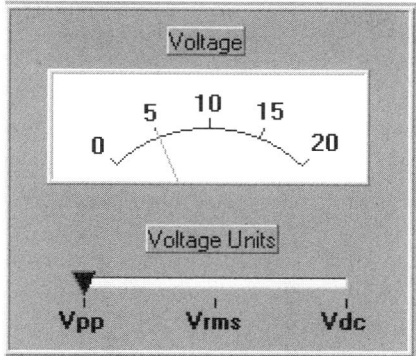

Figure 7–17
Meter and Horizontal Pointer Slide

In Figure 7–15, the *Ring* shown in the decoration box to the right and entitled **Waveform** is called the **Vertical Level Slide.** For clarity, this control is shown individually in Figure 7–16. Again using the *Ring* control, you can only select the individual values shown on the control one at a time. In this example, you have a choice of selecting a sine wave, square wave, triangle wave, or ramp to download to an instrument, such as a waveform generator.

The control below the **Vertical Level Slide** in Figure 7–16 is another kind of *Ring* control called the **Knob.** Allowing the user to select only from the values shown on the knob is an ideal use of this *Ring* control.

Figure 7–17 shows two other kinds of *Ring* controls. The **Meter** is useful if only one of the values shown on the meter controls is acceptable to the user and is given no other choice. For example, the user can only select 0, 5, 10, 15, or 20. If the user wants a value besides these he should not use this control. This may not be the ideal way to use the meter, but this is only an example to introduce you to the various forms of the *Ring* control.

The *Ring* control shown below the **Meter** in Figure 7–17 is called the **Horizontal Pointer Slide.** Here it is used to select a particular type of voltage to configure the equipment: peak to peak, root mean square, or direct current.

There are many other controls in the *Ring* box that you are encouraged to explore. You can find many creative uses of *Rings* in your applications. As mentioned before, *Ring* controls do not have special library functions and use only a sub-set of the *List Box* library functions.

Designing the Test Executive Project

Let us use the information that you have gleaned in the previous sections of this chapter to create a simple **Test Executive** project (`project7.prj`). National Instruments has a very sophisticated **Test Executive tool** called the *TestBench*, which is distributed as an add-on to the base *CVI* program. The purpose of creating a **Test Executive** program is to show you the features of the *List Boxes* and *Ring* controls that were introduced in the previous sections. Through this project, you will learn to use these controls and library functions with all their enhancements.

The basic idea behind creating a **Test Executive** program is to run tests automatically in a pre-selected order and to control the flow of the tests. The user should have the option of running the test a multiple number of times and stopping at the first failure when requested. Consequently, the test data should be logged to an output data file for later review and analysis. Other enhancements can be added, but here you will create a basic test executive with only limited capabilities. This explanation will become clearer as you create this project and analyze the code in detail.

This project will consist of three GUIs, each called by selecting the appropriate command buttons. Start by creating the **TEST SEQUENCE SELECTION** GUI as shown in Figure 7–18. Create a new panel and enter the following attributes. Leave the remaining attributes at their default settings.

Constant Name	TEST
Panel Title	TEST SEQUENCE SELECTION
Menu Bar	No menu bar
Close Control	None
Top	80
Left	224
Height	519
Width	521
Scroll Bars	None

Create two *List Boxes* as shown in Figure 7–18 and set the following attributes:

Chapter 7 • List Boxes & Rings

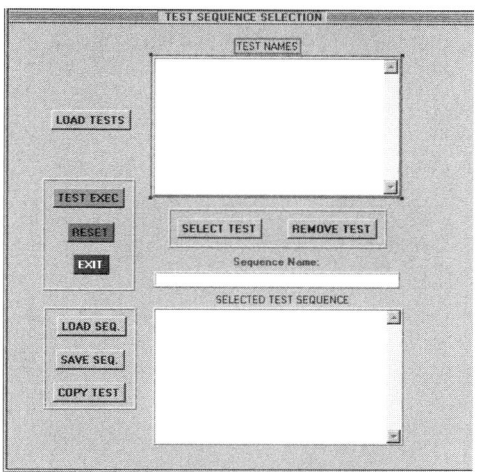

Figure 7–18
Test Sequence Selection GUI

Control Attributes	Upper List Box	Lower List Box
Constant Name	TST_LST	TST_SQU
Control Mode	Hot	Hot
Visible Lines	12	12
Vertical Scroll Bar	Checked	Checked
Label	TEST NAMES	SELECTED TEST SEQUENCE

The **TEST NAMES** list box (upper list box) will contain the names of all the pre-created test files that are loaded from the hard disk. These test files may contain test scripts, which will execute a certain sequence of steps indigenous to that particular test, possibly using a test script file. The **SELECTED TEST SEQUENCE** list box (lower list box) will contain the order of the tests to be run as selected by the user.

Create a **Decoration** Box in between these two *List Boxes* and add two command buttons with the following attributes:

Control Attributes	First Command Button	Second Command Button
Constant Name	`SEL_TST`	`REM_TST`
Callback Function	`SelectTestsCB`	`RemoveTestCB`
Control Mode	`Hot`	`Hot`
Label	`SELECT TEST`	`REMOVE TEST`
Auto Sizing	`Always Auto Size`	`Always Auto Size`

Next to the **TEST NAMES** box, create a command button with **Constant Name** `LOAD_TST`, **Callback Function** `LoadTestNamesCB`, and **Label** `LOAD TESTS`. When this button is selected, the previously created test names are displayed in the **TEST NAMES** box from the hard disk.

The **SELECT TEST** command button will move the highlighted test name from the **TEST NAMES** box to the **SELECTED TEST SEQUENCE** box. This way you can highlight the test you want to select in the **TEST NAMES** box and click the **SELECT TEST** command button. This will add the test name to the **SELECTED TEST SEQUENCE** box. The order of the tests is created by the order in which they are added to the **SELECTED TEST SEQUENCE** box.

If for some reason you want to change the order of the test sequence, you can highlight the test name in the **SELECTED TEST SEQUENCE** box and select the **REMOVE TEST** command button. This will move the test name from the **SELECTED TEST SEQUENCE** box up to the **TEST NAMES** box. You can select another test name and use the **SELECT TEST** command button to move it to the **SELECTED TEST SEQUENCE** box.

Below the **SELECT TEST** and **REMOVE TEST** command buttons, create a string box with **Constant Name** `SEQ_NAME` and **Label** `Sequence Name`. This box displays the name of the selected test case sequence when a pre-existing test sequence is loaded.

To the left of these command buttons, create a **Decoration** box and add three command buttons with the following information:

Chapter 7 • List Boxes & Rings

Control Attributes	Upper Command Button	Middle Command Button	Lower Command Button
Constant Name	TESTEXEC	RESET	EXIT
Callback Function	GoToTestExecCB	ResetSeqCB	ExitCB
Control Mode	Hot	Hot	Hot
Label	TEST EXEC	RESET	EXIT
Auto Sizing	Always Auto Size	Always Auto Size	Always Auto Size

Selecting the **TEST EXEC** command button will bring up the **TEST EXECUTIVE** panel from which you can run the test sequence. The **TEST EXECUTIVE GUI** is shown in Figure 7–19.

Selecting the **RESET** command button in Figure 7–18 will move all the test names from the **SELECTED TEST SEQUENCE** box back to the **TEST NAMES** box to enable you to start your test selection again.

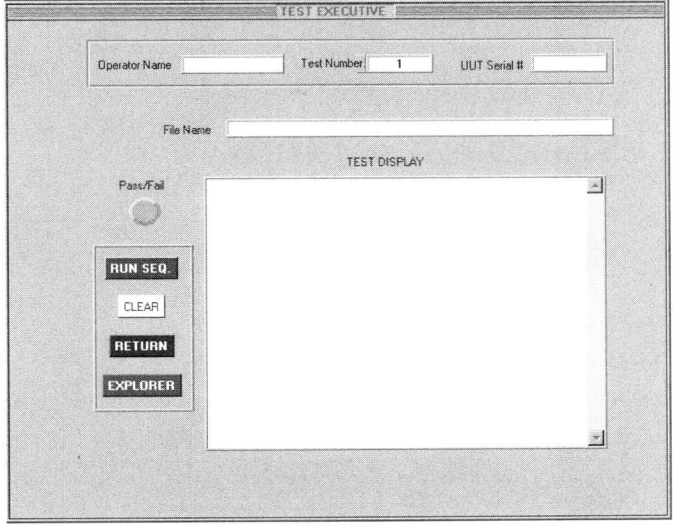

Figure 7–19
Test Executive GUI

To the left of the **Selected Test Sequence** list box, create a **Decoration** box and add three command buttons with the following attributes:

Control Attributes	Top Command Button	Middle Command Button	Bottom Command Button
Constant Name	LOAD_SEQ	SAVE_SEQ	COPY
Callback Function	LoadSeqCB	SaveSeqCB	CopyTestCB
Control Mode	Hot	Hot	Hot
Label	LOAD SEQ.	SAVE SEQ.	COPY TEST
Auto Sizing	Always Auto Size	Always Auto Size	Always Auto Size

The **LOAD SEQ.** command button is used to load a previously created test sequence from the hard disk. The **SAVE SEQ.** command button is used to save the newly created test sequence in a file, and the user will be prompted to enter a file name. The **COPY TEST** command button allows you to add the name of the highlighted test in the **SELECTED TEST SEQUENCE** box in case you want to run that test again. You can copy the test name as many times as you would like to run it by clicking on the **COPY TEST** command button. This will add the test name highlighted in the top list box to the list box below. This completes the **TEST SEQUENCE SELECTION** user interface panel.

Now let us create the **TEST EXECUTIVE** GUI, shown in Figure 7–19, from which the test sequence will be executed. Create a new panel and enter the following attributes:

Constant Name	EXEC
Panel Title	TEST EXECUTIVE
Menu Bar	No menu bar
Close Control	None
Top	84
Left	168
Height	507
Width	649
Scroll Bars	None

Recall that this GUI is displayed when the **TEST EXEC** command button is selected from the **TEST SEQUENCE SELECTION** user interface (Figure 7–18). Create a **Decoration** Box at the top of the panel and add two **String** boxes and a **Numeric** box as shown in Figure 7–19. Set the following control attributes:

Control Attributes	First String Box	Numeric Box	Second String Box
Constant Name	OPER_ID	TEST_NBR	UUT_NUM
Default Value		1	
Data Type		unsigned short int	
Control Mode	Hot	Hot	Hot
Label	Operator Name	Test Number	UUT Serial #

In the **OPER_ID** box, the user can enter the operator name to save it with the test run. The **TEST_NBR** box is an indicator of the number of times this particular test is run, which the user can change. In the **UUT Serial #** box, the serial number of the Unit Under Test is entered to identify uniquely the test run with the unit. All this data is written as a header in the data log file along with the test results.

Create a **String** box below this **Decoration** box and for **Constant Name,** enter FILENAME, for **Control Mode** Hot, and for the **Label** File Name. This box will contain the path of the test being executed as it is selected and run from the test sequence.

Near this box, add an **LED** indicator, which will indicate the pass/fail status of the tests as they are run. Set the following properties for this LED box: **Constant Name**: FAIL_IND; **Control Mode**: Indicator; **Initial State**: OFF; and **Label**: Pass/Fail. Use the *Coloring Tool* to color the LED green.

Create a list box as shown in Figure 7–19 with the **Label**: TEST DISPLAY, **Constant Name**: TE_DISP, **Control Mode**: Hot, **Visible Lines**: 20, and the **Vertical Scroll Bar** box checked.

We will use the list box to display data rather than using the *Text Box* as shown in previous projects. One of the advantages of using the list box is a more aesthetically pleasing data display format, adding vertical lines as columns and adding color to the displayed data. If, for example, you wanted to display a Pass or a Fail with every test on the list box, you may want to set the Pass color to green and the Fail to red. These features are not available in a text box.

Next to the **TEST DISPLAY** list box, create a **Decoration** box with the command buttons shown in Figure 7–20.

Set the following attributes for these command buttons.

Label	RUN SEQ.	CLEAR	RETURN	EXPLORER
Constant Name	RUN_SEQ	CLEAR_LST	RETURN	EXPLORE
Callback Function	SelectTimesToRun	ClearListBoxCB	ReturnCB	ExplorerCB
Control Mode	Hot	Hot	Hot	Hot
Auto Sizing	Always Auto Size	Always Auto Size	Always Auto Size	Always Auto Size

The **RUN SEQ.** command button, as the name suggests, runs the selected test sequence. Selecting this command button brings up the panel shown in Figure 7–21 that shows the **RUN SEQUENCE** user interface. This GUI gives the user the choice of stopping on the first failure and sets the number of times to run the test sequence.

Figure 7–20
Test Executive Command Buttons

Chapter 7 • List Boxes & Rings

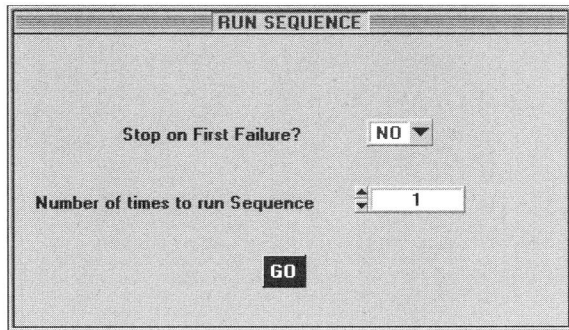

Figure 7–21
Run Sequence GUI

Let us now create the **RUN SEQUENCE** panel as shown in Figure 7–21, with the attributes as shown below:

Constant Name	SEQ_RUN
Panel Title	RUN SEQUENCE
Menu Bar	No menu bar
Close Control	None
Top	120
Left	128
Height	205
Width	375
Scroll Bars	None

The **Stop on First Failure** box is a *Ring* box. Enter the following information in the Edit box for this *Ring*. Enter **Constant Name** FAILURE and **Label** Stop on First Failure?. For the **Label/Value Pairs,** enter the information shown in Figure 7–22. A *List Box* could have been used instead of a *Ring* control. The purpose here is to demonstrate that the *Rings* use the same library functions as the *List Boxes,* as seen in the code listings.

The **Number of times to run Sequence** is a numeric box with the following attributes: **Constant Name** TST_NBR, **Default Value** 1, **Data Type** short int, **Control Mode** Hot, and the **Label** as given above. The rest of the values can be left as defaults.

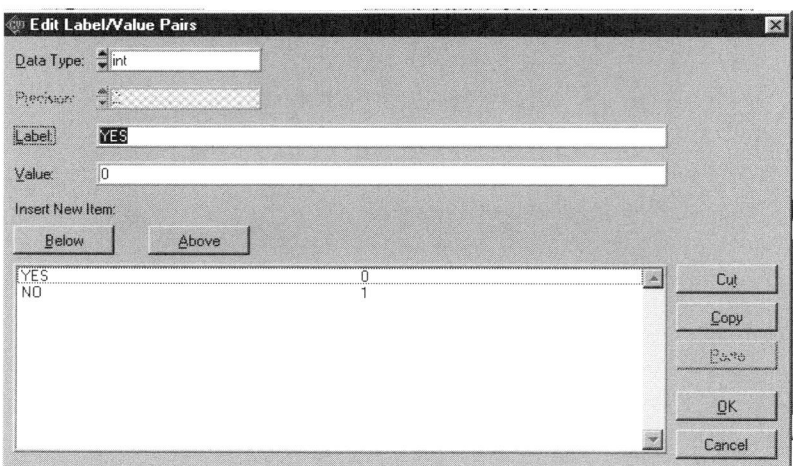

Figure 7–22
Edit Label/Value Pairs

Below these objects add a command button, **GO,** to start the test execution with the following control values. Selecting the **GO** command button will bring up the **TEST EXECUTIVE** panel and start executing the test sequence.

Constant Name	RUNTEST
Callback Function	RunSequenceCB
Control Mode	Hot
Label	GO
Auto Sizing	Always Auto Size

This completes the creation of the GUIs and the setting of their attributes. In the next section, you will look at the code for this project.

Project7 Code Generation and Analysis

Use the *CodeBuilder* to create the template for the source code for each of the three panels and rename the source file to project7.c when the source code is created for the first user interface resource. Then set the same source target file for the other user interface resources using **Code>>Set Target**

File. This will insert all the callback functions into the same source file (`project7.c`).

The code generated for the user interfaces by the *CodeBuilder* will not be shown here. You have already seen the *CodeBuilder* application in previous projects, and you now know how to create the code template. However, the source code for each of the callback functions will be listed and explained step-by-step.

The header files and the function prototype are listed at the top of the program. These header files are the same that you have seen in the previous projects that are included by the *CodeBuilder*. The functions for this project and any library functions not previously described are explained.

Main Function

The *main* function listing for `project7.c` is shown in Figure 7–23. At line 34, the project directory is obtained and saved in the variable `prj_dir`. It is good programming practice to keep as much of the data as possible relative rather than absolute so that if you want to store a project in another directory, you will not have to change the code. The library function *GetProjectDir* locates the project directory.

LoadTestNamesCB Function

The function **LoadTestNamesCB** is called when the **LOAD TESTS** command button is selected from the **TEST SEQUENCE SELECTION** GUI in Figure 7–18. The source code listing for this callback function is shown in Figure 7–24.

This function finds all the test name files (identified by the file extension `.tst`) on the hard-disk and displays them in the **TEST NAMES** box. At line 13, the `Load_File_Flag` variable is checked to determine if the files are already displayed in the **TEST NAMES** box. If they are, then the program executes lines 36 through 39, pops up a message that the test names are already displayed, sets `Load_File_Flag`, and exits the function. Otherwise, lines 16 through 34 are executed.

At line 16, the starting value of the index is set to 0 and incremented with the display of each test name. Line 18 creates the pathname for the test file names. In this project, the test name files were created and saved in a subdirectory `input` and all these file names have the extension `.tst`. This state-

```
1   /*
2   This is a simple Test Executive program, which enables the user to select the tests
3   to create a sequence and save it to a file. Run the test the sequence any number of times,
4   with an option to stop at the first failure or to continue. The test data is also written to
5   a output file, which can be viewed using an ASCII text editor or displayed on the screen.
6   */
7
8   #include <ansi_c.h>
9   #include <utility.h>
10  #include <formatio.h>
11  #include <cvirte.h>        /* Needed if linking in external compiler; harmless otherwise */
12  #include <userint.h>
13  #include "project7.h"
14  #include "Run_Seq.h"
15  #include "TestExecutive.h"
16
17  void CheckFileName (char *FileToCheck, char *extension);//Routine to verify the filename
18
19  static int TestSeqHandle, RunHandle, TEHandle, OutputFileHandle;
20  int stateIndex, Load_File_Flag=0, Reset_Seq_Flag=1,SaveSequenceFlag=0,
21    LoadSequenceFlag=0,Status, FirstFailureFlag=0;
22
23  char prj_dir[260], item_label[260], save_Seq_file[260],inputdir[260],
24  buf[500],Seq_file[260];
25
26  //Main program starts here
27  int main (int argc, char *argv[])
28  {
29          if (InitCVIRTE (0, argv, 0) == 0)
30                  return -1;      /* out of memory */
31          if ((TestSeqHandle = LoadPanel (0, "project7.uir",TEST)) < 0)
32                  return -1;
33          DisplayPanel (TestSeqHandle);
34          GetProjectDir (prj_dir); //Get the project directory
35          //Center the panel on the screen
36          SetPanelPos (TestSeqHandle, VAL_AUTO_CENTER, VAL_AUTO_CENTER);
37          RunUserInterface ;
38          return 0;
39  } //main
```

Figure 7–23
Main Function Source Code

ment, therefore, sets the search path to `prj_dir\input\*.tst` and assigns it to the variable `search_path`.

Note: When you create the project directory, it is advisable to add to it two sub-directories (sub-folders). Name the first folder `input` and include in it all the test files names, and name the second folder `output`, and have the output test data files written to this file during the run.

The *GetFirstFile* library function at line 20 finds the first file in the *search-path* with the `*.tst` extension and puts its search in the variable `result`. This

Chapter 7 • List Boxes & Rings

```
1    //This routine copies the  all the test names into the list box for selection by the user
2    int CVICALLBACK LoadTestNamesCB (int panel, int control, int event,
3                void *callbackData, int eventData1, int eventData2)
4    {
5                int result,index;
6                long size;
7
8                char TC_file_found[260], search_path[260];
9                switch (event)
10               {
11               case EVENT_COMMIT:
12
13                   if (!Load_File_Flag)//Check if the test names are not already loaded
14                   {
15
16                       index=0;//Offset value
17                       //Test file names have ".tst" extension
18                       Fmt(search_path,"%s\\input\\*.tst", prj_dir);
19                       //Find the first test file
20                       result = GetFirstFile (search_path, 1, 1, 0, 1, 1, 0, TC_file_found);
21                       InsertListItem(TestSeqHandle,TEST_TST_LST, index,TC_file_found,
22                                                                                    index);
23                           //Display all the test files
24                           while (result==0) // Loop while files are found
25                           {
26                           result= GetNextFile(TC_file_found);//Find next test file
27                           if (result ==0) //if file name is found then display it
28                               {
29                                   index++;
30                                   InsertListItem(TestSeqHandle, TEST_TST_LST,
31                                                                    index,TC_file_found, index);
32                               }
33                           } //while (result==0)
34                   }//    if (Load_File_Flag!=1)
35
36                   else //(Load_File_Flag==0)
37                       MessagePopup ("TEST FILES LOADED", "The files are already loaded");
38
39                   Load_File_Flag=1; // Load the test files flag set
40           break;
41               }
42       return 0;
43   } //LoadTestsCB
```

Figure 7–24
LoadTestNamesCB Source Code

function has many arguments and needs some explanation. Its prototype follows.

```
int status= GetFirstFile( char searchpath, int NormalAttribute,
int ReadOnlyAttribute,  int SystemAttribute, int HiddenAttribute,
int ArchiveAttribute, int DirectoryAttribute, char filename[ ]);
```

Including a `1` in the appropriate argument enables the file attribute for which you would like to search. A `0` indicates an attribute for which you do not want to search. For example, if you want to search a file that is not a Read-only, System, or Hidden file and that is not a Directory, enter `0` in these arguments and a `1` in the *Normal Attribute* argument. In this particular case, the *ArchiveAttribute* can be either a `1` or `0`. The searched file is returned in the *filename* string, consisting of the file basename and the file extension.

If the *SystemAttribute* argument is '1', then the system file will be found unless the file is hidden and the *HiddenAttribute* was not set. To obtain a hidden file that is also a system file, set both *SystemAttribute* and *HiddenAttribute* to `1`. If multiple attributes are set, then *GetFirstFile* returns the file that meets any one of the attribute criteria.

The `status` of this function consists of the error codes, as shown in the table below:

Codes	Description
0	no errors, file found
-1	no files found matching the arguments set
-3	general I/O error occurred
-4	insufficient memory to complete the search
-5	invalid path name
-6	access denied

Note: Error code -2 does not exist.

Note: The *GetFirstFile* library function can use all the attributes under Windows, but when using Unix, you can only use the **DirectoryAttribute**. A `1` set in this argument finds a directory only and a `0` a non-directory file.

At line 21 of Figure 7–24, the name of the first file found is inserted in the **TEST NAME** box. Lines 26 through 33 are inside a *while* loop and stay in this loop until the files are found, as long as the value of `result` continues to be `0`. The first time, the value of the `result` variable is obtained from the function *GetFirstFile* at line 20. After that, the *GetNextFile* library function at line 26 finds the next files in the *while* loop. If the file is found, then it is listed in the **TEST NAME** box.

The function *GetNextFile* finds the next file in the search path started by *GetFirstFile* for the same attributes. The prototype for this function follows.

```
int result = GetNextFile ( char filename [ ]);
```

The `filename` contains the name of the file found in the search with the basename and extension of the next matching file. The `result` contains the error codes as shown in the table below.

Result	Description
0	no errors, file found
-1	no more files found matching the arguments criteria
-2	search must be started with *GetFirstFile*

SelectTestsCB Function

The *SelectTestsCB* function is invoked when the **SELECT TEST** command button is selected from the GUI shown in Figure 7–18. The source code listing for this callback function is shown in Figure 7–25.

At line 11 of Figure 7–25, the number of test names in the **TEST NAMES** box is obtained from the library function *GetNumListItems* in the variable `count`. If the number is zero, then the program jumps to the *else* part of the *if-else* statement at line 29 and displays a message from the popup panel, indicating the **TEST NAMES** box has no files to insert in the **SELECTED TEST SEQUENCE** box.

If there are test files displayed in the **TEST NAMES** box, the test file name is inserted in the **SELECTED TEST SEQUENCE** box at line 22. The index of the highlighted item is obtained at line 16 and the value and label of the selected test name is obtained at lines 18 and 20 from the index of the highlighted item. At line 23, the variable `insert_index` is incremented each time the test name is inserted. The test name is removed from the **TEST NAMES** box at line 25 by the *DeleteListItem* function. At line 27, the `Reset_Seq_Flag` is set, which is used in *ResetSeqCB* library function listed in Figure 7–29.

RemoveTestCB Function

The *RemoveTestCB* function is called when the **REMOVE TEST** command button is selected from the GUI shown in Figure 7–18. The source code listing for this callback function is shown in Figure 7–26.

The purpose of this function is to remove the test name from the **SELECTED TEST SEQUENCE** box that may have been selected in error and needs to be removed. This function deletes the test name from the **SELECTED TEST SEQUENCE** box and re-inserts the test name in the **TEST NAMES** box.

```
1   //Select the test(s) from the List Box, activated when "SELECT TEST" command
2   //button is selected
3   int CVICALLBACK SelectTestsCB (int panel, int control, int event,
4           void *callbackData, int eventData1, int eventData2)
5   {
6       int item_index, insert_index=0, item_value, count;
7       switch (event)
8       {
9           case EVENT_COMMIT:
10              //Get the number of items in the  Test Case List box
11              GetNumListItems(TestSeqHandle,TEST_TST_LST,&count);
12
13              if (count !=0)   //Are there any files in the Test Case List box?
14              {
15                  //Obtain the index of the highlighted item in the list box
16                  GetCtrlIndex(TestSeqHandle, TEST_TST_LST, &item_index);
17                  //Get the value of the highlighted item from the index
18                  GetValueFromIndex(TestSeqHandle, TEST_TST_LST, item_index, &item_value);
19                  //Obtain the label from the index
20                  GetLabelFromIndex(TestSeqHandle,TEST_TST_LST,item_index, item_label);
21                  //List the file name to the  Selected Test Sequence box
22                  InsertListItem(TestSeqHandle,TEST_TST_SQU, -1, item_label, item_value);
23                  insert_index++;    //increment the index
24                  //Delete the listed item from the Test Case List box
25                  DeleteListItem(TestSeqHandle ,TEST_TST_LST,item_index,1 );
26
27                  Reset_Seq_Flag=0;   //Clear Sequence Flag
28              }
29              else
30                //If all the file names have been moved to the Selected Test Sequence box,
31                //then display this message
32                  MessagePopup ("INSERT SEQUENCE ERROR", "There is nothing to INSERT");
33                  break;
34      }
35      return 0;
36  } //SelectTestsCB
```

Figure 7–25
SelectTestsCB Source Code

As previously discussed, before the test names can be removed, the value of count is obtained from the library function *GetNumListItems* on line 12 and compared to the number of test names in the **SELECTED TEST SEQUENCE** box. If there are no test names in the **SELECTED TEST SEQUENCE** box, then line 32 pops up a message to indicate that there is nothing to be removed. If there is one or more test names in this box, the program continues by obtaining the index at line 16, the index value from line 18, and the label at line 21 of the selected test name. The item is then moved into the **TEST NAMES** box at line 24, and insert_index is incremented at line 26. The *DeleteItemList* at line 28 removes the test name from

Chapter 7 • List Boxes & Rings

```
1   //Remove the test from the selected sequence of tests, activated when the "REMOVE TEST"
2   //command button is selected
3   int CVICALLBACK RemoveTestCB (int panel, int control, int event,
4              void *callbackData, int eventData1, int eventData2)
5   {
6          int item_index, insert_index=0, count, item_value;
7          char item_label[260];
8
9          switch (event)
10         {
11              case EVENT_COMMIT:
12                  GetNumListItems(TestSeqHandle,TEST_TST_SQU,&count);
13                  if (count !=0) //There are items to remove
14                  {
15                  //Get the index of the highlighted file name
16                      GetCtrlIndex(TestSeqHandle, TEST_TST_SQU, &item_index);
17                      //Get the value from the index
18                      GetValueFromIndex(TestSeqHandle, TEST_TST_SQU, item_index,
19                                                              &item_value);
20                      //Get the label from the index
21                      GetLabelFromIndex(TestSeqHandle,TEST_TST_SQU,item_index,
22                                                              item_label);
23                      //Insert the selected file name to Test Case List box
24                      InsertListItem(TestSeqHandle,TEST_TST_LST, -1, item_label,
25                                                              item_value);
26                      insert_index++; //increment the index
27                      //Remove the highlighted item from the Selected Test Sequence box
28                      DeleteListItem(TestSeqHandle ,TEST_TST_SQU,item_index,1 );
29                      Reset_Seq_Flag=0;   //Clear Sequence Flag
30                  }
31                  else //No items to remove
32                      MessagePopup ("REMOVE SEQUENCE ERROR", "There is nothing to
33                                                              REMOVE");
34                  break;
35          }
36          return 0;
37  } //RemoveTestCB
```

Figure 7–26
RemoveTestCB Source Code

SELECTED TEST SEQUENCE box. Notice the similarities between *SelectTestCB* and *RemoveTestCB* function.

SaveSeqCB Function

The *SaveSeqCB* callback function is invoked when the **SAVE SEQ.** command button is selected from the GUI shown in Figure 7–18. The source code listing for this callback function is shown in Figure 7–27.

```
1   //Save selected test sequence to a file, invoked by  "SAVE SEQ." command button
2   int CVICALLBACK SaveSeqCB (int panel, int control, int event,
3           void *callbackData, int eventData1, int eventData2)
4   {
5       int count, status;
6
7       switch (event)
8       {
9                   case EVENT_COMMIT:
10                      //Obtain the number of tests in the Selected Test Sequence box
11                      GetNumListItems(TestSeqHandle,TEST_TST_SQU,&count);
12                      if (count !=0) //There is at least one selection
13                      {
14                          //Enter the sequence file name to save in the pop up menu
15                          PromptPopup ("SAVE SEQUENCE", "Enter File Name\n to save Sequence
16                                                      (no extension)", Seq_file, 30);
17                          //Append the ".sel" extension
18                          Fmt(save_Seq_file,"%s<%s\\input\\%s.sel", prj_dir, Seq_file);
19                          //Check for valid file name
20                          CheckFileName(save_Seq_file,".sel" );
21                          //Check for blank for file name
22                          if ( (CompareStrings(save_Seq_file,0,"", 0,0) ) ==0)
23                          {
24                              //Display message and return to the GUI
25                              MessagePopup ("FILE NAME ERROR",
26                                          "No File Name Entered. Program Halted ");
27                              return -999;
28                          }
29
30                          //Save the  Selected Test Sequence panel to be recalled later
31                          status=SavePanelState(TestSeqHandle,save_Seq_file, stateIndex);
32                          if (status != 0)
33                              {
34                                  //Message if there is some error in saving the panel
35                                  MessagePopup ("PANEL DISPLAY ERROR",
36                                    "Panel cannot be recalled.\nExit and Check Test File.");
37                                      return -999;
38                              }
39
40                          DefaultCtrl(TestSeqHandle,TEST_SEQ_NAME); //Clear the panel
41                                                                                  name
42                          //Write the file name to control on the GUI
43                          SetCtrlVal(TestSeqHandle,TEST_SEQ_NAME, save_Seq_file);
44                          SaveSequenceFlag =1;   //Set Save Sequence Flag
45                      } //if (count !=0)
46                      else
47                          //Popup message if no selection of tests
48                          MessagePopup ("SAVE ERROR", "No Selection has been made. Cannot
49                                                                              Save");
50
51                  break;
52      }
53  return 0;
54  } //SaveSeqCB
```

Figure 7–27
SaveSeqCB Source Code

Chapter 7 • List Boxes & Rings

This function saves the user-selected sequence to a file for later use when recalled and also saves the panel for recall when the panel is loaded by *LoadSequenceCB* callback function.

As before the first step is to check for any test names that can be removed from the **SELECTED TEST SEQUENCE** box at line 11 of Figure 7–27. If not, an error message indicating that no selection has been made is displayed at line 48 in a pop-up panel.

If the `count` value is 1 or greater, then a pop-up message is displayed at line 15 and prompts the user to enter a file name (without an extension) for saving this sequence. The program appends an extension, `.sel`, to all the sequence files saved via the *Fmt* command at line 18, creating the path name as `prj_dir\input` directory. Line 20 calls the user-created *CheckFileName* function to verify that a valid file name has been entered. *SavePanelState* at line 31 is a new function that takes a snapshot of all the items on the panel and their attributes and saves it to a file. The function prototype follows and the function arguments are explained in Table 7–10.

```
int status= SavePanelState (int TestSeqHandle, char save_Seq_file,
                                                   int stateIndex);
```

The `status` of the *SavePanelState* library function is tested for non-zero value at line 32 of Figure 7–27 and displays an error for non-zero value.

Table 7–10 SavePanelState *Function*

Input/Output	Name	Type	Description
Input	TestSeq Handle	integer	the panel handle of the user interface
	save_Seq_file	char	the file name for saving the selected sequence
	stateIndex	integer	assigns a unique state index to each panel so multiple panels can be saved to the same file and over-writes the previous state index in the same file name
	status	integer	refer to Appendix A of *LabWindows/CVI User Interface Library Reference Manual* for error codes or in Online Help; non-zero value creates an error

The *DefaultCtrl* library function clears the panel name at line 40 and then writes the new name to this control box at line 43. A flag, SaveSequence-Flag, is set at line 44 to indicate the test sequence is saved. This flag is used later in the *GoToTestExecCB* callback function (the code listing for this function is shown in Figure 7–31).

LoadSeqCB Function

The next callback function is *LoadSeqCB*, which is invoked when the **LOAD SEQ.** command button is selected. The source code listing for this callback function is shown in Figure 7–28.

At line 9 of Figure 7–28, the directory string is created by the *Fmt* library function for the location of sequence files, and the *FileSelectPopup* library

```
1   //Load the sequence from memory, invoked from "LOAD SEQ." button
2   int CVICALLBACK LoadSeqCB (int panel, int control, int event,
3           void *callbackData, int eventData1, int eventData2)
4   {
5      switch (event)
6        {
7           case EVENT_COMMIT:
8                //Set the  directory where the input files are located
9                Fmt(inputdir, "%s<%s\\input\\", prj_dir);
10               //Display all the files with "*.sel" extension
11               Status = FileSelectPopup (inputdir, "*.sel", "*.*",
12                       "Select Sequence Panel to Load",  VAL_LOAD_BUTTON, 0, 0, 1, 0,
13                                                                   save_Seq_file);
14               //Recall the panel saved with the selected sequence
15
16               Status=RecallPanelState(TestSeqHandle,save_Seq_file, stateIndex); //Recall Panel
17               if (Status != 0)
18               {
19                       //Display message if there is a problem recalling the saved panel
20                       MessagePopup ("PANEL DISPLAY ERROR",
21                                   "Panel cannot be recalled. Exit and Check Test File.");
22                       return -999;
23               }
24               //Display the panel
25               DisplayPanel(TestSeqHandle);
26               DefaultCtrl(TestSeqHandle,TEST_SEQ_NAME); //Clear up the panel name
27               SetCtrlVal(TestSeqHandle,TEST_SEQ_NAME, save_Seq_file);   //Write the file
28                                                                      name to control
29               LoadSequenceFlag=1;   //Set the Load Flag
30               Reset_Seq_Flag=0;
31               break;
32        }
33      return 0;
34  } //LoadSeqCB
```

Figure 7–28
LoadSeqCB Source Code

function at line 11 displays all the files with the *.sel extension for the user's selection.

The *RecallPanelState* library function is called at line 16 and displays the panel with all the control attributes of the user interface resource that were previously saved using the *SavePanelState* function. Its prototype follows and its function arguments are explained in Table 7–11.

```
int status = RecallPanelState (int TestSeqHandle,
                    char sequence_panel,   int stateIndex);
```

Again the status of the *RecallPanelState* is checked for non-zero value at line 17 of Figure 7–28, and an error message is displayed by the pop-up panel. If the panel is found, the recalled panel is displayed at line 25. At line 26, the **Sequence Name** box on the **TEST SEQUENCE** GUI is cleared and the new sequence name written to this control on line 27. On line 29, LoadSequenceFlag is set for later use in the *GoToTestExecCB* and *SelectTimesToRun* callback functions whose listings are shown in Figures 7–31 and 7–32 respectively.

ResetSeqCB Function

The *ResetSeqCB* callback function is invoked when the user selects the **RESET** command button from the GUI shown in Figure 7–18. The source code listing for this callback function is shown in Figure 7–29. This function

Table 7–11 RecallPanelState *Function*

Input/Output	Name	Type	Description
Input	TestSeq Handle	integer	the panel handle of the user interface
	sequence_panel	char	the file name in which the selected sequence panel was saved by the *SavePanelState* function
	stateIndex	integer	the state index with which this panel was saved using *SavePanelState* function
Output	status	integer	refer to Appendix A of *LabWindows/CVI User Interface Library Reference Manual* for error codes or in Online Help; non-zero value creates an error

```
1   // This function restores the sequence select panel to its default state invoked by the
    "RESET" button
2   int CVICALLBACK ResetSeqCB (int panel, int control, int event,
3           void *callbackData, int eventData1, int eventData2)
4   {
5       int result,index,item_index;
6       char TC_file_found[260], search_path[260];
7       long size;
8       switch (event)
9       {
10              case EVENT_COMMIT:
11              if (!Reset_Seq_Flag)   //if (Reset_Seq_Flag=0)
12              {
13                      DefaultCtrl(TestSeqHandle,TEST_SEQ_NAME); //Clear up the panel
14                                                                                name
15                      // Clear the Selected Test Sequence box
16                      ClearListCtrl(TestSeqHandle ,TEST_TST_SQU);
17                      // Clear the  Test Case list box
18                      ClearListCtrl(TestSeqHandle ,TEST_TST_LST);
19
20                      index=0;//Offset value
21                      Fmt(search_path,"%s\\input\\*.tst", prj_dir);
22                      //Find the "*.tst" files
23                      result= GetFirstFile (search_path, 1, 0, 0, 0, 0, 0, TC_file_found);
24                      //Insert the first test in the Test Case list box
25                      InsertListItem(TestSeqHandle,TEST_TST_LST  , index,TC_file_found,
26                                                                                  index);
27                      while (result==0) // Loop while files are found
28                      {
29                       result= GetNextFile(TC_file_found);
30                       if (result ==0)
31                          {
32                                  index++;      //increment index
33                                  //Inser the file names in the Test Case list box
34                                  InsertListItem(TestSeqHandle,TEST_TST_LST,
35                                                                  index,TC_file_found,
36                                                                                  index);
37                          }
38                      } //while (result==0)
39              } //if (Reset_Seq_Flag=0)
40              else
41                      // If there is nothing to RESET then display this message
42                      MessagePopup ("RESET SEQUENCE ERROR", "There is nothing to RESET");
43
44              Reset_Seq_Flag=1;  //Set Reset Sequence Flag
45          break;
46          }
47          return 0;
48  } //ResetSeqCB
```

Figure 7–29
ResetSeqCB Source Code

provides the capability of moving all the test names from the **SELECTED TEST SEQUENCE** box to the **TEST NAMES** box. This command button is handy if you have made extensive test selection errors and you want to start the selection all over again. You can do the same by using the **REMOVE TEST** command button, but that would be very slow and tedious. This function resets the GUI to its state prior to the test name selection and display in the **SELECTED TEST SEQUENCE** box.

At line 11 of Figure 7–29, the `ResetSequenceFlag` is checked for non-zero value, and if the value is `1`, then the function jumps to line 42 and displays a message that there is nothing to reset. Recall that the *ResetSequenceFlag* is set to `0` in the *SelectTestsCB* function (Figure 7–25), indicating that there is a sequence of test names selected.

If the *ResetSequenceFlag* is `0`, the **SELECTED TEST SEQUENCE** box and the **TEST NAMES** box are cleared at lines 16 and 18 by the library function *ClearListCtrl*. Lines 20–21 set the `index` to `0` and create a `search_path` for the location of the test files with extension `*.tst` using the *Fmt* function. The *GetFirstFile* function at line 23 searches for the file attributes indicated and inserts the file name in the **TEST NAMES** box at 0-index location at line 25. Lines 28 to 38 search and insert the files in the **TEST NAMES** box in a *while* loop, each time incrementing the `index`. Lastly, at line 44, the `ResetSequenceFlag` is set for next time this routine is called.

CopyTestCB Function

The **CopyTestCB** callback function is invoked when the **COPY** command button is selected from the GUI shown in Figure 7–18. The source code listing for this callback function is shown in Figure 7–30. This function is used to add one or more copies of the selected test in the **SELECTED TEST SEQUENCE** box if necessary to repeat the same test.

As before the `count` value is obtained using *GetNumListItems* at line 11 of Figure 7–30 for the **SELECTED TEST SEQUENCE** box to check for one or more test names that can be copied. If there are no test names, then the program jumps to line 26 to display an error message that no test names can be copied, indicating that the **SELECTED TEST SEQUENCE** box is empty.

If one or more test names are in this box, then the program continues at line 15. Again, the `item_index`, `item_value`, and `item_label` of the items to be copied are obtained at lines 15, 17, and 21, respectively. The test name is copied below the last test name in the list box using the *InsertItemList* library function at line 23 and `Reset_Seq_Flag` is cleared at line 24.

```
1   //Makes duplicates of the selected test in the selection box, invoked from "COPY" button
2   int CVICALLBACK CopyTestCB (int panel, int control, int event,
3           void *callbackData, int eventData1, int eventData2)
4   {
5       int item_index, item_value, count;
6       char item_label[260];
7       switch (event)
8       {
9               case EVENT_COMMIT:
10              //Get the number of items in the Selected Sequence box
11              GetNumListItems(TestSeqHandle,TEST_TST_SQU,&count);
12              if (count !=0)
13  {
14                  //Get the index of the highlighted item in the Selected Test Sequence box
15                  GetCtrlIndex(TestSeqHandle, TEST_TST_SQU, &item_index);
16                  //Get the value from the index
17                  GetValueFromIndex(TestSeqHandle, TEST_TST_SQU, item_index,
18
19                                                                          &item_value);
20              //Get label from index
21              GetLabelFromIndex(TestSeqHandle,TEST_TST_SQU,item_index, item_label);
22              //Place the test name in the Selected Test Sequence box
23              InsertListItem(TestSeqHandle,TEST_TST_SQU, -1, item_label, item_value);
24                  Reset_Seq_Flag=0;   //Clear Sequence Flag
25              }
26              else
27                  //Error message
28                  MessagePopup ("COPY ERROR", "There is nothing to COPY");
29                  break;
30      }
31      return 0;
32  } // CopyTestCB
```

Figure 7–30
CopyTestCB Source Code

GoToTestExecCB Function

The *GoToTestExecCB* callback function runs the Test Executive menu. The source code listing for this callback function is shown in Figure 7–31. This function is called when the **TEST EXEC** command button is selected on the **TEST SEQUENCE SELECTION** GUI from Figure 7–18.

If the user has not saved the sequence before selecting the **TEST EXEC** button, a pop-up panel at line 11 of Figure 7–31 provides a reminder. The panel is saved on line 17 with the user-supplied name, and its status is checked at line 18. If for some reason the panel is not found, the pop-up panel on line 21 displays an error and the program terminates. The *Hide-Panel* library function at line 26 hides the **TEST SEQUENCE SELECTION** panel but leaves it in memory, and line 28 loads the **TEST EXECUTIVE** GUI

```
1  //Go to the Test Executive menu, invoked by "TEST EXEC" button
2  int CVICALLBACK GoToTestExecCB (int panel, int control, int event,
3                                  void *callbackData, int eventData1, int eventData2)
4  {
5      int status;
6      switch (event)
7      {
8          case EVENT_COMMIT:
9              if ((SaveSequenceFlag !=1) && (LoadSequenceFlag !=1 ) )
10             {
11                 MessagePopup ("SAVE SEQUENCE ERROR", "Save Sequence before
12                                                      running!");
13                 return 0;
14             }
15             //Save the Selected Test Sequence panel to be recalled later
16             status=SavePanelState(TestSeqHandle,save_Seq_file, stateIndex);   //Save Panel
17             if (status != 0)
18             {
19                 //Message if there is some error in saving the panel
20                 MessagePopup ("PANEL DISPLAY ERROR",
21                               "Panel cannot be recalled.\nExit and Check Test File.");
22                 return -999;
23             }
24             //Hide the Test Sequence GUI
25             HidePanel(TestSeqHandle);
26             //Load the Test Executive uir into memory
27             TEHandle = LoadPanel (0, "TestExecutive.uir", EXEC);
28             SetPanelPos (TEHandle, VAL_AUTO_CENTER, VAL_AUTO_CENTER);
29             DisplayPanel(TEHandle); //Display the Test Executive Panel
30             break;
31     }
32     return 0;
33 } //GoToTestExecCB
```

Figure 7-31
GoToTextExecCB Source Code

via the *LoadPanel* function and displays this panel using *DisplayPanel* at line 30. The Test Executive GUI is shown in Figure 7–19.

SelectTimesToRunCB Function

The source code listing for this callback function *SelectTimesToRun* is shown in Figure 7–32. The function verifies that the sequence is either loaded or previously saved and then displays a **RUN SEQUENCE** panel (Figure 7–21).

At line 9 of Figure 7–32, the function checks that `LoadSequenceFlag` or `SaveSequenceFlag` has been set from the *LoadSeqCB* or the *SaveSeqCB* callback functions (Figures 7–28 and 7–27 respectively). This check ensures that either the pre-existing sequence has been loaded or a new sequence of tests has been created and saved. If this is true, then lines 12–15 load and display the **RUN SEQUENCE** GUI as shown in Figure 7–21. If this is not true, the program executes line 20 to display a message via a pop-up panel that no sequence is loaded for running.

The library function *SetPanelPos* at line 13 positions the panel at the coordinates specified in the arguments. Here the panel is positioned at the center

```
1   //Loads the panel to select stop at first failure and the number of times to run the test
2   int CVICALLBACK SelectTimesToRunCB (int panel, int control, int event,
3           void *callbackData, int eventData1, int eventData2)
4   {
5       switch (event)
6       {
7           case EVENT_COMMIT:
8               //Sequence must be loaded or saved before running the test
9               if ( (LoadSequenceFlag) || (SaveSequenceFlag))
10              {
11                  //Load the Run Sequence GUI in memory
12                  RunHandle = LoadPanel (TEHandle, "Run_Seq.uir", SEQ_RUN);
13                  SetPanelPos (TEHandle, VAL_AUTO_CENTER,
14                                          VAL_AUTO_CENTER);
15                  DisplayPanel (RunHandle);   //Display the GUI
16              }
17              else
18                  //Error message
19                  MessagePopup ("SEQUENCE RUN ERROR", "Nothing is selected to
20                          run.\nSave Selection or Load Selection");
21          break;
22      }
23      return 0;
24  }//SelectTimesToRunCB
```

Figure 7–32
SelectTimesToRunCB Source Code

of your screen. The prototype for *SetPanelPos* function follows and its arguments are explained in Table 7–12.

```
int status = SetPanelPos (int panelHandle,
                          int panelTop, int panelLeft);
```

For the top-level panel, (0,0) is the upper-left corner of the screen. For the child-panel, (0,0) is the upper-left corner of the parent panel, directly below the title bar. To center the panel, set both arguments to VAL_AUTO_CENTER.

RunSequenceCB Function

The *RunSequenceCB* function runs the test sequence. This function is called when the **GO** command button is selected from the **RUN SEQUENCE** GUI shown in Figure 7–21. The source code listing for this callback function is shown in Figure 7–33.

Before the test sequence is run, the *RunSequenceCB* function gets the user inputs for the number of times the test sequence has to run and for whether the test should stop at the first failure or continue from the **RUN SEQUENCE** GUI in Figure 7–21. The file header and test data is written to an output data log file with the same base name as the test sequence file but appended by `.log`

Table 7–12 SetPanelPos *Function*

Input/Output	Name	Type	Description
Input	panelHandle	integer	the panel handle of the user interface
	panelTop	integer	the number from the top at which the upper left corner of the panel is placed, directly below the title bar
	panelLeft	integer	the number from the left at which the upper left corner of the panel is placed, directly below the title bar
Output	status	integer	refer to Appendix A of *LabWindows/CVI User Interface Library Reference Manual* for error codes or in Online Help; non-zero value creates an error

```c
//Runs the test sequence, invoked from the "GO" button
int CVICALLBACK RunSequenceCB (int panel, int control, int event,
    void *callbackData, int eventData1, int eventData2)
{
    int StopOnFirstFail, SearchLoc1, SearchLoc2, FirstFailureIndex, FirstFailure=0;
    short NumberTimesToRun;
    int item_index, select_index=0, item_value, count,i, return_value,status, PassFail, red, white,blue,
    black, green;
    unsigned short TestNum;
    char item_label[260], auto_seq_file_name[260],SeqName[260], DataFile[256], FileName[260],
        Operator[30], UUTSerial[30], Result[30], file_buf[500], drive[5], directory[260];
    switch (event)
    {
        case EVENT_COMMIT:

            //Create a file to open for writing to "output" directory with same name
            //as the sequence file name with "log" extension
            GetCtrlVal(TestSeqHandle,TEST_SEQ_NAME,SeqName);

            /*******************************Extract File name from path ******************/
            //Find the file name without the extension from the complete path name
            SearchLoc1 = FindPattern (SeqName, 0, -1, "\\", 0, 1);
            SearchLoc2 = FindPattern (SeqName,SearchLoc1+1 , -1, ".", 0, 1);
            //File name is saved in DataFile string
            CopyString (DataFile, 0, SeqName, SearchLoc1+1, SearchLoc2-SearchLoc1-1);

/*************Alternate and simpler method to extract file name from path ********/

            /*** Split the path name into drive, directory and file name
            SplitPath (SeqName, drive, directory, FileName);
            SearchLoc1 = FindPattern (FileName, 0, -1, ".", 0, 1);     //Get location of "."
            CopyString (DataFile, 0, FileName, 0, SearchLoc1-1);
****************************************************/
            // Append ".log" to the file name
            Fmt(SeqName, "%s<%s\\output\\%s.log", prj_dir, DataFile);

            //Open the file for writing only
            OutputFileHandle = OpenFile (SeqName, VAL_WRITE_ONLY,
                                        VAL_TRUNCATE, VAL_ASCII);

            //Write the file header
            Fmt(buf, "%s<Sequence file name run: %s\n\n", SeqName);
            WriteFile(OutputFileHandle, buf, StringLength(buf));
            //Get the Operator Name, Test number and the UUT Serial number from the GUI
            GetCtrlVal(TEHandle, EXEC_OPER_ID,Operator);
            GetCtrlVal(TEHandle,EXEC_TEST_NBR,&TestNum);
```

```
46      GetCtrlVal(TEHandle,EXEC_UUT_NUM,UUTSerial);
47      //Create a buffer with the above elements
48      Fmt(buf, "%s<Operator Name: %s  Test Number= %d  UUT Serial #=
49                       %s\n", Operator,   TestNum, UUTSerial);
50      //Write this data to the file
51      WriteFile(OutputFileHandle, buf, StringLength(buf));
52
53      //Check if Stop on first fail is selected
54      GetCtrlIndex(RunHandle, SEQ_RUN_FAILURE,&StopOnFirstFail);
55      //Set flag to stop test at first failure
56      if (StopOnFirstFail==0) FirstFailureFlag=1;
57      //Get the number of times to run the sequence
58      GetCtrlVal(RunHandle, SEQ_RUN_TST_NBR,&NumberTimesToRun);
59
60      red   = MakeColor (255, 0,   0);       //red color
61      green = MakeColor (0,255, 0);          //green color
62      blue= MakeColor(0,0,255);              //blue color
63      black = MakeColor (0,0,0);             //black color
64      white = MakeColor (255, 255, 255);  //white color
65
66      //Hide the panels
67      HidePanel (TestSeqHandle);
68      HidePanel(RunHandle);
69
70      //Create the header for test run
71      sprintf(buf, "            TEST NAME            RESULTS");
72      InsertListItem (TEHandle, EXEC_TE_DISP, -1, buf, 0);
73      strcat(buf, "\n");//Add new line to write to file
74      WriteFile(OutputFileHandle, buf, StringLength(buf));
75
76      /       Run the test sequence the number of times selected
77      for ( i=1; i<= NumberTimesToRun;i++)
78          {
79              select_index=0;
80              SetCtrlVal(TEHandle, EXEC_TE_DISP, i); //Display the sequence
                                                                   number
81
82              //Obtain the number of items in the sequence list
83              GetNumListItems(TestSeqHandle,TEST_TST_SQU,&count);
84              while (select_index < count) //Do for all items in the list
85                  {
86                      //Assign random results to tests
87                      PassFail=rand % 3;    // Value can be either 1 or 2
88
```

Figure 7-33
RunSequenceCB Source Code

```c
         DefaultCtrl(TEHandle, EXEC_FILENAME);   //Clear File Name

         //Getlabel from index
         GetLabelFromIndex(TestSeqHandle,TEST_TST_SQU,
                              select_index, item_label);

         //Create the full path name of the file name
         sprintf(auto_seq_file_name, "%s\\input\\%s", prj_dir,
                                     item_label);

         if (PassFail==1)   //Assign 1 to Pass arbitrarily and 2 to Fail
         {
               sprintf(Result, "PASS");
                    //Display the string and write to file
               sprintf(buf,
"\033bg%06X\033fg%06X\033vline\033p180r%s\033fg%06X\033vline\033p300r%s\033fg%06X\033vline
",
                white, blue, item_label, black,  green, Result, black );

               SetCtrlVal(TEHandle, EXEC_FAIL_IND, 0);   //Set LED to green
         }
         else //Fail
         {

                sprintf(Result, "FAIL");
                sprintf(buf,
"\033bg%06X\033fg%06X\033p180r%s\033fg%06X\033vline\033fg%06X\033p300r%s\033fg%06X\033vline
",  white, blue, item_label, black, red, Result, black );
               SetCtrlVal(TEHandle, EXEC_FAIL_IND, 1);    //Set LED to turn red
         }

         InsertListItem (TEHandle, EXEC_TE_DISP, -1, buf, 0);
         //Add linefeed to the buffer for writing to file
         sprintf(file_buf, "%30s!%10s", item_label, Result);//string to write to file
         strcat(file_buf, "\n");
         WriteFile(OutputFileHandle, file_buf, StringLength(file_buf));
         ProcessSystemEvents; //Process everything before going ahead

         //For stopping at first failure
         if( (FirstFailureFlag==1) && (CompareStrings (Result, 0, "FAIL", 0,
                                             1)==0))
         {
```

```
130                    MessagePopup ("FIRST FAILURE", "Found 1st Failure.
131                                                    Aborting!");
132                    return 0;
133            }
134
135            //Increment the test number
136            SetCtrlVal(TEHandle, EXEC_TE_DISP ,select_index+1 );
137
138            //Load up the panel,but don't display
139            RecallPanelState(TestSeqHandle, save_Seq_file,stateIndex);
140            DefaultCtrl( TEHandle,EXEC_FILENAME);
141            //Write the file name to control
142            SetCtrlVal(TEHandle, EXEC_FILENAME, auto_seq_file_name);
143            //Number of list items
144            GetNumListItems(TestSeqHandle,TEST_TST_SQU,&count);
145            select_index++;
146            //Delay 2 seconds between each test to see the LED change states
147            Delay(2.0);
148
149    }//while (count !=0)
150
151    Fmt(buf,"%s< %d Sequence Tests Completed", i);
152    InsertListItem (TEHandle, EXEC_TE_DISP, -1, buf, 0);    //Display
153    strcat(buf, "\n\n");
154    WriteFile(OutputFileHandle, buf, StringLength(buf));//write to file
155    InsertListItem (TEHandle, EXEC_TE_DISP, -1, " ", 0);   //Insert blank line
156
157    }// for ( i=1; i<= Num_Loops;i++)
158    Fmt(buf,"%s< ========= All Test Sequence Completed ==========");
159    InsertListItem (TEHandle, EXEC_TE_DISP, -1, buf, 0); //Display on screen
160    strcat(buf, "\n");
161    WriteFile(OutputFileHandle, buf, StringLength(buf)); //write to file
162    CloseFile(OutputFileHandle); //Close the output data file
163    Reset_Seq_Flag=0; //Clear Reset flag
164
165    break;
166    }
167
168    FirstFailureFlag=0;    //Reset first failure flag
169    return 0;
170  } // RunSequenceCB
```

Figure 7-33
RunSequenceCB Source Code (*continued*)

extension. Random numbers are generated to simulate a pass or fail. If the test fails, the LED control is switched to red; otherwise it is green to indicate pass. The test results are displayed with a header on top of the **TEST DISPLAY** box and in different colors for pass or fail. The results are displayed inside the columns as you saw in the "List Box Example 7.2" section.

Line 18 of the code shown in Figure 7–33 gets the name of the sequence file name from the **Sequence Name** control box on the **TEST SEQUENCE SELECTION** panel. Lines 22–25 perform the string manipulation to extract the base file name from the complete sequence path.

The library function *FindPattern* is called at lines 22 and 23. This function searches `SeqName` for the pattern and returns the zero-based location of the pattern in the integer variable `SearchLoc1`. The function prototype follows and the function arguments are explained in Table 7–13.

```
int SearchLoc1 = FindPattern(char SeqName, int startingIndex,
int NumberOfBytes,  char * pattern, int CaseSensitive, int startFromRight);
```

Using this function the first time, the search is for "\\" from the right side of the string since you know the file name in the path will start with this pattern. For example, if the path is

```
c:\testexec\input\test1.seq
```

in `SeqName`, `SearchLoc1` will contain the index of the pattern \ as searched from the right. In this case the `SearchLoc1` will be 17. This search pattern needs some explanation.

Note: The first \ in "\\" is the escape character and the string is searched for \.

The search index is zero-based, counting the index always from the left even though the search can be started from the right. In the above case if the search were from the left, the first place the string \ would be found is at index value 2. This is because the first \ is found after `c:\`, which makes it a zero-based index of two.

The next *FindPattern* library function at line 22 starts the search at `SearchLoc1` for the index of the pattern ".", and here the `SearchLoc2` value is 23.

Using the library function *CopyString* at line 25, the extracted name of the base file is copied into `DataFile`. The value of `SearchLoc1+1` is used to start copying from `SeqName`, and the number of bytes to copy is `SearchLoc2 – SearchLoc1-1`. Thus, the number of bytes to copy is

$$23 - 17 - 1 = 5$$

Table 7–13 FindPattern *Function*

Input/Output	Name	Type	Description
Input	SeqName	string	the string to search for the pattern
	startingIndex	integer	The position in the search string from where to start the search
	NumberOfBytes	integer	the number of bytes to search; –1 searches the complete string
	pattern	string	the exact pattern to search (within quotes)
	Case Sensitive	integer	case-sensitive mode of search: 0 = Not case-sensitive, Non-zero = case-sensitive only
	startFromRight	integer	indicates whether to start search from left or right of the string and finds the first pattern that meets the conditions set in the other attributes; Zero = start search from left of string, Non-zero = start search from right of string
Output	SearchLoc1	integer	the index in the search in which the pattern is first found; a –1 indicates no pattern is found in the string

and this will copy the five characters of `test1` into the string `DataFile`.

There is a simpler method for extracting the file name from the path name: using the library function *SplitPath* as shown on line 30 and manipulating the string. This function parses the path name into the drive letter, the directory, and the file name. The use of the *FindPattern* function may not be appropriate for searching the file name since the *SplitPath* function performs this task explicitly. The *FindPattern* function was introduced here mainly to show you how to search for string patterns within another string. The prototype of the *SplitPath* function follows and its arguments are described in Table 7–14.

```
int SplitPath ( char pathName, char driveName, char directoryName,
                                                char fileName);
```

Table 7–14 SplitPath *Function*

Input/ Output	Name	Type	Description
Input	pathName	string	pathname to split
Output	driveName	integer	extracted drive name
	directoryName	integer	extracted directory name
	fileName	string	extracted file name

On the *Unix* system, `driveName` is an empty string since there are no drive names.

At line 35, the output data log file path is created using the `DataFile` name with a `.log` extension for

`project directory\output`

in the string `SeqName`. Now you have the complete path name for opening a file for writing the test data.

The file is opened at line 38 and the handle assigned to the open file. At lines, 41 and 42, the sequence file name is written to the output data file at the top of the file. In the next three lines, the data from the **TEST EXECUTIVE** GUI is obtained for the **Operator Name, Test Number,** and the **UUT Serial #.** This data is written in `buf` using the *Fmt* library function. The same buffer is also written to the data log file as a header at line 51.

The control index from the *Ring* in the **RUN SEQUENCE** GUI is obtained to determine if **Stop on First Failure?** is **Yes** or **No.** If it is **Yes,** then the `FirstFailureFlag` is set to 1 at line 56. At line 58, the **Number of times to run the Sequence** is assigned to the variable `NumberTimesToRun`. Lines 67 and 68 hide the **RUN SEQUENCE** and **TEST SEQUENCE SELECTION** panels.

At lines 71 and 72, the header for the test results in the **TEST DISPLAY** box is displayed. Every time there is data shown on the display, the same data must be written to the data file. However in the library function *InsertListItem,* an automatic line feed occurs every time the line of data is output to the monitor. This automatic linefeed does not occur when writing to a file, so you have to be sure that the data line written to the file has a new line character added to the end of the line. This explains the code at lines 41, 48, 73, 122, 153, and 160.

The test sequence is run using two nested loops. At line 77, a *for* loop runs the entire test sequence according to the value of `NumberTimesToRun`. The *while* loop is nested within this *for* loop at line 84 and gets the number of test names in the **SELECTED TEST SEQUENCE** box and loops that many times.

At line 87 for the simulation of pass or fail, the random number generator is used with *modulus* 3 to obtain the value 1 or 2. If the value is 1, **Pass** is assigned to `PassFail`; otherwise **Fail** is assigned. This is just our selection, and you could reverse the values for **Pass** and **Fail** if you wanted. At line 101, the "PASS" string is copied to the **Result** string, and at line 112, the "FAIL" string is copied. The color of the LED is changed appropriately depending on Pass or Fail.

At lines 103–106, and 113–115, the display buffer is created to display the results in color and to insert vertical lines to create a columnar appearance. You learned how to do this in Exercise 7–2. These strings are created in two places: one to display PASS and the other to show FAIL. At line 119, the string is inserted in the **TEST DISPLAY** list box.

The **File Name** is cleared from the **TEST EXECUTIVE** user interface on line 140. The file name is obtained from the **TEST SEQUENCE SELECTION** panel and displayed on the **File Name** control box on line 142.

The *ProcessSystemEvents* library function at line 124 processes all pending systems events. This function is used here to display the data on the monitor as it is executed. Without the use of this function, the data is displayed only after the remaining events have been executed. The *LabWindows/CVI User Interface Reference Manual* cautions you to use this command with care as it can cause other execution problems. See the manual for more details on this function. The *ProcessSystemEvents* function was explained in Chapter 2, "Basics of Creating the Graphical User Interface."

Lines 127 through 133 determine if the `FirstFailureFlag` is set. If the flag is set and the `Result` is "Fail", a pop-up message is displayed indicating that the first failure is found and the program will be aborted.

The *CompareStrings* library function at line 127 compares the two strings with "FAIL" in the string `Result`. Let us look at the prototype for the *CompareStrings* library function. The function arguments are explained in Table 7–15.

```
int result = CompareStrings (char *FirstString, int FirstStringIndex,
      char *SecondString, int SecondStringIndex, int CaseSensitive);
```

The test number is incremented, the test name displayed on the **TEST EXECUTIVE** panel, and the sequence panel is recalled but not displayed

Table 7–15 CompareStrings *Function*

Input/Output	Name	Type	Description
Input	FirstString	string	the first string to compare
	FirstString Index	integer	starting position in first string
	Second String	string	the second string to compare
	Second String Index	string	starting position in second string
	Case Sensitive	integer	case-sensitive mode of search: 0 = Not case-sensitive, non-zero = case-sensitive only
Output	result	integer	the result of comparison as indicated below: –1 = bytes from *FirstString* are less than bytes from *SecondString*, 0 = both strings are identical, 1 = bytes from *FirstString* are greater than bytes from *SecondString*

at line 139. Get the remaining number of list items in the variable `count` at line 144. Increment `select_index` at line 145 for comparison in the *while* loop.

A delay of two seconds is inserted at line 147, between each test in the *while* loop, so the LED indicator switching color can be visible. Otherwise it will switch the Pass/Fail LED very quickly, and you will not be able to see the LED change color between green and red.

The prototype for the library function *Delay* (double `NumberOfSeconds`) library function has one argument, indicating the delay time in seconds of type **double.**

After one sequence of tests is run, a message is displayed and written to the file declaring that all tests in this loop are completed. If `NumberTimesToRun` is more than `1`, the *for* loop repeats the whole sequence over again until all test repetitions are completed and the message is displayed and written to the file. At line 162, the file is closed and the function returns.

ExitCB Function

This function exits the program when the **EXIT** command button is selected. It calls the *QuitUserInterface* library function, which you have seen in the previously.

ClearListBoxCB Function

This function is used when the **CLEAR** button on the **TEST EXECUTIVE** panel is selected. This function uses the *ClearListCtrl* library function to clear all the data from **TEST DISPLAY** box.

ReturnCB Function

This function is invoked when the **RETURN** command button is selected on the **TEST EXECUTIVE** panel. This function hides the **TEST EXECUTIVE** panel and brings up the **TEST SEQUENCE SELECTION** panel. An error message is displayed if the panel is not found.

CheckFileName Function

The *CheckFileName* function is described in Chapter 6, "File Input/Output."

ExplorerCB Function

This function is invoked when the **EXPLORER** command button is selected from the Test Executive GUI. Windows Explorer is called from within *CVI*. The *ExplorerCB* function calls the library function *LaunchExecutable,* which contains the path name of the executable file and the path of the output directory in the project directory. This enables the user to go directly into the output directory instead of moving from the root directory to the desired directory. Including Windows Explorer in this project facilitates viewing the output data file once the test is run. If the results are not acceptable, the test can be run again without exiting the project. The source code listing for *ExplorerCB* is shown in Figure 7–34.

 Any executable application can be launched using the *LaunchExecutable* function, which runs the executable and returns without waiting for the application to exit. You can use the ANSI C library function *system* to run the

```
1   //Run the Windows Explorer, invoked from "EXPLORER" button
2   int CVICALLBACK ExplorerCB (int panel, int control, int event,
3           void *callbackData, int eventData1, int eventData2)
4   {
5       char explore[260], ExplorePath[260];
6       int result, firstindex;
7       switch (event)
8               {
9               case EVENT_COMMIT:
10                      //Launch the Explorer
11                      sprintf( explore, "c:\\Windows\\Explorer.exe %s\\output", prj_dir);
12                      result=LaunchExecutable(explore);
13                      while(result !=0)
14                      {
15                              PromptPopup ("FILE NOT FOUND",
16                              "'Windows Explorer' could not be found.\nPlease enter the full
17                                                      pathname.", ExplorePath, 260);
18                              firstindex=FindPattern (ExplorePath, 0, -1, ":\\", 0, 0);
19
20                      }
21                      break;
22              }
23      return 0;
24  } //ExplorerCB
```

Figure 7–34
ExplorerCB Source Code

executable that waits for the program to exit. In this example, you could use either of these functions.

An enhanced version of the Test Executive demo written using the powerful features of *CVI* is discussed in Appendix E, "CVI Demo Programs." A limited version of this demo program has been included on the CD-ROM with this book to show you the many capabilities of *CVI* programming.

Summary

This chapter showed you how to use *List Boxes* and the associated library functions. Two examples were introduced to give you an idea of the basic features of the *List Boxes*, including moving items from one list box to another, formatting the display by using vertical lines, using different colors to display the data, and checking and clearing list box items. Various *Ring* controls were also introduced and you were shown how to create and use them.

Though they were not covered in detail, some functions of the basic *Ring* controls were described.

The concepts discussed in this chapter and shown in the examples were used to develop the **Test Executive** project. Three user interface panels were created to run the project, and the ways to call and load them on other user interfaces was demonstrated. The method of saving and recalling the panels was introduced in a practical situation. The **Test Executive** project showed you how executables could be called from within the *CVI* environment. You were shown how the results of the tests are logged to a data file with the header information and saved to the `output` folder.

CREATING STAND-ALONE EXECUTABLES AND DISTRIBUTION DISKS

Chapter Highlights

- Creating Stand-Alone Executables and Distribution Disks Overview
- Creating Stand-Alone Executables
- Creating Distribution Disks
- Error Handling for Stand-Alone Executables
- Summary

This chapter describes the systematic procedures for creating the stand-alone executables and distribution disks for *CVI* projects. The procedure of creating distribution disks is explained to enable you to install and run the application on a system devoid of the *CVI* environment. The error handling process for stand-alone executables is discussed to enable the program to abort gracefully when encountering program error(s) without crashing.

Creating Stand-Alone Executables and Distribution Disks Overview

In the previous chapters, you learned how to create and run projects using *CVI*. The projects are executed on your machine, on which *CVI* is installed. By creating the stand-alone executables, you can create projects on your computer and execute them on a different computer (called the target computer), one that lacks *CVI*. You can create distribution disks of your entire project on floppies, take them to the target machine, and install the project on that machine by just clicking on the **Setup** icon on the installation disk. This may become necessary, for example, when you are creating software on one computer and you need to test it with the hardware or instrumentation in the laboratory or at a remote location. This is also required when you want to distribute a software application for commercial purposes (check National Instruments' licensing agreement for the distribution disks licensing agreement).

Before we go into the mechanics of creating a stand-alone executable, let us discuss what making a stand-alone executable file entails. To create a stand-alone executable to run on machines that do not have the *CVI* environment, you need the Run-Time Engine (RTE), which is distributed with *CVI* on the CD-ROM. The Run-Time Engine is automatically included with the distribution kit when you create the distribution disks (unless you choose not to).

The distribution kit includes all the files necessary to run the stand-alone executable except National Instruments' DLLs and the files loaded by *LoadExternalModule*. Let us examine the files included on the distribution disk and their functions. The executable files consist of the compiled, linked files along with instrument driver files linked to the project. The application name and icon resource is on the disk and is registered by the operating system when installed on the target machine.

The Run-Time Engine contains an execute-only version of the *CVI* environment, without the development tools: **Source Code Editor** and the **User Interface Editor;** you cannot use these tools to modify the executable file anyway. If the complete *CVI* environment is installed, the files will become excessively large. The Run-Time Engine is smaller and runs faster than the complete *CVI* environment. Note that one Run-Time Engine installation is sufficient to run all the *CVI* executable files on a target machine.

The stand-alone executable must contain the user interface files, header files, and any graphical image files that are to be displayed on the GUI. Also, the panel state files (if any) that are saved and recalled using the *CVI* functions *SavePanelState* and *RecallPanelState* must be included on the stand-alone executable disks. You must also include the DLL files that are used by the application program, the external .lib and external .obj files loaded by the *LoadExternalModule* function but are not part of the project. Also, any other files that are opened by the project using the file open functions and any data files that your project may use during execution must be included. The files must be in the correct directories on the target machine so the executable program can access them.

Creating Stand-Alone Executables

Let us create a stand-alone executable for project6. Load project6 and select **Build>>Configuration>>Release** to create a **Release** executable version without the debugger options or select **Debug** if you want to include the debugging capability in the executable. The options are shown in Figure 8–1.

Select **Target Settings...** to select the various options for creating the stand-alone executable. The dialog box shown in Figure 8–2 is displayed.

Refer to the **Build Menu** section of Appendix B, "Project Window Environment" for a description of the various options available from this menu.

Set **Release** in the **Application File** box. The **Application File** path name, **Application Title,** and **Application Icon File** are entered into the dialog

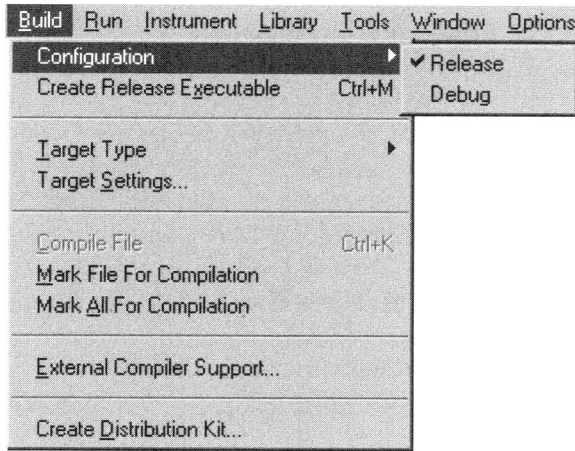

Figure 8–1
Set Configuration Menu

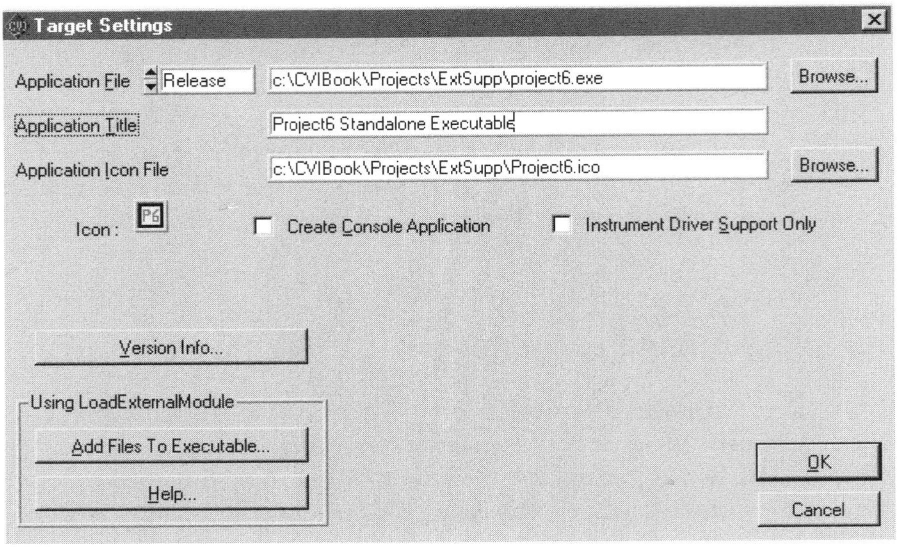

Figure 8–2
Target Settings Dialog Box

boxes for this project as shown in Figure 8–2. The icon for this project is created using the **Icon Editor** as explained in Chapter 4, "Enhancing the User Interface." You can select these files using the **Browse...** command button.

The **Create Console Application** box, if left unchecked, creates your executable as a Windows GUI application. If this box is checked, your executable's output/input is re-directed to/from a Windows console window (Command Prompt or MS-DOS Prompt).

When the **Instrument Driver Support Only** box is checked, your project uses a smaller set of *CVI* libraries. This is particularly useful for creating instrument driver DLLs. For more details, see **Target Setting...** under the **Build** menu in the *LabWindows/CVI User Manual*.

Selecting **OK** will close this dialog box.

Select **Build>>Create Release Executable** from the menu shown in Figure 8–1. The stand-alone executable, project6.exe, is created in the specified pathname using the icon shown above.

Note: If you select the **Release** option, a message will be displayed indicating that the project6.exe file is being built, and later a message will indicate that the build is complete. If you selected the **Debug** option, you will get the same build messages but the executable file name will be `project6_dbg.exe`. These files are added to your project folder. To execute the program in its configured environment (*Release* or *Debug*), you can double-click on either of these files.

Creating Distribution Disks

The **Create Distribution Kit...** option shown in Figure 8–1 is available in Windows only. For *CVI* users in *Unix* environment, the procedure for creating the distribution disks is entirely different. It is best to read the instructions in the "*Unix Compiler/Linker Issues*" section of the *LabWindows/CVI Programmer Reference Manual*. Note that *Unix* is supported on versions earlier than 5.5.

Before creating the distribution kit, the stand-alone executable must be created. Otherwise *CVI* will prompt you to do so. You need to select **Build>>Target Settings,** as in Figure 8–2, and set the appropriate options. As mentioned previously, the distribution disks are useful for loading and running the executable on a target machine that lacks *CVI*. The **Create Distribution Kit...** installs all the files required to run the executable program on another machine except the DLLs for National Instruments hardware and the files loaded with *LoadExternalModule*. The DLL files for the National Instruments hardware can be loaded on the target machine independently of the executable disks. The *LoadExternalModule* files are entered manually using the **Add Group/Edit Group** buttons on the **Create Distribution Kit** dialog box explained here.

To create the distribution kit, select **Build>>Create Distribution Kit...** from the **Project** window, and the dialog box appears as shown in Figure 8–3.

The **Installation Directory** is the default folder (directory) for installing this project on the target computer. In this case, a folder (directory) will be created, `c:\TestExec`, and all the project files will be installed in this directory.

The **Target Path** is the folder (directory) location for building the distribution disk files on your computer. You can build the distribution disks by choosing the floppy drive, but you must include the path to the root directory, e.g., `A:\` where `A` is the floppy drive letter. You can also build the distribution disks on your hard disk and later copy them on to the floppies. When you start to build the stand-alone executable on the floppy disks, the program informs you of the number of formatted floppies required. These floppies must not contain any data.

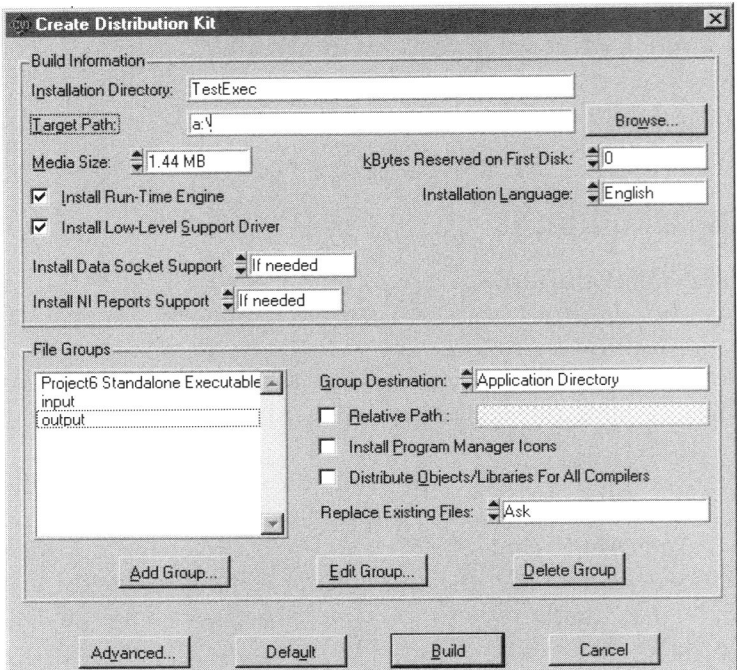

Figure 8–3
Create Distribution Disk Dialog Box

When you selected the **Target Path** to be the floppy disk, the capacity of the floppies is selected automatically for the **Media Size.** The **Media Size** box is disabled and grayed.

You can enter the **kBytes Reserved on First Disk:** if you would like to reserve some space on the first disk of the distribution kit to add extra files. There may be situations in which you would like the user to read the *readme* file before starting the installation. You can reserve the size of the *Readme* file on the disk and copy it onto the disk after the build is complete.

The **Browse** button enables you to select the target path for the distribution disks if you want to create them in a separate folder and later transfer them to the floppies.

The **Install Run-Time Engine** box should be checked if you know that the run-time engine is not already loaded on the target machine so the the run-time engine and the associated files will be included on the distribution kit.

If you already have a run-time engine loaded on that machine, you do not need to do so again.

The **Install Low-Level Support Driver** box will appear for Windows only. Checking this box lets you install the low-level support drivers on the distribution kit. These drivers are required by the *CVI* Utility Library and are loaded automatically at the startup of the executable program. These should be included, unless you are sure that your project does not use these functions. For a list of the Utility Library functions that require the low-level driver functions, see **Create Distribution Kit...** in the *LabWindows/CVI User Manual*.

Install DataSocket Support

Install DataSocket Support should be selected if your application is going to use *CVI* DataSocket Library functions. You have a choice of selecting from **If Needed, Always,** or **Never** when you click on the box next to **Install DataSocket Support.**

Install NI Reports Support

Your application may use *NI-Reports* drivers for report generation features. You can select **If Needed**, **Always,** or **Never** accordingly when you click on the box next to **Install NI Reports Support.**

Both the **Install DataSocket Support** and **Install NI Reports Support** options were not available in *CVI* versions prior to 5.5.

You can set your choice of language for the installation process by selecting from the **Installation Language:** box, which is especially helpful if you are sending the distribution kit to non-English-speaking areas.

In **File Groups,** enter the files that belong together in a group (folder). These files are installed on the distribution kit in the appropriate group and copied in the same folder(s) on the target computer. In this example, there is an `input` folder (directory) with all the test files and an `output` folder (directory) where the output data files are logged. In the distribution disk installation, the same structure is required or the executable program will not find the input files to run the tests nor the `output` folder (directory) for writing the data files.

When the **Create Distribution Kit** dialog box is loaded, the **File Groups** box contains the `project6` files but no other files or folders. Since for `project6` you need the `input` and `output` folders, click on the **Add Group...**

Chapter 8 • Creating Stand-Alone Executables and Distribution Disks

button, and a dialog box will ask you to "Enter name of new group." Enter input the first time and enter output the next time. A window with the directories and file names appears from which you can select the file names to include with the distribution kit. You should select all the files that are required to run your application. Select **Edit Group...** and the **Edit File Group** dialog box is displayed as shown in Figure 8–4.

In this box, you need to group all the files that are in the input folder (directory). Click on the input folder (directory) in the top box, highlight all the files, and click on the **Add** button. Select **OK**. All these files will become part of the input folder (directory) when installed on the target computer.

In the window shown in Figure 8–3, check the **Relative Path** box and enter input, which provides the full path name of the input directory. Do not check the **Relative Path** for the project6 Files because that is specified as TestExec in the **Installation Directory.** The path name for input folder (directory) files will be

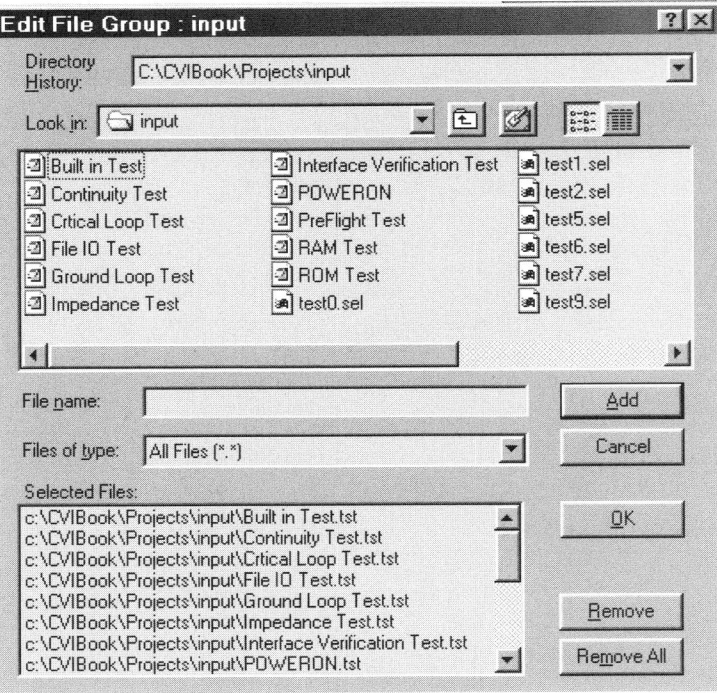

Figure 8–4
Edit File Group Dialog Box

```
c:\TextExec\input
```

Similarly, create the group for the `output` folder (directory) by selecting those files and setting the **Relative Path** to `output`. The pathname for `output` folder (directory) becomes

```
c:\TestExec\output
```

Note: You do not have to add all the output files, as they are created when you run the application. To create the `output` folder (directory), the **Edit File Group** box requires at least one file name in this directory. You can add a dummy file in the directory and add it to the project, or you can use one of the pre-existing data files.

When checked, the **Install Program Manager Icons** option on the **Create Distribution Disk** dialog box (Figure 8–3) creates the icons for the project in the Windows program group. Icon files that can be installed have `.pif`, `.com`, `.txt`, `.wri`, `.bat`, and `.hlp` extensions.

When you installed the *CVI* program, you selected one of the C/C++ compilers from the choices shown in Appendix A:

- Borland C++ 4.51 or higher, Borland C++ Builder 4.0
- Microsoft Visual C++ 2.2 or higher
- Symantec C++ 7.2 or higher
- Watcom C++ 10.5 or higher

If you decide to use a different compiler than the one selected during the installation, you can later re-install *CVI* with the new compiler choice.

When creating a stand-alone executable, you should include the files for all the compatible compilers on the distribution disks by checking the **Distribute Objects/Libraries for All Compilers** box. This feature is useful if you are distributing modules for use with *CVI* or for use by an external compiler. The user is able to obtain the modules for the compatible compiler on the target machine when the user is prompted by the installation program to select one of the external compilers. Checking this box will place the object files, static libraries, and DLL import libraries for all the above compatible compilers in their appropriate directories. *CVI* will create versions for every file in the distribution kit in their respective compiler directories: `msvc`, `borland`, `watcom`, and `symantec`.

If you are distributing the executable file without any external compiler modules, do not check the box next to **Distribute Objects/Libraries for All Compilers.**

Caution: Do not use **Distribute Objects/Libraries for All Compilers** for distributing DLL import libraries for VXI*plug&play* instrument drivers. You have to install two import libraries in the `msc` sub-folder (directory) for Microsoft Visual C/C++ and in `bc` sub-folder (directory) for Borland C/C++.

The **Replace Existing Files** box gives you a choice between the following options: **Never**, **Ask**, **Replace**, **Replace if Newer**, **Always**, and **Check Version**. Select the appropriate choice. It is better to select **Ask** to protect your files from being accidentally replaced.

The **Delete Group** button removes the **File Group** name from the dialog box.

The lower buttons on this dialog box are **Advanced...**, **Default**, **Build**, and **Cancel**.

The **Advanced...** button brings up the **Advanced Distribution Kit Options** dialog box as shown in Figure 8–5.

The top group on this dialog box is **Installation Script File,** in which you can check the box next to **Use Custom Script** if you want to use a customized script file for the distribution kit. Otherwise, you can use the default file name `c:\cvi5\bin\template.inf`, or select another script file using the **Browse** button. Go to the folder (directory) `c:\cvi5\bin` and select from the two customized script files as shown in Figure 8–6. If you are installing the VXI*plug&play*, select the script file `c:\cvi5\bin\vxipnp.inf`, which will include special instructions for VXI*plug&play* installation, which you can read in the `c:\cvi5\bin\vxipnp.doc` file.

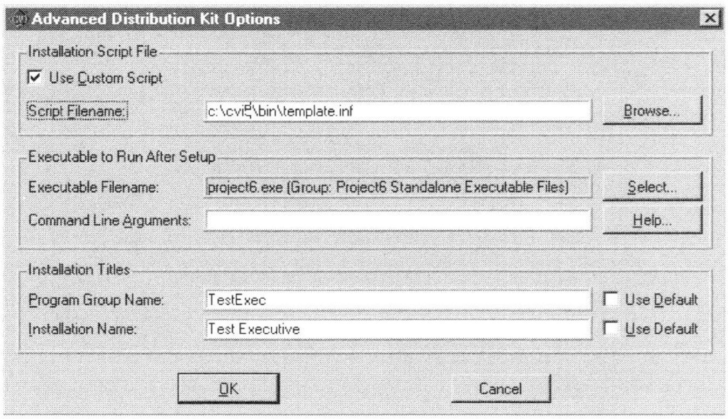

Figure 8–5
Advanced Distribution Kit Options

Chapter 8 • Creating Stand-Alone Executables and Distribution Disks

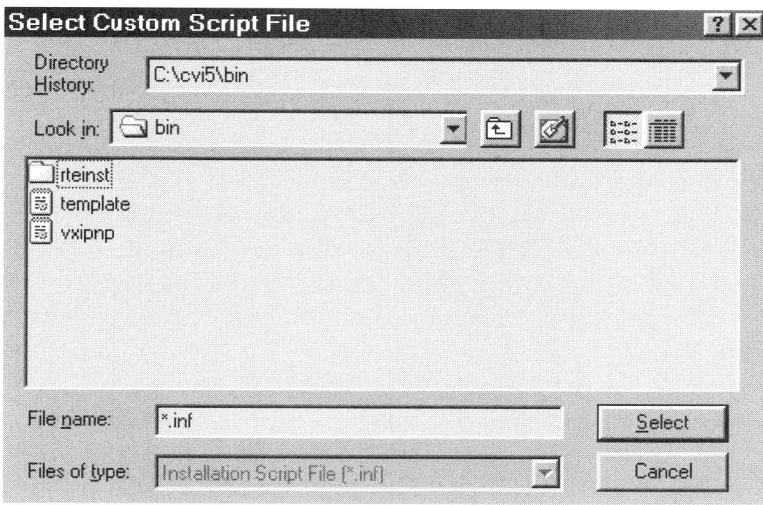

Figure 8–6
Select Custom Script File Dialog Box

The **Executable to Run After Setup** group in the middle section on the **Advanced Distribution Kit Options** dialog box contains two dialog boxes: the **Executable Filename** box and the **Command Line Arguments** box. In the **Executable Filename** box, enter the name of the file that you will want to run immediately after the completion of the installation. Using the **Select…** button can help you enter the file name in this box. If you want to run the executable as soon as the installation is complete, enter the executable file name here. This is useful if you want the project to start executing without letting the user select the execution file name. You may or may not want to use this feature, but it is there as an option.

In the **Command Line Arguments** box, enter the arguments that you need to pass to the executable file after completing the installation. This is required if your *main* function requires the *argc* and *argv[ ]* arguments. For more details on the command line arguments, select the **Help** menu next to this box.

The section at the bottom of this dialog box is **Installation Titles,** in which you can enter the titles for the **Program Group Name** and the **Installation Name.** TestExec and Test Executive, respectively, are chosen here for the installation titles as shown in Figure 8–5. If you do not wish to enter any names, check on the **Use Default** dialog boxes and *CVI* will use the name of the executable file.

When done, select **OK** from the **Advanced Distribution Kit Options** in Figure 8–5, and select **Build** from the **Create Distribution Kit** dialog box (Figure 8–3) to start creating the **Distribution Kit.** The build progress panel is shown in Figure 8–7.

The **Default** button in Figure 8–3 changes all the controls on the **Create Distribution Kit** dialog box to the default values. To return the values of the controls prior to selecting the **Default** button, select **Cancel.** The **Cancel** button also stops the **Create Distribution Kit** operation.

If you are creating and copying the distribution disks to floppies, you are prompted to insert the disks, as shown in Figure 8–8; otherwise messages for the build are displayed.

When the build is complete, you will get a message indicating so and including the location of the files on the hard disk or the floppy.

Take the disks to the target computer and click on the setup file. The installation will start as shown in Figure 8–9. Once the program is installed on

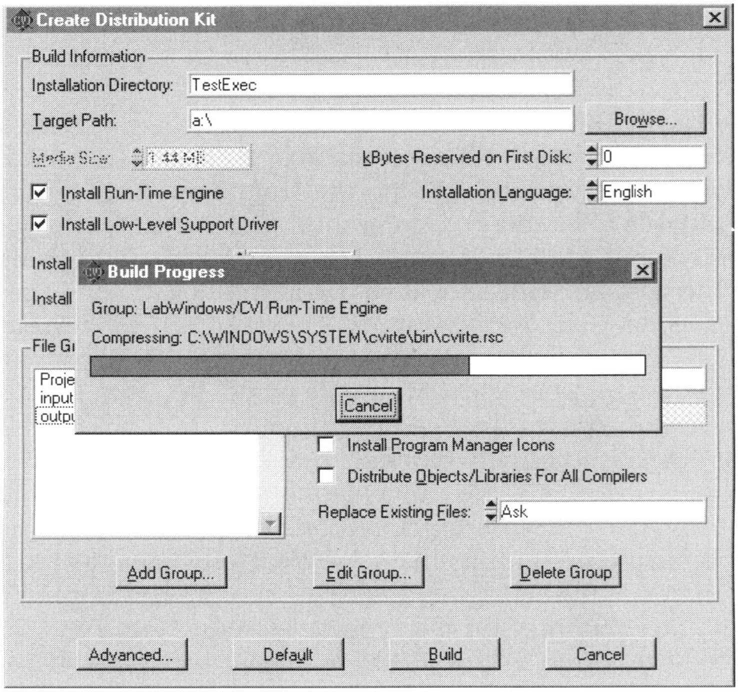

Figure 8–7
Build Progress Panel

Chapter 8 • Creating Stand-Alone Executables and Distribution Disks

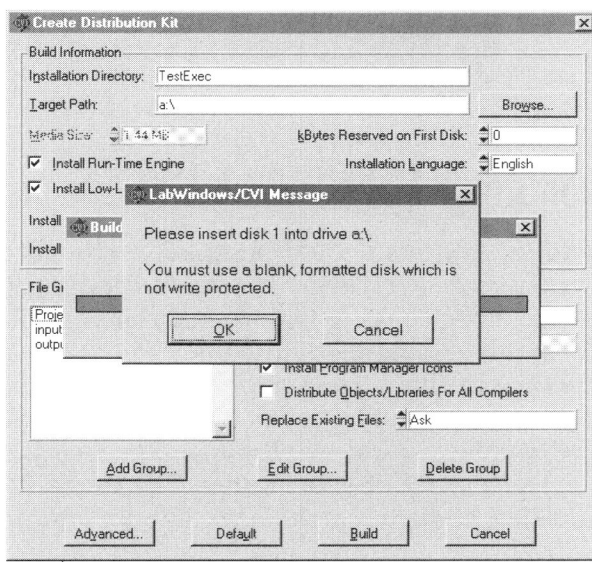

Figure 8–8
Insert Disk Message

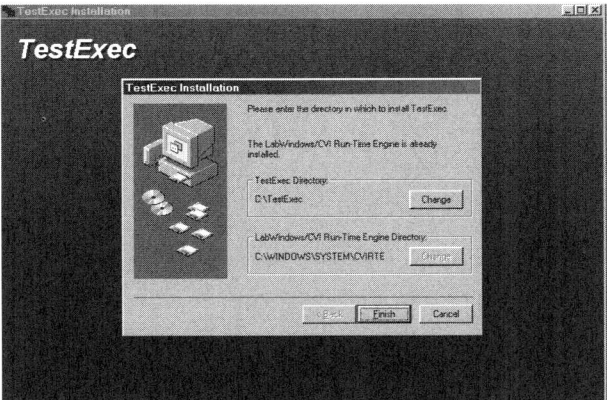

Figure 8–9
Test Exec Installation

the target machine, clicking on the *TestExec* icon can run the program. The `input` and `output` directories and their files will be installed appropriately.

Error Handling for Stand-Alone Executables

The error handling process should be given thorough consideration when creating a stand-alone executable. The error handling must be creative. If you find an error while running the program on your machine, you can use the debugger, set the breakpoints to debug your code, and compile, build, and re-run the project until it runs error-free. When you create a stand-alone executable, debugging features are not available, and no compilers or linkers are there to re-compile your code. It is possible that your stand-alone executable will be tested at an off-site location that may not be easily accessible. If the user runs into problems with the execution, finding the error(s), fixing them, and returning the corrected stand-alone executable may take time.

There may be situations in which the errors occur due to problems external to the software execution environment, such as running out of memory, though the same executable file worked on your system. Also you may try to load a file that was inadvertently excluded from the executable, generating a "File open error" or a *LoadPanel* error if the *user interface* file is not found.

What you, as a software designer, should do, is ensure that the software continues to run if the error(s) is(are) not fatal. If the error is fatal, then the program should abort gracefully. Built-in error handling routines should accomplish this objective. If that is not at all possible, a descriptive error message should be displayed by means of a pop-up panel, informing the user of the error and what can be done to resolve it. This error message can be communicated back to the designer to facilitate deciphering and fixing the problem quickly.

There may be situations in which you designed your GUI on a high-resolution monitor, using 1024 by 768 pixels or higher on a twenty- or twenty-one-inch monitor for example. When the executable is loaded on the target machine with a fifteen-inch monitor and using a resolution of 640 by 480 pixels, the user will see only part of the GUI, with the rest disappearing off the sides of the screen. In such situations, you will have to be careful to include functions to programmatically resize the panels by using the library functions *SetSystemAttribute*, *GetPanelAttribute*, *SetPanelAttribute*, *SetPanelPos*, and *SetPanelSize*. These functions are described in the *LabWindows/CVI User Interface Reference Manual* or online.

Another method of resolving this situation is to design the GUI with the small monitor size and low resolution in mind. Center the panels on the screen and size them proportionally to fit both large and small monitors to reduce the chance of the panels being cut off at the edges. Avoid **Bold** type face fonts in your GUIs and use only *metafonts* for the **Fonts Types** from the control attribute dialog box to prevent scaling problems on different-sized monitors.

The whole idea of this discussion was to make you understand that stand-alone executables should be created more robustly with abundant error handling and smooth termination in case of errors.

Summary

In this chapter, you learned how to create a stand-alone executable and its distribution disks. Once the distribution disks are created, they can be installed and run on the target computer without installing *LabWindows/CVI* on that machine. All the items in the Create Stand-alone Executable dialog box were explained using an example project. You learned the reasons for and the methods for avoiding errors in stand-alone executables.

CREATING AND USING DYNAMIC LINK LIBRARIES

Chapter Highlights
- Dynamic Link Libraries (DLL) Overview
- Creating Dynamic Link Libraries
- Creating DLLs without a User Interface
- Creating DLLs with the User Interface
- Summary

This chapter introduces you to the concept of *Dynamic Link Libraries (DLLs)*. A systematic process to create a *DLL* in one *CVI* project and to call it from another project is demonstrated. Two different examples are given here for you to understand the notion of *DLL* creation, export, and use in another project. In the first example, you are shown how to create a simple *DLL* to export two functions, without the *user interface*. In the second example, the functions and the *user interface DLL* are created and called from another project.

Dynamic Link Libraries (DLL) Overview

Before creating the *DLLs*, you must understand the concept of a *DLL*. When you program using high-level programming languages, you first create the source code in the compiler for that language. All the appropriate headers are included that contain library functions of the routines that you are calling in the source code. The compiler compiles the source code, and the linker combines the object files of these library functions into an executable file. These library functions become a part of the executable file. If any library function is not found during linking, the linker flags an error. This method of linking where the library errors and conflicts are resolved during the linking process is called static linking.

Dynamic linking is different from static linking; dynamic linking allows the Windows applications to link the libraries during run-time. These library applications are stand-alone executable files and are not linked with the application's executable. They are only linked to an application when it is loaded and executed at run-time. When an application uses DLLs, the operating system loads the *DLL* into memory and resolves the references to functions in the *DLL*, enabling the application program to use these functions. The *DLL* is unloaded when it is no longer needed by the application.

There are advantages of using *DLLs*: the application executable can be smaller compared to the static linked code because the *DLL* files are not loaded at link time and are not part of the executable. They appear only during the run-time, do their job, and unload when not needed by the program. The *DLL* applications are modular, meaning they can be worked independently of the application program that uses the *DLL*. The new *DLL* can be linked in the same way without any modification to the application executable. Many executable programs can use the same *DLL*, thus saving further memory by avoiding multiple copies of the *DLL*'s code. In static linking, each executable will have its own copy of the functions, thus using more memory.

There are a few disadvantages of using *DLLs*. The program becomes more complex to develop. The executable program is not one executable file as it is for statically linked files. The user has to make certain that the required *DLLs* are present on the target system before running the program.

Other issues are involved if you are working in a Windows 3.1 environment with *CVI* versions prior to 5.5. Windows 3.1 is a sixteen-bit environment and *CVI* for Windows works in thirty-two-bits. *CVI* requires a special code, called the glue code, that does the address conversion from thirty-two-bit to sixteen-bit and from sixteen-bit to thirty-two-bit. The sixteen-bit to thirty-two-bit conversion is called thunking. There are various issues with the glue code features for which you should refer to the *LabWindows/CVI Programmer Reference Manual*. Note that only *CVI* versions prior to 5.5 support Windows 3.1.

You need to understand that sixteen-bit *DLL*s are not supported in *CVI* for Windows, which is thirty-two-bits. To use a sixteen-bit *DLL*, you must obtain a thirty-two-bit to sixteen-bit thunking *DLL* and a thirty-two-bit import library.

Creating Dynamic Link Libraries

Let us now look at the mechanics of creating a *DLL*. To create a stand-alone executable, the **Build>>Target Type** is set to **Executable,** whereas to create the **Dynamic Link Library,** the **Target Type** has to be set to **Dynamic Link Library.**

The creation of a *DLL* will be demonstrated using two examples. The first example will create a *DLL* to export two functions without any user interface file. The second example will show you how to export the functions along with the user interface. Using the procedures shown in the examples, you can create *DLL*s for as many functions as you need.

Creating DLLs without a User Interface

In this example, you will create a *DLL* to export two functions. After the *DLL*s are created, you will create another project to demonstrate how these functions are loaded, exported, and then unloaded.

To explain how this is accomplished, let us create the source code for `SimpleDll.c`, listed in Figure 9–1. We cannot use *CodeBuilder* to create the source skeleton code because *CodeBuilder* creates the code from a user interface only. You, therefore, have to rely on entering code the conventional way.

Examine the code listed in Figure 9–1.

In the *DllMain* function listed between lines 8 through 24 in Figure 9–1, there are a number of items to observe. Notice that there is no *main* function when creating a *DLL;* the *main* function is replaced by *DLLMain,* whose prototype follows.

```
            int _stdcall DllMain( HINSTANCE hinstDLL, DWORD fwdReason,
                                               LPVOID lpvReserved);
```

```
1    //Include files needed for SimpleDll
2    #include <stdio.h>
3    #include <stdlib.h>
4    #include <ansi_c.h>
5    #include <utility.h>
6
7    //Entry point for DLL routines
8    int __stdcall DllMain (HINSTANCE hinstDLL, DWORD fdwReason, LPVOID lpvReserved)
9    {
10       /* The DllMain function is called when ever the DLL is loaded and    */
11       /* unloaded. Place Initialization code for the DLL in this function. */
12       if (fdwReason == DLL_PROCESS_ATTACH) {
13         /* Place any initialization which needs to be done when the DLL */
14           /* is loaded here. */
15           printf("Export Routines loaded\n");
16       } else if (fdwReason == DLL_PROCESS_DETACH) {
17           /* Place any clean-up which needs to be done when the DLL */
18           /* is unloaded here. */
19           printf("Export Routines un-loaded\n");
20           Delay(5.0);           //Pause so you can see the message
21       }
22
23       return 1;
24   }
25
26   //Mult function is exported using the _export qualifier
27   int __stdcall __export Mult(int a,int b)
28   {
29     int Result;
30
31     Result = a*b;
32     printf("In Mult function\n");
33     return Result;
34   }  //Mult
35
36   //Add function is exported using the _export qualifier
37   int __stdcall __export Add(int a,int b)
38   {
39     int Result;
40
41     Result = a + b;
42     printf("In Add function\n");
43     return Result;
44   }  //Add
```

Figure 9–1
SimpleDLL.c Code Listing

The *DllMain* function has the `__stdcall` keyword before it. This is the calling convention for the *DLLMain* function. The calling convention is used to specify how the machine code is going to be generated to place the function call arguments on the stack.

In Windows, there are two calling conventions: in C, the *calling convention* is __cdecl, and standard *calling convention* is __stdcall.

Either of these calling conventions can be used depending on how your *DLL* is going to be used. If you intend your *DLL* to be used by C or C++ or the Watcom stack-based calling convention to declare the functions you want to export, use the __cdecl convention. If you want the *DLL* to be used in other environments such as Microsoft Visual Basic, __stdcall should be used. Most often, __stdcall is used by Windows function calls.

The arguments of *DllMain* are listed here.

- hinstDll—This is the instance handle of the program. The handle uniquely identifies the program.
- fwdReason—This argument is one of the following reasons why Windows is calling the DLL:
  - DLL_PROCESS_ATTACH—DLL has been attached to the process.
  - DLL_PROCESS_DETACH—DLL is no longer needed by the process.
  - DLL_THREAD_ATTACH—DLL process is created as a new thread.
  - DLL_THREAD_DETACH—DLL thread has terminated.
- lpvReserved—This argument is reserved by the system.

For more information on the *DLLMain* function and the meaning of its arguments, check the reference *Programming Windows* mentioned in the Bibliography.

Line 12 of Figure 9–1 uses *DLL_PROCESS_ATTACH*, causing the imported *DLL* function to be attached between lines 13 through 15. A message is displayed at line 15 as a marker to indicate that the functions have been loaded here. At lines 17 through 20, in the *else-if* part of the statement, *DLL_PROCESS_DETACH* unloads the *DLL*(s) when they have completed processing, and a message is displayed to that effect.

At lines 26 through 34, a multiplication function, *Mult*, is defined and will be exported in the DLL. This function performs the multiplication of two integers and returns the Result. The function prototype includes the __stdcall __export qualifier before the function name. The __export qualifier exports this function.

Similarly between lines 37 through 44, the *Add* function is defined, which adds two integers and returns the Result. This function is also exported and has the __export qualifier before the function name.

Add the SimpleDll.c file to the project and save the project as SimpleDll.prj. From the **Project** window, select **Build>>Target Type** as **Dynamic Link Library** as shown in Figure 9–2.

Chapter 9 • Creating and Using Dynamic Link Libraries

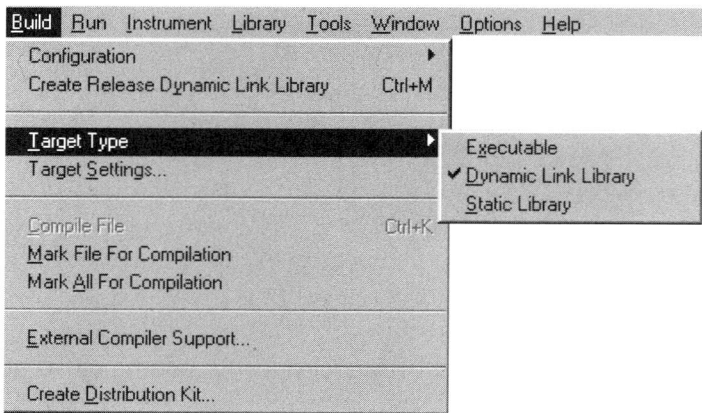

Figure 9–2
Setting Dynamic Link Library as Target Type

From the **Build** menu, select **Target Settings...** and a dialog box will be displayed as in Figure 9–3. The dialog boxes functionalities are described in the following list.

- **DLL File**—You can change the configuration to **Release** or **Debug** in the ring box. The box next to it will contain the name of the file to be created with the .dll extension. The existing file can be selected from the **Browse...** button. You can enter another name of your choice if you do not like the automatically loaded function base name.
- **Import Library Base Name**—This is the same name as the base name of the DLL but with a .lib extension. You can choose a different name if you prefer by removing the check mark from the **Use Default** box.
- **Where to copy DLL**—This gives you a choice of copying the *DLL* in either the Windows System directory or the VXI*plug&play* directory or of not copying the DLL. Use **Do not copy** for this example.
- **Instrument Driver Support Only**—This has the same use as was discussed in Chapter 8, "Creating Stand-Alone Executables and Distribution Disks."
- **Version Info**—Refer to the **Build Menu** section of Appendix B, "Project Window Environment" for an explanation.
- **Import Library Choices**—This gives you the choice of creating the *DLL* for only the selected compiler or for each of the compatible compilers. Selecting this button brings up Figure 9–4. If you choose the latter, the subdirectories with the names of the compilers are created and each

Chapter 9 • Creating and Using Dynamic Link Libraries

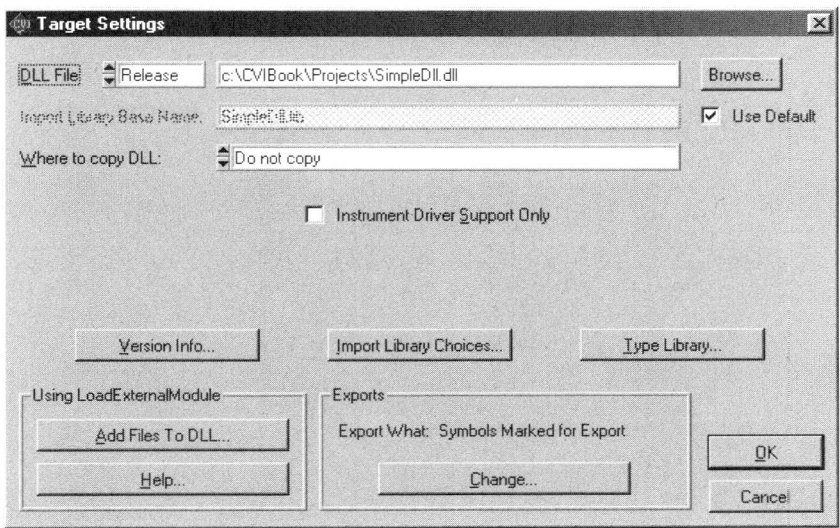

Figure 9–3
Target Settings for Dynamic Link Library Dialog Box

DLL is placed in the appropriate directory. For this example, you can select the option for current compatible compiler.

- **Type Library**—You can add a Type Library resource to the DLL by selecting the appropriate box as shown in Figure 9–5. This is useful when

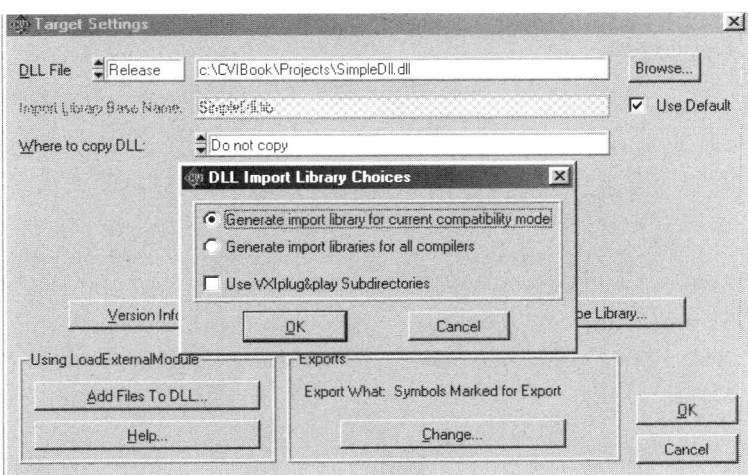

Figure 9–4
DLL Import Library Choices Dialog Box

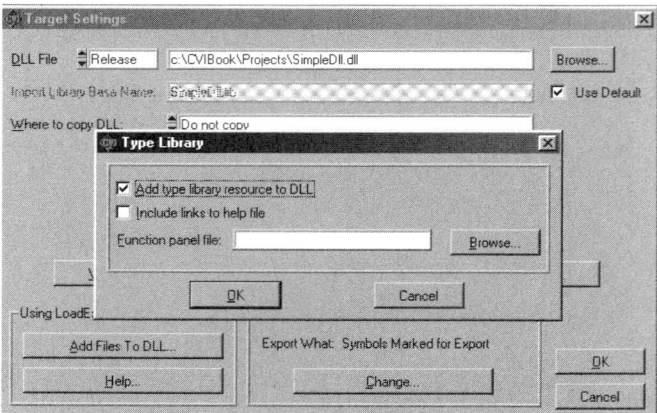

Figure 9–5
Type Library Dialog Box

you want your *DLL* to be used from Visual Basic. For more details, refer to the *LabWindows/CVI User Manual*.

- **Using LoadExternal Module**—This group consists of the following two buttons.
    - **Add Files to DLL**—This lets you add additional files to your DLL by using *LoadExternalModule* function, which loads an external object file. There is a difference in whether this function is used for a Windows 3.1, Windows, or Unix environment. In the Windows 3.1, you can use the object file (.obj), a library file (.lib), or a DLL file in this function. These files can be created only using either the Watcom C compiler for Windows or a *CVI* compiler. For Windows, you can use an object file (.obj), a library file (.lib) or a *DLL* import (.lib) file. A *DLL* cannot be loaded directly. In this environment, you can use either a compatible external compiler or the *CVI* compiler to compile object and library modules. For Unix users, the object file with (.o) extension or a statically linked library (.a) can be used. There are certain rules for the loadable compiled modules, which are discussed in Chapter 8, "Creating Stand-Alone Executables and Distribution Disks" in the *LabWindows/CVI Programmer Reference Manual* for *CVI* versions prior to 5.5.
    - **Help**—This button has information on using *LoadExternalModule* and **Add Files to DLL.**
- **Exports** group—**Export What: Symbols marked for Export** displays your current choice for exporting from the *DLL* to the users of this DLL. You can use the **Change...** button to select the method to use for deter-

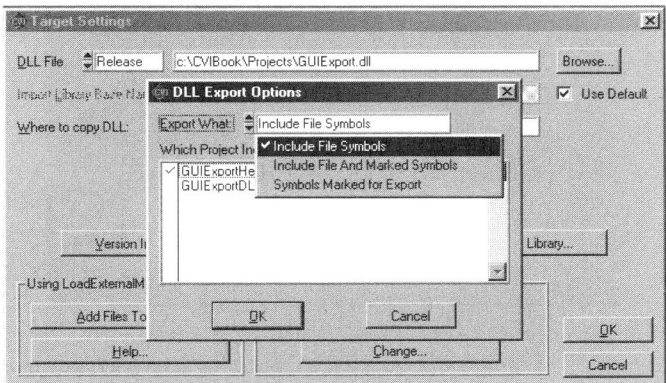

Figure 9–6
DLL Export Options

mining which symbols in the *DLL* to export to the *DLL* users. You can select from one of the three methods, as shown in Figure 9–6. The first choice, **Include File Symbols,** is used when you are using the include file to export the global symbols in the DLL. Leave it set to **Include File Symbols.** The last choice, **Symbols Marked for Export,** exports all symbols that are defined in the *DLL* with the qualifier declspec(dllexport) or export. The middle choice, **Include File And Marked Symbols,** is now self-explanatory.

- **OK**—The selection is accepted and the *DLL* creation starts.
- **Cancel**—Operation is cancelled and the **Create Dynamic Link Library** dialog box is removed.

With this introduction, let us start to create the *DLL*. Select **Build>>Create Release Dynamic Link Library** or **Build>>Create Debuggable Dynamic Link Library** depending on whether you selected **Release** or **Debug** from **Build>>Configuration.** A message is displayed indicating that *CVI* is building the libraries. The export files with the project base name SimpleDLL and extension .dll and .lib will be created as shown in Figure 9–7.

In the next section, you will see how the DLL is used to export the functions to another project.

Using the DLL

You will create a project here to call the exported functions and verify that the *DLL* created above for the *Add* and *Mult* functions are loaded and unloaded appropriately after they have performed the necessary functions. To

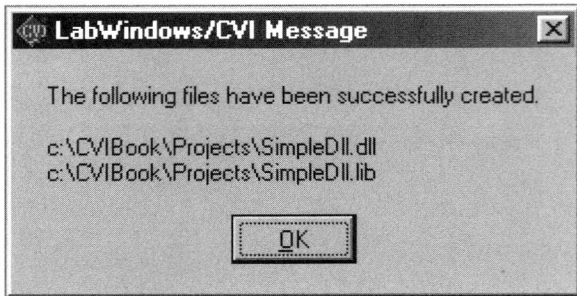

Figure 9–7
Creation of SimpleDLL.dll and .lib Files for Current Compiler

do so, create a new source code file naming it UseSimpleDLL.c that will call these two functions as shown in Figure 9–8.

Notice that in the listing in Figure 9–8, the export function prototypes and the arguments along with the prototype variables are declared globally. The values of the two function arguments are assigned as 6 and 2, and the results

```
#include <ansi_c.h>
//UseSimpleDll.c
//Function top call the SimpleDll functions
// Include the export function prototypes
int __stdcall __export Mult(int ,int );
int __stdcall __export Add(int ,int );
int first, second, Calc;

//Export functions are called in main
main
{

// Assign values to the two arguments
  first =6;
  second =2;

//Call the Add function
  Calc= Add(first, second);
  printf("Add= %d\n", Calc);
//Call the Mult function
  Calc=Mult(first, second);
  printf("Mult= %d\n", Calc);

}//main
```

Figure 9–8
UseSimpleDLL.c Listing to Call the Export Functions

Chapter 9 • Creating and Using Dynamic Link Libraries

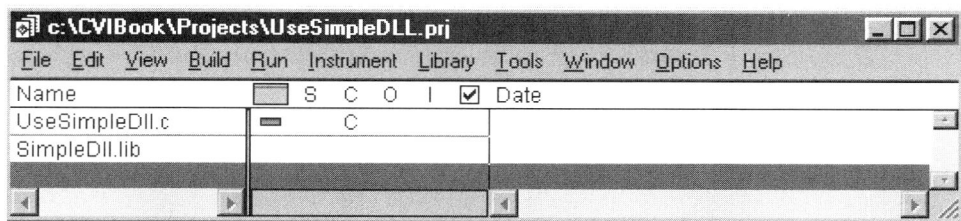

Figure 9–9
UseSimpleDLL Project Files

are assigned to the variable `Calc` and displayed to the *CVI* Standard Output window or to the Console window depending on your **Target Settings.**

Create a new project and call it `UseSimpleDLL.prj`. Add `UseSimpleDll.c` and `SimpleDll.lib` that were created previously. The project window list will now look like Figure 9–9.

From the **Build** menu, set the **Target Type** to **Executable** and the **Configuration** to **Release.** Leave the **Create Console Application** box unchecked in the **Target Settings** dialog box to display the output in **Standard Output.** Select **Build>>Create Release Executable** to create the executable. **Run** the project, and the **Standard Output,** as shown in Figure 9–10, displays the results.

Let us analyze the output data. The first message displayed shows `Export Routines loaded` from line 15 of Figure 9–1. This is where the *DLL* gets loaded by `DLL_PROCESS_ATTACH`. The next message is from line 42 of the same listing in which the *printf* statement displays the *Add* function being called and the result of addition. Similarly, the *Mult* function message is dis-

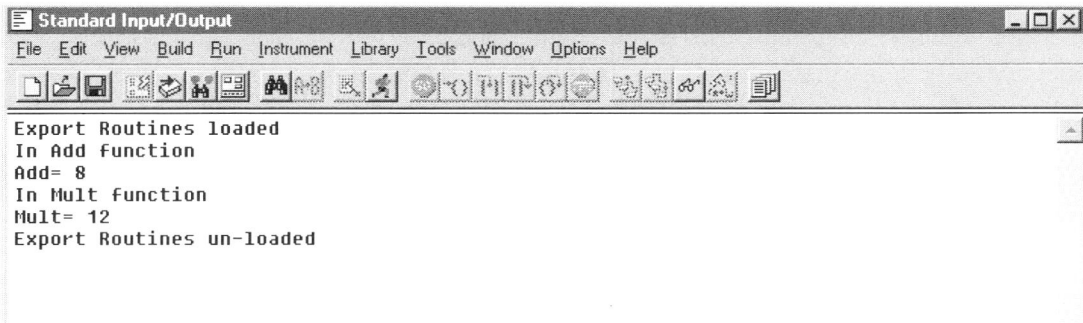

Figure 9–10
UseSimpleDLL Project Output

played from line 32, and the result of multiplication displayed. Lastly, the functions are unloaded by `DLL_PROCESS_DETACH` as their work is done, and the message is displayed at line 19.

Creating DLLs with the User Interface

As you saw in the previous chapter when creating the code for the stand-alone executables, *CodeBuilder* is a very valuable tool in creating the skeleton code for the functions, and this is also true for creating the *DLLs*.

Let us start by creating a user interface and then create the skeleton code using *CodeBuilder*. Create a simple user interface as shown in Figure 9–11, which will display the messages in the string box on this GUI.

Set the **Constant Name** of the panel to `DLL_PNL` and the **Panel Title** to `DLL TESTING`. The rest of the attributes are not significant here and can be left as defaults. Create a **String Control** box in the middle of the panel, with the **Constant Name** `STATUS` and the **Label** `IMPORT ROUTINES STATUS`. Create the **String Control** box sufficiently long so the messages can be displayed without being cut off at the edges. You can set the length of this box by experimenting with the displayed messages in this box.

Name the user interface file `GUIExportDLL.uir` from the **File>> Save As...** menu of the **User Interface Editor.** After the file is renamed, select **Add File to Project.** Also, add the header file to the project, using the same base name. Bring up the **Project** window and notice that these files have been added to the **Project** window. From the **File** menu of the **Project** window, select **Save As...** and name the project `GUIExportDLL.prj`.

In previous projects when you created the project using *CodeBuilder,* you set **Build>>Target Type** to **Executable.** *CodeBuilder* creates the *main* function

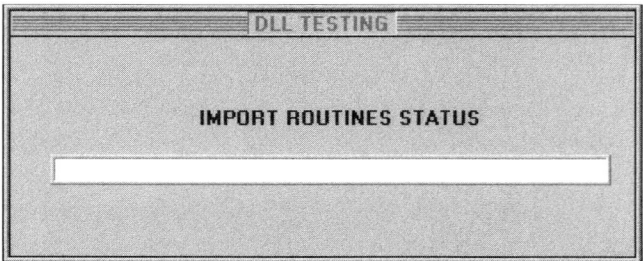

Figure 9–11
DLL User Interface Resource

in the code. To create a *DLL* function, change **Build>>Target Type** to **Dynamic Link Library** from the **Project** window. This will inform *CodeBuilder* that you will be creating a *DLL*.

Return to the **User Interface** window and select **Code>>Generate>>AllCode.** In previous projects, you assigned *QuitUserInterface* to the *Exit* callback function. In this user interface, there is no exit function, and therefore you cannot assign the *QuitUserInterface* function to any exit callback function. Select **OK** to create the source code.

CodeBuilder will name the file `untitled1.c`. From the **Source Editor** window, select **Save As...** and name the file `GUIExportDLL.c`. From the **Source Editor** window, select **File>>Add File to Project.** This will add `GUIExportDLL.c` to the **Project** window list.

Let us examine the code created by *CodeBuilder* for the *DLL* as listed in Figure 9–12. At line 9, the *DLLMain* function acts as the entry point of *DLL*, just as *main* does for C language.

At line 11 of Figure 9–12, the `switch` statement causes the imported *DLL* function to be attached between lines 13 through 17. It is here that you should include the *InitUIForDLL* function, which is defined between lines 41 through 52, to load the user interface and display it. The *InitUIForDLL* function will be described below. The imported functions are unloaded between line 20 and 28 by being detached when no longer needed by the program. The *DiscardUIObjectsForDLL* function at line 22 unloads the exported routine(s).

For those readers who are familiar with C++, you can consider *InitUIForDLL* as the *constructor* since it creates all the initializations and loads the objects and resources required. *DiscardUIObjectsForDLL* can be thought of as the *destructor*; objects not needed are removed and all the resources are un-allocated.

At lines 34, the *DllEntryPoint* function is the entry point of the exported routine, which is used by the Borland compiler. When using the Borland compiler, this is where you will call the *InitUIForDLL* function. This function loads and displays the panel and runs user interface events (if any) from the library function *RunUserInterface* called at line 50. In this project, there are no callback functions, so no events will be generated. Therefore, the line with the *RunUserInterface* function will be left commented. Between lines 53 and 58, the *DiscardUIObjectsForDLL* function is defined. This function is called in the *DLLMain* to discard the panel and release the resources loaded by the *InitUIForDLL* function.

A new library function, *LoadPanelEx,* is introduced at line 45. This function is very similar to *LoadPanel*, which was discussed in an earlier chapter, with one exception. The callback functions you reference in your `.uir` file can be defined in the *DLL* that contains the call to *LoadPanelEx*. This difference only

```c
#include <cvirte.h>    /* Needed if linking in external compiler; harmless otherwise */
#include <userint.h>
#include "GUIExportDLL.h"

int InitUIForDLL (void);
void DiscardUIObjectsForDLL (void);
static int pnl;

int __stdcall DllMain (HINSTANCE hinstDLL, DWORD fdwReason, LPVOID lpvReserved)
{
  switch (fdwReason)
  {
        case DLL_PROCESS_ATTACH:
                /* Needed if linking in external compiler; harmless otherwise */
                if (InitCVIRTE (hinstDLL, 0, 0) == 0)
                        return 0;        /* out of memory */
                break;
        case DLL_PROCESS_DETACH:
                /* Do not call CVI functions if cvirte.dll has already been detached. */
                if (!CVIRTEHasBeenDetached)
                {
                        DiscardUIObjectsForDLL ;     /* Discard the panels loaded in
                                                                        InitUIForDLL */

                        CloseCVIRTE ;                /* Needed if linking in external compiler;
                                                                        harmless otherwise */
                }
          break;
      }

  return 1;
}

int __stdcall DllEntryPoint (HINSTANCE hinstDLL, DWORD fdwReason, LPVOID
lpvReserved)
{
  /* Included for compatibility with Borland */
   return DllMain (hinstDLL, fdwReason, lpvReserved);
}

int InitUIForDLL (void)
{
  /* Call this function from the appropriate place in your code */
  /* to load and display startup panels.                        */
  if ((pnl = LoadPanelEx (0, "GUIExportDLL.uir", DLL_PANEL, __CVIUserHInst)) < 0)
        return -1;
DisplayPanel (pnl);
  /* Uncomment this call to RunUserInterface or call it from elsewhere */
  /* in your code to begin issuing user interface events to panels     */
  /* RunUserInterface ; */
  return 0;
}
void DiscardUIObjectsForDLL (void)
{
  /* Discard the panels loaded in InitUIForDLL */
  if (pnl > 0)
        DiscardPanel (pnl);
}
```

Figure 9–12
GUIExportDLL Source Code Created by *CodeBuilder*

applies to Windows. In other operating systems, there is no difference between *LoadPanel* and *LoadPanelEx*. The prototype of this function follows and

```
int panelHandle = LoadPanelEx (int parentPanelHandle, char filename,
                int panelResourceID, void *callingModuleHandle);
```

the function arguments are explained in Table 9–1.

Add the code shown in Figure 9–13 to the bottom of the source code listing created by *CodeBuilder* and shown in Figure 9–12 to create two functions for export, namely *ExportFunction* and *SwapNumbers.* Name the source file as `GUIExportDLL.c`.

Add the code shown in Figure 9–14 to the *DLLMain* function under `case DLL_PROCESS_ATTACH`.

The *InitUIForDLL* function loads and displays the user interface. It is included under `DLL_PROCESS_ATTACH` to call this function from where the

Table 9–1 LoadPanelEx *Function*

Input/Output	Name	Type	Description
Input	parentPanelHandle	integer	handle of the panel on which to load the panel as a child panel; use 0 to load the panel at top-level
	filename	string	name of the user interface file (`.uir`) that contains the panel
	panelResourceID	integer	the panel name assigned in the **User Interface Editor**
	callingModuleHandle	void pointer	usually, the module handle of the calling DLL; to use the module handle of the *DLL,* use _CVIUserHInst; if 0 is used in this argument, it behaves like the function *LoadPanel*
Output	panelHandle	integer	this is the identifier used to refer to this panel and negative values indicate error; for error codes, see Appendix A of the *LabWindows/CVI User Interface Manual* or in On-line Help

```
//This function is also exported
void ExportFunction(void)
{
            Fmt(message, "%s< 'ExportFunction' being exported ");
            SetCtrlVal(pnl, DLL_PNL_STATUS ,message);
            Delay(2.0);
}
//This exported function swaps values of two integers
void SwapNumbers(int a, int b)
{
            int temp;
            Fmt(message, "%s<'In SwapNumbers' routine is exported ");
            SetCtrlVal(pnl, DLL_PNL_STATUS ,message);
            //Swap a and b
            temp=a;
            a=b;
            b=temp;
            //Show a and b values on the panel
            Fmt(message, "%s<a= %d     b=%d", a, b);
            SetCtrlVal(pnl, DLL_PNL_STATUS ,message);
            Delay(2.0);
}
```

Figure 9–13
ExportFunction Callback Function

process will be attached. The *SetCtrlVal* displays the message to give you visibility into where the exported routine(s) are loaded and unloaded. These messages are unnecessary when you are running the actual DLLs to be used in some application.

Add the code shown in Figure 9–15 at line 14 in Figure 9–12 beneath `case DLL_PROCESS_DETACH`.

Add `char message[260]` above the *DLLMain* to declare `message` as a global variable. The source file is now complete and you can compile it. When you do, it will ask if you want the `<formatio.h>` and `<utility.h>` headers added; select **Yes.** The complete listing should now look like the code in Figure 9–16.

The code before line 70 has already been explained, so let us examine the code starting from there. There are two functions that are created for export:

```
            InitUIForDLL;    //Display the user Interface Panel
            Fmt(message, "%s<In the Main program being loaded");
            SetCtrlVal(pnl,DLL_PNL_STATUS ,message);
            Delay(2.0);
```

Figure 9–14
Code Fragment Added to *DLLMain*

Chapter 9 • Creating and Using Dynamic Link Libraries

```
                    Fmt(message, "%s<In the Main program being unloaded");
                    SetCtrlVal(pnl,DLL_PNL_STATUS ,message);
                    Delay(2.0);
```

Figure 9–15
Code Fragment Added to DLL_PROCESS_DETACH

```
1    //This function creates the DLL with the GUI
2
3    #include <formatio.h>
4    #include <utility.h>
5    #include <cvirte.h>    /* Needed if linking in external compiler; harmless otherwise */
6    #include <userint.h>
7    #include "GUIExportDLL.h"
8
9    char message[260];
10   int InitUIForDLL (void);
11   void DiscardUIObjectsForDLL (void);
12   static int pnl;
13
14   //DLL Main to load and unload the DLLs
15   int __stdcall DllMain (HINSTANCE hinstDLL, DWORD fdwReason, LPVOID lpvReserved)
16   {
17     int a,b;
18     switch (fdwReason)
19     {
20     case DLL_PROCESS_ATTACH:
21           InitUIForDLL;    //Display the user Interface Panel
22           Fmt(message, "%s<In the Main program export routines being loaded");
23           SetCtrlVal(pnl,DLL_PNL_STATUS ,message);
24           Delay(2.0);
25           if (InitCVIRTE (hinstDLL, 0, 0) == 0)    /* Needed if linking in
26                                                      external compiler; harmless otherwise */
27           return 0;        /* out of memory */
28           break;
29     case DLL_PROCESS_DETACH:
30           Fmt(message, "%s<In the Main program export routines being unloaded");
31           SetCtrlVal(pnl,DLL_PNL_STATUS ,message);
32           Delay(2.0);
33           if (!CVIRTEHasBeenDetached)    /* Do not call CVI functions if
34                                             cvirte.dll has already been detached. */
35           {
36                 DiscardUIObjectsForDLL ;/* Discard the panels
37                                                                 loaded in InitUIForDLL */
38                 CloseCVIRTE ;/* Needed if linking in external compiler;
39                                                                 harmless otherwise */
40           }
41           break;
```

Figure 9–16
GUIExportRoutine.c Listing

```
42              } //switch (fdwReason)
43              return 1;
44      }
45      int __stdcall DllEntryPoint (HINSTANCE hinstDLL, DWORD fdwReason, LPVOID pvReserved)
46      {
47        /       * Included for compatibility with Borland */
48              return DllMain (hinstDLL, fdwReason, lpvReserved);
49      }
50      //Loads the panel
51      int InitUIForDLL (void)
52      {
53        /* Call this function from the appropriate place in your code */
54        /* to load and display startup panels.                        */
55        if ((pnl = LoadPanelEx (0, "GUIExportDLL.uir", DLL_PNL, __CVIUserHInst)) < 0)
56              return -1;
57        DisplayPanel (pnl);
58        /* Uncomment this call to RunUserInterface or call it from elsewhere */
59        /* in your code to begin issuing user interface events to panels    */
60        // RunUserInterface ;
61        return 0;
62      }
63      //Discards the panel
64      void DiscardUIObjectsForDLL (void)
65      {
66        /* Discard the panels loaded in InitUIForDLL */
67              if (pnl > 0) DiscardPanel (pnl);
68      }
69      //This function is also exported
70      void ExportFunction(void)
71      {
72              Fmt(message, "%s< Inside 'ExportFunction'");
73              SetCtrlVal(pnl, DLL_PNL_STATUS ,message);
74              Delay(2.0);
75      }
76      //This exported function swaps values of two integers
77      void SwapNumbers(int a, int b)
78      {
79              int temp;
80              Fmt(message, "%s<Inside 'SwapNumbers' function ");
81              SetCtrlVal(pnl, DLL_PNL_STATUS ,message);
82
83              //Swap a and b
84              temp=a;
85              a=b;
86              b=temp;
87              //Show a and b values on the panel
88              Fmt(message, "%s<a= %d    b=%d", a, b);
89              SetCtrlVal(pnl, DLL_PNL_STATUS ,message);
90              Delay (2.0);
91      }
```

Figure 9–16
GUIExportRoutine.c Listing (*continued*)

```
//Prototypes of the functions being exported
void SwapNumbers(int a, int b);
void ExportFunction(void);
```

Figure 9–17
GUIExportHeader File

ExportFunction, which displays the message at line 72 when exported, and SwapNumbers(int a, int b), which swaps two numbers, a and b. This function is shown on lines 77 through 91. When this function is imported by the *DLL* calling project, the swapped values are displayed on the GUI at line 88 as you will see later when this *DLL* is called from another project.

Create a new header file consisting of the lines in Figure 9–17, name this file GUIExportHeader.h, and add this header file to the project.

This file will export the function prototype headers for the two exported functions. The **Project** window should contain the files shown in Figure 9–18.

To create the *DLLs*, select **Build>>Target Type>>Dynamic Link Library**, and **Build>>Create Release Dynamic Link Library** or **Create Debuggable Dynamic Link Library** depending on the **Release** or **Debug Configuration.**

Select **Target Settings…** from the **Build** menu. Leave all the settings to their default values, but select the **Change** button. Figure 9–19 is displayed. From the dialog box, check the GUIExportHeader.h to export from the dialog box that is shown in Figure 9–19.

Select **OK** from the **DLL Export Options** dialog box and then **OK** again to exit the **Target Settings** dialog box. From the **Build** menu, select **Create Release Dynamic Link Library.** A *CVI* message is displayed indicating that

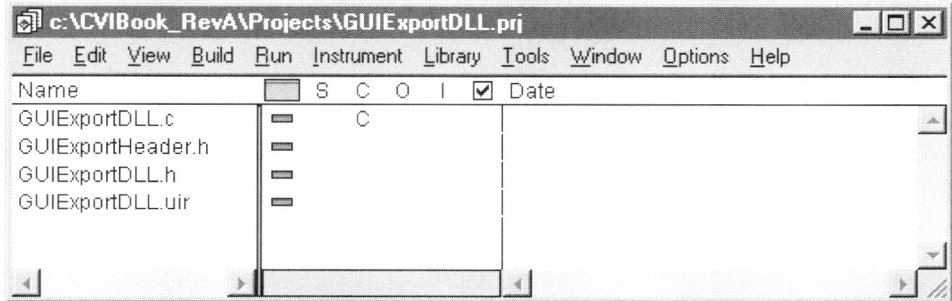

Figure 9–18
GUIExportDLL Project Window

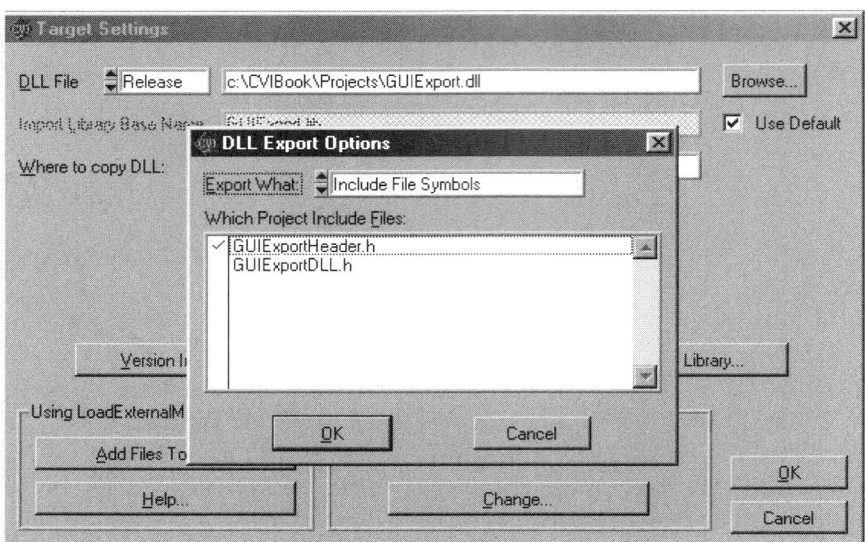

Figure 9–19
GUIExport Create DLL Dialog Box

GUIExport.dll and GUIExport.lib have been successfully created as shown in Figure 9–20.

In the next section, you will see how to call these *DLLs* from another project. As in the previous example, the .lib file will be included in the calling project.

Using the GUI DLL

To call the *DLL*, create a new project and name it UseGUIExportDLL.prj. Add the GUIExport.lib created above to this **Project** window list. Create a source file UseGUIExport that includes the exported header file GUIExportHeader.h with source code shown in Figure 9–21.

Name this file UseGUIExport.c and add it to the **Project** window list. The **Project** window list should now consist of the files shown in Figure 9–22.

In the **Project** window, set the **Build>>Target Type** to **Executable** and select **Created Release Executable** from the **Build** menu. A *CVI* message indicating that the executable has been created is displayed. Now run the project and notice the following messages are displayed in sequence. Figure 9–23 displays the message in the *DLLMain* indicating that the *DLL* is loaded. The

Chapter 9 • Creating and Using Dynamic Link Libraries

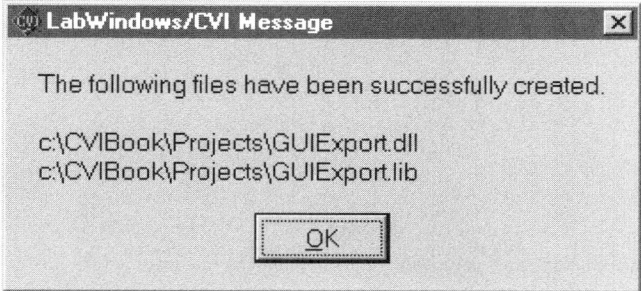

Figure 9–20
GUIExport.dll and GUIExport.lib Created

```
#include "GUIExportHeader.h"
//This routine will run the two exported functions
main
{
        int a, b;

        a=300;
        b=400;
        SwapNumbers(a,b);
        ExportFunction;
}   //  main
```

Figure 9–21
UseGUIExport Listing

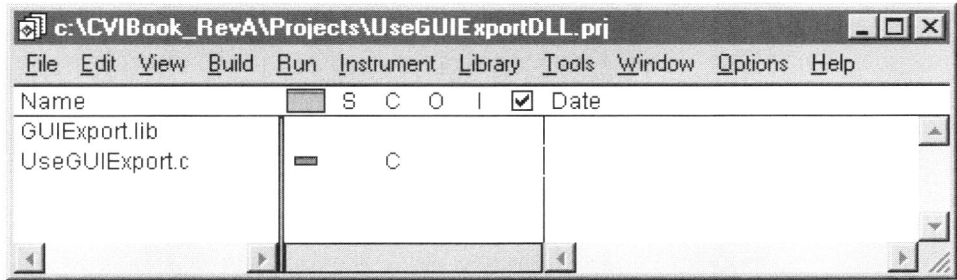

Figure 9–22
UseGUIExportDLL Project Window

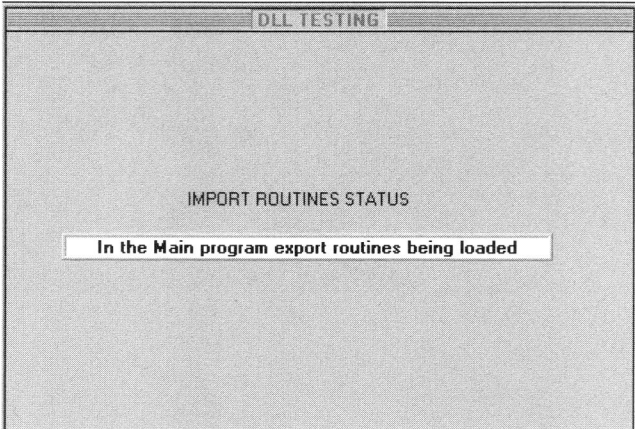

Figure 9–23
Functions Loaded in Main

message in Figure 9–24 shows that the values of a and b have been switched by the *SwapNumbers* function (before the swap, a=300 and b=400). The third message is from inside the ExportFunction as shown in Figure 9–25. Figure 9–26 displays the message from the DLL_PROCESS_DETACH section of *DLLMain* function to indicate that all export routines are unloaded.

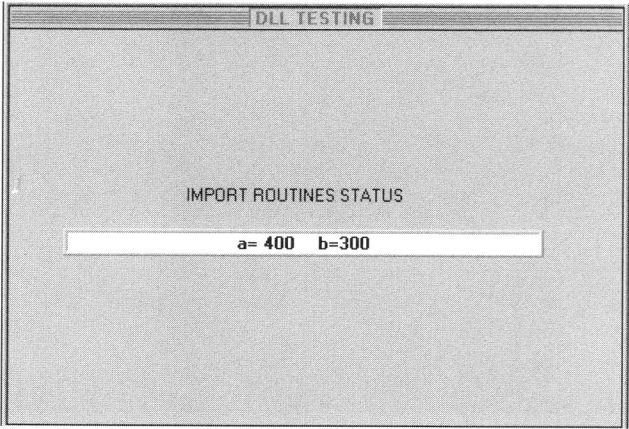

Figure 9–24
Values of a and b Swapped

Chapter 9 • Creating and Using Dynamic Link Libraries 339

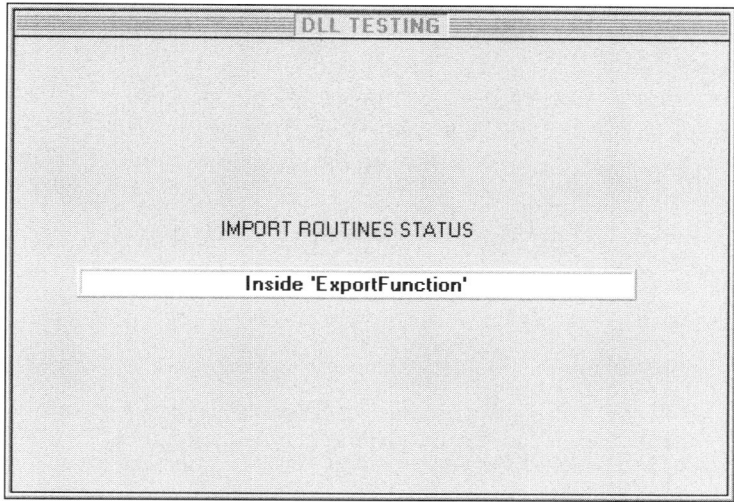

Figure 9–25
Inside ExportFunction

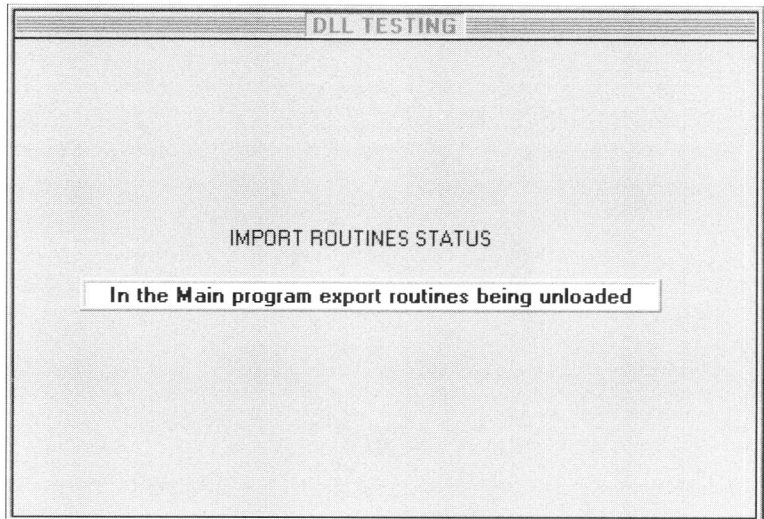

Figure 9–26
Exported Functions Unloaded from DLLMain

Summary

This chapter introduced you to the concepts of DLLs. The first project created a *DLL* that exported a function and displayed the results on Standard Output when the *DLL* was called from another project. In the other project, a *DLL* was created that used a GUI, exported the *DLL,* and displayed the results on the GUI when called from another project.

10
EXTERNAL COMPILER SUPPORT

Chapter Highlights
- External Compiler Support
- Creating an Object File for Use in an External Compiler
- Creating DLL Containing Callback Functions
- Summary

The topic of external compiler support is an advanced feature of *CVI* and actually belongs in a book for advanced users. This topic is introduced for the benefit of the users who are more comfortable using the external compiler of their choice but still want to use the numerous powerful *CVI* features. This way the user can take advantage of the computing power of *CVI* analysis libraries and the ease of creating GUIs, callback functions, instrument drivers, and function panels in the *CVI* environment. Once the user has created the application using *CVI*, it can be imported and used in a favorite external compiler.

Two ways of using *CVI*-created files in external compilers are introduced here via projects: first, by creating and using the callback function *object* file that is called by the external compiler, and then by creating the *DLL* in *CVI* and calling it from an external compiler.

Beginner readers who are not interested in using this feature of *CVI* can skip this chapter without any loss of continuity in the text.

External Compiler Support

As mentioned in Appendix A, *CVI* supports five different compatible compilers that can be used with *CVI*:

- Borland C++ 4.51 or higher, Borland C++ Builder 4.0,
- Microsoft Visual C++ 2.2 or higher,
- Symantec C++ 7.2 or higher,
- Watcom C++ 10.5 or higher.

Situations may occur in which you are more comfortable using a compiler of your choice, but you would still prefer to use CVI and its rich libraries to develop the GUIs. The user interface consists of objects that have callback functions associated with them that are used specifically by the *CVI* compiler and user interface and do not form part of the external compiler. To use the *CVI* libraries, you must include the appropriate header files at the top of the code. Also in your external compiler, add the path of the *CVI include* files so the compiler knows where to find these files.

You must also add the *CVI* support libraries to the external compiler so that the function calls can be resolved. The library files `cvisupp.lib`, `cviwmain.lib`, and `cvirt.lib` contain all the *CVI* functions from the User Interface, Utility, TCP, DDE, RS232, and Format I/O libraries. Any other special library files, such as for communicating with instruments, must likewise be included and will be explained later in this chapter.

You can use the callback functions in an external compiler in one of two ways:

1. Define the *user interface resource* (UIR) callback functions in the external compiler,
2. Define the *user interface resource* callback functions in *CVI* and compile as *DLL* to be linked into the external compiler.

Both methods are discussed here.

Creating an Object File for Use in an External Compiler

Before creating a callback using the external compiler, you should understand what is required for the creation of the callbacks by an external compiler.

CVI keeps a table of non-static functions in your project when you link the program. When a panel or a menu bar is loaded, the **User Interface Library** uses this table to find the callback functions associated with objects loaded from the `.uir` file.

Such a table is not created when you link using an external compiler. Therefore, the external compiler will not be able to resolve the callbacks associated with the objects. To do this, you must create an object file from the **Project** window by selecting **Build>>External Compiler Support** to create such a table.

An example will help you better understand this process. Create a folder (directory) `ExtSupp` under the **Projects** window. The complete pathname for this project is

```
c:\CVIBook\Projects\ExtSupp
```

This is where you will store the files needed for the external compiler callback creation. Let us use the previously created `project6` files for demonstrating this feature. Copy the entire set of `project6` files to the `\ExtSupp` folder to avoid over-writing the original files.

Open the `project6.uir` file and from the **Code** menu, select **Code >>Set Target File...** and select `project6.c`. You want to use the already created source code and make minimal changes to make the program work through the external compiler. One of the major changes required, the *WinMain* function must be used rather than the *main*. It's easiest to let *CodeBuilder* create the skeleton code for you.

The `project6.c` source code file is using the *main* function, and you need to replace the *main* function with the *WinMain* function. For that reason, you must remove the highlighted lines shown in Figure 10–1 from the *main* function and leave your cursor near the top of the *main* program. *CodeBuilder* will create code with the *WinMain* function arguments.

In C language, the entry point for the code is through the *main* function, and similarly, the starting point for Windows programming is through the *WinMain* function. This is where the initializations for the project are performed. To read about the *WinMain* function, you can refer to any good Windows programming book.

Bring up the GUI, select **Code>>Preferences,** and remove the check mark from **Always Append Code to End.** This way the code will be inserted at the cursor location from where you removed the highlighted lines in the *main* function. Select **Code>>Generate >>All Code** and check the *ExitPanelCB* in the bottom dialog box to indicate the *QuitUserInterface* function being called from this callback function. Also be sure to check the **Generate WinMain instead of main** box shown on the top of the dialog box, as shown in Figure 10–2. Select **OK,** and *CodeBuilder* will start to generate the code. If it finds an already exist-

Chapter 10 • External Compiler Support

```
//Main program for project6.
int main (int argc, char *argv[])
{
if (InitCVIRTE (0, argv, 0) == 0)  /* Needed if linking in external compiler; harmless otherwise */
         return -1;          /* out of memory */
if ((TestHandle = LoadPanel (0, "project6.uir", TST)) < 0)
         return -1;


DisplayPanel (TestHandle);
GetProjectDir (prj_dir); //Get the project directory
SetCtrlVal(TestHandle, TST_DATE, DateStr);
Fmt (buf, "%s<Time      Temperature   Result    Pressure    Result");
InsertTextBoxLine (TestHandle, TST_RESULTS, -1, buf);
SetPanelPos (TestHandle, VAL_AUTO_CENTER, VAL_AUTO_CENTER);

RunUserInterface ;
return 0;
}// main
```

Figure 10–1
Remove Lines from Project6 Main Function

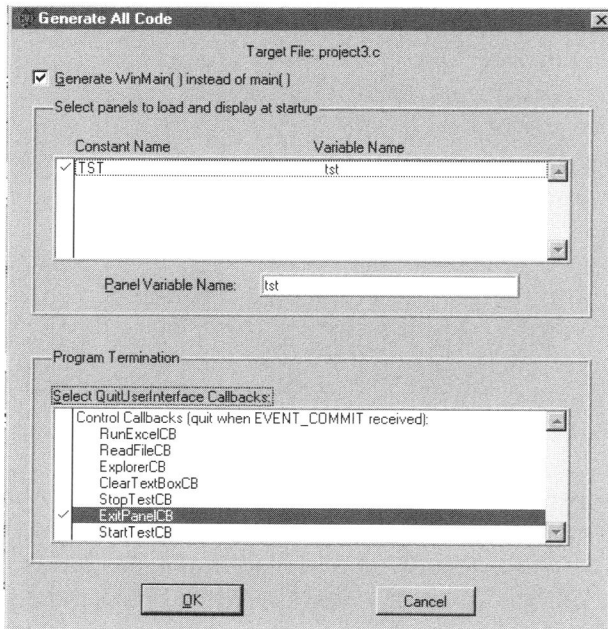

Figure 10–2
Generate All Code Dialog Box

Chapter 10 • External Compiler Support

ing callback function, it will prompt you with the various options, which you saw in previous chapters. Select **Stop** since you do not want your code to be over-written by *CodeBuilder*. The only new code created by *CodeBuilder* is that for the *WinMain* function. The rest of the `project6.c` code is to remain unaltered. The source code listing should now look as shown in Figure 10–3.

Remove the highlighted lines in Figure 10–3 and paste them to make your code look like Figure 10–4. Change `tst` to `TestHandle` in *DiscardPanel* as shown on line 20 below. Your complete code for *WinMain* function should now look similar to Figure 10–4.

The function *InitCVIRTE* is called in *WinMain* so that *CVI* run-time libraries can initialize the list of names from the object file.

Compile the source code and build and run the project. It will run as before. Now you are ready to create the project that can be run using the external compiler.

From the **Build** menu in the **Project** window, select **External Compiler Support**. The **External Compiler Support** dialog box will appear as shown in Figure 10–5. Each of the dialog box items are explained in the following list.

```
//Main program for project6
int __stdcall WinMain (HINSTANCE hInstance, HINSTANCE hPrevInstance,
    LPSTR lpszCmdLine, int nCmdShow)
{
        if (InitCVIRTE (hInstance, 0, 0) == 0)
                return -1;       /* out of memory */
        if ((tst = LoadPanel (0, "project6.uir", TST)) < 0)
                return -1;
        DisplayPanel (tst);
        RunUserInterface ;
        DiscardPanel (tst);
        return 0;
}
        if ((TestHandle = LoadPanel (0, "project6.uir", TST)) < 0)
                return -1;

        DisplayPanel (TestHandle);
        GetProjectDir (prj_dir); //Get the project directory
        SetCtrlVal(TestHandle, TST_DATE, DateStr);
        Fmt (buf, "%s<Time      Temperature    Result    Pressure    Result");
        InsertTextBoxLine (TestHandle, TST_RESULTS, -1, buf);
        SetPanelPos (TestHandle, VAL_AUTO_CENTER, VAL_AUTO_CENTER);

        RunUserInterface ;

        return 0;
```

Figure 10–3
CodeBuilder-Created *WinMain* for Project6

```
1   //Main program for project6.
2   int __stdcall WinMain (HINSTANCE hInstance, HINSTANCE hPrevInstance,
3          LPSTR lpszCmdLine, int nCmdShow)
4   {
5   if (InitCVIRTE (hInstance, 0, 0) == 0) /* Needed if linking in external compiler; harmless
6                                    otherwise */
7          return -1;       /* out of memory */
8
9   if ((TestHandle = LoadPanel (0, "project6.uir", TST)) < 0)
10          return -1;
11  DisplayPanel (TestHandle);
12  GetProjectDir (prj_dir); //Get the project directory
13  SetCtrlVal(TestHandle, TST_DATE, DateStr);
14  Fmt (buf, "%s<Time     Temperature   Result   Pressure   Result");
15  InsertTextBoxLine (TestHandle, TST_RESULTS, -1, buf);
16  SetPanelPos (TestHandle, VAL_AUTO_CENTER, VAL_AUTO_CENTER);
17
18
19  RunUserInterface();
20  DiscardPanel (TestHandle);
21  return 0;
22  }
```

Figure 10–4
WinMain Function Created for Project6

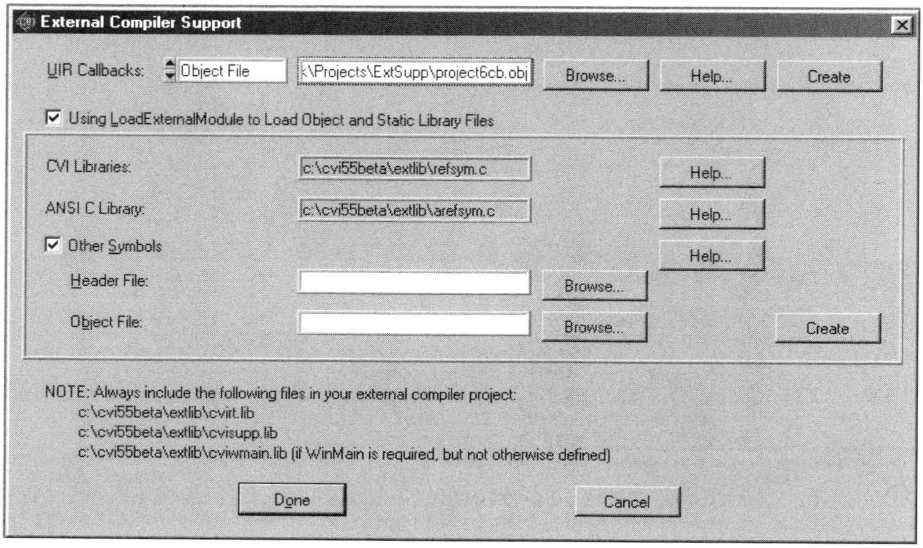

Figure 10–5
External Compiler Support Dialog Box

- **UIR Callbacks**—This box gives you a choice of **None, Object File,** and **Source File.** When the box on this line is checked, the dimmed dialog box line is available for you to enter the path and the file name. If you select **Object File,** then the object file created is linked to the executable or *DLL*. This object file contains the list of the callback functions that you had specified in the user interface resource (.uir) files used in the project. This list references the objects on the panel or the menu bar to the callback functions in your executable or *DLL*. For this example, enter c:\CVIBOOK\Projects\ExtSupp\project6cb.obj.

- **Using LoadExternalModule to Load Object and Static Library Files**—This box is checked when external object modules are to be loaded using the *LoadExternalModule* function to load object or static library files. When a check mark is placed in this box, the middle section of the dialog box becomes available for entering data.

- **CVI Libraries**—The object file name displayed in this dialog box is used by the *LoadExternalModule* function, which is used by the run-time module reference symbols in any of the following *CVI* libraries: User Interface, RS-232, DDE, TCP, Formatting I/O, Utility. The object file in the indicator box must be included in your external project if any of these libraries are used.

- **ANSI C Library**—The object file name displayed in this dialog box is used by the *LoadExternalModule* function, which is used by the run-time module reference symbols in the ANSI C library.

- **Other Symbols**—Select this box if you need to include functions or variables that you define globally in your executable or *DLL* to which your object or static library run-time modules are going to link. The two indicator boxes that are enabled by selecting this box are **Header File** and **Object File.**
    - **Header File**—In this indicator box, insert the name of the include file. This file contains the entire symbol names and declarations to resolve references from the run-time modules.
    - **Object File**—In this indicator box, enter the name of the object file to create. The **Create** button will create this file, which must be included in your external compiler project.

Notice that at the bottom of this dialog box the names of all the files that will be added to the external compiler are listed. You will see later how to add these files.

Leave the **Using LoadExternalModule to Load Object and Static Library Files** box unchecked for this demonstration. When the **Create** button is selected, the object file is created in the assigned path. In this case, `project6cb.obj` will be created in the folder (directory) `c:\CVIBook\Projects\ExtSupp`. Select the **Done** command button to make the **External Compiler Support** dialog box disappear. *CVI* has created the object file required for the external compiler. You can now exit *CVI*.

The *Microsoft* Visual C++ 5.0 compiler will be used here to show you how to run the *CVI* project by calling it from this external compiler. Any of the compatible compilers could be used to run the *CVI* project with a few modifications unique to that compiler.

Load the *Microsoft* Visual C++ 5.0 compiler on your system, and select **File>>New** from the Visual C++ compiler menu bar. A dialog box shown in Figure 10–6 is displayed. Select the **Projects** tab and highlight **Win32 Application** from the list. For the **Project Name,** enter `ExtCallback` (for External Callback). Enter the pathname `c:\CVIBook\Projects\ExtSupp\` in the **Location** box. Select **OK**.

Select **Project>>Add To Project>>Files…** as shown in Figure 10–7. Go one level up to the **ExtSupp** directory and be sure the `project6.c` and `project6cb.obj` are included in this directory. For these files to be visible, you need to change the **Files of type** to `All Files (*.*)`.

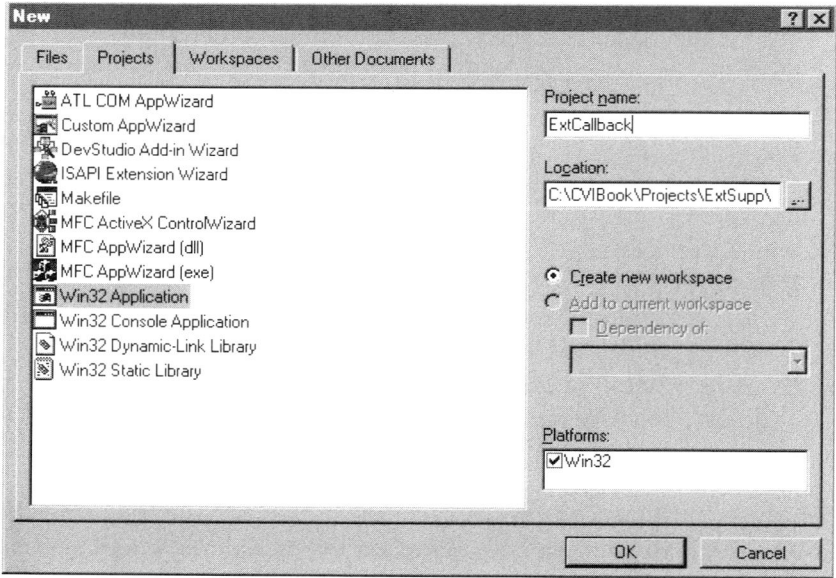

Figure 10–6
Microsoft Visual C++ File New Dialog Box

Chapter 10 • External Compiler Support

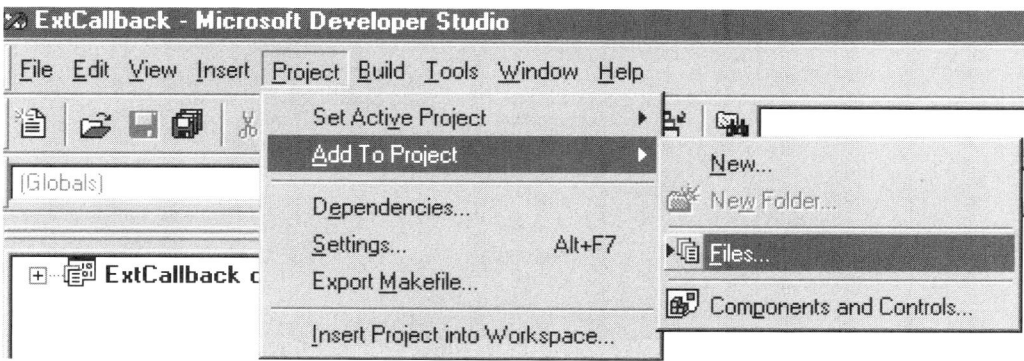

Figure 10–7
Microsoft Visual C++ Add to Project Menu Item

Highlight `project6.c` and `project6cb.obj` and select **OK,** as shown in Figure 10–8, adding these two files to the external compiler project.

Similarly, add the `cvirt.lib`, `cvisupp.lib`, and `cviwmain.lib` library files to the project from the `cvi5\extlib` directory by highlighting these files and selecting **OK,** as shown in Figure 10–9. You will have to set the **Files of type** to `All Files (*.*)` again to view all the file types. Recall you were re-

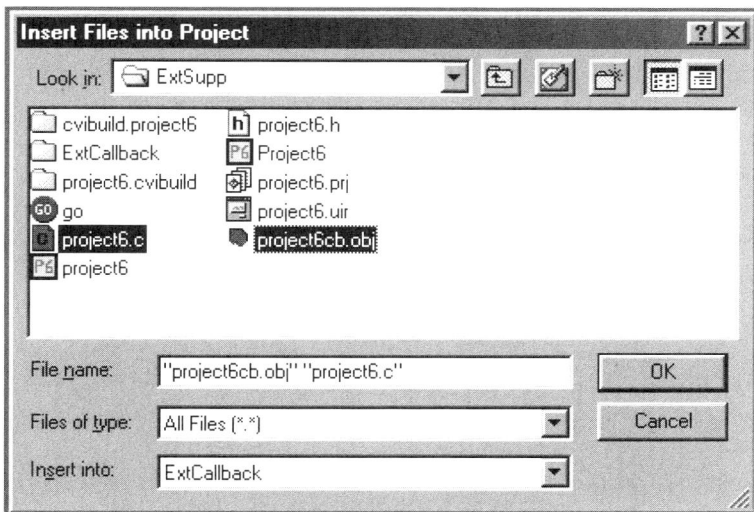

Figure 10–8
Microsoft Visual C++ Insert Files into Project

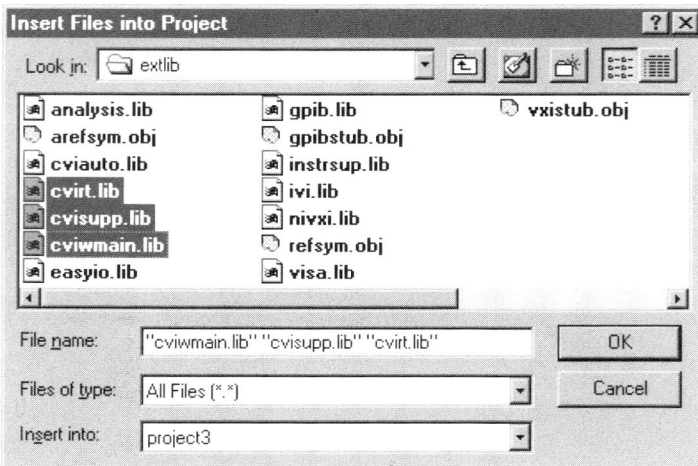

Figure 10–9
Microsoft Visual C++ Insert Library Files into Project

minded to add these files to the external compiler project at the bottom of the dialog box in Figure 10–5.

Select **Project>>Settings** to display a dialog box as shown in Figure 10–10. In the **Settings For** display box, select Win32Debug.

On the right side under the **General** tag, you will see Debug under **Intermediate Files** and **Output Files.** This will create a Debug folder (directory) under the project folder (directory) to store the intermediate and output files if you are creating a debug version of the executable. It easier to work with files in the same folder (directory) for this example. Therefore, you should remove Debug from both the **Intermediate Files** and **Output Files** dialog boxes.

Similarly, select the **Project>>Settings** to display Figure 10–11. In the **Settings For** display box, select Win32Release. On the right side under the **General** tag, you will see Release under **Intermediate Files** and **Output Files.** As described in the previous paragraph, this will create a Release folder (directory) under the project directory. Again to keep everything in the same directory, remove Release from both the **Intermediate** and **Output Files** indicator boxes.

Go to the main menu of the MS Visual C++ compiler and from the **Options** menu under **Tools,** select **Directories** and add c:\cvi5\include as shown in Figure 10–12. This gives the external compiler the pathname of the *CVI* include directory.

Chapter 10 • External Compiler Support

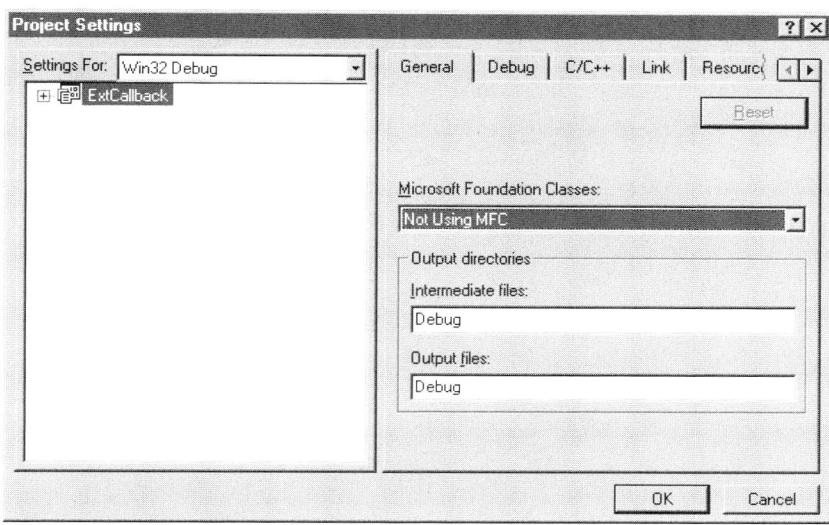

Figure 10–10
Microsoft Visual C++ Project Settings for Win32 Debug

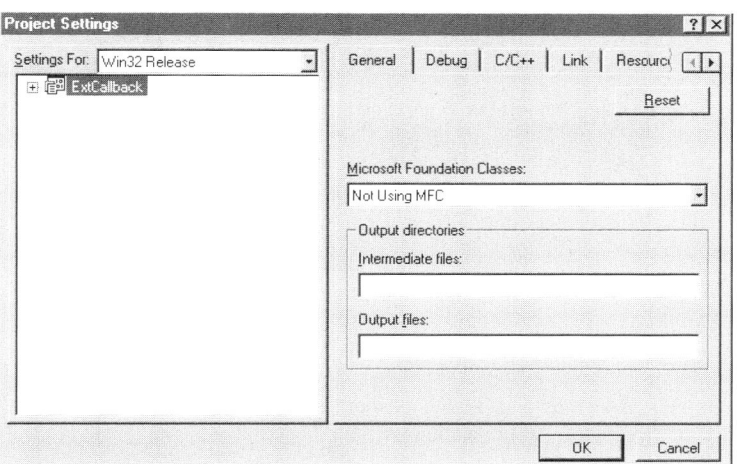

Figure 10–11
Microsoft Visual C++ Project Settings Win32 Release

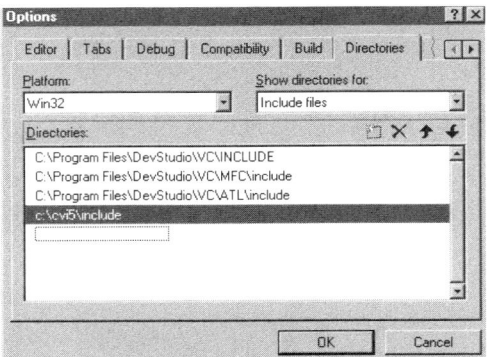

Figure 10–12
Microsoft Visual C++ Directory Options

The external compiler project will look for the `project6.uir` and `project6.h` files in the same folder (directory) where the executable is kept, so you must copy `project6.uir` and `project6.h` to the `ExtCallback` directory. This can be done using Windows Explorer. Build the project using the **Build** menu as shown in Figure 10–13. The files in these project directories are shown in Figure 10–14 and 10–15.

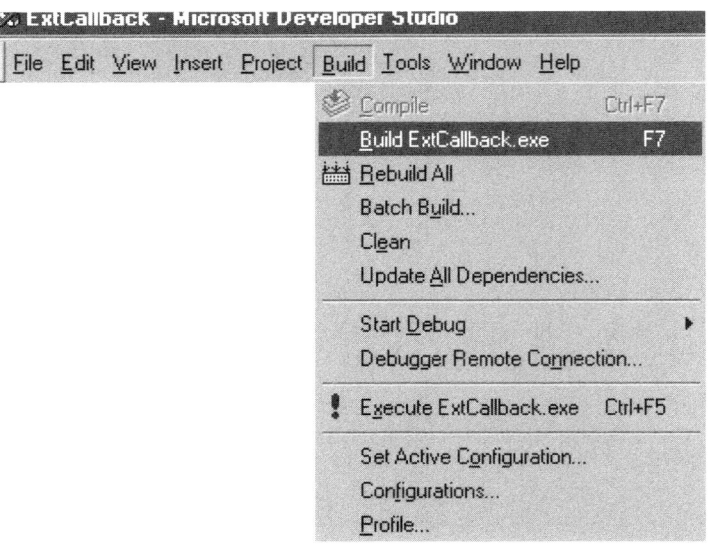

Figure 10–13
Microsoft Visual C++ Project Build Menu

Chapter 10 • External Compiler Support

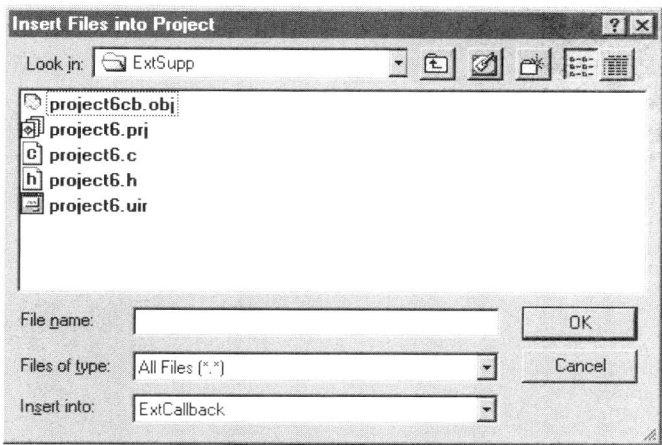

Figure 10–14
Microsoft Visual C++ ExtCallback Directory Listing

If all these files exist in the indicated directories, you can start to build the project. From the **Build** menu, select **Build ExtCallback.exe** or <F7>. If there no errors, select **Execute ExtCallback.exe** or <Ctrl-F5> from the **Build** menu. The program will start to execute and a panel as shown in Figure 10–16 will be displayed, which shows the same GUI as when the project was run from

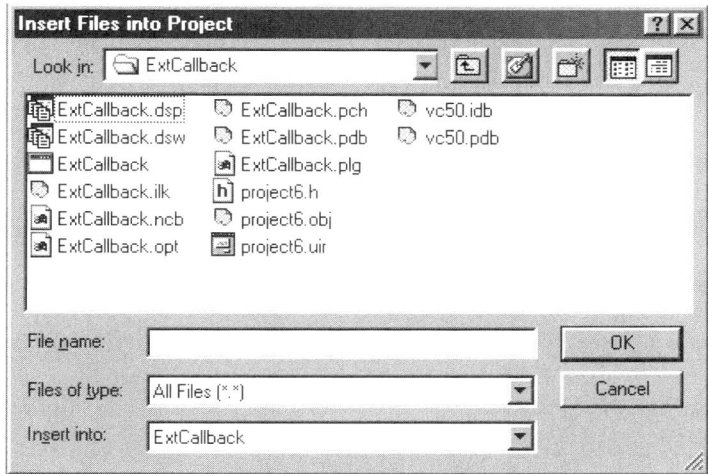

Figure 10–15
Microsoft Visual C++ ExtSupp Directory Listing

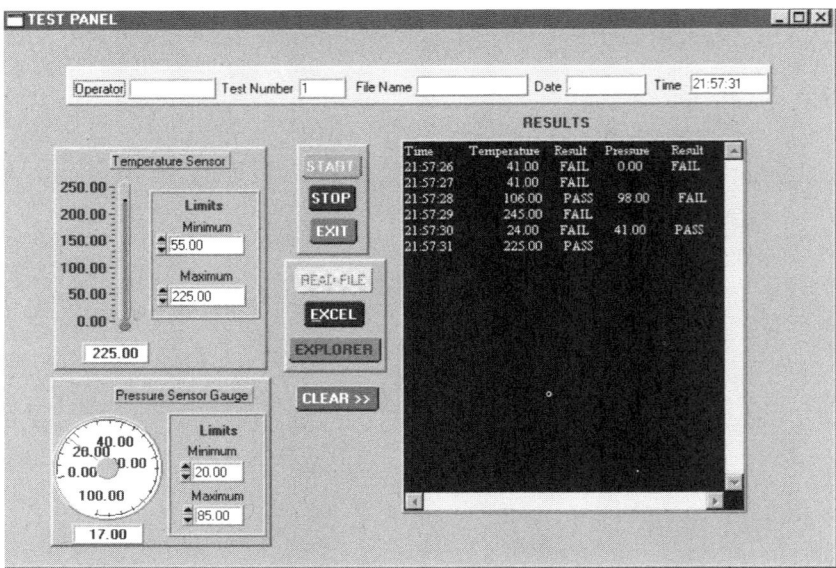

Figure 10–16
Project Run Using Microsoft Visual C++ as External Compiler

CVI. Check that all callbacks functions are operational by clicking on the command buttons.

Creating DLL Containing Callback Functions

In the last section, you saw how the *object* file for the callbacks could be created and run from an external compiler. In this section, the second method, creating a *CVI* program as a *DLL* and calling it from an external compiler, is discussed. For comparison purposes, a *DLL* for the same project, `project6`, will be created again.

Create a new directory with the path name: `C:\CVIBook\Projects\ExtDLL`. Copy all the `project6` files (except `project6.prj` file) from the `Project` folder (directory) into this folder (directory). As before you will not be using the *main* function, but will create a *DllMain* function, by using the *CodeBuilder*.

Create a new *CVI* project, and add `project6.uir`, `project6.h` and `project6.c` files to the `Project` window list. Save this as `project6DLL.prj`. In the `project6.c` file, delete all the code in the *main* function to enable *CodeBuilder* to create the *DllMain*. Leave your mouse cursor at the location from where you removed the *main* function for the *CodeBuilder* to insert the *DllMain* there.

Save the source file as `Project6DLL.c`. Do not rename the `project6.uir` and `project6.h` files. They remain the same. From the **Project** window, select **Build>>Target Type>>Dynamic Link Library.** Open `project6.uir` and from the menu bar, select **Code>> Set Target File** to `Project6DLL.c`. Select **Code>> Generate >>All Code.** As before from the **Generate All Code** dialog box, select the *ExitPanelCB* for the *QuitUserInterface* callback. Be sure to select **Skip** for all the callback functions and do not select **Stop** during the *CodeBuilder* generation.

The *DllMain*, *InitUIForDLL*, and *DiscardUIObjectsForDLL* functions that load, display and then discard the *DLL* are created by *CodeBuilder*. Refer to Chapter 9, "Creating and Using Dynamic Link Libraries" to refresh your memory on the purpose of these functions.

Examine the code generated by *CodeBuilder* and given in Figure 10–17. Lines 11 through 16 are copied from the *main* function shown in Figure 10–4 and added to the *InitUIForDLL* function. Also, *CodeBuilder* will create the panel handle as `tst`, which you should rename to `TestHandle` to be consistent with the rest of the program.

The *InitUIForDLL* function will use the *RunUserInterface* function, which must be un-commented for the function to run the user interface *events*. The operations of the *DllMain* and *DllEntryPoint* functions were explained in the previous chapter. You should refer back to this chapter if you are uncertain of their use.

Compile the source file from the **Build** menu and add the necessary header files (if prompted). The *InitUIForDLL* function will be called from the external compiler to load the user interface file, display the panel, execute code in this function, and start to run the user interface events. The *InitUIForDLL* function is exported to the compatible external compiler. Create a new header (`.h`) file from the *CVI* **Project** window. In this header file, type `int InitUIForDLL(void);`. Save this file as `ExportProject6.h` and add it to the project.

To create the *DLL*, you will use the same procedure as shown in Chapter 9, "Creating and Using Dynamic Link Libraries."

You have a choice to build for **Release** or a **Debug** configuration by selecting **Build>>Configuration.** For this example select **Release.** From the **Project** window select **Build>>Target Settings...** to display a panel as shown in Figure 10–18. You have an option of selecting **Release** or **Debug** in

```c
int __stdcall DllMain (HINSTANCE hinstDLL, DWORD fdwReason, LPVOID lpvReserved)
{
switch (fdwReason)
    {
        case DLL_PROCESS_ATTACH:
          if (InitCVIRTE (hinstDLL, 0, 0) == 0)    /* Needed if linking in external
                                                      compiler; harmless otherwise */
            return 0;        /* out of memory */
            break;
        case DLL_PROCESS_DETACH:
          if (!CVIRTEHasBeenDetached)    /* Do not call CVI functions if cvirte.dll has
                                            already been detached. */
            {
               DiscardUIObjectsForDLL ;    /* Discard the panels loaded in
                                              InitUIForDLL */
               CloseCVIRTE ;    /* Needed if linking in external compiler; harmless
                                   otherwise */
            }
        break;
    }
return 1;
}

int __stdcall DllEntryPoint (HINSTANCE hinstDLL, DWORD fdwReason, LPVOID lpvReserved)
{

/* Included for compatibility with Borland */

 return DllMain (hinstDLL, fdwReason, lpvReserved);
}

int InitUIForDLL (void)
{
/* Call this function from the appropriate place in your code */
/* to load and display startup panels.              */

if ((TestHandle = LoadPanelEx (0, "project6.uir", TST, __CVIUserHInst)) < 0)
        return -1;
DisplayPanel (TestHandle);
GetProjectDir (prj_dir); //Get the project directory
SetCtrlVal(TestHandle, TST_DATE, DateStr);
Fmt (buf, "%s<Time     Temperature   Result   Pressure   Result");
InsertTextBoxLine (TestHandle, TST_RESULTS, -1, buf);
SetPanelPos (TestHandle, VAL_AUTO_CENTER, VAL_AUTO_CENTER);

/* Uncomment this call to RunUserInterface or call it from elsewhere */
/* in your code to begin issuing user interface events to panels   */
 RunUserInterface ;
return 0;
}

void DiscardUIObjectsForDLL (void)
{
/* Discard the panels loaded in InitUIForDLL */
if (TestHandle > 0)
        DiscardPanel (TestHandle);
}
```

Figure 10–17
Completed DLL Creation Code

Chapter 10 • External Compiler Support

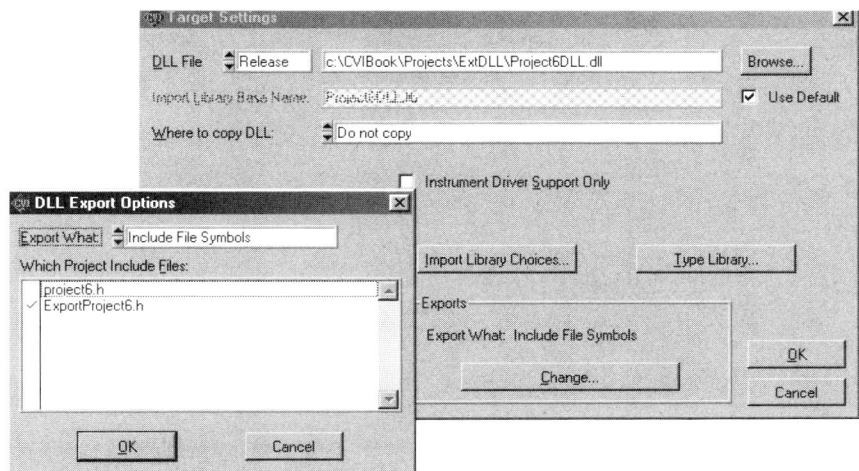

Figure 10–18
Create Dynamic Link Library for Project6DLL

the **DLL File** box. In the box next to it, enter the DLL file path name and the file name as `Project6DLL.dll` (if not already displayed).

Leave the rest of the dialog boxes to their default settings. Select the **Change** command button under the **Export What:Include File Symbols** section. A **DLL Export Options** dialog is displayed. In **Export What,** select `Include File Symbols` and in the **Which Project Include Files** list box, check `ExportProject6.h`. This is the header file that you created above consisting of the function *InitUIForDLL* that will be used by the external compiler. Select **OK** to exit the **DLL Export Options** dialog box. From the **Target Settings** dialog box, select **OK** to close this dialog box. From the **Build** menu, select **Create Release Dynamic Link Library** to create the *DLL* and the import library. The DLL libraries created are shown in Figure 10–19.

Note that because you did not select all the compilers in this **Target Setting** dialog box, you will create the import library for the current compatibility mode, which is Microsoft Visual C++. If you are using a different compatible compiler, you can select the **Generate import libraries for all compilers** option from the **DLL Import Library Choices** dialog box, shown in Figure 10–20.

Next, create a source file `importdll.c` for this project that will import the function *InitUIForDLL*. Add the lines shown in Figure 10–21 to `importdll.c`.

Before using this *DLL* in an external compiler, it is good practice to try running it from *CVI*. Create a new project `importdll.prj` and add the import library `Project6DLL.lib` to the project.

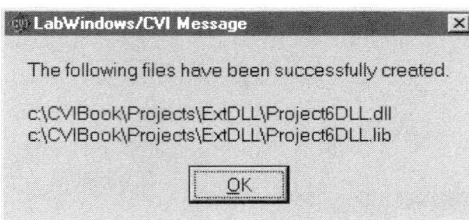

Figure 10–19
Created Dynamic Link Library Files

Add this source file to the project. Save the project as `importdll.prj`. The **Project** window for `importdll.prj` is shown in Figure 10–22.

Build and **Run** the project. It will run the same as before, as shown in Figure 10–16.

With a successful run you have now established that the *DLL* and the import library files are working, so you can now run it from the Microsoft Visual C++ external compiler.

You can now close *CVI* and start the Microsoft Visual C++ 5.0 compiler on your system, and select **File>>New** from the menu bar. A panel as shown in Figure 10–23 is displayed. Select the **Projects** tab and highlight `Win32 Appli-`

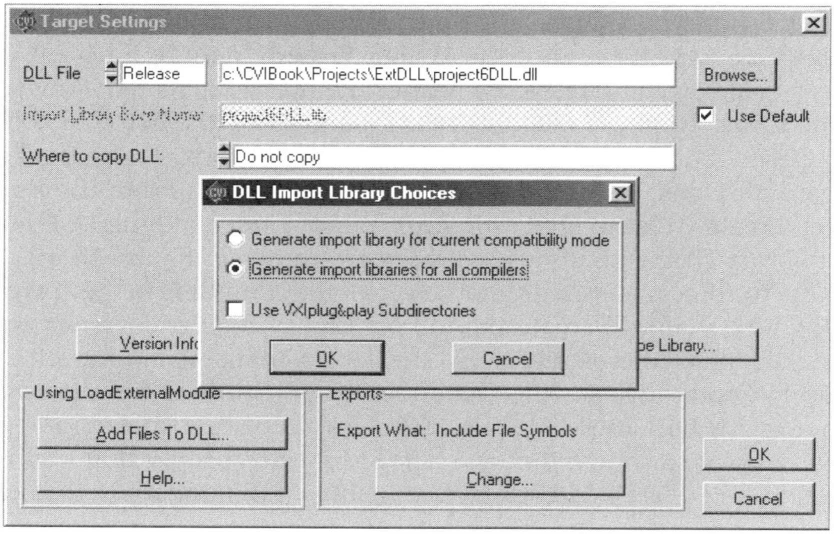

Figure 10–20
DLL Import Library Choices

Chapter 10 • External Compiler Support

```
#include "ExportProject6.h"
main
{
InitUIForDLL;
return 0;
}
```

Figure 10–21
Importdll.c Source Code

cation. For the **Project Name,** enter DLLCall. Enter the appropriate pathname (c:\CVIBook\Projects\ExtDLL\ for this example) in the **Location** box and select **OK.**

Select **Project>>Add To Project>>Files...** and you will see the **Insert Files into Project** box shown in Figure 10–24.

Go to the ExtDLL directory, and be sure that project6DLL.lib and importdll.c are included in this directory. For these files to be visible, you need to change **Files of type** to **All Files (*.*)**. Select the files Project-6DLL.lib and importdll.c as shown in Figure 10–24. Select **OK,** adding these two files to the external compiler project. Note that to add these files to the project you must highlight these files and select **OK.**

Similarly, add the cvirt.lib, cvisupp.lib, and cviwmain.lib library files to the project from the cvi5\extlib directory (or the appropriate path) by highlighting these files and selecting **OK**. For these files to be visible in the list box, change **Files of type** to **All Files (*.*),** as illustrated in Figure 10–25.

Select **Project>>Settings** as displayed in Figure 10–26. In the **Settings For** pull-down list box, select Win32Debug.

On the right side under the **General** tag, you will see Debug under **Intermediate Files** and **Output Files.** This will create a Debug folder (directory)

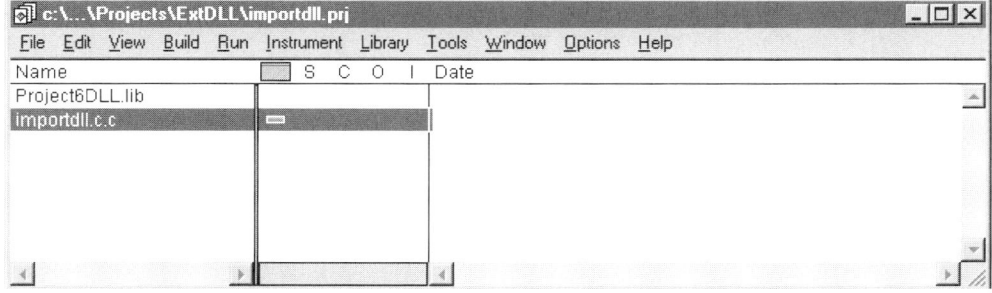

Figure 10–22
Project Window for importdll.prj

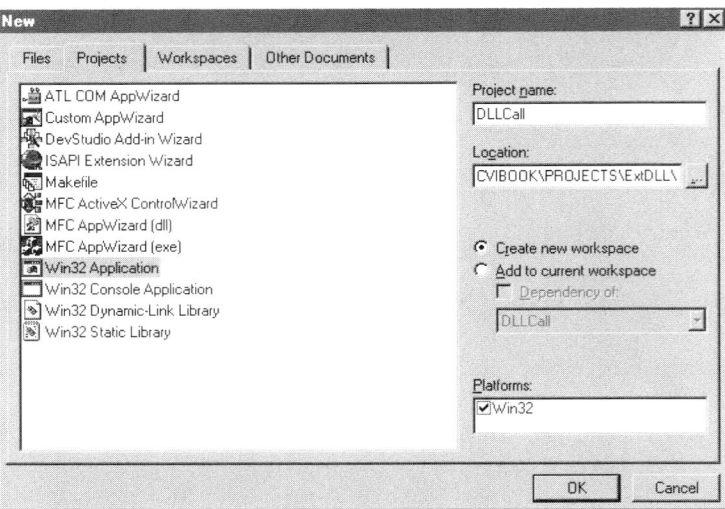

Figure 10–23
Microsoft Visual C++ DLL File New Dialog Box

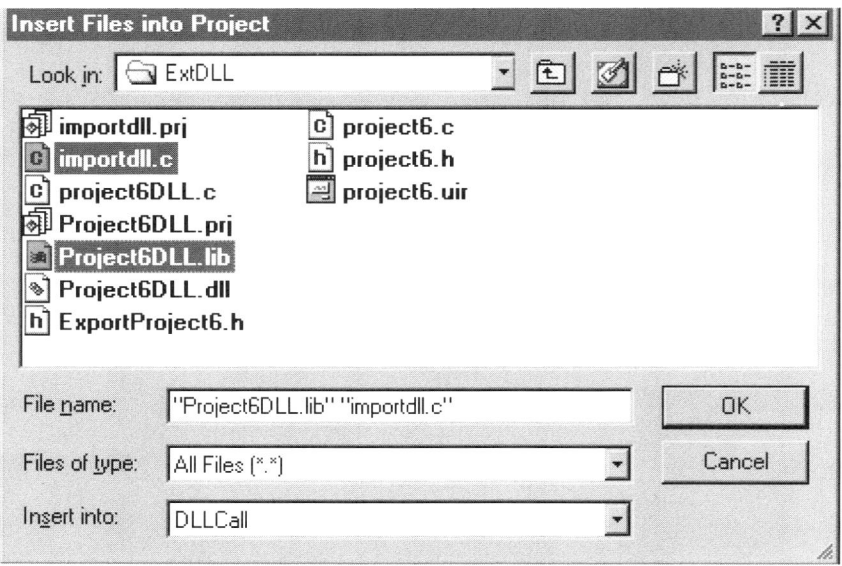

Figure 10–24
Microsoft Visual C++ Insert Files into DLL Project

Chapter 10 • External Compiler Support

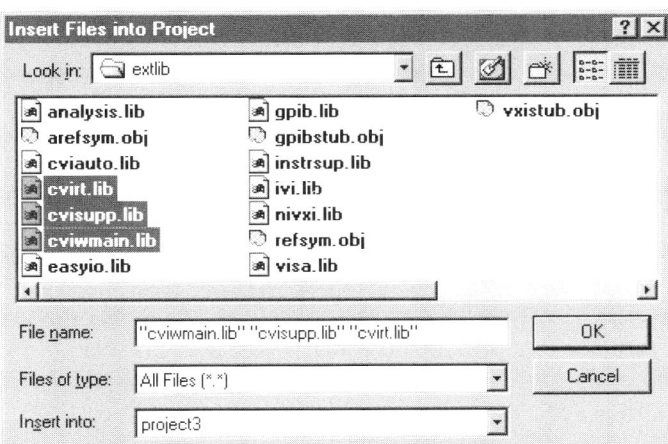

Figure 10–25
Microsoft Visual C++ Insert Library Files into DLL Project

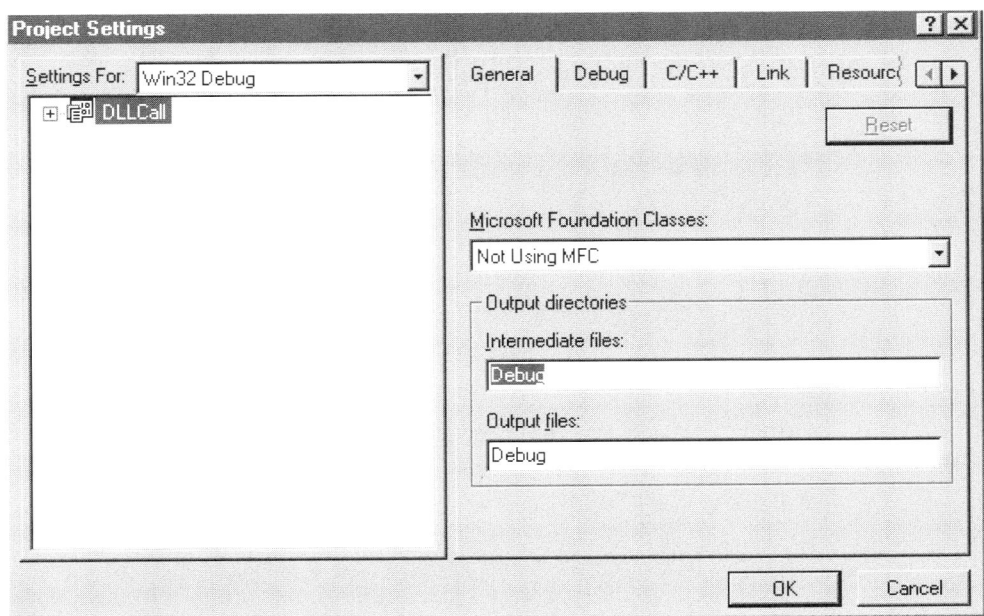

Figure 10–26
Microsoft Visual C++ DLL Project Settings for Win32 Debug

under the project folder (directory) to store the intermediate and output files if you are creating a debug version of the executable. It easier to work with files in the same folder (directory). Therefore, I recommend you remove `Debug` from both the **Intermediate Files** and **Output Files** indicator boxes. You should always create the project using the `Debug` version to enable the use of the Debugger (if required).

Similarly, select **Project>>Settings** to display a dialog box as shown in Figure 10–27. In the **Settings For** display box, select `Win32Debug`. On the right side under the **General** tag, you will see `Release` under **Intermediate Files** and **Output Files.** As described above, this will create a `Release` folder (directory) under the project directory. Again, you would like to keep everything in the same directory, so delete `Release` from both the **Intermediate Files** and **Output Files** dialog boxes.

When the project is finalized and you are ready to distribute the file, the `Release` version should be created. The `Debug` version is larger and runs slower. The `Release` version does not have any of the debug features, and so it is smaller and runs faster.

From the **Options** menu under **Tools,** select **Directories** and add `c:\cvi5\include` (or an equivalent path) as shown in Figure 10–28. This gives the external compiler the path name of the *CVI include* directory.

The external compiler project will look for the `project6.uir` and `project6.h` files in the same folder (directory) where the executable is created, so

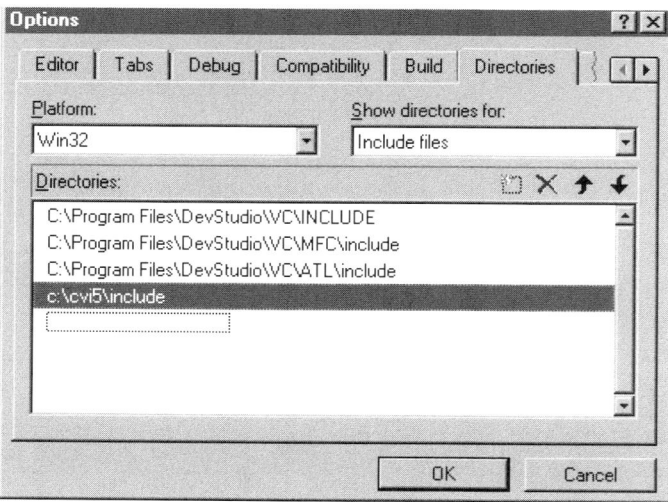

Figure 10–27
Microsoft Visual C++ DLL Project Settings Win32 Release

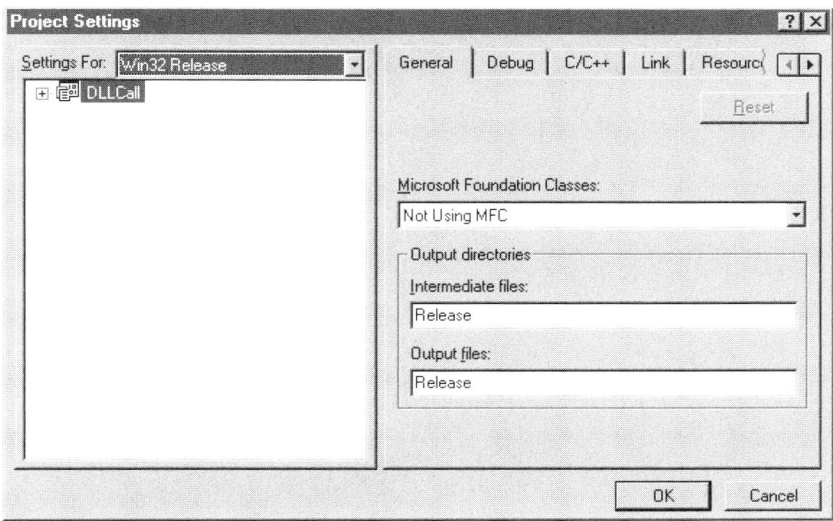

Figure 10–28
Microsoft Visual C++ Directory Options

you must copy `project6.uir` and `project6.h` to the `DLLCall` folder. Also, copy `project6DLL.lib` and `project6DLL.dll` from the `ExtDLL` folder to the `DLLCall` folder. Build the project using **Build>>Build DLLCall.exe** command or <F7> as shown in Figure 10–30. The `DLLCall` folder should now contain the files shown in Figure 10–29.

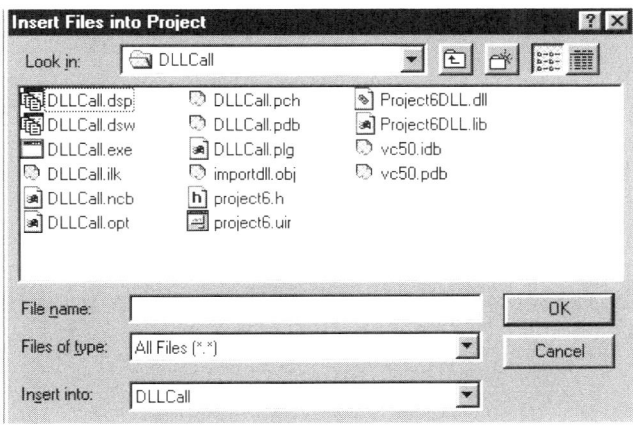

Figure 10–29
Files in DLLCall Folder

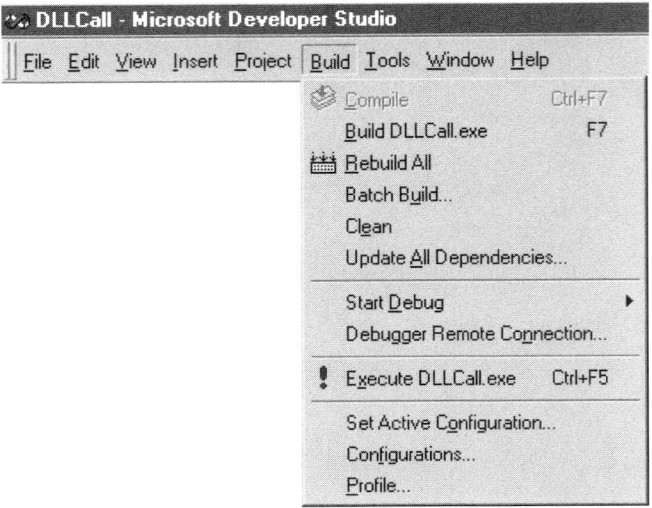

Figure 10–30
Microsoft Visual C++ Project Build Menu

Run the project from the **Build** menu by selecting **Execute ExtCall.exe** or <Ctrl+F5>.

The same GUI, as shown in Figure 10–16, displays as it did when the project was run from *CVI*. Check that all callbacks functions are operational by clicking on all the GUI command buttons.

Summary

The topic of external compiler support was introduced, using two methods of defining the user interface callback functions from an external compiler. The first method dealt with creating an *object* file of the user interface callbacks. The second method consisted of creating a *DLL* for the callbacks. These callback files were then loaded and built in the Microsoft Visual C++ compiler, and the *CVI* project was run from the external compiler. Any of the other compatible compilers could have been used to execute the callbacks.

11

GPIB COMMUNICATIONS

Chapter Highlights

- GPIB Communication Introduction and History
- GPIB Communication
- Getting Started with GPIB
- GPIB Communication Using the NI-488 Routines
- WIN32 Interactive Control Utility
- NI Spy Utility
- Using the IEEE-488.2 Standards
- Using GPIB Instrument Drivers
- Summary

The real power and flexibility of *CVI* is shown when using *CVI* to communicate with devices, facilitated by numerous communication drivers available via the instrument function panel. This can be achieved using the *General Purpose Interface Bus* (GPIB) interface protocol.

In this chapter, the GPIB communication and its implementation, using the various IEEE 488 standards, is explained. A short history of the GPIB evolution is given to familiarize you with the various standards. Projects using the IEEE-488.1 and IEEE-488.2 routines are created, and the board-level and device-level routines are explained.

To make the debugging task easier, the *Win32 Interactive Utility* is explained. It enables you to send the commands interactively through the keyboard.

The use of *NI Spy* is explained through an example. *NI Spy* is a debugging tool that is loaded with the GPIB drivers. It monitors and records the interface communication between the GPIB interface board and the GPIB and *Virtual Instrumentation Software Architecture* (VISA) interfaces.

A project using the multimeter is demonstrated using the instrument drivers to give the user a flavor of communicating with interface boards and devices using the device driver functions.

GPIB Communication Introduction and History

The GPIB is a bus protocol for controlling instruments and transferring data between computers and the instruments (devices). Protocol refers to an exact sequence of bits, characters, and control codes used for transferring data between computers and devices on the communication channel (interface). GPIB is a very popular, powerful, and comparatively fast communication protocol, transferring upwards of 800 Kbytes/sec. This communication interface was originally developed by Hewlett-Packard in 1965 to control and transfer data between the computer and the test equipment and was called *Hewlett-Packard Interface Bus* (HP-IB). Hewlett-Packard still refers to it as HP-IB.

The Institute of Electrical and Electronic Engineers (IEEE) standardized the electrical, mechanical, and functional specification for GPIB protocol and it became known as IEEE-488 standard. In 1987, IEEE renamed this standard to IEEE-488.1-1987. The International Electrotechnical Commission (IEC) adopted this as an international standard and called it IEC 625-1, and it is now referred to as the IEC Bus in some countries.

Even with this new standard, there was much confusion regarding instrumentation communication. In 1990, some of the instrumentation companies defined the *Standard Commands for Programmable Instrumentation* (SCPI) that further standardized the instrumentation commands.

SCPI provides a standard instrument vocabulary to control and command test and measurement equipment. This enables different manufacturers' instruments to be interchanged without changing the software considerably, since the same commands and device responses control corresponding instrument functions in the SCPI compatible equipment.

The SCPI is structured in a hierarchical command tree. Functions, sub-functions, and sub-sub-functions are separated by a colon (:). For instance, if you want to send a command to the voltmeter to configure it to read DC voltage, the SCPI command would be `MEAS:VOLT:DC?`.

The SCPI standard was further refined to standardize the way instruments and controllers operate. In 1992, this standard became known as the IEEE-488.2-1992. Almost all recently designed equipment follows this convention.

In 1993, a high-speed data transfer protocol for the IEEE-488, called the HS488, was introduced, which increases the speed of communication to about ten times the original GPIB speed to 8 Mbytes/sec.

The GPIB IEEE-1394 standard, called the Firewire, controls the instruments over the IEEE-1394 bus, using the VXIbus system. This bus can communicate with speeds of up to 400 Mb/sec. **VXI** stands for **V**ME (Versa-Modular Eurocard) e**X**tensions for **I**nstruments, a very high-performance system for instrumentation. The VXI test and measuring instruments are on individual cards with no front panels for display and control as in the conventional instruments, and are plugged into a VXI chassis. GPIB communication for VXI devices will not be discussed here since it is a topic for an advanced book.

GPIB Communication

Data is communicated between the PC and the device(s) using a GPIB interface board or a GPIB box connected externally to the computer. In this book, reference made to the GPIB interface board will also mean the GPIB box. To communicate over the GPIB, you must have the GPIB device drivers installed and the device(s) connected using the GPIB cable. The GPIB interface board is usually plug-and-play and comes with the installation software, a complete set of installation instructions, and user manuals.

The GPIB communication is established using the interface board connected to your computer. You can install more than one GPIB interface board on your computer and connect them to different devices. Be sure to assign different addresses to the GPIB interface boards. Only one board can be operated at a time, and the board in operation is called the Controller-In-Charge (CIC). You cannot issue commands to several interface boards simultaneously.

You can control up to fifteen devices on one bus (including the interface board) or up to thirty-one devices using the GPIB *bus extender*. Each GPIB-based device on the bus, including the GPIB interface board, must have a unique address. The interface board is usually assigned 0 address, and the devices can be assigned the remaining addresses (1 through 30). Most devices come with their addresses factory-installed. You can change the GPIB address on the device if you wish from either the switch setting at the back of the instrument or the front panel controls. Be sure that each device on the same bus has a unique address.

The GPIB devices can be Talkers, Listeners, or Controllers. The Talker sends out the data messages, the Listeners receive the messages, and the Controller manages the flow of information on the bus. Some instruments can be both Talkers and Listeners if they are receiving commands and sending back data to the application through the interface board.

The devices are connected to each other and to the computer using a special GPIB cable. This cable has a similar connector on both ends that consists of a plug on one side and a receptacle on the backside of the same connector for connecting compatibility. The GPIB bus is a twenty-four-conductor parallel, or bus-based, interface. There are eight data lines that transfer one byte (eight bits) at a time, making it possible to transfer data in ASCII strings. The five Interface Management Lines manage the flow of data across the bus. The three hardware handshake lines asynchronously control the transfer of message bytes between devices. These lines take care of the send and receive message bytes on the data lines without any data collision. The remaining eight lines are ground lines.

You can connect the devices in a linear configuration, by connecting the computer to the first device, connecting the first device to the second device, and so on. In the star configuration, the computer is connected to one device, and to that device, all the remaining devices are connected.

As good as the GPIB communication may sound, there are a few limitations. First, as you saw above, some special hardware is required to communicate over this interface. The physical connection between the devices is limited to a distance of four meters. The average distance of two meters over the entire bus is required, and the total cable length should not exceed twenty meters. A bus extender should be used if you are exceeding the GPIB specified cable length. A further restriction requires at least two thirds of the devices on the bus should be powered on.

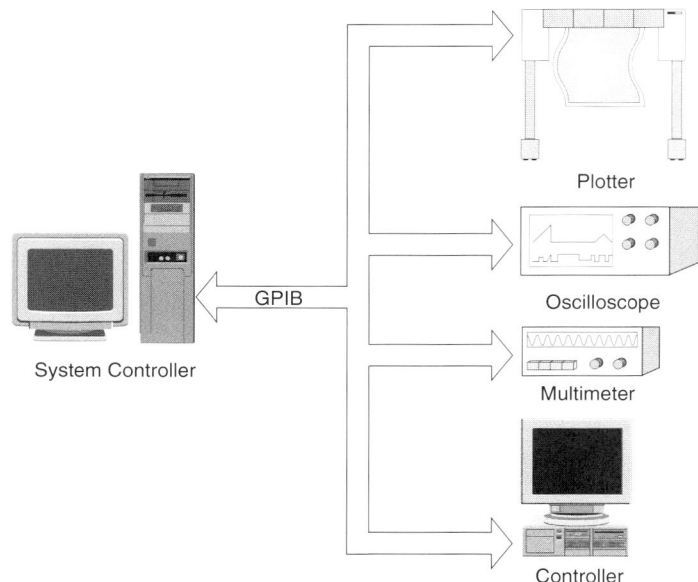

Figure 11–1
Typical GPIB System

Figure 11–1 shows one of the ways the GPIB devices can be connected to the computer.

Getting Started with GPIB

The GPIB interface board comes with the National Instruments standard GPIB DLL for Windows. For Sun users, the standard Solaris-based GPIB drivers are used.

The manufacturer of the interface board recommends that you install the software first and then the GPIB interface board. When the system is booted, the plug-and-play card is detected and installed automatically. Once you have the interface board and the software installed on your computer, you are ready to communicate using the GPIB interface. You can communicate with GPIB-compatible devices using the IEEE-488.2 routines. Some of the earlier devices do not accept the IEEE-488.2 routines and can only communicate using the IEEE-488 routines. On the other hand, all GPIB-based devices can converse using the IEEE-488 routines. Most of the newer instruments have the instrument drivers written using the VISA format to facilitate communication. VISA is a topic for an advanced book and will not be discussed here. The newer devices also follow the SCPI protocol. The GPIB drivers can easily be downloaded from the manufacturers' or National Instruments' web site.

The GPIB library routines can be broadly categorized into device-level and board-level routines. Device-level routines handle the bus management operations, such as configuring the device, reading and writing data to the device, obtaining the status of the device, performing serial and parallel polling of the device, and many more operations.

The board-level routines handle the basic GPIB interface board control related routines, such as placing the interface board online and taking it offline, changing the primary and secondary addresses, checking the presence of the device on the bus, and returning the status of the eight GPIB control lines. These board-level routines are sometimes necessary when the high-level routines cannot achieve the programming needs.

The GPIB library routines are described in the *LabWindows/CVI Standard Libraries Reference Manual* and in the *NI-488.2M Function Reference Manual for Win32*. The *NI-488.2M Function Reference Manual for Win32* is part of the manual set that is shipped with the interface board from National Instruments. To go through the complete list of these routines and describe their usage is redundant. Instead, a *CVI* project will be created, and the purpose of the IEEE-488 routines will be explained as encountered in the code.

GPIB Communication Using the NI-488 Routines

Take a look at the sequence of steps that are usually required to communicate with the interface board and the device(s). You may want to refine these steps or just follow them depending on your requirements.

1. Initialize the interface board.
2. Open the device for communication.
3. Check the device for GPIB communications at the specified address.
4. Configure the instrument(s) by writing configuration data.
5. Read data from the instrument.
6. Close the device and take the interface board offline.

It is always a good practice to check for possible errors whenever communicating with the hardware (interface board or device(s)). Doing so acts as an effective debugger and will prevent you from wasting much time in locating possible communication errors. Error checking is described below as we go through the code listing.

The addressing scheme used in GPIB communication is discussed briefly here. All GPIB interface boards and the devices connected to the GPIB bus must have a unique address to establish communication. The interface board is usually at address 0. You can have more than one interface board on the same computer, but these boards must have different addresses. Only one board at a time can be the Controller in-Charge (CIC) and can communicate with devices on the same bus.

A GPIB address consists of a sixteen-bit address consisting of the primary address and an optional secondary address. The primary address is a decimal number in the range of 0 to 30. The device address is pre-set at the factory, though the user can change it programmatically or otherwise to any unique address. The instrument manual for the device gives the primary device address and is usually accessible from the front panel of the instrument. The secondary address is seldom used and only provides flexibility in accessing devices uniquely when, for example, they reside on the same primary address mounting rack. This may occur in situations where a VXI chassis is connected via a GPIB-VXI/C interface board and the entire chassis is given a single primary GPIB address, and each instrument card given the secondary address. The secondary address is a decimal number between 96 and 126.

Chapter 11 • GPIB Communications

The Controller uses the primary address and/or the secondary address to establish communication with the device. The primary address is stored in the lower byte of the address and the secondary address in the upper byte. If the device uses both the primary and secondary address, then the full address of the device is in the form `0xsspp`, where `0x` represents the hexadecimal format in ANSI C, `ss` is the secondary address, and `pp` is the primary address.

You will create a *CVI* project (`project11-1.prj`) using the NI-488 routines to communicate with the GPIB-based devices and the interface board. These routines are explained in detail in the *GPIB NI-488.2M Function Reference Manual for Win32*.

The purpose of `project11-1` is to establish basic communication between a generic GPIB-based device and to explain the procedures involved with the interface board and the device communication.

Figure 11–2 shows the GPIB communication user interface for a generic device, a device for which the device addresses can be specified in the GUI by the user, and the GPIB communication is established with that device by writing appropriate instrument dependent commands. Only the IEEE-488.1 routines are used to communicate with the device. These routines will be described as you build the project.

The description of the project is given in the code listing comments shown in Figure 11–3, and the code description follows.

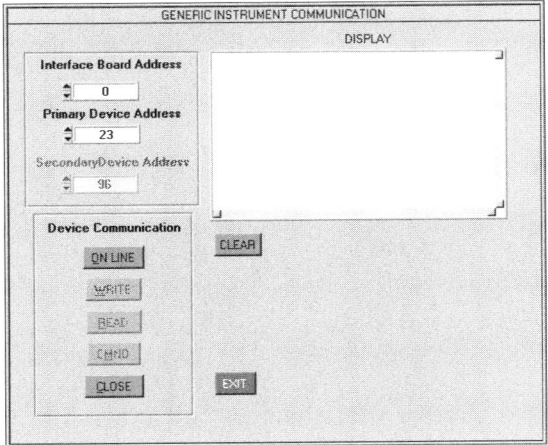

Figure 11–2
Project11–1 User Interface

```
/*******************************************************************
Project Name: project11-1.prj

This project establishes communication with a generic GPIB-based device.
The user selects the GPIB interface board address: the primary address,
and optionally a secondary device addresses. The board is brought on-line,
and after the device is no longer in-use, is taken off-line. The user can
write specific data to the device and read back the responses.
The data written and read is displayed on the GUI. Error checking is
performed whenever data is written to the interface board or the device.

Author: Shahid F. Khalid

*******************************************************/

#include <ansi_c.h>
#include "project11-1CMD.h"
#include <gpib.h>
#include <formatio.h>
#include <cvirte.h>     /* Needed if linking in external compiler; harmless otherwise */
#include <userint.h>
#include "project11-1.h"

static int genericHandle, CommandHandle;
char buf[260], write_buf[30], write_cmd[260];

int deviceDesc, BoardOffLine, WriteFlag;
unsigned short BoardAddress,DeviceAddress;

int display_error( int *iberr);

//Main function to load the panel and place interface
//board on-line
int main (int argc, char *argv[])
{
    if (InitCVIRTE (0, argv, 0) == 0)/* Needed if linking in external compiler; harmless otherwise */
        return -1;  /* out of memory */
    if ((genericHandle = LoadPanel (0, "project11-1.uir", GENERIC)) < 0)
        return -1;

    DisplayPanel (genericHandle);

    ibonl(BoardAddress,1);      //Put interface board on line
    if (ibsta & ERR)        //Check for error
    {
        Fmt( buf, "%s<Error Opening device");
```

```
46              InsertTextBoxLine(genericHandle, GENERIC_DISP, -1, buf);
47              return -1; //abort program
48          }
49      RunUserInterface ;
50      return 0;
51  } //main
52
53
54  //Place device on line, invoked by "ONLINE" command button
55  int CVICALLBACK IbdevCB (int panel, int control, int event,
56              void *callbackData, int eventData1, int eventData2)
57  {
58      short DevicePresent;
59      switch (event)
60      {
61          case EVENT_LEFT_CLICK:
62              //Obtain the user-specified interface and device addresses
63              GetCtrlVal(genericHandle, GENERIC_BRD_ADD,&BoardAddress);
64              GetCtrlVal(genericHandle, GENERIC_DEVICE_ADD,&DeviceAddress);
65              //Set up the device descriptor based on the device address
66              deviceDesc = ibdev (BoardAddress, DeviceAddress, NO_SAD, T3s, 1, 0);
67              if (deviceDesc <0)
68              {
69                  Fmt( buf, "%s<Cannot open device addressed");
70                  InsertTextBoxLine(genericHandle, GENERIC_DISP, -1, buf);
71                  return -1; //abort program
72              }
73              //Find device on line
74              ibln(deviceDesc,DeviceAddress,NO_SAD, &DevicePresent);
75              //Check for errors
76              if (ibsta & ERR) display_error( &iberr);
77
78              if ( DevicePresent >0)  //Device found
79              {
80                  Fmt( buf, "%s<Found device at this address");
81                  InsertTextBoxLine(genericHandle, GENERIC_DISP, -1, buf);
82              }
83              else //Device not found
84              {
85                  Fmt( buf, "%s<No device at this address");
86                  InsertTextBoxLine(genericHandle, GENERIC_DISP, -1, buf);
87                  return -1; //abort program
```

Figure 11-3
Project11-1 Code Listing

```
90                }
91                break;
92        }
93        SetInputMode (genericHandle, GENERIC_WRITE_DEV, 1);   //Enable the command button
94        SetInputMode (genericHandle, GENERIC_CLOSE_BRD, 1);   //Enable the command button
95        SetInputMode (genericHandle, GENERIC_CMD, 1);          //Enable the command button
96        BoardOffline=0; //GPIB interface board is on-line
97        return 0;
98 } //IbdevCB
99
100
101 //Write the GPIB device commands. Invoked from the "WRITE" command button
102 int CVICALLBACK WriteToDeviceCB (int panel, int control, int event,
103                                   void *callbackData, int eventData1, int eventData2)
104 {
105        switch (event)
106        {
107            case EVENT_LEFT_CLICK:
108                //Load the GUI to write device data
109                if ((CommandHandle = LoadPanel (0, "project11-1CMD.uir", WRT)) < 0)
110                    return -1;
111                DisplayPanel(CommandHandle);
112                WriteFlag=1;//Set Flag to indicate "Write"
113                break;
114        }
115        return 0;
116 } //WriteToDeviceCB
117
118 //Hide the write device data panel and enable the "READ" command button.
119 //Invoked by "BACK" command button
120 int CVICALLBACK BackCB (int panel, int control, int event,
121                         void *callbackData, int eventData1, int eventData2)
122 {
123        switch (event)
124        {
125            case EVENT_COMMIT:
126                HidePanel(CommandHandle);
127                SetInputMode (genericHandle, GENERIC_READ_DEV, 1); //Enable the
128                                                                    command button
129                break;
130        }
131        return 0;
132 } //BackCB
```

```
134
135  //Write the device command, and display on GUI.
136  int CVICALLBACK SendCommandData (int panel, int control, int event,
137                                    void *callbackData, int eventData1, int eventData2)
138  {
139      switch (event)
140      {
141          case EVENT_COMMIT:
142              if (WriteFlag)       //Write data to device
143              {
144                  //Get the user specified device command
145                  GetCtrlVal(CommandHandle,WRT_GPIB_CMD, write_buf);
146                  //Display command to write
147                  Fmt( buf, "%s<Written: %s to device ",write_buf);
148                  InsertTextBoxLine(genericHandle, GENERIC_DISP, -1, buf);
149                  //Write the data to the device
150                  ibwrt(deviceDesc,write_buf, StringLength(write_buf));
151                  if (ibsta & ERR) display_error( &iberr);
152
153                  SetInputMode (genericHandle, GENERIC_READ_DEV, 1); //Enable
154                                               the "READ" command button
155              if (WriteFlag);    //Write data to device
156          }
157          else   //Send GPIB command
158          {
159              //Get the user specified device command
160              GetCtrlVal(CommandHandle,WRT_GPIB_CMD, write_cmd);
161              //Display command to write
162              InsertTextBoxLine(genericHandle, GENERIC_DISP, -1, write_cmd);
163              //Write the data to the device
164              ibcmd(BoardAddress,write_cmd, StringLength(write_cmd));
165              if (ibsta & ERR)
166              {
167                  Fmt( buf, "%s<Error... Could not send the GPIB command");
168                  InsertTextBoxLine(genericHandle, GENERIC_DISP, -1, buf);
169                  return -1; //abort program
170              }
171              Fmt( buf, "%s<Written %s GPIB Command ",write_cmd);
172              InsertTextBoxLine(genericHandle, GENERIC_DISP, -1, buf);
173              SetInputMode (genericHandle, GENERIC_READ_DEV, 1); //Enable
174                                               READ" command button
175          }
176
```

Figure 11-3
Project11-1 Code Listing *(continued)*

```
            break;
        }
        return 0;
}//GoCmdCB

//Read data a from the device
int CVICALLBACK ReadFromDeviceCB (int panel, int control, int event,
        void *callbackData, int eventData1, int eventData2)
{
    char read_data[51], read_buf[260];
    int i;
    
    switch (event)
    {
        case EVENT_LEFT_CLICK:
            //Read data from device
            ibrd (deviceDesc, read_data, 50);
            
            if (ibsta & ERR) display_error( &iberr);
            
            Fmt(read_buf, "%s<Data read from device: %s",read_data);
            InsertTextBoxLine(genericHandle, GENERIC_DISP, -1, read_buf);
            SetInputMode (genericHandle, GENERIC_READ_DEV, 0); //Disable the
                                                                READ" command button
            break;
    }
    return 0;
}//ReadFromDeviceCB

//Routine to send interface messages to configure the GPIB state.
//Called from "CMND" command button.
int CVICALLBACK GPIBCmdCB (int panel, int control, int event,
        void *callbackData, int eventData1, int eventData2)
{
    char cmd_buf[260];
    
    switch (event)
    {
        case EVENT_COMMIT:
        
            if ((CommandHandle = LoadPanel (0, "project11-1CMD.uir", WRT)) < 0)
                return -1;
            
            DisplayPanel(CommandHandle);
```

```
222                    WriteFlag=0;                    //Reset flag to indicate GPIB command mode
223                    break;
224         }
225         return 0;
226 } //GPIBCmdCB
227
228 //Take the device off-line. Invoked by "CLOSE" command button
229 int CVICALLBACK IbonlOFFCB (int panel, int control, int event,
230                 void *callbackData, int eventData1, int eventData2)
231 {
232         switch (event)
233         {
234             case EVENT_LEFT_CLICK:
235
236                 ibonl (deviceDesc, 0);              //Take device off-line
237                 if (ibsta & ERR) display_error( &iberr);
238
239                 BoardOffline=1; //Board is taken off-line Flag
240
241                 SetInputMode (genericHandle, GENERIC_WRITE_DEV, 0);   //Disable the
242                                                                       command button
243                 SetInputMode (genericHandle, GENERIC_READ_DEV, 0);    //Disable the
244                                                                       command button
245                 SetInputMode (genericHandle, GENERIC_CLOSE_BRD, 0);   //Disable the
246                                                                       command button
247                 SetInputMode (genericHandle, GENERIC_CMD, 0);         //Enable the command
248                                                                       button
249                 break;
250         }
251         return 0;
252 } //IbonlOFFCB
253
254 //Clear the display
255 int CVICALLBACK ClearDisplayCB (int panel, int control, int event,
256                 void *callbackData, int eventData1, int eventData2)
257 {
258         switch (event)
259         {
260             case EVENT_LEFT_CLICK:
261                 DefaultCtrl(genericHandle, GENERIC_DISP);
262                 break;
263         }
264
```

Figure 11-3
Project11-1 Code Listing (*continued*)

```
265             return 0;
266     } //ClearDisplayCB
267     //Exit the project
268     int CVICALLBACK ExitCB (int panel, int control, int event,
269             void *callbackData, int eventData1, int eventData2)
270     {
271         switch (event)
272         {
273             case EVENT_COMMIT:
274                 if (BoardOffLine ==0)  //Has board been taken off-line?
275                 {                      // If no, take it off-line
276                     ibonl(deviceDesc, 0); //Good practice to take device off-line when
277                                                                                  exiting
278                     if (ibsta & ERR) display_error( &iberr);
279                 }
280                 QuitUserInterface (0);
281                 break;
282         }
283         return 0;
284     } //ExitCB
285     //Isolate and display error messages
286     int display_error( int *iberr)
287     {
288         switch (*iberr)
289         {
290             case 0:
291                 Fmt( buf, "%s<Error Mnemonic: EDVR iberr value= %d  System error",iberr);
292                 break;
293             case 1:
294                 Fmt( buf, "%s<Error Mnemonic: ECIC iberr value= %d  Function requires GPIB
295                                                                       board to be CIC",iberr);
296                 break;
297             case 2:
298                 Fmt( buf, "%s<Error Mnemonic: ENOL iberr value= %d No Listeners on GPIB",iberr);
299                 break;
300             case 3:
301                 Fmt( buf, "%s<Error Mnemonic: EADR iberr value= %d GPIB board not addressed
302                                                                            correctly",iberr);
303                 break;
304             case 4:
305                 Fmt( buf, "%s<Error Mnemonic: EARG iberr value= %d  Invalid argument to function
306                                                                            call",iberr);
```

```
311            break;
312    case 5:
313            Fmt( buf, "%s<Error Mnemonic: ESAC iberr value= %d GPIB board not System
314                                                             Controller as required",iberr);
315            break;
316    case 6:
317            Fmt( buf, "%s<Error Mnemonic: EABO iberr value= %d I/O operation aborted
318                                                             (timeout)",iberr);
319            break;
320    case 7:
321            Fmt( buf, "%s<Error Mnemonic: ENEB iberr value= %d Nonexistent GPIB
322                                                             board",iberr);
323            break;
324    case 8:
325            Fmt( buf, "%s<Error Mnemonic: EDMA iberr value= %d DMA error",iberr);
326            break;
327    case 9:
328    case 13:
329    case 17:
330    case 18:
331    case 19:
332            Fmt( buf, "%s<iberr value= %d is un-defined",iberr);
333            break;
334    case 10:
335            Fmt( buf, "%s<Error Mnemonic: EOIP iberr value= %d Asynchronous I/O in
336                                                             progress",iberr);
337            break;
338    case 11:
339            Fmt( buf, "%s<Error Mnemonic: ECAP iberr value= %d No capability for
340                                                             operation",iberr);
341            break;
342    case 12:
343            Fmt( buf, "%s<Error Mnemonic: EFSO iberr value= %d File system error",iberr);
344            break;
345    case 14:
346            Fmt( buf, "%s<Error Mnemonic: EFSO iberr value= %d GPIB bus error",iberr);
347            break;
348    case 15:
349            Fmt( buf, "%s<Error Mnemonic: ESTB iberr value= %d Serial poll status byte queue
350                                                             overflow",iberr);
351            break;
```

Figure 11-3
Project11-1 Code Listing (*continued*)

```
352        case 16:
353            Fmt( buf, "%s<Error Mnemonic: ESRQ iberr value= %d  SRQ stuck in ON
354                                                                position",iberr);
355            break;
356
357        case 20:
358            Fmt( buf, "%s<Error Mnemonic: ETAB iberr value= %d  Table problem",iberr);
359            break;
360        default:
361            Fmt( buf, "%s<Illegal iberr value= %d ",iberr);
362            break;
363        }
364
365        InsertTextBoxLine(genericHandle, GENERIC_DISP, -1, buf);
366
367    return -1;
368
369
370 }//display_error
371
```

Figure 11-3
Project11-1 Code Listing *(continued)*

Chapter 11 • GPIB Communications

In this project, only the primary address will be used to communicate with the device. The control box for setting the secondary address is shown dimmed in Figure 11–2 and can be enabled and used if your device requires the secondary address.

The *main* function loads and displays the panel, `project11-1.uir.`, at lines 37 and 40 respectively.

At line 42, the GPIB routine *ibonl* is used to place the interface board on-line. The prototype of *ibonl* follows, and the routine arguments are explained in Table 11–1.

```
int ibonl( int device/board address, int online/offline);
```

The GPIB routines are available by selecting **Library>>GPIB/GPIB 488.2** as shown in Figure 11–4.

From the GPIB/GPIB 488.2 Library Function Panel, as shown in Figure 11–5, you can select the appropriate routine.

For example, the function panel for *ibonl* routine is displayed as shown in Figure 11–6.

Line 78 of the code calls the error detection, isolation, and displaying routine: *display_error*. Before discussing how the errors are isolated, let me explain the status word *ibsta*. The GPIB global status word *ibsta* contains the state of the GPIB hardware and the interfaces. This status word is updated by all the IEEE-488 routines except *ibfind* and *ibdev*. The status word bits are described in Appendix B of the *GPIB NI-488.2M Function Reference Manual for Win32*. The status word *ibsta* is a sixteen-bit word in which a set bit in any of the positions indicates that a particular state is enabled. You can examine the meaning of a

Table 11–1 ibonl *Routine*

Input/Output	Name	Type	Description
Input	device/board address	integer	board or device descriptor
	online/offline	integer	non-zero places online, 0 takes it off line
Return Value	value of *ibsta*	integer	return status of the routine call; For Status Word Conditions, see Appendix B of *GPIB, NI-488.2M Function Reference Manual for Win32*

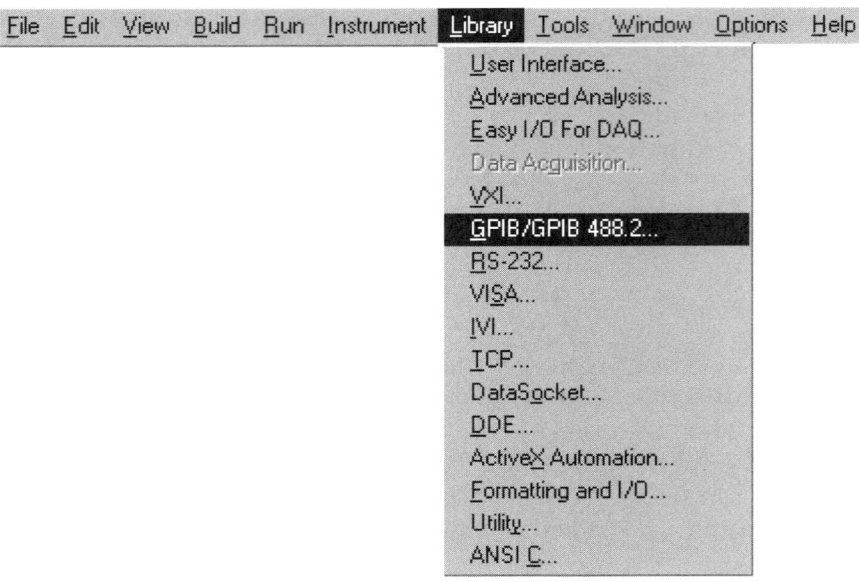

Figure 11–4
Accessing the GPIB Library Routines

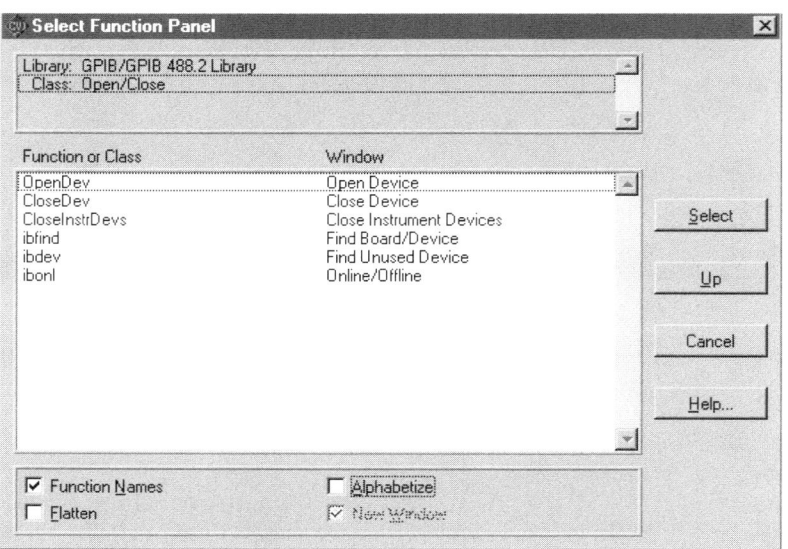

Figure 11–5
GPIB/GPIB 488.2 Function Panel

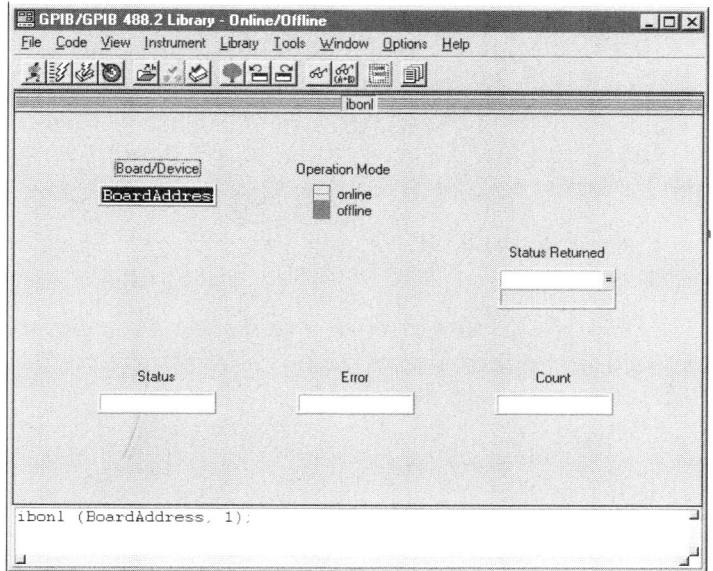

Figure 11–6
Function Panel for *ibonl* Routine

particular bit by performing a bitwise AND with the bit mnemonic indicated in the Status Word Layout in Appendix B of the *GPIB NI-488.2M Function Reference Manual for Win32*. To check for errors in GPIB communications, perform a bitwise AND of *ibsta* and ERR. It is advisable to perform error checking whenever any GPIB messages are sent/received to/from the device(s) or interface board as shown at line 43 in the code.

The global variable *iberr* is a sixteen-bit variable that returns the GPIB errors. This value is then passed to the *display_error* function that displays the appropriate error string on the user interface. Instead of referring to the numbers in Appendix C of the *GPIB NI-488.2M Function Reference Manual for Win32*, the error mnemonics and the messages are displayed on the GUI.

There are two other global variables, *ibcnt* and *ibcntl*, that are updated with each GPIB call. These are thirty-two-bit integers in Win32 applications that give the number of bytes transferred. The global variable *ibcnt* is a sixteen-bit integer that is used in MS-DOS applications, and *ibcntl* is a thirty-two-bit integer variable that is used in Windows. For compatibility between various systems, it is advisable to use the variable *ibcntl*. These variables are useful when you want to know the number of bytes transferred over the bus.

Between lines 55 and 99, the callback function *IbdevCB* is invoked when the **ON LINE** command button is selected. Lines 65 and 66 obtain the user-specified interface board and device addresses. If you choose, you can include the *GetCtrlVal* function here to also obtain the secondary address from the user interface control. This function sets up the device descriptor, a temporary integer value used by the handler. It is used in all subsequent device-level calls. The prototype of *ibdev* routine follows and the arguments for this routine are explained in Table 11–2.

```
int ibdev ( int board address, int pad, int sad, int tmo, int eot,
                                                              int eos);
```

At line 76, the *ibln* routine queries if there is a device at the specified primary and secondary addresses on the bus. The prototype for this routine follows and the arguments are described in Table 11–3.

Table 11–2 ibdev *Routine*

Input/ Output	Name	Type	Description
Input	board address	integer	board address to which device is connected
	pad	integer	primary GPIB address of device
	sad	integer	secondary GPIB address of device; if secondary address is not used, enter NO_SAD
	tmo	integer	time-out value for I/O communication with device; T followed by the time-out value in seconds, e.g., T10s for ten seconds.
	eot	integer	if enabled, specifies the END message is sent automatically with the last byte
	eos	integer	specifies the end of string character and the data transfer termination method; refer to the function panel for *ibdev* for more details
Return Value	device descriptor	integer	unique descriptor to communicate with the device

Chapter 11 • GPIB Communications

Table 11–3 ibln *Routine*

Input/Output	Name	Type	Description
Input	board address	integer	board address to which device is connected
	pad	integer	primary GPIB address of device
	sad	integer	secondary GPIB address of device; if secondary address is not used, enter NO_SAD.
Return Value	listen	short	0 means no device found; non-zero means device found

```
int ibln ( int board address, int pad, int sad, short *listen);
```

The appropriate message indicating the presence or absence of a device is displayed between lines 80 through 91.

Lines 94–96 enable the command buttons **WRITE**, **CLOSE** and **CMND** that are shown dimmed in Figure 11–7. Their attributes are set *dimmed* at the initiation of the project. This prevents the user from trying to write or read data to

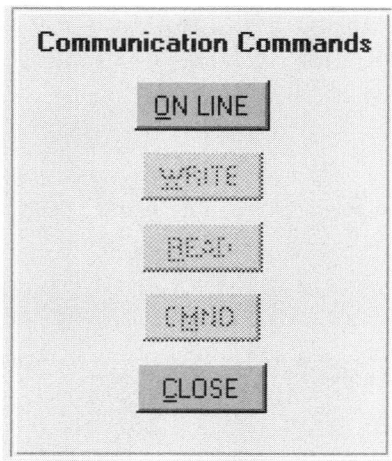

Figure 11–7
Communication Command Buttons

the board or the device before the appropriate GPIB setup is established. The **READ** button is not enabled until the string is written to the GPIB interface.

At line 103, the *WriteToDeviceCB* function is called when the **WRITE** command button, shown in Figure 11–7, is selected. This callback function loads and displays the user interface shown in Figure 11–8. You can write a valid GPIB string command to transmit to the interface or send data to the device.

After entering the string in the dialog box, press <Enter> to send the string to the device or the interface board. By invoking <Enter>, the callback function *SendCommandData* is called, which is the callback function associated with this dialog box. The *SendCommandData* function obtains the string command entered in the dialog box, displays it on the GUI, and writes to the GPIB device via the *ibwrt* routine on line 151 or to the *ibcmd* routine at line 165. The `WriteFlag` value at line 143 determines which GPIB function is called (*ibwrt* or *ibcmd*). The `WriteFlag` value is set in the *WriteToDeviceCB* or in the *GPIBCmdCB* function depending upon whether the data is written to the device or sent to the interface respectively. The routine prototype for *ibwrt* follows and its arguments are shown in Table 11–4.

```
int ibwrt ( int Device Descriptor, void *write_cmd, long count );
```

The routine prototype of *ibcmd* follows, and its arguments are shown in Table 11–5.

```
   int ibcmd ( int board unit descriptor, void *cmd, long count );
```

The *ReadFromDeviceCB* callback function at lines 183 through 205 is enabled when the **READ** command button is selected. The GPIB function *ibrd* reads the data from the device. The *ibrd* function prototype follows and its arguments are shown in Table 11–6.

Figure 11–8
Enter GPIB Device Command User Interface

Chapter 11 • GPIB Communications

Table 11–4 ibwrt *Routine*

Input/Output	Name	Type	Description
Input	Device Descriptor	integer	board or device descriptor
	write_cmd	void	address of the write buffer
	count	integer	length of the string written
Return Value	value of *ibsta*	integer	return status of the routine call; for Status Word Conditions, see Appendix B of the *GPIB NI-488.2M Function Reference Manual for Win32*

```
int ibrd ( int Device Descriptor, void *read_data, long count );
```

The callback function *GPIBCmdCB* is invoked when the **CMND** command button is selected from Figure 11–7. Lines 218 and 221 load and display the panel, which is shown in Figure 11–8.

The *IbonlOFFCB* function takes the device offline, setting the last argument of *ibonl* routine to 0, and disables the command buttons indicated on lines 230–253.

Table 11–5 ibcmd *Routine*

Input/Output	Name	Type	Description
Input	board unit descriptor	integer	board descriptor
	cmd	void	address of the write command string to configure the GPIB state
	count	integer	length of the string written
Return	value of *ibsta*	integer	return status of the routine call; for ValueStatus Word Conditions, see Appendix B of the *GPIB NI-488.2M Function Reference Manual for Win32*

Table 11–6 ibrd *Routine*

Input/ Output	Name	Type	Description
Input	Device Descriptor	integer	board or device descriptor
	read_cmd	void	address of the read buffer
	count	integer	length of the string read
Return Value	count	long	the number of bytes to read, the number actually read are updated in the global variable *ibcntl*

The *display_error* function is listed between lines 290 and 370. This function converts the error code into the appropriate error message and displays it on the user interface.

The *ClearDisplayCB* and *ExitCB* callback functions are not discussed here since we have seen them many times in previous projects.

Win32 Interactive Control Utility

The *Win32 Interactive Control Utility* is used for communicating with the GPIB interface board and the GPIB-based devices by sending the commands interactively through the keyboard. This utility is useful if you want to learn about the device or to troubleshoot for possible communication errors in your software. This utility is installed when you install the GPIB interface board drivers.

To access the *Win32 Interactive Control Utility*, select **Start>>Programs>> GPIB Software>>Win32 Interactive Control** and Figure 11–9 is displayed. If you would like help on a particular IEEE-488 routine, you can type `help` followed by the routine name, as shown in Figure 11–10, or just type `help` to learn more about this utility. For example, Figure 11–10 shows the `help` for the *ibdev* routine.

To establish interactive communication with the device, enter `ibdev` at the : prompt. The *Win32 Interactive Control Utility* prompts you for the information required by this routine, as shown in Figure 11–11. Note that you can also enter all the routine arguments on the same line in the correct order

Chapter 11 • GPIB Communications

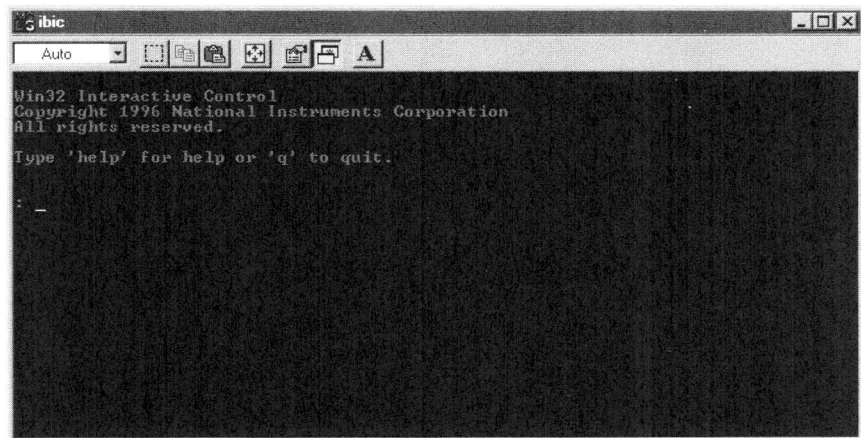

Figure 11–9
Win32 Interactive Control Utility Start Screen

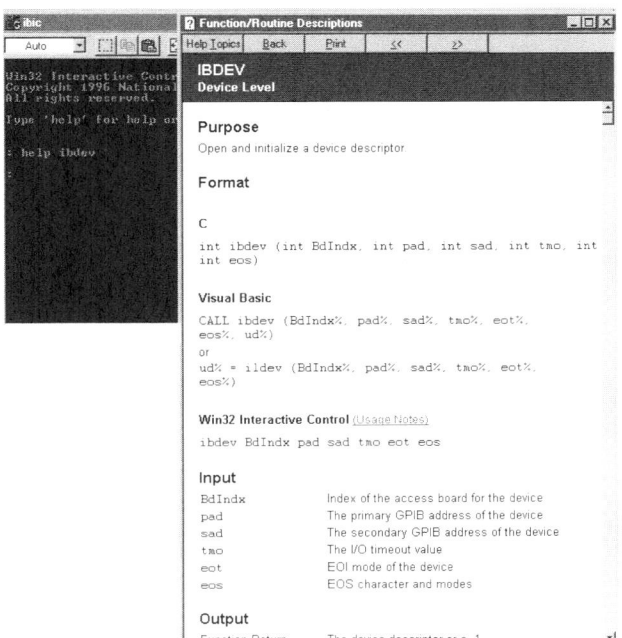

Figure 11–10
Win32 Interactive Utility help ibdev

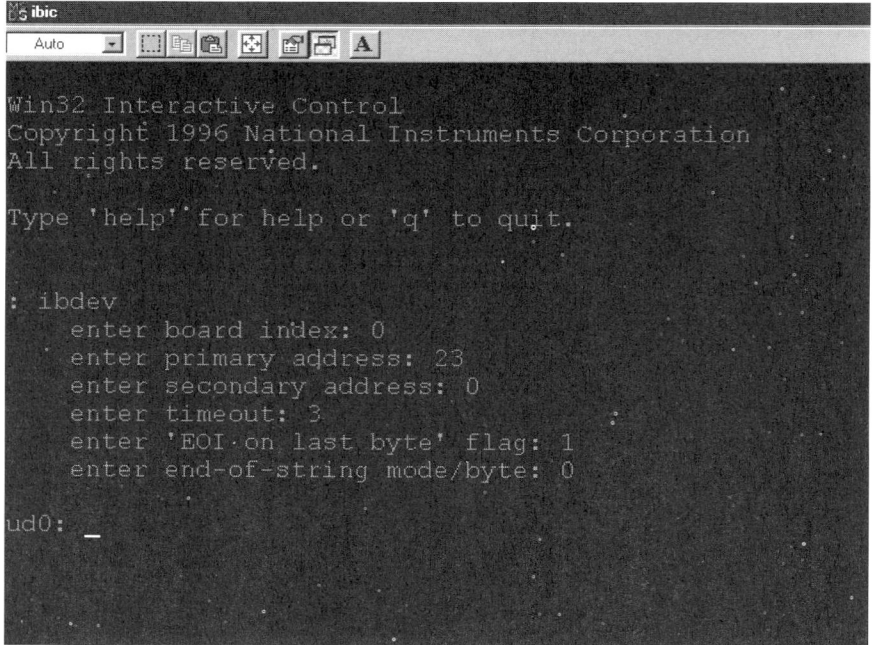

Figure 11–11
Win32 Interactive Utility ibdev data

following the routine name, instead of waiting for the *Win32 Interactive Control Utility* to prompt you for each argument value.

When all the information is entered correctly, the device handle is returned. In this example **ud0** is returned as the device handle and signifies that the device is ready to communicate and to accept further commands.

You can enter any of the routine commands followed by their arguments, possibly in the same order as your code if you are trying to use *Win32 Interactive Control Utility* for debugging. After you have made all the routine calls, you can enter ibonl to close the device and the **ud0** prompt is replaced by :.

A list of caveats when using *Win32 Interactive Control Utility* follow.

1. Unlike the GPIB routine calls, the board or device descriptors are never included as part of the routine arguments.
2. The number of bytes written or read to/from the device are returned automatically.
3. The routine return values are handled by the *Win32 Interactive Control.*

Chapter 11 • GPIB Communications

4. The numbers can be entered in decimal or hexadecimal format. The hexadecimal numbers are prefixed by `0x`.
5. The ASCII character strings must be enclosed within quotation marks.
6. When hex byte is entered as a string, you must use a backslash character and `x` followed by the hex value.
7. Instruments that require special end-of-string (EOS) characters must have these termination characters included within quotes. Some of the more common EOS characters are the carriage return, represented by `\r`, and the linefeed, by `\n`.
8. The primary and secondary addresses follow the same addressing rules explained previously.

NI Spy Utility

NI Spy is a debugging tool that monitors and records the interface communication between the GPIB interface board and the GPIB and VISA interfaces. This utility records all the calls made to the driver routines in chronological order with time stamps on each call. The results of the status word, errors, and the bytes transferred during communication are stored in the global variables *ibsta*, *iberr*, and *ibcntl* as explained previously.

The *NI Spy* software comes as part of the interface board drivers software and is automatically loaded on the system with the installation of the GPIB interface board software.

To run *NI Spy*, select **Start>>Programs>>GPIB Software>>NI Spy.** The main *NI Spy* window will be visible as shown in Figure 11–12. To record the results of the run, start the *NI Spy* by clicking on the blue arrow on the tool bar. Start the GPIB project that you want monitored by *NI Spy*. As each IEEE-488 and 488.2 call is made to the interface, the results are logged in the *NI Spy* window.

Figure 11–13 shows the monitoring of the GPIB calls with *NI Spy* when `project11-1` is run with various data inputs. The data input is shown on the `project11-1` DISPLAY window, but the *NI Spy* gives more details of the GPIB interface calls.

You can obtain further details of each of the monitored calls in the *NI Spy* main window by double-clicking on the line for which you require additional information. Double-clicking on a GPIB interface call in the *NI Spy* main window brings up the **NI Spy Property Sheet** for that line, as shown in Figure 11–14. On the **Property Sheet,** you have the choice of selecting

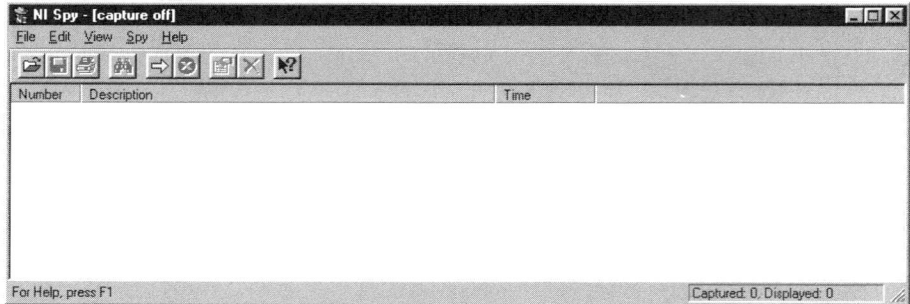

Figure 11–12
NI Spy Main Window

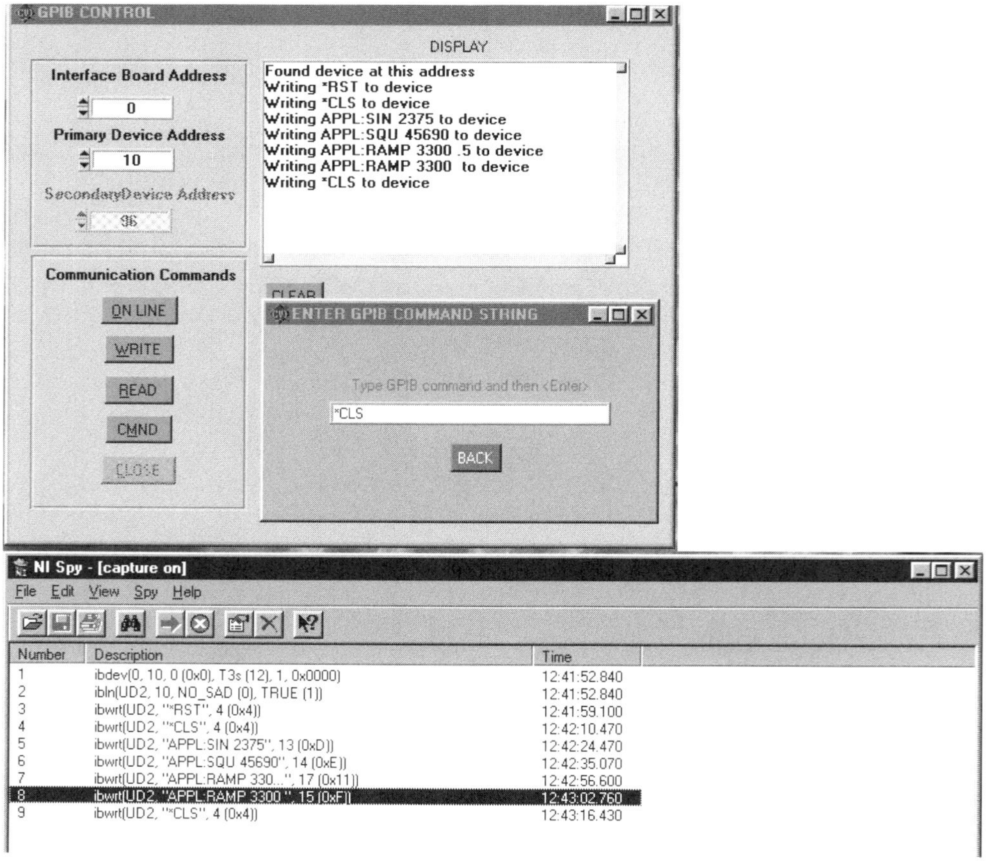

Figure 11–13
Project11–1 Interactive Monitoring with NI Spy

Chapter 11 • GPIB Communications

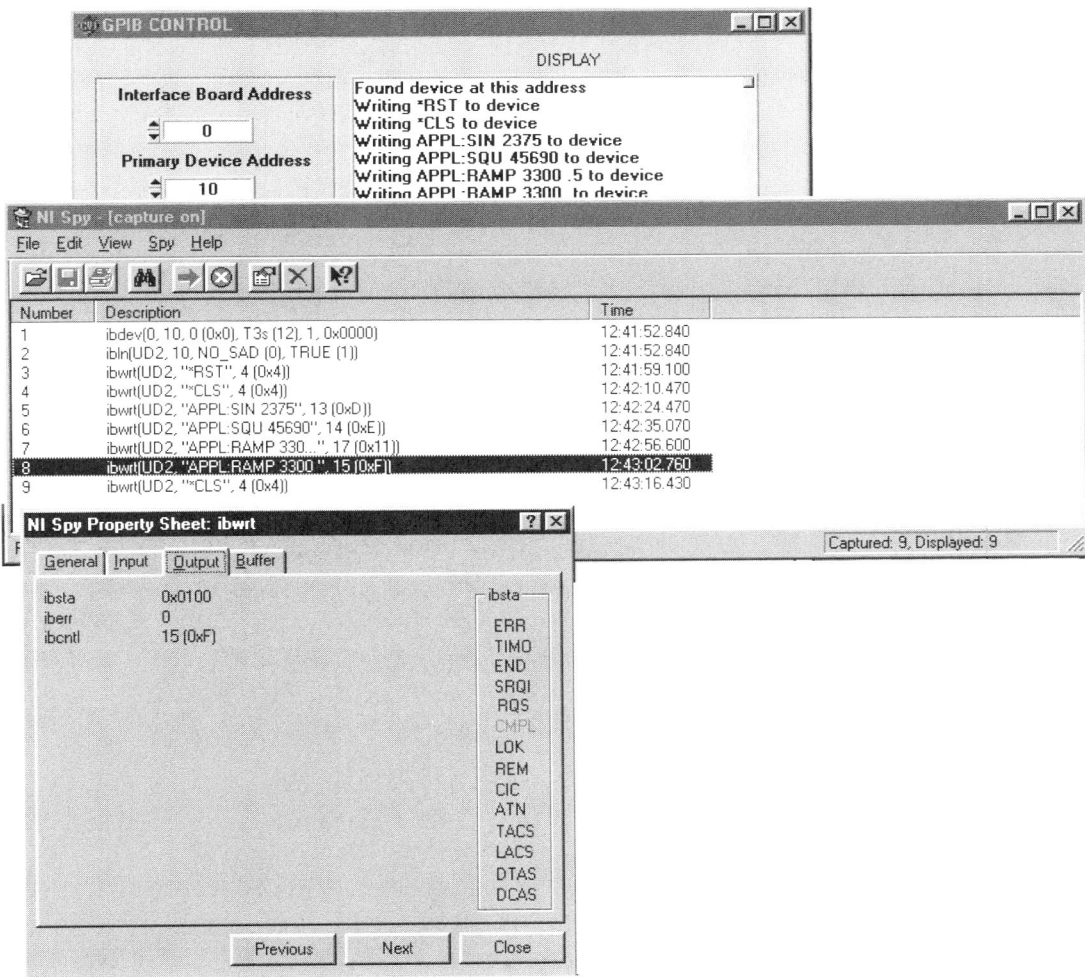

Figure 11–14
Getting Communication Call Details with NI Spy

from **General, Input, Output,** or **Buffer** properties tabs to examine more data attributes for that call. The condition of the *ibsta* word is shown highlighted on the right of the **Property Sheet.** The *iberr* value, the *ibcntl* for the number of words transmitted, and *ibsta* are shown here.

You can step through all the lines by clicking the **Previous** or **Next** command buttons. Select the **General, Input,** and **Buffer** tabs to examine the various information for the call.

The session run shown in Figure 11–14 was recorded communicating with a HP 33120A Function Generator/Arbitrary Waveform Generator. Some of the commands used to communicate with this instrument are SCPI commands. For information on the meaning of the commands used, refer to the *HP 33120A Users Manual* referred to in the bibliography.

The price paid for using *NI Spy* is the slowing of the GPIB application that you are monitoring. If time is not a constraint in your application, this tool is invaluable for troubleshooting.

To avoid losing the collected data, it is advisable to save the recorded session. First click on the red **X** button on the tool bar (Figure 11–14) to stop the recording, and then select **File>>Save As** to save the data in a file with a .spy extension. You can exit *NI Spy* by selecting **File>>Exit**.

Using the IEEE-488.2 Standards

The IEEE-488.2 standard was an improvement over the IEEE-488.1 standards with more added features. These features come in the form of common instrument commands and better error handling and controller functionality, specifying data formats and status reporting. The IEEE-488.2 compatible device can understand and communicate with the IEEE-488.1 standard.

All IEEE-488.2 compatible devices must accept a certain set of commands that are required to facilitate the basic GPIB communication. These commands and their meanings are listed in Table 11–7.

The GPIB 488.2 library routines in *CVI* consist of high-level routines that are used in the next project (`project11-2`) to modify and demonstrate `project11-1` functions. Additional features are added to `project11-2` and will be explained as we go through the code listing shown in Figure 11–15.

Notice that at line 27, the address variables are of type *Addr4882_t,* which are defined in the C language interface header as of type *short*. Using the IEEE-488.2 GPIB calls requires that the address variable must be declared of type *Addr4882_t* for proper communication.

The *SendIFC* routine at line 43 is used to reset the GPIB by sending an interface clear message. It makes the interface board Controller-In-Charge (CIC) of the GPIB and resets all the devices on the bus. The routine prototype for *SendIFC* is

```
void SendIFC(int boardID);
```

Chapter 11 • GPIB Communications

Table 11-7   IEEE-488.2 Commands

Mnemonics	Description
*IDN?	request device identification
*RST	reset device
*TST?	perform self test
*OPC	operation complete
*OPC?	operation complete request
*WAI	wait for completion
*CLS	clear the status
*ESE	event status enable
*ESE?	event status enable request
*ESR?	event status register request
*SRE	service request enable
*SRE?	service request enable request
*STB?	read status byte request

Note: In the above commands the ? following the command always implies a query.

The *boardID* is the GPIB interface board number and is usually 0 unless you have installed multiple interface boards on your system.

The *IbdevCB* function between lines 57 and 105 is called when the **ON-LINE** command button is selected. This function finds all the devices on the bus and displays their addresses. A user interface panel is displayed prompting the user to enter the primary address of the device to establish GPIB communication. The function *ReturnFromAddressCB*, between lines 125 and 156, inquires via a pop-up panel if the user wishes to enter for a secondary address. If the answer is **Yes,** another pop-up panel appears in which the user can enter the secondary address. At lines 69 and 72, an array of valid primary addresses (0 to 30) is generated, with the last array element set to the constant NOADDR. This constant signifies the end of the array and is defined in the header file. Note that for IEEE-488.2-compliant devices, multiple devices can be opened and addressed and data can be sent and received.

```c
/*********************************************************************
Project Name: project11-2.prj

This project performs the same functions as project 11-1, using the IEEE-488.2
library functions.

Author: Shahid F. Khalid

**********************************************************************/

#include <ansi_c.h>
#include "project11-1CMD.h"
#include <gpib.h>
#include <formatio.h>
#include <cvirte.h>            /* Needed if linking in external compiler; harmless otherwise */
#include <userint.h>
#include "project11-2.h"
#include "AddressSelect.h"
#include "project11-2CMD.h"
//#include "Decl-32.h"

static int genericHandle, CommandHandle, AddressHandle;
char buf[260];
char  write_cmd[260], WriteBuf[260];
int   BoardAddress, SecondaryAddr, BoardOffLine, WriteFlag=0;

/*Addr4882_t - Address type (for 488.2 calls) */
Addr4882_t device[32],PAddr, SAddr,PrimaryAddr,found[31];        //typedef short Addr4882_t

//Main function to load the panel and place interface
//board on-line
int main (int argc, char *argv[])
{
    if (InitCVIRTE (0, argv, 0) == 0)  /* Needed if linking in external compiler;
                                                              harmless otherwise */
        return -1;        /* out of memory */
    if ((genericHandle = LoadPanel (0, "project11-2.uir", GENERIC)) < 0)
        return -1;

    DisplayPanel (genericHandle);
    //Obtain the user-specified interface and device addresses
    GetCtrlVal(genericHandle, GENERIC_BRD_ADD,&BoardAddress);

    SendIFC (BoardAddress);  //Intialize the GPIB interface and make the board as CIC
    if (ibsta & ERR)             //Check for error
    {
```

```
46        Fmt( buf, "%s<Could not send IFC");
47        InsertTextBoxLine(genericHandle, GENERIC_DISP, -1, buf);
48        return -1; //abort program
49     }
50
51     RunUserInterface ;
52     return 0;
53  } //main
54
55  //Place device on line, invoked by "ONLINE" command button and find
56  //the devices present
57  int CVICALLBACK IbdevCB (int panel, int control, int event,
58                           void *callbackData, int eventData1, int eventData2)
59  {
60     short i, number;
61     switch (event)
62     {
63         case EVENT_LEFT_CLICK:
64
65  //The array is created in "device" with GPIB addresses from 1 to 30. This array of
66  //addresses is an argument in FindLstn function to compare against the addresses
67  //of the listeners on the bus, if any.
68
69         for (i=1; i<=30;i++)
70              device[i]=i;           //Initialize the array
71
72         device[31]=NOADDR;    //NOADDR signifies the end of the array. This is
73                                //   defined in "decl.h" header file.
74
75  //Find device on line
76         FindLstn (BoardAddress, device, found, 31);   //The listener's address is stored in
77                                                        //   "found"
78         if (ibsta & ERR) //Check for errors
79         {
80              Fmt( buf, "%s<No device found");
81              InsertTextBoxLine(genericHandle, GENERIC_DISP, -1, buf);
82              return -1; //abort program
83         }
84
85         number = ibcntl -1; //The number found is in global variable ibcntl. The interface
86                              //   board is also counted by 'FindLstn'. Hence, 1 is sbutracted to
87                              //   get the actual devices on the bus.
88
```

Figure 11-15
Project11-2 Code listing

```c
89              //Display the number of devices on the bus
90              Fmt( buf, "%s<Number of devices present on bus= %d",number);
91              InsertTextBoxLine(genericHandle, GENERIC_DISP, -1, buf);
92              //Display address of device(s) found
93              for (i=1; i<=number; i++)
94                  {
95                  Fmt( buf, "%s<Found Device at address: %d",found[i]);
96                  InsertTextBoxLine(genericHandle, GENERIC_DISP, -1, buf);
97                  }
98              break;
99
100         }//switch (event)
101
102         SetInputMode (genericHandle, GENERIC_ADDR_ENTER, 1); //Enable SELECT
                                                                    ADDR.button
103         return 0;
104
105 } //IbdevCB
106
107 int CVICALLBACK EnterAddressesCB (int panel, int control, int event,
108         void *callbackData, int eventData1, int eventData2)
109 {
110     switch (event)
111         {
112         case EVENT_COMMIT:
113             //Load and display the Primary Address selection panel
114             if ((AddressHandle = LoadPanel (0,"AddressSelect.uir", ADDR_PNL)) < 0)
115                 return -1;
116             DisplayPanel(AddressHandle);
117             break;
118
119         }
120     return 0;
121
122 }// EnterAddressesCB
123
124 //Inquire for Secondary Address
125 int CVICALLBACK ReturnFromAddressCB (int panel, int control, int event,
126         void *callbackData, int eventData1, int eventData2)
127 {
128     int Yes_no;
129     char SAddrString[12];
130     switch (event)
131         {
132         case EVENT_LEFT_CLICK:
133             GetCtrlVal(AddressHandle, ADDR_PNL_DEVICE_ADD,&PrimaryAddr);
```

```
134             HidePanel(AddressHandle); //Hide the address panel
135             Yes_no = ConfirmPopup ("SECONDARY ADDRESS",
136                          "Do you want to enter Secondary Address?");
137
138             if (Yes_no)   //If Yes
139             {
140                     PromptPopup ("ENTER SECONDARY ADDRESS",
141                         "Enter Secondary Address of device", SAddrString, 10);
142                     Scan(SAddrString, "%s>%i", &SecondaryAddr);
143             }
144
145             break;
146     }
147     //Enable the command buttons
148     SetInputMode (genericHandle, GENERIC_WRITE_DEV, 1);
149     SetInputMode (genericHandle, GENERIC_CLOSE_BRD, 1);
150     SetInputMode (genericHandle, GENERIC_CMD, 1);
151
152     BoardOffLine=0; //GPIB interface board is on-line
153
154     return 0;
155
156 } // ReturnFromAddressCB
157
158
159
160
161 //Write the GPIB device commands. Invoked from the "WRITE" command button
162 int CVICALLBACK WriteToDeviceCB (int panel, int control, int event,
163                     void *callbackData, int eventData1, int eventData2)
164 {
165     switch (event)
166     {
167         case EVENT_LEFT_CLICK:
168             //Load the GUI to write device data
169             if ((CommandHandle = LoadPanel (0, "project11-2CMD.uir", WRT)) < 0)
170                     return -1;
171             DisplayPanel(CommandHandle);
172             WriteFlag=1; //Flag Set
173             break;
174         }
175     return 0;
176 } //WriteToDeviceCB
```

Figure 11-15
Project11-2 Code Listing (*continued*)

```
//Write to device or send interface command, and display on GUI
int CVICALLBACK SendCommandData (int panel, int control, int event,
                void *callbackData, int eventData1, int eventData2)
{
    switch (event)
    {
        case EVENT_COMMIT:
            if (WriteFlag)      //Write data to device
            {
                //Get the user specified device command
                GetCtrlVal(CommandHandle,WRT_GPIB_CMD, WriteBuf);
                
                //Write data to the device
                Send (BoardAddress,PrimaryAddr,WriteBuf,
                                    StringLength(WriteBuf),2);
                if (ibsta & ERR)
                {
                    Fmt( buf, "%s<Error... Could not write to device");
                    InsertTextBoxLine(genericHandle, GENERIC_DISP, -1,
                                                           buf);
                    return -1; //abort program
                }
                //Display command to write
                Fmt( buf, "%s<Writing %s to device at address: %d ",WriteBuf,
                                                            PrimaryAddr);
                InsertTextBoxLine(genericHandle, GENERIC_DISP, -1, buf);
                SetInputMode (genericHandle, GENERIC_READ_DEV, 1);
                                                //Enable the
                                                 "READ" command button
            }
            else  //send GPIB command
            {
                //Get the user specified device command
                GetCtrlVal(CommandHandle,WRT_GPIB_CMD, write_cmd);
                //Send command to the GPIB interface
                SendCmds (BoardAddress, write_cmd, StringLength(write_cmd));
                if (ibsta & ERR)
                {
                    Fmt( buf, "%s<Error... Could not send command");
                    InsertTextBoxLine(genericHandle, GENERIC_DISP, -1,
                                                            buf);
                    return -1; //abort program
                }
```

```
222                 ( buf, "%s<Writing %s to GPIB interface ",write_cmd);
223                 InsertTextBoxLine(genericHandle, GENERIC_DISP, -1, buf);
224                 SetInputMode (genericHandle, GENERIC_READ_DEV, 1);
225                                             //Enable the
226                                                   "READ" command button
227             }
228             break;
229
230     }
231     return 0;
232 }//SendCommandData
233
234 //Routine to send interface messages to configure the GPIB state.
235 //Called from "CMND" command button.
236 int CVICALLBACK GPIBCmdCB (int panel, int control, int event,
237                 void *callbackData, int eventData1, int eventData2)
238 {
239     char cmd_buf[260];
240
241     switch (event)
242     {
243         case EVENT_COMMIT:
244
245             if ((CommandHandle = LoadPanel (0, "project11-2CMD.uir", WRT)) < 0)
246                 return -1;
247
248             DisplayPanel(CommandHandle);
249             WriteFlag=0;    //Reset Flag
250
251             break;
252     }
253     return 0;
254 } //GPIBCmdCB
255
256 //Hide the write device data panel and enable the "READ" command button.
257 //Invoked by "BACK" command button
258 int CVICALLBACK BackCB (int panel, int control, int event,
259                 void *callbackData, int eventData1, int eventData2)
260 {
261     switch (event)
262     {
263         case EVENT_COMMIT:
264             HidePanel(CommandHandle);
```

Figure 11-15
Project11-2 Code Listing *(continued)*

```
                    break;
            }
        return 0;
    } //BackCB

//Read data from the device
int CVICALLBACK ReadFromDeviceCB (int panel, int control, int event,
        void *callbackData, int eventData1, int eventData2)
{
    char read_data[201], read_buf[201];
    int i;

    switch (event)
        {
        case EVENT_LEFT_CLICK:
            for (i=0;i<200;i++) //initialize the arrays
                {
                    read_data[i]=0;
                    read_buf[i]=0;
                }

            //Read data from device until number of bytes in "read_data" or the STOPend is detected
            Receive (BoardAddress,PrimaryAddr, read_data, StringLength(read_data), 256);
            if (ibsta & ERR)
                {
                    Fmt( buf, "%s<Error... Could not read from device");
                    InsertTextBoxLine(genericHandle, GENERIC_DISP, -1, buf);
                    return -1; //abort program
                }
            Fmt(read_buf, "%s<Data read from device: %s",read_data);
            InsertTextBoxLine(genericHandle, GENERIC_DISP, -1, read_buf);
            break;
        }
    return 0;
} //ReadFromDeviceCB

//Take the device off-line. Invoked by "CLOSE" command button
int CVICALLBACK IbonlOFFCB (int panel, int control, int event,
        void *callbackData, int eventData1, int eventData2)
{
    switch (event)
        {
        case EVENT_LEFT_CLICK:
```

```
310             ibonl (BoardAddress, 0);                    //Take device off-line
311             if (ibsta & ERR)
312             {
313                 Fmt( buf, "%s<Error Closing device");
314                 InsertTextBoxLine(genericHandle, GENERIC_DISP, -1, buf);
315                 return -1; //abort program
316             }
317             BoardOffline=1;  //Board is taken off-line Flag
318
319             SetInputMode (genericHandle, GENERIC_WRITE_DEV, 0);   //Disable the
320                                                                    command button
321             SetInputMode (genericHandle, GENERIC_READ_DEV, 0);   //Disable the
322                                                                    command button
323             SetInputMode (genericHandle, GENERIC_CLOSE_BRD, 0);  //Disable the
324                                                                    command button
325             SetInputMode (genericHandle, GENERIC_CMD, 0);        //Enable the command
326                                                                                button
327         break;
328         }
329         return 0;
330
331 } //IbonlOFFCB
332
333
334 //Clear the display
335 int CVICALLBACK ClearDisplayCB (int panel, int control, int event,
336                 void *callbackData, int eventData1, int eventData2)
337 {
338     switch (event)
339     {
340         case EVENT_LEFT_CLICK:
341             DefaultCtrl(genericHandle, GENERIC_DISP);
342         break;
343     }
344     return 0;
345 } //ClearDisplayCB
346
347
348 //Exit the project
349 int CVICALLBACK ExitCB (int panel, int control, int event,
```

Figure 11-15
Project 11-2 Code Listing (*continued*)

```
350          void *callbackData, int eventData1, int eventData2)
351 {
352    switch (event)
353    {
354       case EVENT_COMMIT:
355
356          if (BoardOffline ==0)   //Has board been taken off-line?
357          { // If no, take it off-line
358                ibonl(BoardAddress, 0);  //Good practice to take device off-line
359                                         when exiting
360                if (ibsta & ERR)
361                {
362                   Fmt( buf, "%s<Error Closing device");
363                   InsertTextBoxLine(genericHandle, GENERIC_DISP, -1,
364                                                                 buf);
365                   return -1; //abort program
366                }
367          }
368          QuitUserInterface (0);
369          break;
370       }
371    return 0;
372
373 }//ExitCB
```

Figure 11-15
Project 11-2 Code Listing (continued)

The *FindLstn* routine at line 76 finds all the listeners on the GPIB bus. This routine checks against the values in the array device. If the device with the primary address is found, it is placed in the found array. If no primary address of the device(s) is found, all the secondary addresses are searched. If there are device(s) with secondary addresses, they are stored in the found array. The last argument sets the limit for the maximum number of addresses to be searched. The routine prototype for *FindLstn* follows and its arguments are defined in Table 11–8.

```
void FindLstn ( int BoardAddress, Addr4882_t *device,
                                  Addr4882_t*found, int limit);
```

The *FindLstn* routine updates the *ibcntl* variable to signify the actual number of devices found, including the GPIB interface board. Line 85 subtracts one from the *ibcntl* value to indicate the number of devices present excluding the interface board.

The address of the devices found is displayed by the *for* loop at lines 93 through 97. The **AddressSelect** user interface is loaded and displayed at line 114. The user can enter the primary address in the dialog box to communicate with the device as shown in Figure 11–16. The **RETURN** command button invokes the callback function *ReturnFromAddressCB*, and this function was previously discussed.

The *WriteToDeviceCB* function is defined between lines 162 and 176 and performs the same function as defined in `project11-1`.

The *SendCommandData* function starts at line 179. Depending on the value of `WriteFlag`, either lines 185 through 207 or lines 208 through 227 are executed. If the value of `WriteFlag` is set to 1, the WRITE command button was

Table 11–8 FindLstn *Routine*

Input/Output	Name	Type	Description
Input	BoardAddress	integer	board number, usually zero
	device	Addr 4882_t	a list of all the valid primary addresses
	limit	integer	array limit for found
Output	found	Addr 4882_t	an array consisting of all the listening devices found

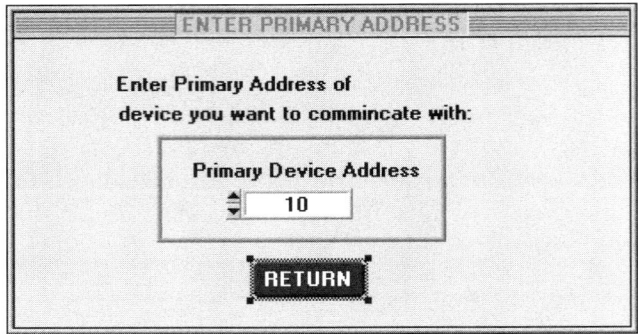

Figure 11–16
Enter Primary Address User Interface

selected to write data to the device. Line 188 obtains the user-entered data from the dialog box shown in Figure 11–8. The GPIB routine *Send*, at line 191, is used to write data to the device. The prototype for the *Send* routine follows and the arguments are defined in Table 11–9.

```
void Send (int BoardAddress, Addr4882_t PrimaryAddr,
           void *write_cmd,long buffer_length, int eotmcde);
```

Table 11–9 Send *Routine*

Input/ Output	Name	Type	Description
Input	BoardAddress	integer	board number, usually zero
	PrimaryAddr	Addr 4882_t	primary address of device to be send data
	write_cmd	char	buffer consisting of data to write to device
	buffer_length	long	length of data to write to device
	eotmode	integer	one of the three data termination modes: NULLend = 0, Nlend = 1, DABend = 2

Chapter 11 • GPIB Communications

The *eotmode* informs the interface of how the last byte of data transmission will be sent. *NULLend* signifies that the GPIB bus management line "End or Identify" (EOI) will not be asserted. *Nlend* sends a newline character (\n) with the EOI line asserted after the last byte of data. *DABend* asserts the EOI with the last data byte. The *eotmode* value entered in the *Send* routine argument identifies which mode is used.

When the **CMND** command button is selected, the GPIB command bytes are sent to the interface. The *GPIBCmdCB* function at line 236 is called and loads the same dialog box as shown in Figure 11–8. This sets the value of `WriteFlag` in the *SendCommandData* function to transmit the GPIB command bytes, and lines 208 through 227 are executed. At line 211, the user-entered command bytes are obtained from the dialog box in Figure 11–8 and sent over the interface using the *SendCmds* routine. The prototype for the *SendCmds* routine follows and its arguments defined are in Table 11–10.

```
void SendCmds(int BoardAddress, void *write_cmd, long length);
```

The callback function *ReadFromDevice*, starting at line 271, is called when the **READ** command button is selected. The data arrays are initialized to clear the arrays at lines 279–283 in a *for* loop. The routine *Receive* reads the device data at line 287, checks for errors, and displays the data read back.

We have been using error checking in the projects above by performing a bitwise AND of *ibsta* and *ERR*. When an error is encountered, the GPIB error handler shows the error number in a pop-up panel and gives you the choice of selecting **Break** or **Continue.** If you select **Continue,** the error messages that you supplied using the *InsertTextBoxLine* or decoded using *display_error* function will be displayed. If **Break** is selected, the project aborts without any explanation of the error. Always select **Continue** if you want to see the error message.

Table 11–10 SendCmds *Routine*

Input/Output	Name	Type	Description
Input	BoardAddress	integer	board number, usually zero.
	write_cmd	char	buffer consisting of command bytes to send over the interface.
	length	long	length of command bytes string.

The two projects created above show the IEEE-488.1 and IEEE-488.2 routine calls used in GPIB communications. You saw a few of the IEEE-488.2 routines, and their prototypes were demonstrated by means of examples in the code. These routines are listed in the *LabWindows/CVI Standard Libraries Reference Manual* and explained in *GPIB NI-488.2M Function Reference Manual for Win32*.

Using GPIB Instrument Drivers

In the previous sections, you saw how you can communicate with devices using the IEEE-488.1 and IEEE-488.2 routines. In actual practice, most of the newer devices on the market today come with GPIB device drivers (also called the instrument module), which are high-level instrument control functions designed specifically to communicate with a particular device. Using these device drivers makes the communication task between the controller and the devices much simpler. All you have to do is to use the appropriate GPIB functions and enter the correct argument variables in the function, and the task is completed when the driver function is called. This task is even made easier if you use the instrument driver function panels. These drivers communicate with the low-level routines that are usually transparent to the user and take care of the device-specific details to configure, control, send, and receive information appropriately.

All the device drivers come with a set of documentation explaining the function's purpose, the prototype, and the arguments' description and use. Most of the device drivers can be downloaded from the manufacturer's or from National Instruments' web site if you want the latest version.

The purpose of this section is to show you, by example, how to use the function device driver. `Project11-3` demonstrates controlling the Hewlett Packard 3478A digital multimeter (DMM) from your computer, using the GPIB device drivers. The instrument module provides the GPIB functions for opening, configuring, taking measurements from, and closing the instrument.

In this project, you will be able to issue almost all the commands by clicking the buttons on the user interface, configuring the DMM, and reading back the data from the instrument without even touching the instrument's front panel. This is an example of *virtual instrumentation*. Your computer

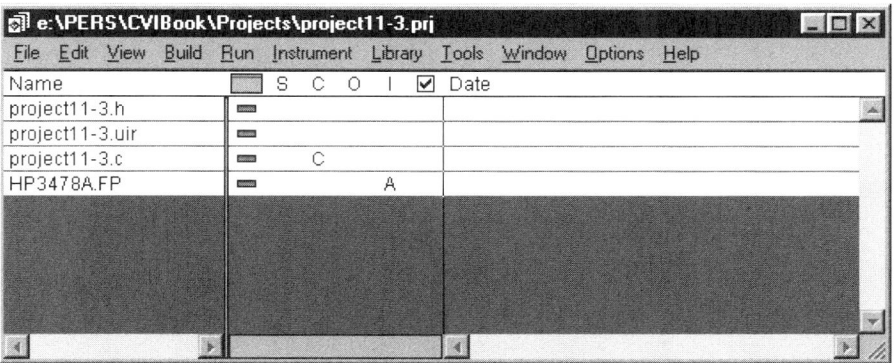

Figure 11–17
Project11-3 Project Window

monitor acts as the instrument panel from which you can command, control, and send and receive data to/from the device.

Project11-3 consists of the files shown in Figure 11–17. A noteworthy observation is that the instrument driver of the device is added to the **Project** window, though it could also have been loaded by selecting **Instrument>>Load**.

Let us look at some of the salient features of project11-3 code listing in Figure 11–18. As mentioned before, it is good practice to check for errors when reading or writing to the interface or device. Line 54 checks the global variable hp_3478a_err for any non-zero value and reports the error. The hp_3478a_err variable is part of the instrument module that records the communication errors.

In the *StartCB* callback function starting at line 69, the *hp3478a_init* device driver function is called at line 80. This device driver function initializes the HP3478A and resets it to its power-on default values. The function prototype for this device driver function follows.

```
void hp3478a_init( int DeviceAddress);
```

The argument for this function, DeviceAddress, is an input argument that gives the address of the device as an integer.

The argument for *hp3478a_init* consists of an *integer* variable in which you input the address of the device to be initialized.

The device is configured at line 87 using the device driver function *hp3478_config*, which configures the function, range, and triggering mode of

```c
/*******************************************************
  File Name: project11-3.c

  Purpose:  This project communicates with the HP 3478A DMM
  via the GPIB. This project uses the device drivers to control
  the configuration and to read the acquired data by selecting
  the controls on the GUI to activate and run the DMM. Where possible
  the device driver are used, along with the IEEE-488.1 routines.


  Written By:    Shahid F. Khalid

********************************************************/

#include <utility.h>
#include <formatio.h>
#include <userint.h>
#include <ansi_c.h>
#include <stdio.h>
#include <stdlib.h>
#include <string.h>
#include <gpib.h>
#include "project11-3.h"
#include "HP3478A.h"

#define dmmAddr   23      //Address of the DMM

int hp3478a_write_data (char *, int);
int hp3478a_read_data(char *, long);
void DMMoff(void);   //Turn Off the DMM
void ClearDisplay(void);

int dmm, menuHandle, error,     /* file descriptor for digital multimeter */
    i,      dmmHandle,  /* loop control variable      */
                        // Handle of the DMM uir
ShiftFlag=0, AutoZeroFlag=0, AutoManCBFlag=0;

char buf[300];

double reading;

int hp_3478a_err;

//Main function to load the panel and the menu bar
void main
```

```
46  {
47      dmmHandle = LoadPanel (0, "project11-3.uir", MULT); //Load the panel in memory
48      DisplayPanel(dmmHandle);          //Display the panel
49      menuHandle = LoadMenuBar (dmmHandle, "project11-3.uir",DMM_MENU ); //Load the
50                                                                menu bar in memory
51
52      ibonl(0,1);//put interface board on line, board address= 0
53      if(hp_3478a_err !=0)//Display errors
54      {
55          Fmt(buf, "%s<IFC NOT SENT");
56          SetCtrlVal (dmmHandle,  MULT_DISP, buf);
57      }
58      if ((dmm = ibdev(0, dmmAddr, 0, T10s, 1, 0)) < 0)
59
60      {
61          Fmt(buf, "%s<DMM NOT OPEN");
62          SetCtrlVal (dmmHandle, MULT_DISP, buf);
63      }
64
65      RunUserInterface;
66
67  } //main

//This function is activated when the ON/OFF line is selected on the GUI
int CVICALLBACK StartCB (int panel, int control, int event,
        void *callbackData, int eventData1, int eventData2)
{
    int count, LineOnOff, fun, trigger;

    double ac_curr;

    switch (event)
    {
        case EVENT_COMMIT:

            hp3478a_init (dmmAddr);    //Initialize the DMM
            if(hp_3478a_err !=0)       //Check for errors
            {
                Fmt(buf, "%s<CANNOT INITIALIZE DMM");
                SetCtrlVal (dmmHandle, MULT_DISP, buf);
            }
            // Set the power up mode for the DMM
            hp3478a_config (2, -3, 1);  //AC Voltage, Autoranging, internal trigger
```

Figure 11-18
Project11-3 Code Listing

```
 88     //   The hp3478a_config routine could have been replaced by the following two lines:
 89     //       Fmt(buf, "%s<F2RAT1");
 90     //       hp3478a_write_data (buf, NumFmtdBytes);
 91
 92              Fmt(buf, "VAC");  //Display the VAC on the screen
 93              SetCtrlVal (dmmHandle, MULT_CONFIG_DISP, buf);
 94
 95              //Display errors
 96          if(hp_3478a_err !=0)
 97          {
 98              Fmt(buf, "%s<CANNOT WRITE");
 99              SetCtrlVal (dmmHandle, MULT_DISP, buf);
100          }
101
102          GetCtrlVal (dmmHandle, MULT_LINE, &LineOnOff);  //Get value of Line
103                                                                       switch
104          if (LineOnOff)  //Line is ON
105              SetCtrlVal (dmmHandle, MULT_LINE, 1);  //Turn ON the switch
106          else
107          {
108              DMMOff;  //Turn off DMM
109              return 0;
110          }
111
112      break;
113  }
114  return 0;
115 }//StartCB]
116
117 //This function is activated from the DC Volts command button
118 int CVICALLBACK SetDCVoltConfigCB (int panel, int control, int event,
119              void *callbackData, int eventData1, int eventData2)
120 {
121     switch (event) {
122         case EVENT_COMMIT:
123             ClearDisplay;
124
125             hp3478a_config(1,-3,1);  //DC Voltage, Autoranging, internal trigger
126
127             if(hp_3478a_err !=0)
128             {
129                 Fmt(buf, "%s<CONFIG ERROR");
130                 SetCtrlVal (dmmHandle, MULT_DISP, buf);
```

```
133            }
134            Fmt(buf, "%s<V DC");
135            SetCtrlVal (dmmHandle,  MULT_CONFIG_DISP, buf);
136
137            hp3478a_measure(&reading); /* Read data from dmm */
138            if(hp_3478a_err !=0)    //Check for errors in reading
139                {
140                Fmt(buf, "%s<CANNOT READ");
141                SetCtrlVal (dmmHandle, MULT_DISP, buf);
142                }
143
144
145            Fmt(buf, "%f\n", reading);// Display the reading on the screen
146            SetCtrlVal (dmmHandle, MULT_DISP, buf);
147            break;
148
149        return 0;
150    }//SetDCVoltConfigCB
151
152    //This function is activated from the AC Volts command button
153    int CVICALLBACK SetACVoltConfigCB (int panel, int control, int event,
154            void *callbackData, int eventData1, int eventData2)
155    {
156        switch (event) {
157            case EVENT_COMMIT:
158            ClearDisplay;
159
160            hp3478a_config(2,-3,1); //AC Volts, autoranging,internal trigger
161            // The hp3478a_config routine does the same as the two commented lines below:
162            //         Fmt(buf, "%s<F2RAT1");
163            //         hp3478a_write_data (buf, NumFmtdBytes);
164            Fmt(buf, "%s<V AC");
165            SetCtrlVal (dmmHandle,  MULT_CONFIG_DISP, buf);
166
167            /* Read data from dmm */
168            hp3478a_measure(&reading);
169            // Check for errors
170            if(hp_3478a_err !=0)
171                {
172                Fmt(buf, "%s<CANNOT READ");
173                SetCtrlVal (dmmHandle,  MULT_DISP, buf);
174                }
175            Fmt(buf, "%f\n", reading);
```

Figure 11-18
Project11-3 Code Listing (*continued*)

```
176                 SetCtrlVal (dmmHandle, MULT_DISP, buf); //Display the data on DMM
177                 Delay (.25);
178             break;
179         }
180      return 0;
181 }//SetACVoltConfigCB
182
183
184 //Measure 2-wire resistance, invoked from the "2 Wire" command button
185 int CVICALLBACK Resis_2WireCB (int panel, int control, int event,
186             void *callbackData, int eventData1, int eventData2)
187 {
188     switch (event) {
189         case EVENT_COMMIT:
190             ClearDisplay;
191
192             Fmt(buf, "%s<OHMS");
193             SetCtrlVal (dmmHandle, MULT_CONFIG_DISP, buf);
194             hp3478a_config(3,-3,1); //2-wire resistance, auto-ranging, internal trigger
195
196 // The hp3478a_config routine does the same as the two commented lines below:
197 //          Fmt(buf, "%s<F3RAT1");
198 //          hp3478a_write_data (buf, NumFmtdBytes);
199
200             //Check for write errors
201             if(hp_3478a_err !=0)
202             {
203                 Fmt(buf, "%s<CANNOT WRITE");
204                 SetCtrlVal (dmmHandle, MULT_DISP, buf);
205             }
206
207             Fmt(buf, "%s<OHMS");
208             DefaultCtrl(dmmHandle,  MULT_CONFIG_DISP); //Display "OHMS"
209             SetCtrlVal (dmmHandle,  MULT_CONFIG_DISP, buf);
210
211
212             /* Read data from dmm */
213             hp3478a_measure(&reading);
214             //Check for read errors
215             if(hp_3478a_err !=0)
216             {
217                 Fmt(buf, "%s<CANNOT READ");
218                 SetCtrlVal (dmmHandle,  MULT_DISP, buf);
219             }
```

```
220              Fmt(buf, "%f\n", reading);//Display the read data
221              SetCtrlVal (dmmHandle, MULT_DISP, buf);
222           break;
223        }
224        return 0;
225     }
226   }  // Resis_2WireCB
227
228   //Function to measure the 4-ire resistance
229   int CVICALLBACK Resis_4WireCB (int panel, int control, int event,
230        void *callbackData, int eventData1, int eventData2)
231   {
232
233      switch (event) {
234         case EVENT_COMMIT:
235            ClearDisplay;
236            hp3478a_config(4,-3,1);
237   // The hp3478a_config routine does the same as the two commented lines below:
238   //          Fmt(buf, "%s<F4RAT1");
239   //          hp3478a_write_data (buf, NumFmtdBytes);
240            if(hp_3478a_err !=0) //Check for errors
241            {
242               Fmt(buf, "%s<CANNOT WRITE");
243               SetCtrlVal (dmmHandle, MULT_DISP, buf);
244            }
245
246            Fmt(buf, "%s<OHMS");
247            SetCtrlVal (dmmHandle, MULT_CONFIG_DISP, buf);
248
249            /* Read data from dmm */
250            hp3478a_measure(&reading);
251
252            if(hp_3478a_err !=0) //Check for errors
253            {
254               Fmt(buf, "%s<CANNOT READ");
255               SetCtrlVal (dmmHandle, MULT_DISP, buf);
256            }
257
258            Fmt(buf, "%f\n", reading);
259            SetCtrlVal (dmmHandle, MULT_DISP, buf); //Display results
```

Figure 11-18
Project11-3 Code Listing (continued)

```
                break;
            }
            return 0;
    } //Resis_4WireCB

    //This function is activated when the DC Current button is selected
    int CVICALLBACK CurrentDCCB (int panel, int control, int event,
            void *callbackData, int eventData1, int eventData2)
    {
        switch (event) {
            case EVENT_COMMIT:
                ClearDisplay;
                hp3478a_config(5,-3,1);   //DC Current, Auto-ranging, internal trigger
                //The hp3478a_config routine does the same as the two commented lines below:
                //              Fmt(buf, "%s<F5RAT1");
                //              hp3478a_write_data (buf, NumFmtcBytes);
                Fmt(buf, "%s< ADC");
                SetCtrlVal (dmmHandle,  MULT_CONFIG_DISP, buf);

                /* Read data from DMM */
                hp3478a_measure(&reading);
                //Check errors
                if(hp_3478a_err !=0)
                {
                    Fmt(buf, "%s<CANNOT READ");
                        SetCtrlVal (dmmHandle,  MULT_DISP, buf);
                }
                Fmt(buf, "%f\n", reading);
                SetCtrlVal (dmmHandle, MULT_DISP, buf);  //Display data
                Delay (.25);

                break;
            }
            return 0;
    } //CurrentDCCB

    //This function is activated when the AC Current button is selected
    int CVICALLBACK CurrentACCB (int panel, int control, int event,
            void *callbackData, int eventData1, int eventData2)
    {
        switch (event) {
            case EVENT_COMMIT:
                ClearDisplay;
                hp3478a_config(6,-3,1);   //AC current, auto-ranging, internal trigger
```

```
308        // The hp3478a_config routine does the same as the two commented lines below:
309
310        //          Fmt(buf, "%s<F6RAT1");
311        //          hp3478a_write_data (buf, NumFmtdBytes);
312        if(hp_3478a_err !=0)   //Check for errors
313        {
314               Fmt(buf, "%s<CANNOT write data");
315               SetCtrlVal (dmmHandle, MULT_DISP, buf);
316        }
317        Fmt(buf, "%s< AAC");
318        SetCtrlVal (dmmHandle, MULT_CONFIG_DISP, buf);
319
320        /* Read data from dmm */
321        hp3478a_measure(&reading);
322        //Check error
323        if(hp_3478a_err !=0)
324        {
325               Fmt(buf, "%s<CANNOT READ");
326               SetCtrlVal (dmmHandle, MULT_DISP, buf);
327        }
328        Fmt(buf, "%f\n", reading);
329        SetCtrlVal (dmmHandle, MULT_DISP, buf); //Display reading
330        Delay (.25);
331
332        break;
333
334     }
335     return 0;
336
337 } //CurrentACCB
338
339 //Activated when the Blue key (Shift key) is selected
340 int CVICALLBACK ShiftKeyCB (int panel, int control, int event,
341            void *callbackData, int eventData1, int eventData2)
342 {
343     switch (event) {
344         case EVENT_COMMIT:
345             ShiftFlag=1;
346             break;
347      }
348      return 0;
349 } //ShiftKeyCB
350
```

Figure 11-18
Project11-3 Code Listing *(continued)*

```c
351  //Auto/Manual selector
352  int CVICALLBACK AutoManCB (int panel, int control, int event,
353                             void *callbackData, int eventData1, int eventData2)
354  {
355      switch (event)
356      {
357          case EVENT_COMMIT:
358              if (AutoManCBFlag==1)    //Auto Ranging
359              {
360                  hp3478a_write_data("RA",2);    //Set Autoranging mode
361                  if(hp_3478a_err !=0)   //Check for errors
362                  {
363                      Fmt(buf, "%s<CANNOT write data");
364                      SetCtrlVal (dmmHandle, MULT_DISP, buf);
365                  }
366
367              AutoManCBFlag=0;   // Set Auto Ranging Flag
368              }// if (AutoManCBFlag==0)
369              else   // Manual Ranging
370              {
371                  AutoManCBFlag=1;   // Manual Ranging Flag
372              }
373          break;
374      } // switch (event)
375
376      return 0;
377  }//AutoManCB
378
379  /*This function will move the DMM to the next higher range every time
380  the UP Arrow is selected
381  */
382  int CVICALLBACK UpRangeCB (int panel, int control, int event,
383                             void *callbackData, int eventData1, int eventData2)
384  {
385      switch (event)
386      {
387          case EVENT_LEFT_CLICK:
388              //Add code to select up range based on the DMM function
389              break;
390      }
391      return 0;
392  } //UpRangeCB
```

```
/*This function will move the DMM to the next lower range every time
the DOWN Arrow is selected
*/
int CVICALLBACK DownRangeCB (int panel, int control, int event,
        void *callbackData, int eventData1, int eventData2)
{
    switch (event)
    {
        case EVENT_LEFT_CLICK:
            //Add code to select down range based on the DMM function
            break;
    }
    return 0;
} //DownRangeCB

//This function performs auto trigger, and turns on and off the auto zeroing function
//when the Shift key is selected
int CVICALLBACK IntTrigCB (int panel, int control, int event,
        void *callbackData, int eventData1, int eventData2)
{
    switch (event) {
        case EVENT_COMMIT:
            ClearDisplay;
            if ( ShiftFlag==0) //Shift key not selected, select Internal Trigger mode
            {                                                // for fastest triggering
                hp3478a_write_data("T1",2);       //Turn on internal trigger
                if(hp_3478a_err !=0)  //Check for errors
                {
                    Fmt(buf, "%s<CANNOT write data");
                    SetCtrlVal (dmmHandle, MULT_DISP, buf);
                }
            }
            if ((ShiftFlag==1) && (AutoZeroFlag==0))
            {
                hp3478a_write_data("Z1",2);  //Turn on auto zeroing
                if(hp_3478a_err !=0)  //Check for errors
                {
                    Fmt(buf, "%s<CANNOT write data");
                    SetCtrlVal (dmmHandle, MULT_DISP, buf);
                }
```

Figure 11-18
Project11-3 Code Listing (*continued*)

```
                    AutoZeroFlag=1;
                    ShiftFlag=0;
                }
            if ((ShiftFlag==1) && (AutoZeroFlag==1))
                {
                hp3478a_write_data("Z0",2);    //Turn Off auto zeroing
                if(hp_3478a_err !=0)   //Check for errors
                    {
                    Fmt(buf, "%s<CANNOT write data");
                    SetCtrlVal (dmmHandle, MULT_DISP, buf);
                    }
                AutoZeroFlag=0;
                ShiftFlag=0;
                }

            break;
            }
    return 0;
} //IntTrigCB

/*This function sets the DMM tothe Single trigger mode,
to take a reading every time button is selected.
If the Shift key and this button is selected,
then Self test is performed.
*/
int CVICALLBACK SignTrigCB (int panel, int control, int event,
        void *callbackData, int eventData1, int eventData2)
{
    switch (event) {
        case EVENT_COMMIT:
            ClearDisplay;
            if(ShiftFlag==1)
                {
                hp3478a_init (dmmAddr); //Do self test
                if(hp_3478a_err !=0)   //Display error message
                    {
                    Fmt(buf, "%s<CANNOT INIT");
                    SetCtrlVal (dmmHandle, MULT_DISP, buf);
                    }
                ShiftFlag=0;    //Reset the shift key
                }
            else //Shift key not selected
                {
                hp3478a_write_data("T3",2);    //Select Single Trigger mode
```

```
483             if(hp_3478a_err !=0)    //Check for errors
484                 {
485                 Fmt(buf, "%s<CANNOT write data");
486                 SetCtrlVal (dmmHandle, MULT_DISP, buf);
487                 }
488         }//else
489
490         break;
491         }
492     return 0;
493
494
495 } //SignTrigCB
496
497 /*When Shift key is slected it displays the DMM address.
498 This routine does not read the address over the bus. I am just
499 writing the value to the display. I could not find a GPIB command
500 to read the address.
501 When shift is not enabled then the SRQ is performed
502 */
503 int CVICALLBACK SRQCallback (int panel, int control, int event,
504                              void *callbackData, int eventData1, int eventData2)
505 {
506     switch (event) {
507         case EVENT_COMMIT:
508             ClearDisplay;
509             if(ShiftFlag==1)
510                 {
511                 Fmt(buf, "%s<HPIB ADRS. %d",dmmAddr );
512
513                 SetCtrlVal (dmmHandle, MULT_DISP, buf);
514                 ShiftFlag=0;  //Reset the shift key
515                 }
516             else  //Shift key is not enabled, perform SRQ
517                 {
518                 hp3478a_write_data("M20",3);   //Enable SRQ
519                 if(hp_3478a_err !=0)   //Check for errors
520                     {
521
522                     Fmt(buf, "%s<CANNOT request service");
523                     SetCtrlVal (dmmHandle, MULT_DISP, buf);
524                     }
525
```

Figure 11-18
Project 11-3 Code listing (*continued*)

```
                break;
        }
        return 0;
} //SRQCallback

//Select the LOCAL mode
int CVICALLBACK LocalModeCB (int panel, int control, int event,
                void *callbackData, int eventData1, int eventData2)
{
        switch (event) {
            case EVENT_COMMIT:
                ClearDisplay;
                SetCtrlVal (dmmHandle, MULT_LINE, 0); //Turn OFF the switch
                hp3478a_close; //Take off line
                if(hp_3478a_err !=0)
                {
                        Fmt(buf, "%s<CANNOT CLOSE");
                        SetCtrlVal (dmmHandle, MULT_DISP, buf);
                }
                break;
        }
        return 0;
} //LocalModeCB

//Exit the GUI, after closing the DMM
void CVICALLBACK ExitCB (int menuBar, int menuItem, void *callbackData,
        int panel)
```

```
556     {
557         ClearDisplay;
558         hp3478a_close;        /* Take DMM off-line */
559         QuitUserInterface(0);
560     }
561
562
563     //Clear the display
564     void ClearDisplay(void)
565     {
566         DefaultCtrl(dmmHandle,    MULT_CONFIG_DISP);    //Clear the box
567         DefaultCtrl (dmmHandle,   MULT_DISP);
568     } // ClearDisplay
569
570     //Turn Off the DMM
571     void DMMOff
572     {
573         ClearDisplay;
574         SetCtrlVal (dmmHandle,    MULT_LINE, 0);  //Turn Line switch OFF
575         DefaultCtrl(dmmHandle,    MULT_DISP);//Clear the display
576         DefaultCtrl(dmmHandle,    MULT_CONFIG_DISP);\
577
578         hp3478a_close;        /* Take dmm off-line */
579         if(hp_3478a_err !=0) //Check for errors
580         {
581             Fmt (buf, "%s<DMM OFF LINE");
582             SetCtrlVal (dmmHandle, MULT_DISP, buf);
583         }
584
585     } //DMMOff(
```

Figure 11-18
Project 11-3 Code Listing (continued)

the device. The function prototype for this device driver function follows and its arguments are defined in Table 11–11.

```
void hp3478a_config ( int function, int range, int trigger);
```

This configuration could also be achieved by using the instrument-specific GPIB commands listed in the *HP 3478A Users Manual*. If you prefer not to use the device driver function, you can achieve the same by using the commented

Table 11–11   hp3478a_config *Routine*

Input/Output	Name	Type	Description
Input	function	integer	the HP3478a is configured to the following meter functions depending on the argument value: • 1—DC Volts • 2—AC Volts • 3—2 wire resistance • 4—4 wire resistance • 5—DC Current • 6—AC Current • 7—Extended Ohms
	range	integer	select from the following measurement range: • -3—Autoranging • -2—30 mV/mA • -1—300 mV/mA • 0—3 V/A • 1—30 V/Ohm • 2—300 V/Ohm • 3—3 K Ohms • 4—30 K Ohms • 5—300 K Ohms • 6—3 M Ohms • 7—30 M Ohms
	trigger	integer	select from the following trigger modes: • 1—internal trigger • 2—external trigger • 3—Single trigger • 4—Trigger hold • 5—Fast trigger

Chapter 11 • GPIB Communications

code on lines 89 through 90 instead. The instrument-specific GPIB commands are shown as an alternate to the device driver function(s) throughout the code listing. It may be advisable to experiment with some of these commands by commenting out the driver function and un-commenting the instrument commands. Both of these commands will produce the same results.

The *hp3478a_measure* driver function at line 168 takes the reading from the multimeter. The prototype of this function follows.

```
void hp3478a_measure(double *reading);
```

The argument of this function, `reading`, is a variable of type *double* that receives the instrument value.

You can write directly to the instrument using the function *hp3478a_write_data* as indicated on line 361. This sets the instrument to the auto ranging mode by writing RA to the device. The last argument in this function refers to the number of bytes in the string.

When communication with the multimeter is no longer required, the device driver function *hp3478a_close* at line 540 closes the instrument and returns it to the local mode.

The objective of this section has been to give you a flavor of the device driver functions and how to use them in GPIB communication. Every device comes with its own set of instrument driver functions, which can contain from a few to over half a score of driver functions, depending on the features of the instrument.

Run this project and when the GUI in Figure 11–19 is displayed, click on the **Line** switch to turn **On** the power to the multimeter. You can take the

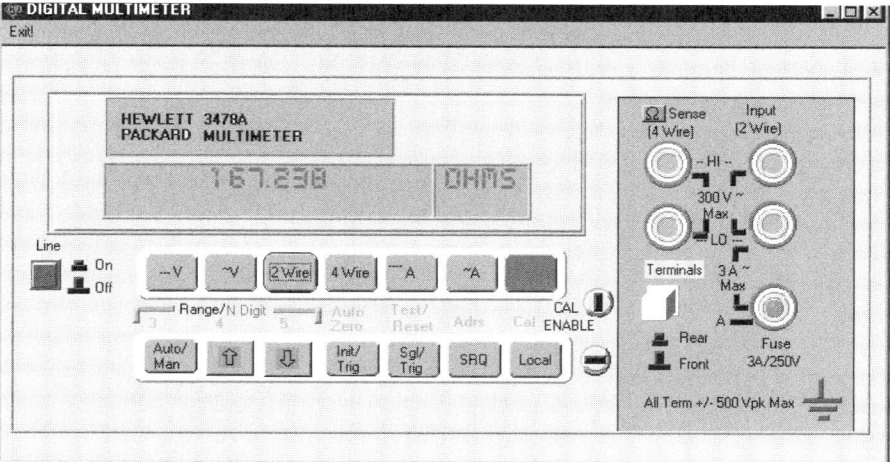

Figure 11–19
Project11–3 Run

same measurements from the *virtual instrument* on your computer monitor as by pushing the panel buttons on the instrument's panel.

Summary

This chapter introduced you to the GPIB evolution history and to the various IEEE-488 standards and the differences between them. It explained the GPIB communication between the interface board and the various GPIB-based devices on the bus, and the addressing and bus requirements.

Three projects were developed: two using different IEEE-488 standards and the third using the multimeter instrument drivers. You were made aware of the differences between them and the advantages of using one over the other.

The use of *Win32 Interactive Control Utility* and *NI Spy* were explained and their usefulness in debugging and obtaining GPIB routine details was discussed as were the global variables.

12

RS-232 SERIAL COMMUNICATION

Chapter Highlights

- Overview of Serial Communication
- Serial Communication Project
- Configuring the RS-232
- Error Checking Functions
- Writing and Reading Data
- Sending and Receiving Data Packets
- Serial Communication Utility Functions
- Communicating with RS-485
- Summary

This chapter explains the hardware and software aspects of serial communication. The protocol for transferring serial data using the character frame is explained. The various pin connectors with their pin numbers and signal names as well as their purpose is shown. You are shown by means of a project how to configure, send, and receive serial data using the *CVI* RS-232 Library functions. Various ways of transferring data are explained using the library functions and their prototypes. Error checking and serial interface utility functions are discussed and demonstrated.

Overview of Serial Communication

Serial communication refers to transferring the data over a single wire, one bit at a time as a series in a queue between the connecting devices. The data is entered as character strings on your computer, and the serial interface communication drivers convert the data into a standard bit serial format that is transmitted out of the serial connector. A clock or timing reference is used to control the flow of data and decide when to send and read each bit. There are two types of serial-data formats: synchronous and asynchronous. In synchronous transmission, all devices use a common clock and the bits transmitted and received are synchronized to this clock. This kind of transmission requires an extra line for the clock signal.

Asynchronous transmission does not include a clock line because each of the connecting devices provides its own clock frequency, but each must match the others very accurately. Data is sent to the serial port using a character frame that makes up a single character. The frame is defined as the characters from a *Start* bit to the last *Stop* bit, inclusively. The *Start* bit is a '0' that precedes each character to indicate to the receiving system that the following numbers are data bits. One or two bits are used as *Stop* bits to indicate to the receiving device that the character has been sent. Within the frame, you can select the number of data bits and the parity type.

The RS-232 ports on PCs use asynchronous transmission to communicate with other devices. An RS-232 interface can also transfer synchronous data, but it is mainly used for asynchronous transmissions. In this chapter, only asynchronous transmission will be discussed.

The *Universal Asynchronous Receiver/Transmitter* (UART) chip controls the serial data. When sending data out to the serial port, the UART converts the parallel data on the PC system bus to the serial data for transmitting. After the communication protocol, data rate, and other settings have been established for a communication port, the data byte is written to the transmit buffer of the selected port. The UART sends the data one bit at a time with the *Start*, *Stop,* and *Parity* bits as needed. The data is written to the receive buffer of the receiving device. The UART on the receiving system converts the received serial data to parallel data. The CPU reads this data on the PC system bus.

Two examples of the character frame format are shown. Figure 12–1 shows the frame for seven data bits using a *Parity* bit and two *Stop* bits. Figure 12–2 shows eight data bits with no *Parity* and one *Stop* bit.

Adding the *Start* and *Stop* bits to the data increases the overhead of transmitting the data by about 20% but is necessary for serial communication. The number of *Stop* bits are specified when you set up the serial port config-

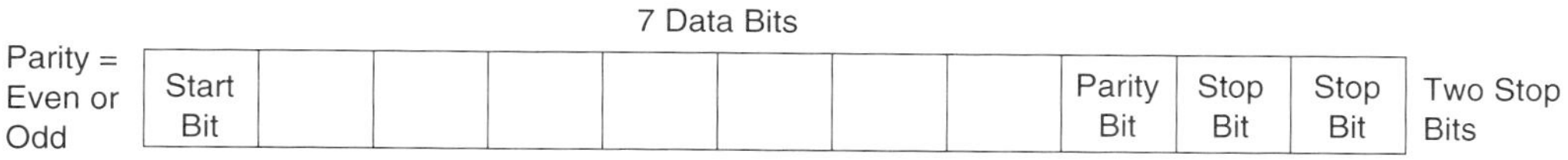

Chapter 12 • RS-232 Serial Communication

Parity = Even or Odd	Start Bit			7 Data Bits					Parity Bit	Stop Bit	Stop Bit	Two Stop Bits

Figure 12–1
Character Frame Format for Seven Data Bits, One Parity Bit, Two Stop Bits

uration, and they must match the configuration of the receiving device. Configuring the serial port(s) is explained in the *Configuring the RS-232* section.

Many instruments and controllers come equipped with a serial port that enables you to set up communication without the use of any additional hardware. Every computer system has at least one serial port; some later models come with two serial ports. If you need additional serial ports on your system, you can install multi-port serial cards that usually come with two to thirty-two serial ports.

All serial devices fall into one of the two types: Data Transmission Equipment (DTE) or Data Communication Equipment (DCE). IBM-compatible PCs are DTE devices. Modems' serial ports are configured as DCEs. The types determine the meaning assigned to the pins. For example, looking at Tables 12–1 and 12–2, TD (Transmit Data) carries data from the DTE to the DCE, so it is an output on DTE and an input on DCE. Likewise, RD (Receive Data) carries data from the DCE to the DTE, and it is an input to DTE and an output to DCE. The Direction of Data Flow in these tables is from the perspective of the DTE.

Connectors and Signals

To communicate with a RS-232 serial device, you need a serial cable to connect between the computer and the device. The serial cable consists of wires, or lines that join two connectors. The serial port has two types of connectors:

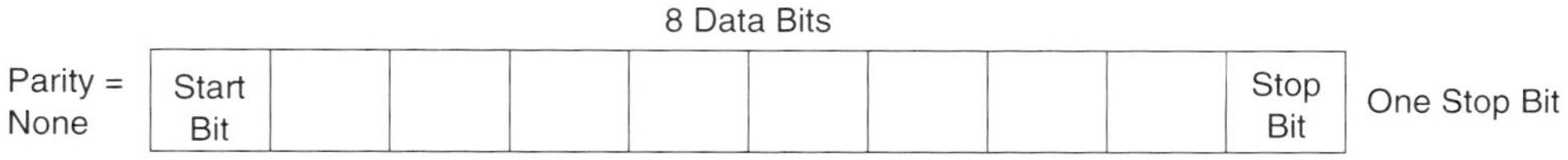

Parity = None	Start Bit			8 Data Bits						Stop Bit	One Stop Bit

Figure 12–2
Character Frame Format for Eight Data Bits, No Parity Bit, One Stop Bit

Table 12–1   DB-9 Connector Pin Signals

Pin	Signal	Signal Description	Direction of Data Flow
1	CD	Carrier Detect	In
2	RD	Receive Data	In
3	TD	Transmit Data	Out
4	DTR	Data Terminal Ready	Out
5	SG	Signal Ground	-
6	DSR	Data Set Ready	In
7	RTS	Request To Send	Out
8	CTS	Clear To Send	In
9	RI	Ring Indicator	In

Table 12–2   DB-25 Connector Pin Signals

Pin	Signal	Description	Direction of Data Flow
1	-	Chassis Ground	-
2	TD	Transmit Data	Out
3	RD	Receive Data	In
4	RTS	Request To Send	Out
5	CTS	Clear To Send	In
6	DSR	Data Set Ready	In
7	SG	Signal Ground	-
8	CD	Carrier Detect	In
9	-	+Transmit current loop return	Out
11	-	−Transmit current loop data	Out
18	-	+Receive current loop data	In
20	DTR	Data Terminal Ready	Out
22	RI	Ring Indicator	In
25	-	−Receive current loop return	In

Chapter 12 • RS-232 Serial Communication

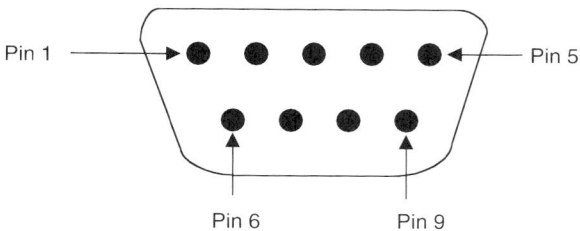

Figure 12–3
DB-9 (Male) Connector Pin Locations

nine-pins, called DB-9, and twenty-five-pins, called DB-25. These can be either male or female type connectors or receptacles. A male connector has pins inside the connector shell and a female connector has holes inside the connector shell. The male connectors are shown in Figures 12–3 and 12–4.

You can use either type of connector depending on the connector receptacle on your device. The pin numbers, the signal names, the signal descriptions, and the direction of data flow for these connectors are shown in Tables 12–1 and 12–2.

At times, you may need a serial cable that has to connect between a DB-9 and a DB-25 receptacle on the communicating devices. You can use the *adapter* on the serial cable that connects the nine-pin signals to the appropriate twenty-five-pin signals. The adapter connections for the nine-pin and the twenty-five-pin are shown in Table 12–3.

After configuring both the communicating devices (which is explained later), you can start the serial communication by sending the character string from your computer to the receiving device.

There are advantages and disadvantages to using serial port communication. The advantage of using serial port communication is the lack of ad-

Figure 12–4
DB-25 (Male) Connector Pin Locations

Table 12–3   DB-9 to DB-25 Adaptor Connections

DB-9 Pin	DB-25 Pin	Signal	Signal Description
1	8	CD	Carrier Detect
2	3	RD	Receive Data
3	2	TD	Transmit Data
4	20	DTR	Data Terminal Ready
5	7	SG	Signal Ground
6	6	DSR	Data Set Ready
7	4	RTS	Request To Send
8	5	CTS	Clear To Send
9	22	RI	Ring Indicator

ditional hardware requirements. One main disadvantage of using serial communication is its slower pace, moving one data bit at a time, compared to the GPIB, which transfers data in parallel and is much faster.

The serial communication interface standards defined by the Electronics Industries Association (EIA) are: RS-232C, RS-422A, RS-432A, and RS-485A. These standards include the specifications for cabling, connector pinouts, signal levels, and timing.

The most commonly used standard supported by your computer and serial instruments is the RS-232C. This standard supports using a twenty-five-pin D-style connector or a nine-pin version on IBM PC. The pin numbers that send and receive data, perform handshaking, and connect to ground are described in the EIA standard. The RS-232C is electrically single-ended, which results in poor noise reduction and limits the serial cable length to a hundred meters and the maximum serial data transfer rate to 19,200 baud (bits per second).

RS-422A is similar to RS-232 but has a better noise rejection ratio since it uses differential drivers and receivers, requiring two pairs of wires and a ground as a minimum connection. RS-422 can operate up to speeds of 9600 baud on a cable length of twelve hundred meters and up to 1 Mbaud using shorter cable lengths.

RS-432A has better electrical characteristics than RS-232C and supports faster baud rates over longer cable lengths. If you want to use multi-drop systems (multiple talkers and listeners on one set of wires), then RS-485A is ideal for you. You will need an adapter box connected to your computer to support these serial protocols. The RS-232 library functions can be used with RS-485 boards from various manufacturers, such as with National Instruments RS-485 serial board.

The RS-232 Library is supported under Unix for *CVI* versions prior to 5.5, but the Unix kernel must support asynchronous I/O functions, e.g., *aioread* and *aiowrite*. You can do this by building your Unix kernel as *Generic* instead of *Generic Small*.

This section gave you an overview of the serial communication basics. For more details on serial communication, refer to the books listed in the bibliography.

Serial Communication Project

The serial communication project (`project12.prj`) will demonstrate serial port communication from one serial port of your PC (called COM1) to the other serial port (called COM2) of the same computer. To establish such communication, you must have a computer with two serial ports and connect them by means of a special type of cable, called a null-modem (laplink) cable, which is a special cable that connects the transmit data pin to the receive data pin on each serial port connector. Such a cable is available in computer stores, or you can build your own cable if you prefer per the wiring diagram shown in the Tables 12–4 and 12–5 and using either DB-9 (9-pin) or DB-25 (25-pin) connectors.

If this hardware is not available, you will not be able to run this project. However, by reading the text and following the source code, you will be able to understand the use of *CVI* RS-232 library functions.

`Project12.prj` will establish communication between the serial ports COM1 and COM2 on your PC by configuring them to the same settings. This project will show you how to send and receive data bytes from a buffer, how to send and receive a single byte of data, and how to transfer data using files.

Table 12–4 DB-9 to DB-9 Null Modem Pin Connections

COM1			COM2
RD	2	3	TD
TD	3	2	RD
SG	5	5	SG

Table 12–5 DB-25 to DB-25 Null Modem Pin Connections

COM1			COM2
RD	2 ←→	3	TD
TD	3 ←→	2	RD
SG	7 ←→	7	SG

Almost all the serial interface library functions are included in the project. Some of the other library functions that are not included in the project will be discussed as necessary.

Load `project12.prj` from the `Projects` directory and display the `project12Main.uir` GUI as shown in Figure 12–5. This is the main GUI for this

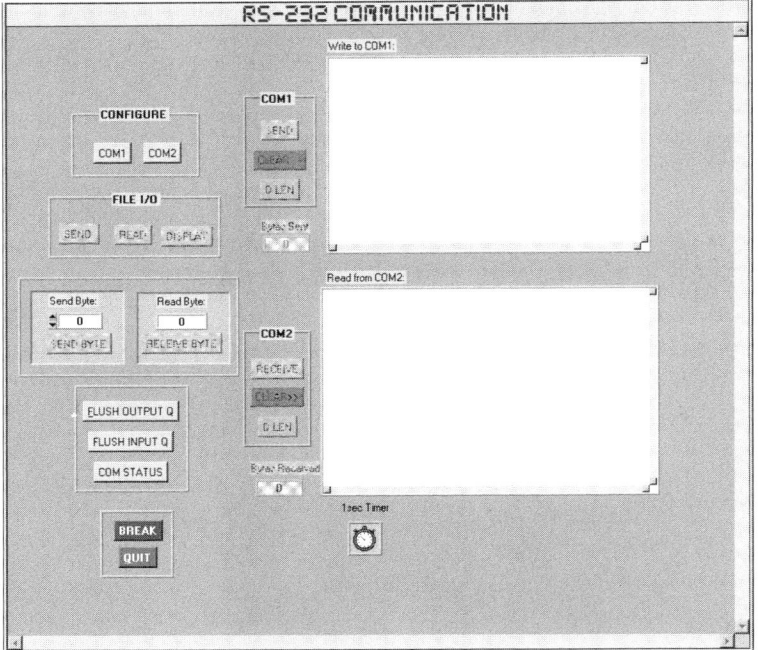

Figure 12–5
Project12 Main GUI

project, and it is titled **RS-232 COMMUNICATION**. You will setup the configuration for the COM ports, send and receive data, display data, and perform various serial interface utility functions.

Before transmitting the data, both ports must be configured by first selecting the **COM1** command button in **CONFIGURE** group on the GUI and then selecting the **COM2** command button. The configuration for the COM ports is discussed next.

Configuring the RS-232

To communicate, the serial communication devices must have exactly the same configuration. The port parameters can be hard-coded if you know in advance that they are never going to be changed by the user. It is better, however, to give the user the flexibility to change these parameters by means of a serial interface configuration GUI by selecting the **COM1** and **COM2** command buttons in Figure 12–5. A sample configuration GUI is shown in Figure 12–6. This configuration GUI is used in `project12.prj` to

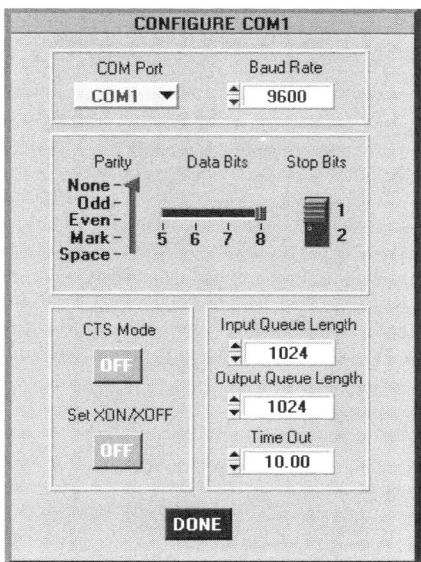

Figure 12–6
Serial Communication Configuration GUI

setup both the COM ports. A discussion of each of the configuration parameters follows.

COM Port—This is the communication port number on your computer system that you will use to communicate with the other device. On most computer systems, it is COM1 and COM2 with port number values assigned as '1' and '2,' respectively. If you are using a multi-port serial card, then you will have access to more COM ports with numbers assigned to these ports sequentially.

Baud Rate—The serial port will send data in bits per second at the rate entered in this box. The possible baud rates are visible when you click inside the box, but not all the baud rates are supported by PCs and SPARC stations. PCs do not support a 150 baud rate. Some PC serial drivers do not support 115200, 128000, and 256000 baud rates. SPARC stations do not support 14400, 28800, 56000, 57600, 115200, 128000, and 256000 baud rates.

Parity—The *Parity* bit can be set as None, Odd, Even, Mark, and Space. Odd *Parity* bit is calculated by setting the value of the *Parity* bit to a 1 or 0 such that the number of 1s in the data plus the *Parity* bit total to an odd number of 1s. For example, if you were sending eight data bits using odd *Parity*, the format of data transmission would be as shown in Figure 12–7.

The total number of data bits with 1s in the above example is four, which is an even number. The *Parity* bit must contain a 1 to make the total number of 1s an odd number to make an odd *Parity*.

If you were sending eight data bits using even *Parity*, the format of data transmission would be as shown in Figure 12–8. The total number of data bits with 1s in the above example is three, which is an odd number. The *Parity* bit must contain a 1 to make the total number of 1s an even number to make up an even *Parity*.

Mark parity refers to having a value of 1 always in the *Parity* bit. *Space* parity refers to having a value 0 always in the *Parity* bit. The *Mark* and *Space* parities are not used for error checking but use this bit to differentiate between addresses and data. If the *Parity* bit is a 1, the byte is considered an address, if 0, data. This method is called the 9-bit format and uses the hardware for most of the decoding. In this method, the serial port sends and receives eleven-bit words, including the *Start* bit, eight data bits, a ninth bit, and one *Stop* bit.

Data bit number	0	1	2	3	4	5	6	7	8	Parity bit
Data bit value	1	0	0	1	0	1	0	1	0	1

Figure 12–7
Eight Data Bits Using Odd Parity

Data bit number	0	1	2	3	4	5	6	7	8	Parity bit
Data bit value	0	0	0	1	0	1	0	0	1	1

Figure 12–8
Eight Data Bits Using Even Parity

Data Bits—The number of bits of data to transmit that are between the *Start* and the *Stop* bits. You have a choice of five, six, seven, or eight data bits.

Stop Bits—Depending on the device with which you are communicating, you have a choice of using one or two *Stop* bits.

Before explaining **CTS Mode** and **Set XON/XOFF,** let us look at the purpose of these two controls and why they are necessary. Sometimes in serial communication, the receiver loses the data that the sender transmits. This is typically because the receiver is unable to empty its input queue quickly enough. Handshaking is used to resolve this problem by preventing the overflow of the input queue. The RS-232 library has two types of handshaking: hardware and software. You can use one or the other to ensure that your application synchronizes the data transfer between the devices.

CTS Mode—This enables hardware handshaking by using the RS-232 library function *SetCTSMode*. For hardware handshaking to work, the serial devices must use the same lines for the handshake and assign the same meanings to these lines. In addition, the serial cable that connects the two devices must include the lines required to support the protocol. Most *CTS* and *RTS* lines perform hardware handshaking and use the *DTR* line to signal online presence to other device. Some serial devices also use the *DTR* line for hardware handshaking, similar to the *CTS* line. The *SetCTSMode* function can handle either of these modes. This is explained later when you use this function to configure the serial ports.

Set XON/XOFF—This enables software handshaking by using the RS-232 library function *SetXMode*. Software handshaking is to be used to transfer ASCII data or text to or from a serial device that uses software handshaking. This option should not be enabled when transmitting binary data. The special XON/XOFF characters can occur as part of the data stream that the receiver can mistake for control codes. Hardware handshaking is independent of the type of data you transfer. Software handshaking requires no special cable configuration. The *SetXMode* will be explained when we use this function for configuring the serial ports.

Input Queue Length—You can specify the maximum length of the data you expect to receive. Under Windows, there is no maximum limitation on the queue size. Under Windows 3.1, the maximum queue size is 65,535 bytes. However, some serial drivers have a maximum of 32,767 bytes. Therefore, National Instruments recommends that you use a queue size no greater than 32,767 bytes.

Output Queue Length—You can specify the maximum length of data you expect to send. The limitations on the data size are the same as given in the **Input Queue Length** paragraph.

Time Out—You specify the maximum time to send and receive the data for the operations. For receiving data, this is the maximum time from the start of the transfer to the arrival of the first byte. This is also the elapsed time between the arrival of any two consecutive bytes. If the data is not received within this timeout limit, a timeout error is displayed.

For sending data, this is the maximum time before the first byte is transferred to the output queue. This is also the elapsed time between sending any two consecutive bytes to the output queue. If for some reason the hardware or software handshaking does not occur within the timeout limit, a timeout error is displayed. The timeout is set through RS-232 library function *SetComTime*. This function will be explained when you use it later in the project.

To setup the COM ports, make the changes to the configuration controls on this GUI (or leave them at their default setting) and select the **DONE** command button. This will invoke the *SetUpConfig1CB* callback function, whose source code is listed in Figure 12–9.

The library function for configuring the RS-232 communication parameters will be explained here. Lines 12–18 obtain the selected configuration values from the GUI for the COM port, baud rate, parity, data bits, stop bits, and the input and output queue lengths. Line 19 obtains the index for the selected COM port. This index is used at line 21 to obtain the COM port label. These configuration parameters are input to the RS-232 library function *OpenComConfig* at line 25. This function opens the selected COM port and sets the selected configuration parameters for this port. The prototype for the *OpenComConfig* function follows and its parameters are explained in Table 12–6.

```
int result = OpenComConfig (int ComportNumber, char Device,
       long BaudRate, int Parity, int databits, int stopbits,
                     int In_Q_Length, int Out_Q_Length);
```

```
1   //Invoked with the "DONE" command button for COM1 configuration
2   int CVICALLBACK SetUpConfig1CB (int panel, int control, int event,
3                   void *callbackData, int eventData1, int eventData2)
4   {
5   int CTS_On_Off1, X_Mode1;
6   float Timeout1;
7
8       switch (event)
9       {
10      case EVENT_LEFT_CLICK:
11          //Get the Communication parameters from the GUI
12          GetCtrlVal (SetupHandle1, CONFIG1_COM_PORT, &ComportNumber1);
13          GetCtrlVal (SetupHandle1, CONFIG1_BAUD, &Baud1);
14          GetCtrlVal (SetupHandle1, CONFIG1_PARITY, &Par1);
15          GetCtrlVal (SetupHandle1, CONFIG1_DATA_BITS, &databits1);
16          GetCtrlVal (SetupHandle1, CONFIG1_STOP_BITS, &stopbits1);
17          GetCtrlVal (SetupHandle1, CONFIG1_INPUT, &In_Q1);
18          GetCtrlVal (SetupHandle1, CONFIG1_OUTPUT, &Out_Q1);
19          GetCtrlIndex (SetupHandle1, CONFIG1_COM_PORT, &ComPortIndex1);
20
21          GetLabelFromIndex (SetupHandle1, CONFIG1_COM_PORT,
22                                  ComPortIndex1,Device1);
23
24          //Open the COM port and send the configuration parameters
25          status = OpenComConfig (ComportNumber1, Device1, Baud1, Par1,
26                      databits1, stopbits1, In_Q1, Out_Q1);
27          if (status <0) //Check for communication errors
28          {
29              *ErrorMessage = GetRS232ErrorString(ReturnRS232Err); //Return the Error
30                                                                                    String
31              InsertTextBoxLine (mainhandle, SER_SEND_BOX, -1, *ErrorMessage);
32          }
33          //Enable or disable hardware handshaking
34          GetCtrlVal(SetupHandle1, CONFIG1_CTS_MODE, &CTS_On_Off1);
35          status = SetCTSMode (ComportNumber1, CTS_On_Off1);
36          if (status <0) //Check for communication errors
37          {
38              *ErrorMessage = GetRS232ErrorString(ReturnRS232Err); //Return the Error
39                                                                                    String
40              InsertTextBoxLine (mainhandle, SER_SEND_BOX, -1, *ErrorMessage);
41          }
42
43      //Enable or disable software handshaking
44      GetCtrlVal(SetupHandle1, CONFIG1_XON_XOFF, &X_Mode1);
45      status= SetXMode(ComportNumber1, X_Mode1);
46      if (status <0) //Check for communication errors
47      {
48              *ErrorMessage = GetRS232ErrorString(ReturnRS232Err); //Return the Error
49                                                                                    String
50              InsertTextBoxLine (mainhandle, SER_SEND_BOX, -1, *ErrorMessage);
51      }
52
53      //Set the timeout
54      GetCtrlVal(SetupHandle1, CONFIG1_TIME_OUT, &Timeout1);
55      status = SetComTime (ComportNumber1, Timeout1);
56      if (status <0) //Check for communication errors
57      {
58              *ErrorMessage = GetRS232ErrorString(ReturnRS232Err); //Return the Error
```

Figure 12–9
RS-232 Configuration Source Code

```
59                                                                            String
60                InsertTextBoxLine (mainhandle, SER_SEND_BOX, -1, *ErrorMessage);
61         }
62
63         HidePanel(SetupHandle1);
64         Com1Set = 1;    //Set flag
65         SetInputMode (mainhandle, SER_CONFIGURE1, 0);   //Disable once configured
66         if (Com1Set==1 && Com2Set==1)
67            EnableControls;
68      break;
69      }
70      return 0;
71 } //SetUpConfig1CB
```

Figure 12–9
RS-232 Configuration Source Code (*continued*)

Table 12–6 OpenComConfig *Function*

Input/Output	Name	Type	Description
Input	Comport Number	integer	the number assigned to the COM port, can be between 1 and 1,000
	Device	character	name of the COM port, e.g. COM1, COM2, etc.
	BaudRate	long	baud rate settings; the acceptable values for Windows and SPARC stations were explained earlier
	Parity	integer	*Parity* bit settings are: 0 = no parity 1 = odd parity 2 = even parity 3 = mark parity 4 = space parity
	databits	integer	the number of data bits to transmit between the *Start* and *Stop* bits are either five, six, seven, or eight
	stopbits	integer	*Stop* bits set to 1 or 2
	In_Q_Length	integer	length of input queue to receive data
	Out_Q_Length	integer	length of output queue to send data
Output	result	integer	error codes (negative values indicate error), refer to Table 5–10 in the *CVI Standard Libraries Reference Manual* or in Online Help

Chapter 12 • RS-232 Serial Communication

The *SetCTSMode* library function is invoked at line 35 and enables or disables the hardware handshaking. The function prototype follows and its mode arguments are explained in Table 12–7.

```
int result = SetCTSMode( int ComportNumber, int mode);
```

Mode	Description
0	Hardware handshaking is disabled, CTS line ignored. The RTS and DTR lines are raised while the port is open.
1	Hardware handshaking is enabled, and CTS line is monitored. The RTS and DTR lines are used for handshaking.
2	Hardware handshaking is enabled and CTS line is monitored. RTS line is used for handshaking and the DTR line is raised while the port is open.

The *SetXMode* library function, as mentioned above, is invoked at line 45 and enables or disables software handshaking by turning On or Off the XON/XOFF sensitivity. The function prototype folows and its arguments explained in Table 12–8.

```
int result = SetXMode( int ComportNumber, int mode);
```

Caution: Be sure that you have opened the port by calling *OpenComConfig* before using the *SetCTSMode* or the *SetXMode* functions or you will get an error.

Table 12–7 SetCTSMode *Function*

Input/Output	Name	Type	Description
Input	Comport Number	integer	the number assigned to the COM port, can be between 1 and 1,000
	mode	integer	zero is used to disable hardware handshaking and non-zero to enable it; these modes are described later
Output	result	integer	error codes (negative values indicate error), refer to Table 5–10 in the *CVI Standard Libraries Reference Manual* or in Online Help

Table 12–8 SetXMode *Function*

Input/Output	Name	Type	Description
Input	Comport Number	integer	the number assigned to the COM port, can be between 1 and 1,000
	mode	integer	zero is used to disable software handshaking and non-zero to enable it
Output	result	integer	error codes (negative values indicate error), refer to Table 5–10 in the *CVI Standard Libraries Reference Manual* or in Online Help

The *SetComTime* library function, as mentioned earlier, is invoked at line 55. This sets the timeout for sending and receiving data. The function prototype follows and its arguments are explained in Table 12–9.

```
int result = SetComTime( int ComportNumber, double timeout);
```

Do the same by selecting the **COM2** command button from the **CONFIGURE** group, making sure all the parameters are the same as those for COM1, but select **COM2** from the **COM Port** pull-down menu.

Table 12–9 SetComTime *Function*

Input/Output	Name	Type	Description
Input	Comport Number	integer	the number assigned to the COM port, can be between 1 and 1,000
	timeout	integer	timeout for sending and receiving data as explained earlier
Output	result	integer	error codes (negative values indicate error), refer to Table 5–10 in *CVI Standard Libraries Reference Manual* or in Online Help

Error Checking Functions

When communicating with a device, it is always advisable to perform error checking to establish that the device performed the expected task. For example, it is important to know that the data sent out on the serial port was actually sent to the output queue. For this reason, you will notice that every RS-232 serial function call in the *SetUpConfig1CB* function code listing in Figure 12–9 (and other functions in this project) is checked for errors.

The RS-232 library function always returns negative values when an error occurs. In addition, the global variable `rs232err` is updated after each function call to the RS-232 library. If the function returns no errors, `rs232err` is set to zero; otherwise it is set to an error code. The library function *ReturnRS232Error* returns the same error code as `rs232err` except it keeps track of separate error codes for each thread in your application when running a multithreading application. Hence, if you are running a multithreading application in Windows, use the *ReturnRS232Err* function instead of `rs232err`. Multithreading will not be discussed in this text and is a topic for an advanced book.

Let us examine the error checking code in the configuration setup listing as extracted from Figure 12–9 and shown in Figure 12–10.

The results of the *OpenComConfig* function are returned in the variable `status` at line 1. If the value of `status` is negative, the value of the error code obtained from the function *ReturnRS232Err* is used in another library function, *GetRS232ErrorString*, using line 5. The *GetRS232ErrorString* converts the error code into the error message string `ErrorMessage` that is displayed in the text box. Table 5–10 in the *LabWindows/CVI Standard Libraries Reference Manual* lists the error messages associated with the error codes generated by this function.

```
1      status = OpenComConfig (ComportNumber1, Device1, Baud1, Par1,
2                              databits1, stopbits1, In_Q1, Out_Q1);
3      if (status <0) //Check for communication errors
4      {
5           *ErrorMessage = GetRS232ErrorString(ReturnRS232Err); //Return the Error
6                                                                                 String
7           InsertTextBoxLine (mainhandle, SER_SEND_BOX, -1, *ErrorMessage);
8      }
```

Figure 12–10
Error Checking Code Fragment

Writing and Reading Data

The RS-232 Input/Output Library support various functions with which data can be written to or read from a serial port. Data can be sent/received using a buffer, a single byte, or to/from a file. These different methods of serial data transmission are explained in the following sections.

Sending Data Buffer

The *ComWrt* function is used for writing a specified number of data bytes to the output queue of the serial port. To send the data from COM1 to COM2, enter the data in the text box **Write to COM1,** shown in Figure 12–5, and select the **SEND** command button. The *SendSerialCB* callback function is invoked, and its code listing is shown in Figure 12–11.

At line 11, the data written to **Write to COM1** box is copied into the string buffer `send_data`. The data is sent to the serial port COM1 using the library function *ComWrt* shown on line 12.

```
1   //Send serial data from COM1
2   int CVICALLBACK SendSerialCB (int panel, int control, int event,
3                void *callbackData, int eventData1, int eventData2)
4   {
5        char send_data[1024];
6
7        switch (event)
8                {
9                case EVENT_LEFT_CLICK:
10                       FlushOutQ(ComportNumber1);
11                       GetCtrlVal(mainhandle,SER_SEND_BOX, send_data);
12                       bytes_sent = ComWrt (ComportNumber1, send_data, StringLength(send_data));
13                       if (bytes_sent <0) //Check for communication errors
14                       {
15                               *ErrorMessage = GetRS232ErrorString(ReturnRS232Err); //Return the
16                                                                           Error String
17                               InsertTextBoxLine (mainhandle, SER_RECEIVE_BOX, -1,
18                                                                           *ErrorMessage);
19                       }
20                       else
21                               SetCtrlVal(mainhandle, SER_SENT_BYTES, bytes_sent);
22                       break;
23                }
24        return 0;
25  }//SendSerialCB
```

Figure 12–11
SendSerialCB Function Source Code

Using the *ComWrt* function, you can specify the number of bytes written to the output port. The *ComWrt* function returns the number of bytes actually placed in the output queue in the variable. A negative return value indicates an error. The error message at line 15 is trapped and displayed in a text box at line 17. The prototype for *ComWrt* function follows and its arguments are described in Table 12–10.

```
int bytes_sent = ComWrt(int ComportNumber, char send_data[ ],
                                                   int BytesToSend);
```

Receiving Data Buffer

Serial port data is read from COM2 by selecting the **RECEIVE** command button in Figure 12–5, which invokes the callback function *ReceiveSerialCB*. The listing for this callback function is shown in Figure 12–12.

At line 13, the library function *ComRd* is used to read the data into the buffer `receive_data` and the number of bytes read from the port are written to `bytes_received`. Error checking between lines 14 and 18 is done as previously explained. Line 23 displays the received data in the **Read from COM2:** text box. The number of bytes received are written to the **Bytes Received** indicator box on line 24. The prototype for the *ComRd* function follows and its arguments are described in Table 12–11.

```
int bytes_received = ComRd(int ComportNumber,
                       char receive_data[ ],int BytesToRead);
```

Table 12–10 ComWrt *Function*

Input/Output	Name	Type	Description
Input	Comport Number	integer	the number assigned to the COM port, can be between 1 and 1,000
	send_data	string	buffer containing the data string to send
	BytesTo Send	integer	number of bytes to send to the output queue
Output	bytes_ sent	integer	number of bytes placed in the queue (negative value indicates error), refer to Table 5–10 in the *LabWindows/CVI Standard Libraries Reference Manual* or in Online Help

```
1   //Read serial data from COM2
2   int CVICALLBACK ReceiveSerialCB (int panel, int control, int event,
3                   void *callbackData, int eventData1, int eventData2)
4   {
5   int bytes_received,i;
6   char receive_data[1024];
7
8           switch (event)
9               {
10              case EVENT_LEFT_CLICK:
11                  for (i=0; i<1024;i++) receive_data[i]=0; //clean the buffer
12
13                  bytes_received = ComRd (ComportNumber2, receive_data, bytes_sent);
14                  if (bytes_received <0)  //Check for communication errors
15                  {
16                      *ErrorMessage = GetRS232ErrorString(ReturnRS232Err); //Return the
17                                                                          Error String
18                      InsertTextBoxLine (mainhandle, SER_RECEIVE_BOX, -1,
19                                                          *ErrorMessage);
20                  }
21                  else
22                  {
23                      InsertTextBoxLine (mainhandle, SER_RECEIVE_BOX, -1, receive_data);
24                      SetCtrlVal(mainhandle, SER_REC_BYTES, bytes_received);
25                  }
26              break;
27              }
28      return 0;
29  }//ReceiveSerialCB
```

Figure 12–12
ReceiveSerialCB Function Source Code

Table 12–11   ComRd *Function*

Input/Output	Name	Type	Description
Input	Comport Number	integer	number assigned to the COM port, can be between 1 and 1,000
	receive_data	string	buffer to read the data string
	BytesTo Read	integer	number of bytes to receive
Output	bytes_received	integer	number of bytes read from the input queue (negative value indicates error), refer to Table 5–10 in the *LabWindows/CVI Standard Libraries Reference Manual* or in Online Help

Sending a Data Byte

There may be times when you may want to send a single byte of data to the serial port instead of a string buffer. The RS-232 library has a function called *ComWrtByte* just for this purpose. This function writes the low-order byte of the integer and returns a 1 if the operation is successful and a negative value if unsuccessful.

In the GUI as shown in Figure 12–5, enter the byte to send in the **Send Byte** box and select **SEND BYTE** to place the byte in the output queue on the COM1 port. The callback function *SendByteCB* is invoked, and its listing is shown in Figure 12–13.

The data obtained from the SendByte box at line 11 is sent to the output queue using the RS-232 library function *ComWrtByte* at line 12. The function prototype follows and its arguments are described in Table 12–12.

```
int status = ComWrtByte(int ComportNumber, int byte);
```

```
1   //Send data byte to output queue
2   int CVICALLBACK SendByteCB (int panel, int control, int event,
3                void *callbackData, int eventData1, int eventData2)
4   {
5   unsigned short  send_byte;
6
7
8       switch (event)
9           {
10              case EVENT_LEFT_CLICK:
11                  GetCtrlVal (mainhandle, SER_SEN_BYTE_BOX, &send_byte);
12                  bytes_sent = ComWrtByte (ComportNumber1, send_byte);
13                  if (bytes_sent <0)   //Check for communication errors
14                  {
15                      *ErrorMessage = GetRS232ErrorString(ReturnRS232Err); //Return the
16                                                                          Error String
17                      InsertTextBoxLine (mainhandle, SER_RECEIVE_BOX, -1,
18                                                                  *ErrorMessage);
19                  }
20                  break;
21          }
22          return 0;
23  } //SendByteCB
```

Figure 12–13
SendByteCB Function Source Code

Table 12–12 ComWrtByte *Function*

Input/Output	Name	Type	Description
Input	Comport Number	integer	number assigned to the COM port, can be between 1 and 1,000
	byte	integer	low order byte to send to output queue
Output	status	integer	0= timeout error, 1= one byte placed in output queue (negative value indicates error), refer to Table 5–10 in the *LabWindows/CVI Standard Libraries Reference Manual* or in On-line Help

Reading a Data Byte

The data byte sent out on the serial port is read in the **Read Byte** control box when the **RECEIVE BYTE** command button is selected from the GUI, as shown in Figure 12–5. The callback function *ReceiveByteCB* is listed in Figure 12–14.

```
1   //Receive data byte
2   int CVICALLBACK ReceiveByteCB (int panel, int control, int event,
3                   void *callbackData, int eventData1, int eventData2)
4   {
5   char ReadByte;
6           switch (event)
7                   {
8                   case EVENT_LEFT_CLICK:
9                           ReadByte = ComRdByte(ComportNumber2);
10                          if (ReadByte <0)   //Check for communication errors
11                          {
12                                  *ErrorMessage = GetRS232ErrorString(ReturnRS232Err); //Return the
13                                                  Error String
14                                  InsertTextBoxLine (mainhandle, SER_RECEIVE_BOX, -1,
15
16                                                                                  *ErrorMessage);
17                          }
18                          SetCtrlVal (mainhandle, SER_REC_BYTE_BOX, ReadByte);
19                          break;
20                  }
21          return 0;
22  } //ReceiveByteCB
```

Figure 12–14
ReceiveByteCB Function Source Code

Table 12–13 ComRdByte *Function*

Input/Output	Name	Type	Description
Input	Comport Number	integer	number assigned to the COM port, can be between 1 and 1,000
Output	byte	integer	low order byte contains the byte read (negative value indicates error), refer to Table 5–10 in the *LabWindows/CVI Standard Libraries Reference Manual*; –99= timeout error

At line 9, the *ComRdByte* library function reads the data byte from the COM2 port. This function returns an integer whose low-order byte contains the byte read and displayed using line 14. Error checking for *ComRdByte* is performed as before. The prototype for this function follows and its arguments are described in Table 12–13.

```
int byte = ComRdByte(int ComportNumber);
```

Reading Data Using a Termination Byte

The function *ComRdTerm* is used for reading data from the input queue until a certain termination character in the data buffer is read. Many times when communicating with certain instruments, you would like to know when to stop receiving data. Looking for the end of data transmission character would give you such an indication. The prototype for this function follows and its arguments are described in Table 12–14.

```
int bytes_received = ComRdTerm (int COMPort, char receive_data[ ],
                                int count, int terminationByte);
```

The *ComRdTerm* function reads the data until it finds the **terminationByte** character in its input queue, reaches the **count** number of bytes, or encounters an error. This function neither writes the **terminationByte** to the buffer nor includes it in **count.**

The *ComRdTerm* function times out if the input queue remains empty during the entire timeout period. When time out occurs, bytes_received are the

Table 12–14 ComRdTerm *Function*

Input/Output	Name	Type	Description
Input	Comport Number	integer	number assigned to the COM port, can be between 1 and 1,000
	receive_data	string	buffer containing the data
	count	integer	number of bytes to read
	termination Byte	integer	low byte contains the numeric equivalent of the terminating byte
Output	bytes_received	integer	number of bytes read from the input queue (negative value indicates error), refer to Table 5–10 in the *LabWindows/CVI Standard Libraries Reference Manual* or in Online Help

number of bytes read up to the time before the time out occurred. The global variable `rs232err` error flag is set to –99.

Writing Data Files

The RS-232 library contains functions that can send and receive data from/to files over the serial port(s). The callback function *SendFileCB* is invoked when you select the **SEND** command button in the **FILE I/O** group in the GUI shown in Figure 12–5. You select the file to send from the file select pop-up dialog box, open the selected file as read only, and send it to the output queue of the serial port. The listing for *SendFileCB* function is shown in Figure 12–15.

The RS-232 Library function *ComFromFile* is called at line 22. It reads from the specified file and writes to the output queue of the selected port. Error checking for the *ComFromFile* function is performed as explained for other functions. The number of bytes sent from the file are written to **Bytes Sent** indicator box shown in the GUI in Figure 12–5. The prototype for this function follows and its arguments are explained in Table 12–15.

```
1   //Transmit the File over COM1
2   int CVICALLBACK SendFileCB (int panel, int control, int event,
3                  void *callbackData, int eventData1, int eventData2)
4   {
5
6        switch (event)
7        {
8             case EVENT_COMMIT:
9
10                 DefaultCtrl(mainhandle, SER_REC_BYTES);
11                 FileSelectPopup (c_prj_dir, "*.txt", "*.txt",
12                 "Select File to write to COM1 port", VAL_LOAD_BUTTON,
13                                            0, 0, 1, 0, pathname);
14
15                 sprintf(buf, "File %s sent on COM1:\n",pathname);
16                 InsertTextBoxLine (mainhandle, SER_SEND_BOX, -1, buf);
17
18                 // open file for reading only
19                 ToFileHandle = OpenFile (pathname, VAL_READ_ONLY, VAL_APPEND,
20                                                           VAL_ASCII);
21                 //Sent data and close file
22                 bytes_sent= ComFromFile(ComportNumber1, ToFileHandle, 0, -1); // Write file
23                                                                    to COM port
24                 if (bytes_sent <0)    //Check for communication errors
25                 {
26                      *ErrorMessage = GetRS232ErrorString(ReturnRS232Err); //Return the
27                                                                    Error String
28                      InsertTextBoxLine (mainhandle, SER_RECEIVE_BOX, -1,
29                                                           *ErrorMessage);
30                 }
31                 SetCtrlVal(mainhandle, SER_SENT_BYTES, bytes_sent);
32                 CloseFile(ToFileHandle);
33
34                 SetInputMode (mainhandle, SER_READFILE, 1);  //Enable the "Read" command
35                                                                    button
36                 SetInputMode (mainhandle, SER_SENDFILE, 0);  //Disable the "Send" command
37                                                                    button
38             break;
39        }
40        return 0;
41  } //SendFileCB
```

Figure 12–15
SendFileCB *Function Source Code*

```
int bytes_sent = ComFromFile(int ComportNumber, int fileHandle,
                             int count, int terminationByte);
```

To send the entire file, set **count** to 0 and **terminationByte** to -1. The *ComFromFile* function will send all the data until the EOF or an error is encountered. This function times out anytime it does not write any bytes from the output queue to the COM port during the set timeout period. On a timeout,

Table 12–15 ComFromFile *Function*

Input/Output	Name	Type	Description
Input	Comport	integer	the number assigned to the COM port, can be between 1 and 1,000
	fileHandle	integer	file handle of the file to send
	count	integer	number of bytes to send; if 0, terminates on **terminationByte,** End of File (EOF), or on error
	terminationByte	integer	specified termination byte to stop sending data; enter –1 for the function to ignore this argument and terminate on **count,** bytes written, EOF, or on error
Output	bytes_sent	integer	number of bytes written to the output queue (negative value indicates error), refer to Table 5–10 in the *LabWindows/CVI Standard Libraries Reference Manual* or in Online Help

the value returned in `bytes_sent` is the number of bytes actually read from the COM port. The global variable `rs232err` is set to -99.

Reading Data Files

To read the file from the COM port and write it to a file, select the **READ** command button in the **FILE I/O** group shown in the GUI in Figure 12–5. The callback function *ReadFileCB,* listed in Figure 12–16, is invoked.

The file is written using the same file name as the sent file but with a .wrt extension. Lines 9 through 15 perform these manipulations by extracting the file name using the function *SplitPath*. This extension is removed and the new extension concatenated to the same file name. Recall that the *SplitPath* function was explained in the "Run SequenceCB Function" section of Chapter 7.

A file handle is created for opening a file for reading at line 17. This file handle is used in the library function *ComToFile* at line 28 to read the input queue of the specified COM port and write the data to this file. The

```
1    //read the File from COM2
2    int CVICALLBACK ReadFileCB (int panel, int control, int event,
3                  void *callbackData, int eventData1, int eventData2)
4    {
5         switch (event)
6                 {
7                 case EVENT_COMMIT:
8                      //Split the path name into drive, directory and file name
9                      SplitPath (pathname, drive, directory, Split_FileName);
10                     //Find the location of "." to remove the file extension
11                     ExtensionLocation= FindPattern (Split_FileName, 0, -1, ".", 0, 0);
12                     //Copy the file name without the extension
13                     CopyString (WriteFileName, 0, Split_FileName, 0, ExtensionLocation);
14                     //Append ".wrt" as extension to the file name
15                     strcat(WriteFileName,".wrt");   //Append the ".wrt" extension
16
17                     FromComFileHandle = OpenFile (WriteFileName, VAL_READ_WRITE,
18                                                   VAL_TRUNCATE, VAL_ASCII);
19
20                     sprintf(buf, "Reading file on COM2............\n");
21                     InsertTextBoxLine (mainhandle, SER_RECEIVE_BOX, -1, buf);
22                     ProcessSystemEvents;
23
24                     sprintf(buf, "File received on COM2:%s\\%s\n",c_prj_dir,Split_FileName);
25                     InsertTextBoxLine (mainhandle, SER_RECEIVE_BOX, -1, buf);
26
27                     //Write data received to file and close the fil
28                     bytes_read= ComToFile(ComportNumber2, FromComFileHandle, 0, -1);
29                     if (bytes_read <0) //Check for communication errors
30                          {
31                          *ErrorMessage = GetRS232ErrorString(ReturnRS232Err); //Return the
32                                                                      Error String
33                          InsertTextBoxLine (mainhandle, SER_RECEIVE_BOX, -1,
34                                                                *ErrorMessage);
35                          }
36
37                     //Change the Label programmatically
38                     SetCtrlAttribute (mainhandle, SER_REC_BYTES, ATTR_LABEL_TEXT,
39                                                        "Bytes Received");
40                     SetCtrlVal(mainhandle, SER_REC_BYTES, bytes_read);
41                     CloseFile(FromComFileHandle);
42                     SetInputMode (mainhandle, SER_READFILE, 0);  //Disable the "Read" command
43                                                                                        button
44
45                     SetInputMode (mainhandle, SER_DISPLAY, 1);  //Enable the "Display" command
46                                                                                        button
47                     SetInputMode (mainhandle, SER_SENDFILE, 1); //Enable the "Send" command
48                                                                                        button
49
50                 break;
51                 }
52          return 0;
53   }//ReadFileCB
```

Figure 12–16
ReadFileCB Function Source Code

ComToFile function reads the data until the specified number of bytes are read from the input queue, the termination byte is encountered, the function times out, or an error occurs. Lines 29 through 35 perform the error checking. The number of bytes written to the file is written to the **Bytes Received** indicator using line 40. The prototype for this function follows and its arguments are explained in Table 12–16.

```
int bytes_received = ComToFile(int ComportNumber,
            int fileHandle, int count, int terminationByte);
```

To read the entire file, set the **count** to 0 and the **terminationByte** to –1 in the *ComToFile* function. If the **terminationByte** is set to –1, *ComToFile* ignores this argument and terminates on **count** bytes. If **terminationByte** is not -1, the function stops reading the data when this byte is encountered. The *ComToFile* function removes the termination byte and does not write it to the file. If the **terminationByte** is a carriage return (CR or decimal 13) and if the character immediately following CR is a linefeed (LF or decimal 10), *ComToFile* discards the LF and the CR characters. If the **terminationByte** is a LF and if the character immediately following LF is CR, *ComToFile* discards the CR and the LF characters.

Table 12–16 ComToFile *Function*

Input/Output	Name	Type	Description
Input	Comport Number	integer	number assigned to the COM port, can be between 1 and 1,000
	fileHandle	integer	file handle of the file to write data
	count	integer	number of bytes to read; if 0, terminates on **terminationByte** or on error
	termination Byte	integer	specified termination byte to stop reading data; enter –1 for the function to ignore this argument and terminate on **count** bytes read or on an error
Output	bytes_received	integer	number of bytes written to the file (negative value indicates error), refer to Table 5–10 in the *LabWindows/CVI Standard Libraries Reference Manual* or in Online Help

This function times out if the input queue is empty during the set timeout period. On a timeout, the value returned in `bytes_received` is the number of bytes actually written to the file. The global variable `rs232err` is set to –99.

Sending and Receiving Data Packets

For reliable communications, data must be sent and received error-free over the serial interface. There are various methods used to achieve error checking. In one method, the sender sends the same data twice and the receiver verifies it to be the same data both times. In another method, the error-checking byte is sent along with the data as a checksum by performing an arithmetic or logical operation on the bytes in the message. A more reliable type of error-checking is to use *Cyclic Redundancy Code* (CRC), which uses a more complex mathematical algorithm than the checksum method. The common error-checking protocols used in file transfers are *Kermit*, *XModem*, *YModem*, and *ZModem*.

The RS-232 library supports *XModem* functions that transfer files using a data transfer protocol. This protocol uses a file transfer technique with error checking. These functions transfer the data packets from the files along with the error checking and synchronization information. The *XModem* functions calculate the checksum and retransmit the data automatically when they detect any errors. The *XModem* functions are different from the *ComFromFile* and *ComToFile* functions since *ComFromFile* and *ComToFile* do not perform any such error checking.

To transfer a file, you must first set the *XModem* configuration parameters for both the COM ports using the *XModemConfig* function. To send data packets, use the *XModemSend* function. The *XModemReceive* function receives the data packet over a COM port and writes it to a file.

Configuring *XModem*

The *XModemConfig* function is used for configuring the COM port(s) when you want to send/receive data packets. The same baud rate and the input/output queue size that you specified in the *OpenComConfig* function are used. However, the data bits, parity, and the stop bit settings of the *OpenComConfig* function are ignored, and the *XModem* functions always use eight bits, no parity, and one stop bit. In addition, the timeout value specified in the *SetComTime* function is not used, but the *XModem* functions use a one-second timeout between data bytes. The prototype for the *XModemConfig* function follows and a description of its parameters is given in Table 12–17.

Table 12–17 XModemConfig *Function*

Input/Output	Name	Type	Description
Input	Comport Number	integer	number assigned to the COM port, can be between 1 and 1,000
	startDelay	double	initial time to establish communication between the sender and the receiver (setting 0.0 selects the default value of 10.0 seconds)
	maximum#ofRetries	integer	maximum number of retries to send the data packet (0 selects the default value of 10)
	waitPeriod	double	time between transfer of two packets (0.0 selects the default value to 10.0 seconds; National Instruments recommends > 5.0 seconds)
	packetSize	integer	packet size in bytes (0 sets the default value to 128)
Output	result	integer	0 means the transfer of data packet is successful (negative value indicates error), refer to Table 5–10 in the *LabWindows/CVI Standard Libraries Reference Manual* or in Online Help

```
int result = XModemConfig (int ComportNumber, double startDelay,
    int  maximum#ofRetries, double waitPeriod, int packetSize);
```

The arguments for the *XModemConfig* function should be explained in more detail. The **startDelay** is the time required for establishing the initial connection between the sender and the receiver. When you are receiving the data packet, **startDelay** is the time in which to send the initial negative acknowledgment character to the transmitter. *XModemConfig* sends that character every **startDelay** seconds, up to the **maximum#ofRetries** times. When you are sending the data packet, **startDelay** specifies the interval in seconds, during which the sender waits for the initial negative acknowledgment. The sender waits for up to **startDelay** times the **maximum#ofRetries** seconds.

The **maximum#ofRetries** argument is the maximum number of times the sender retries to send the data packet to the receiver due to an error condition.

The **waitPeriod** parameter is the time between the transfer of two data packets. When you are the sender, you wait for up to **waitPeriod** seconds for an acknowledgement before re-sending the data packet. When you are receiving the data, you wait for up to **waitPeriod** seconds for the next data packet after you send out an acknowledgment for the current data packet. If you do not receive the packet within **waitPeriod** seconds, you re-send the acknowledgment and wait again for up to **maximum#ofRetries** times.

The **PacketSize** parameter sets the packet size in bytes. The value must be less than or equal to the input and output queue size specified in *OpenComConfig* function. The standard XModem protocol defines packet size as 128 or 1024. The default packet size is 128.

Sending Data Packets

The *XModemSend* function reads the data from a file and sends it as a data packet over the COM port using the *XModem* file transfer protocol. You do not need to open the data file as in the *ComFromFile* function and close it after the data is transferred. The *XModemSend* function does that for you automatically. It opens the file in binary mode and sends all the special characters including carriage returns and line feeds to the receiver without translation.

The standard *XModem* protocol supports only 128 or 1024 as packet sizes, though *CVI* attempts to support any packet size that is specified in the *XModemConfig* function. If you send a file that is not an even multiple of the packet size, the last data packet is padded with ASCII NUL bytes. The prototype for the *XModemSend* function follows and a description of its parameters is given in Table 12–18.

```
int result = XModemSend (int ComportNumber, char filename[ ]);
```

Receiving Data Packets

Use the *XModemReceive* function to receive the data packets. The data packet received is written to the file specified in the *XModemReceive* function. Like the *XModemSend* function, it opens the file in binary mode and does not treat the linefeed and carriage return as special characters.

Table 12–18 XModemSend *Function*

Input/Output	Name	Type	Description
Input	Comport Number	integer	number assigned to the COM port, can be between 1 and 1,000
	filename	string	the file path name containing the data
Output	result	integer	0 means transfer is successful (negative value indicates error), refer to Table 5–10 in the *LabWindows/CVI Standard Libraries Reference Manual* or in Online Help

The sender and the receiver of the data packets must negotiate on the error checking protocol. This is done at the beginning before the data packets are sent, which can cause a delay before the data packet transfer begins. The *XModemReceive* function tries ((maximum#ofRetries + 1)/2 times) to negotiate a Cyclic Redundancy Check (CRC) error check transfer. If there is no response, the *XModemReceive* function tries to negotiate a check sum transfer (up to (maximum#ofRetries − 1)/2 times). Again, if the size of the file is not an even multiple of the packet size, the file received is padded with ASCII NUL bytes. The prototype for the *XModemReceive* function follows and a description of its parameters is given in Table 12–19.

```
int result = XModemReceive (int ComportNumber, char filename[ ]);
```

Table 12–19 XModemReceive *Function*

Input/Output	Name	Type	Description
Input	Comport Number	integer	number assigned to the COM port, can be between 1 and 1,000
	filename	string	the file path name receiving the data
Output	result	integer	0 means transfer is successful (negative value indicates error), refer to Table 5–10 in the *LabWindows/CVI Standard Libraries Reference Manual* or in Online Help

Serial Communication Utility Functions

Besides the library functions that set up the configuration of the serial ports and send and receive data as either individual bytes, strings of data, or a file, there are functions that perform tasks to support other serial communication needs. These functions are discussed in this section.

Flush Output Queue

Sometimes you may want to discard the data sent to the output port and load another set of data to the output queue. To perform this task, the RS-232 library function *FlushOutQ* is used. It removes all the characters from the specified COM port. In this project, when the **FLUSH OUPUT Q** command button is selected from the GUI shown in Figure 12–5, the function *FlushOutputQCB* is called. The source code listing for this function is given in Figure 12–17.

At line 8, the function *FlushOutQ* is called and its returned value checked for possible errors. The prototype for this function follows and its arguments are explained in Table 12–20.

```
int status = FlushOutQ( int ComportNumber);
```

```
1   //This function flushes the output queue on COM port 1
2   int CVICALLBACK FlushOutputQCB (int panel, int control, int event,
3               void *callbackData, int eventData1, int eventData2)
4   {
5       switch (event)
6           {
7           case EVENT_COMMIT:
8               status = FlushOutQ (ComportNumber1);
9               if (status <0) //Check for communication errors
10              {
11                  *ErrorMessage = GetRS232ErrorString(ReturnRS232Err); //Return the
12                                                                      Error String
13                  InsertTextBoxLine (mainhandle, SER_SEND_BOX, -1, *ErrorMessage);
14              }
15              sprintf(buf, "Output Queue Flushed");
16              InsertTextBoxLine (mainhandle, SER_SEND_BOX, -1, buf);
17              break;
18          }
19      return 0;
20  } //FlushOutputQCB
```

Figure 12–17
Flush OutputQCB Callback Function

Table 12–20 FlushOutQ *Function*

Input/Output	Name	Type	Description
Input	Comport Number	integer	COM port number to flush the output queue, can be between 1 and 1,000
Output	status	integer	negative value indicates error, refer to Table 5–10 in the *LabWindows/CVI Standard Libraries Reference Manual* or in Online Help

Flush Input Queue

Similarly, if for some reason you want to remove all the characters from the input queue, you can use the RS-232 library function *FlushInQ*. In this project, selecting the command button **FLUSH INPUT Q** from the GUI shown in Figure 12–5 invokes the callback function *FlushInputQCB*, whose listing is given in Figure 12–18.

```
1   //This function flushes the output queue on COM port 2
2   int CVICALLBACK FlushInputQCB (int panel, int control, int event,
3                   void *callbackData, int eventData1, int eventData2)
4   {
5       switch (event)
6           {
7           case EVENT_COMMIT:
8               status = FlushInQ (ComportNumber2);
9               if (status <0) //Check for communication errors
10      {
11                      *ErrorMessage = GetRS232ErrorString(ReturnRS232Err); //Return the
12                                                                          Error String
13                      InsertTextBoxLine (mainhandle, SER_RECEIVE_BOX, -1,
14                                                                      *ErrorMessage);
15              }
16              sprintf(buf, "Input Queue Flushed");
17              InsertTextBoxLine (mainhandle, SER_RECEIVE_BOX, -1, buf);
18              break;
19          }
20      return 0;
21  } // FlushInputQCB
```

Figure 12–18
FlushInputQCB Callback Function

Table 12-21 FlushInQ *Function*

Input/Output	Name	Type	Description
Input	Comport Number	integer	COM port number to flush the input queue; can be between 1 and 1,000 and is the port to be flushed
Output	status	integer	negative value indicates error, refer to Table 5-10 in the *LabWindows/CVI Standard Libraries Reference Manual* or in Online Help

The function *FlushInQ* removes all the characters from the input queue of the COM port as shown on line 8. The prototype for this function follows and its arguments are explained in Table 12-21.

```
int status = FlushInQ( int ComportNumber);
```

Queue Length

To obtain the number of characters in the output queue, you can use the RS-232 library function *GetOutQLen*. There may be several reasons for you to use this function. You can use this function to check if all the characters in the output queue have been sent before you close the COM port. Alternatively, you may want to know the number of characters sent on one COM port and compare them to the number received on the other COM port. This function is invoked by selecting the **Q LEN** command button from the GUI shown in Figure 12-5. The listing for the output Q length callback function is given in Figure 12-19.

The prototype for this function follows and its arguments are explained in Table 12-22.

```
int len = GetOutQLen( int ComportNumber);
```

Similarly, the RS-232 library function *GetInQLen* is used to obtain the number of characters in the input queue. The prototype of this function is similar to *GetOutQLen*.

```
//Obtain the output Q length
int CVICALLBACK GetOutQLengthCB (int panel, int control, int event,
            void *callbackData, int eventData1, int eventData2)
{
    int Out_Length;
    switch (event)
        {
            case EVENT_LEFT_CLICK:
                Out_Length= GetOutQLen(ComportNumber1);
                sprintf(buf, "Output Q length = %d",Out_Length);
                InsertTextBoxLine (mainhandle, SER_SEND_BOX, -1, buf);
                break;
        }
    return 0;
}//GetOutQLengthCB
```

Figure 12–19
GetOutQLengthCB Callback Function

Obtaining COM Port Status

To obtain the status of the COM port, select the **COM STATUS** command button from the GUI shown in Figure 12–5. The callback function *GetCOM-StatusCB* is invoked, and its listing is shown in Figure 12–20.

The bits returned by the *GetComStat* function indicate the status of the COM port. They are explained in Table 5–5 in the *LabWindows/CVI Standard Libraries Reference Manual*. The *DisplayComStatus* function, called after *GetComStat*, translates the status bit information into a meaningful status definition. The *DisplayComStatus* function is copied in its entirety from the

Table 12–22   GetOutQLen *Function*

Input/ Output	Name	Type	Description
Input	Comport Number	integer	the number assigned to the COM port, can be between 1 and 1,000
Output	len	integer	number of characters in the output queue (negative value indicates error), refer to Table 5–10 in the *LabWindows/CVI Standard Libraries Reference Manual* or in On-line Help

Chapter 12 • RS-232 Serial Communication

```
//Obtain the Status for the COM port
int CVICALLBACK GetCOMStatusCB (int panel, int control, int event,
                void *callbackData, int eventData1, int eventData2)
{
        int PortNumber;
        char ComPortNumber[3];
        switch (event)
                {
                case EVENT_COMMIT:
                        PromptPopup ("STATUS FOR COM PORT NUMBER",
                                        "Enter COM Port number", ComPortNumber, 2);
                        Scan(ComPortNumber, "%s>%d", &PortNumber);
                        com_status =GetComStat(PortNumber);
                        DisplayComStatus ;
                        break;
                }
        return 0;
}//GetCOMStatusCB
```

Figure 12–20
GetCOMStatusCB Callback Function

rs232 samples project that comes with *CVI*. The prototype for this function follows and its arguments are explained in Table 12–23.

```
int status = GetComStat( int ComportNumber);
```

Break Function

The RS-232 library function *ComBreak* generates a break signal for a specified number of milliseconds. This function causes the serial communication to be interrupted and the send or receive operations aborted. A break signal is

Table 12–23   GetComStat *Function*

Input/Output	Name	Type	Description
Input	Comport Number	integer	the number assigned to the COM port, can be between 1 and 1,000
Output	status	integer	bits indicating the COM port status (negative value indicates error), refer to Table 5–10 in *LabWindows/CVI Standard Libraries Reference Manual* or in Online Help

```
//Loop every second to check if Break invoked
int CVICALLBACK LoopEverySecondCB (int panel, int control, int event,
         void *callbackData, int eventData1, int eventData2)
    {

         if ((BreakSignal == 1) && (event == EVENT_TIMER_TICK))
         {
                 ComBreak (ComportNumber1, 0);
                 ComBreak (ComportNumber2, 0);
                 sprintf(buf, "Break triggered on COM1 and COM2");
                 InsertTextBoxLine (mainhandle, SER_SEND_BOX, -1, buf);
                 BreakSignal=0;


         }
      return 0;
} //LoopEverySecondCB
```

Figure 12–21
LoopEverySecondCB Callback Function

the transmission of a special character on the communication line for a period longer than the transmission time for one character and its framing bits.

When the **BREAK** command button is selected from the GUI in Figure 12–5, the break flag is set. It causes the processing of the *ComBreak* function inside the timer loop function, as shown in Figure 12–21.

The prototype for *ComBreak* follows and its arguments are explained in Table 12–24.

```
int status = ComBreak( int ComportNumber, int BreakTimeMsec);
```

Table 12–24 ComBreak *Function*

Input/Output	Name	Type	Description
Input	Comport Number	integer	the number assigned to the COM port, can be between 1 and 1,000
	Break Time Msec	integer	range 1–255 (use 0 to select 250 milliseconds)
Output	status	integer	negative value indicates error, refer to Table 5–10 in the *LabWindows/CVI Standard Libraries Reference Manual* or in On-line Help

Communicating with RS-485

As mentioned earlier in the chapter, you can use all the RS-232 library functions with the RS-485 AT-Serial board. The *ComSetEscape* function enables you to control the transceiver mode of the board. Using this function, you can carry out extended serial port operations by setting the escape codes in the function to control different line signals. All escape codes may not be valid and may be dependent upon the serial interface driver. The function prototype follows and its arguments are explained in Table 12–25.

```
int result = ComSetEscape( int ComportNumber, int escapeCode);
```

The following values of the *escapeCode* can be used, though not all of them may be supported by all device drivers.

escapeCode	Purpose
CLRDTR	clears the DTR (data-terminal-ready) signal
CLRRTS	clears the RTS (request-to-send) signal
GETMAXCOM	returns the maximum COM port identifier the system supports, ranges from 0x00 to 0x7F, e.g., 0x00 corresponds to COM1, 0x01 to COM2, etc.
STDTR	sends the DTR signal

Table 12–25 ComSetEscape *Function*

Input/Output	Name	Type	Description
Input	Comport Number	integer	the number assigned to the COM port, can be between 1 and 1,000
	escape Code	integer	specifies the escape code
Output	status	integer	negative value indicates error (–1 = Unknown System Error—device driver does not support a particular escape code), refer to Table 5–10 in the *LabWindows/CVI Standard Libraries Reference Manual* or in Online Help

SETRTS	sends the RTS signal
SETXOFF	causes the port to act as if has received an XOFF character
SETXON	causes the port to act as if it has received an XON character

The following values can only be used with the National Instruments RS-485 serial driver.

WIRE_4	Sets the transceiver to four-wire mode
WIRE_2_ECHO	Sets the transceiver to two-wire DTR controlled with echo mode
WIRE_2_CTRL	Sets the transceiver to two-wire DTR controlled without echo
WIRE_2_AUTO	Sets the transceiver to two-wire auto TXRDY controlled mode

Summary

In this chapter, a brief introduction to serial communication was given in which you were shown how to configure the serial ports and how to send and receive data bytes, data strings, ASCII files, and data packets with error checking. The relevant RS-232 library functions were explained and demonstrated in the project. You can use the RS-232 routines in this project with modifications to suit your own needs. RS-485 communication was mentioned to carry out extended serial communication operations.

If you would like to learn more about serial communications, refer to the books referenced in the bibliography.

LABWINDOWS/CVI INSTALLATION

Chapter Highlights

- Installing *CVI*
- Installing the Projects Directory

Installing CVI

The installation of *CVI* is convenient and automatic using the CD-ROM and needs no detailed explanation here. Select the compatibility mode from among the five C/C++ compilers when asked by the installation screen. You have a choice of five *CVI* compatible compilers from four manufacturers:

- Borland C++ 4.51 or higher, Borland C++ Builder 4.0,
- Microsoft Visual C++ 2.2 or higher,
- Symantec C++ 7.2 or higher,
- Watcom C++ 10.5 or higher.

This is also useful when creating stand-alone executables and *DLLs* if you want to create them only for your selected compiler rather than for all the above compilers. Chapter 8, "Creating Stand-Alone Executables and Distribution Disks," discusses this topic in detail.

During the installation process, you will be asked for the destination location for installing the *CVI* application and its supporting libraries. For the purpose of this text, *CVI* will be installed in the `c:\cvi5` folder and referred to this location throughout the text.

Creating a *CVI* icon on the desktop is the easiest way to start running the program, and spares you from searching through **Start>>Programs >> National Instruments CVI** to find the program name. To create an icon on the desktop, open the **Explorer** and go to the directory in which the program was installed. Scroll down through the files until you see a `CVI.exe` icon. Right-click on this icon with the mouse and drag it onto the desktop. A menu will open, asking if you want to create a shortcut here; select **Yes.** The name of the icon can be changed to `CVI5` if you want.

To run the program, double-click on the **CVI5** icon. The program starts, and for the first time user, the screen in Figure A–1 is displayed.

Enter your name, your company name (if any), and the serial number from the software package in the dialog boxes. It is very important to enter this information now because this program has a habit of nagging for this information every time you start the program. Also, keep the serial number with the CD in case you want to re-install it.

Later *CVI* versions can run the projects created in the earlier versions of this tool. However, projects created in the newer versions of *CVI* cannot be run using the older *CVI* versions. If for some reason you would like to use

Figure A–1
CVI Registration

the older version(s) of *CVI*, you should keep them in separate directories and not over-write the older version's files with the newer version of *CVI*.

Note that the installation screens may be different for different versions of *CVI*, but this should not make a difference in the installation of the program.

Installing the Projects Directory

The *CVI* projects used in this text are included in the `Projects` folder on the CD-ROM. It is advisable to copy the `Projects` folder to the folder where you installed *CVI*. Some projects write data to the files that are created in the `Projects` directory during their execution and will cause problems if these projects try to write data to the CD-ROM. In addition, your application(s) will execute much faster when loaded on the hard disk drive.

Sometimes when you copy the files from the CD-ROM, you may find the files have the `RA` attributes set. The `R` refers to the read-only attribute and `A` to the archive attribute. These attributes must be removed from all the files, including the files in the `input` and `output` folders, for the project to run. Open the `Projects` folder, select all the files and right-click the mouse button. Select the `Properties` folder and remove the check marks from the **Read-only** and the **Archive** boxes. Do the same for the `input` and `output` folders.

PROJECT WINDOW ENVIRONMENT

Chapter Highlights

- Starting the Project
- Overview of the Project Window
- Using the File Menu
- Using the Edit Menu
- Using the View Menu
- Using the Build Menu
- Using the Run Menu
- Using the Instrument Menu
- Using the Library Menu
- Using the Tools Menu
- Using the Window Menu
- Using the Option Menu
- Using the Help Menu
- Summary

This appendix gives you an overview of the **Project** window, explaining the menu and sub-menu commands. It is very easy to get overwhelmed with all the capabilities and functions of the **Project** window, but there is no need to be. You probably will not use all of them at once, but you will at least learn their functions and where to find them. The **Project** window menus and sub-menus are explained here to enable you to understand the basics required to create projects.

Starting the Project

As mentioned in the previous chapter, when you start *CVI*, you are presented with a new **Project** window screen as shown in Figure B–1 (or with the files of a project that you have opened previously). You always start a project from the **Project** window environment.

Before the various menus and menu items in the **Project** window are explained, you should know that there are two other *CVI* environments: the **User Interface Editor** and the **Source Editor**. An understanding of these environments is vital to building *CVI* projects. The **User Interface Editor** environment is explained in Chapters 2, 3, and 4, and the menu commands are explained in Appendix C, "User Interface Editor Environment." The **Source Code Editor** and **Debugging Techniques** are explained in Chapter 5. In this appendix only the **Project** window menu commands will be discussed.

The **Title Bar,** pointed to in Figure B–1, is the horizontal bar located on the top of the screen. It displays the name of the project loaded in the **Project** window. If this is a new project, then the `Untitled1.prj` default name is used. Notice that the project name has the three-letter extension `.prj` appended to it. Projects will always have the `.prj` extension appended to the

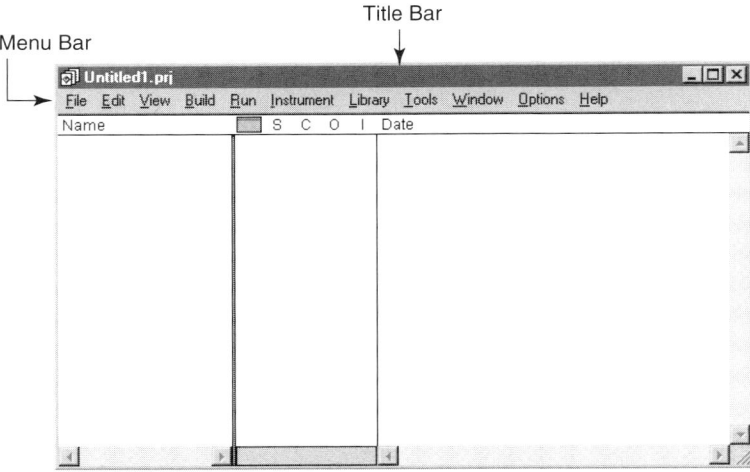

Figure B–1
New Project Window

Appendix B • Project Window Environment

file name to differentiate them from other files in the same project (such as files with extensions .c, .uir, .h, and .fp).

The **Menu Bar** is just below the **Title Bar** and lists all the pull-down menus used in the **Project** window. In the next section, an overview of the **Project** window is given, explaining the various menu commands.

Overview of the Project Window

As you may know from working with Windows applications, having a *menu bar* enhances your accessibility into controlling, executing, and manipulating the features of an application. The same is true in the *CVI* environment. Menu items needed to setup, load, develop, create, debug, run, and save the project are all available to the user.

Let us now look at the various **Project** window menus and menu items.

Using the File Menu

The **File** menu contains the commands to open and close files, load various project files, and save and print them. Figure B–2 shows the menu items under the **File** menu.

New

New is the first command in the **File** menu. The **New** command consists of the sub-menu shown in Figure B–3.

If you want to create a new source file, a file with a .c extension, select **Source (*.c)**. It will open a new source window to create a .c file.

If you want to add a new **Include** file (a file with a .h extension), also known as the header file, select **Include (*.h)**. It will open a new source window to create a .h file.

To create a new user interface file (GUI), select **User Interface (*.uir)**. A new **User Interface Editor** window opens with a .uir extension.

On selecting **Project (*.prj)**, a new blank project window is loaded.

If you choose **Function Tree**, a new **Function Tree Editor** window appears in which you can create a new .fp file. The .fp extension refers to

Appendix B • Project Window Environment

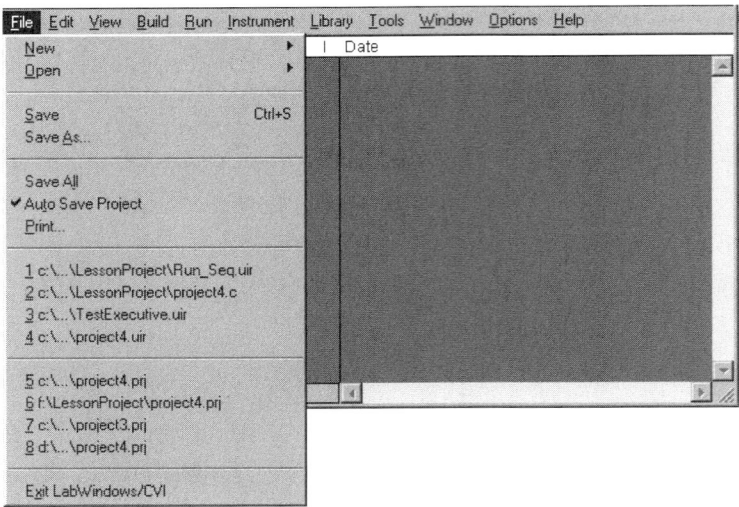

Figure B–2
Project Window File Menu

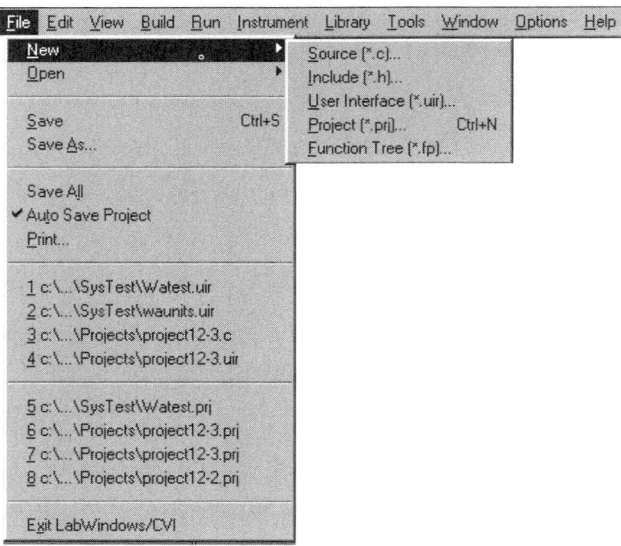

Figure B–3
Project Window New Sub-Menu

Appendix B • Project Window Environment

function panel, which performs certain specialized tasks related to instrumentation files or utility functions. The function panels were used many places in the previous chapters. In particular, the function panel was defined in Chapter 3, "More Graphical User Interface" and its sample shown in Figure 3–11.

Open

The **Open** command has a sub-menu, as shown in Figure B–4. It enables you to load the pre-existing files, depending on the item selected from the sub-menu. Selecting any of these sub-menu command shows an **Explorer**-like folder (directory) structure showing the files. You can move around to the appropriate folder (directory) and select the .c file if **Source (*.c)** is selected.

If the **Include** sub-menu command is selected, a folder (directory) structure is shown with header files (.h extensions).

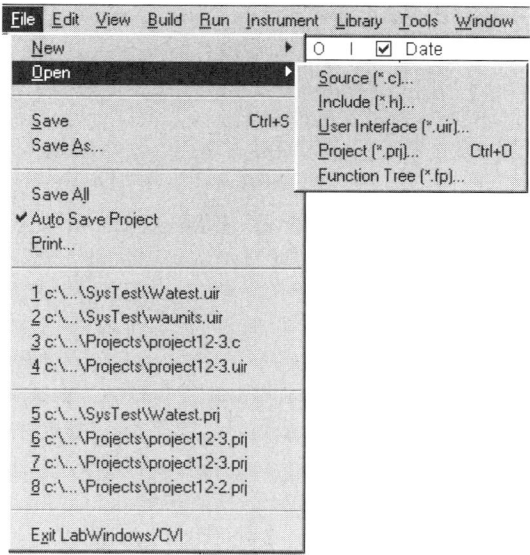

Figure B–4
Project Window Open Sub-Menu

Similarly for **User Interface,** the .uir files are shown. These are the files that consist of the user interface resources layout consisting of buttons, controls, indicators, graphs, charts, or other controls (objects).

The **Project** command on this sub-menu similarly opens files with a .prj extension.

If you select the **Function Tree,** a new **Function Tree Editor** window is displayed with your specified .fp file. Refer to *LabWindows/CVI Instrument Driver Developers Guide* for information about using the **Function Tree Editor** window.

Save

The **Save** command saves all the files of the project to the disk. It is necessary to keep the same type of extensions as *CVI* uses by default and not to create your own extensions.

Save As

The **Save As** command is similar to any in the Windows **File** menu, enabling the user to change the file name and/or the destination of the file before saving it.

Save All

The **Save All** command saves all the files to the disk.

Auto Save Project

When the **Auto Save Project** command is enabled, the project files are automatically saved each time you make changes to your project files. If this item is not checked, you will be prompted to save the file before running the project. Also when you exit *CVI* and files have not been saved, you will be prompted to save the files if you do not have **Auto Save Project** enabled. It is advisable to leave the **Auto Save Project** command enabled.

Print

The **Print** command prints the selected file on the default printer.

Most Recently Closed Files

In addition to the commands shown in **File** menu, there are two blocks of the four most recently closed files followed by a list of the four most recently closed project files. This handy feature of *CVI* saves you time by allowing you to select the file from this menu rather than going through **File >>Open**, searching the project/file folder (directory), and selecting the file to load. Clicking on the file name from the **File** menu loads the file/project instantaneously.

Exit LabWindows/CVI

Exit LabWindow/CVI is used to exit *CVI* and close all opened files. If any files are unsaved, you are prompted to save them before exiting.

Using the Edit Menu

The **Edit** menu is used to edit the project files listed in the **Project** window. It is here that you add and remove files from the project, select all the files in the project, exclude files from build, and change the order in which the files are displayed in the **Project** window list.

The menu items of the **Edit** menu are shown in Figure B–5 and are explained here.

Add Files To Project...

The **Add Files to Project...** command is the first on the **Edit** menu and has a sub-menu associated with it, as shown in Figure B–6. This command is used when you open a project and want to include files to the project. It is here that you select the **Source, Include, Object, Library, User Interface,** or **Instrument** files. These sub-menu commands only show the files with the following extensions `.c`, `.h`, `.obj`, `.lib`, `.uir`, and `.fp` respectively. If you select **All Files,** all the files are displayed, enabling you to select the necessary files to include in the project.

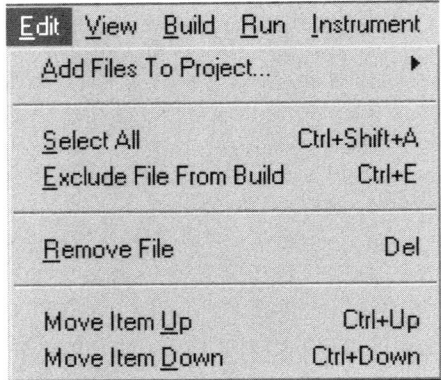

Figure B–5
Project Window Edit Menu

Often as you are building your project, you keep adding new files to the project, and it may become necessary to use the **Add Files To Project** command.

Select All

The **Select All** command is used to select all the files in the project. This enables you to select all the files in the project and execute a command on all the files simultaneously. This is the same as using <Ctrl-Shift-A> from the keyboard.

Note the following shortcuts.

- Use <Ctrl-Left Click> to add the selected file to the currently selected files.

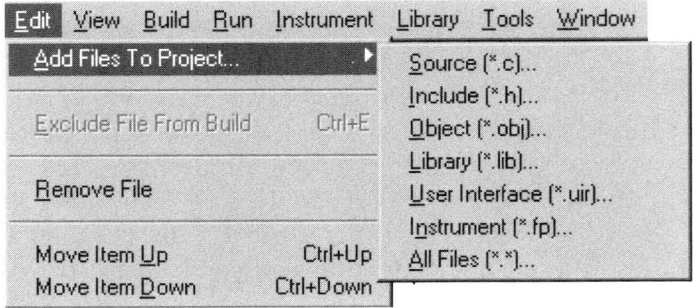

Figure B–6
Add Files to Project Sub-Menu

- Use <Shift-Left Click> to add all the files between the selected file and the file you clicked on to the currently selected files.
- Use <Shift-Down Arrow> to add the file below the currently selected file to the list of selected files.
- Use <Shift-Up Arrow> to add the file above the currently selected file to the list of selected files.

Exclude File From Build

Use **Exclude File From Build** when you do not want to include a file in your project build. Highlight the file(s) you do not want to include in the project (with the exception of `.h`, `.fp`, or `.uir` files). The excluded file will be dimmed on the project window. The file(s) will not be compiled or linked into the project. Note that when the file is excluded, this command dynamically changes to **Include File to Build,** giving you an opportunity to add the same or different files to the project.

Remove File

The **Remove File** command permanently removes the highlighted file from the **Project** window list. Sometimes this is necessary when you are building projects.

Move Item Up

The **Move Item Up** command moves the selected file up one line at a time in the project list as many times it is selected. <Ctrl-Up Arrow> can also do this. This command is only activated if **No Sorting** is checked in the **View** menu. The purpose of this command is to arrange the files to your preference. Some users would like to keep all the `.uir files` together and all the header files together, whereas others prefer to have the files with the same base name listed together.

Move Item Down

The **Move Item Down** command moves the highlighted file one line down in the project list. <Ctrl-Down Arrow> can also do this. Again, this command is only activated if **No Sorting** is selected from the **View** menu.

Remember the shortcuts listed below.

Command	Shortcut
Add the selected file to the currently selected files	<Ctrl-Left Click>
Add all the files between the selected file and the file you clicked on	<Shift-Left Click>
Add the file below the currently selected file to the list of selected files	<Shift-Down Arrow>
Add the file above the currently selected file to the list of selected files	<Shift-Up Arrow>
To move the file up in **Project** window	<Ctrl-Up Arrow>
To move the file down in **Project** window	<Ctrl-Down Arrow>

Using the View Menu

The **View** menu controls the way the files are displayed using the various date formats and sorting features in the **Project** Window list. The **View** menu items are shown in Figure B–7 and the commands are described here.

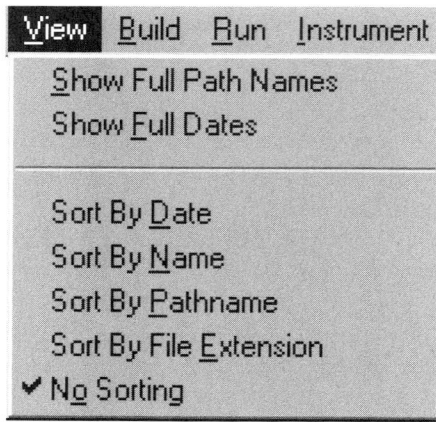

Figure B–7
Project Window View Menu

Show Full Path Names

The **Show Full Path Names** command is used to display the full path names of the files or the base file names. If the command is checked, the full path names will be shown. If not, the base file names are shown.

Show Full Dates

The **Show Full Dates** command is used for displaying the last date the file was saved. The date can appear in full-length form, e.g., `Mon, Sep 14, 2025`, or short form, e.g., `9/14/25`.

Sort By Date

The **Sort By Date** command sorts the project list files by modified date and displays the files in chronological order in the project list.

Sort By Name

The **Sort By Name** command sorts the files by name in alphabetical order in the project list.

Sort By Pathname

The **Sort By Pathname** command sorts by the path names in alphabetical order in the project list.

Sort By File Extension

The **Sort By File Extension** command sorts the file names in alphabetical order by the file extension.

No Sorting

If the **No Sorting** command is highlighted in the **View** menu, the files can be listed in any order in the **Project** window list by the **Move Item Up** and **Move Item Down** commands in the **Edit** menu as previously discussed.

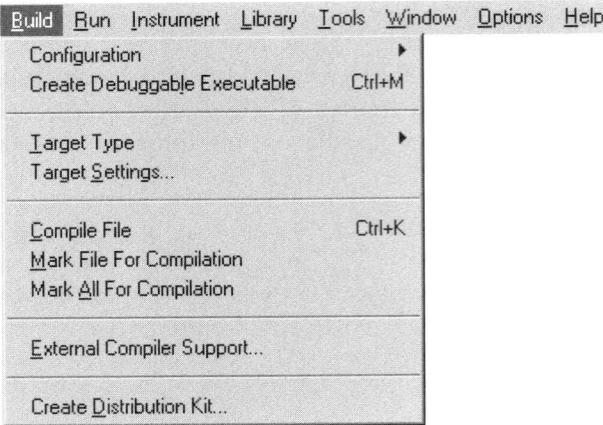

Figure B–8
Project Window Build Menu

Using the Build Menu

The **Build** menu shown in Figure B–8 has the commands to compile files, build and link projects, mark files for compilation, and create DLLs and distribution disks. The sub-menus also provide for external compiler support and configuration and creation of **Debug** or **Release** executables.

Configuration

The **Configuration** menu item gives you the choice of creating the executable or DLL with the **Debug** option or the **Release** option, as shown in the sub-menu in Figure B–9.

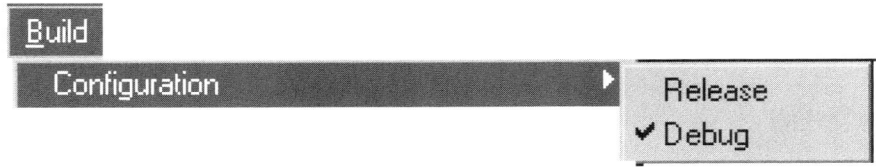

Figure B–9
Configuration Menu Commands

The **Release** configuration should be selected when you know that your project is working error-free and you do not want to use the debugging features. Using this configuration, the executable code is smaller and executes faster but all run-time memory errors are no longer enabled. The **Release** configuration should only be used when you are certain that your executable, DLL, or static library is working.

The **Debug** configuration is selected when you want the debugging features enabled, allowing you the ability to perform the level of debugging set from the **Options>>Build Options** menu shown later in this appendix.

Target Type

The next item on the menu **(Create)** can be better explained after you look at **Target Type,** since this governs what item is displayed and operational on the **Build** menu. The **Target Type** sub-menu is shown in Figure B–10.

The **Target Type** gives you a choice of creating an executable, dynamic link library, or static library file. The name of the sub-menu below **Configuration** (**Create** sub-menu) changes according to the **Configuration** and the **Target Type** selected. For example, if you selected **Release** from the **Configuration** sub-menu and **Dynamic Link Library** from **Target Type,** then the menu item below **Configuration** will be **Create Release Dynamic Link Library.**

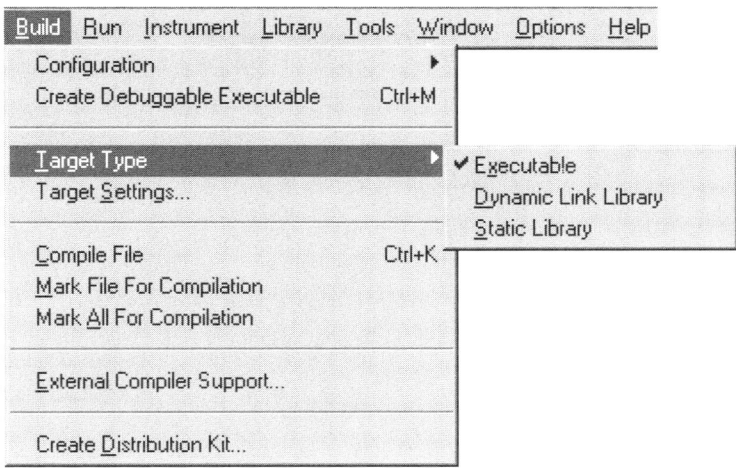

Figure B–10
Target Type Sub-Menu

Target Settings...

The **Target Settings** box is shown in Figure B–11. The control settings in this dialog box are initially dependent on the selection made in the **Target Type** sub-menu.

When you click on the ring counter in the **Application File** dialog box, the **Debug** changes to **Release,** allowing the application file name to change accordingly. This application file name with the complete path is displayed in the box to the right. You can edit the file name in this box or select the **Browse...** command button that opens a folder in the project directory from which you can select the file name. For **Debug,** the pathname of the application file name is always appended with _dbg.

When creating an application for distribution and using **Create Distribution Kit...** in the **Build** menu, enter the title in the **Application Title** box as you would like it to appear in the Windows **Start** menu. This is also the name registered by the operating system during installation.

The **Application Icon File** specifies the icon that you would like to use for this application. To select one of the pre-existing icons, select **Browse...** or double-click on the displayed **Icon** and the project folder with existing icon files (if any) are displayed. The icon files use the .ico extension. If no icon file is selected, then the default *CVI* icon is used as shown in Figure B–11. You can create your own icons using the **Icon Editor** from the **Start** menu as

Figure B–11
Target Settings Dialog Box

Appendix B • Project Window Environment

explained in the "Creating and Using Toolbars" section of Chapter 4, "Enhancing the User Interface."

When the **Create Console Application** box is checked, it creates the executable as a console application. Console applications create a Windows console window (Command Prompt or MS-DOS Prompt) and default the standard I/O port to the console. If this box is left unchecked, your executable is created as a Windows GUI application.

This will become clear with an example. Load `project2-1.prj` and from the **Project** Window, select **Build>>Configuration>>Debug**. From **Build>> Target Settings,** check the **Create Console Application** box to display the data on the Windows console window. To run the project, select <Shift-F5>. The output is displayed as shown in Figure B–12.

Likewise, leave the **Create Console Application** box unchecked and run the project. Your executable is now created as a Windows GUI application. The output is now displayed in a **Standard Input/Output** window as shown in Figure B–13. The **Standard Input/Output** window, like the **Source Editor** window, can contain up to one million lines of code, 254 characters per line.

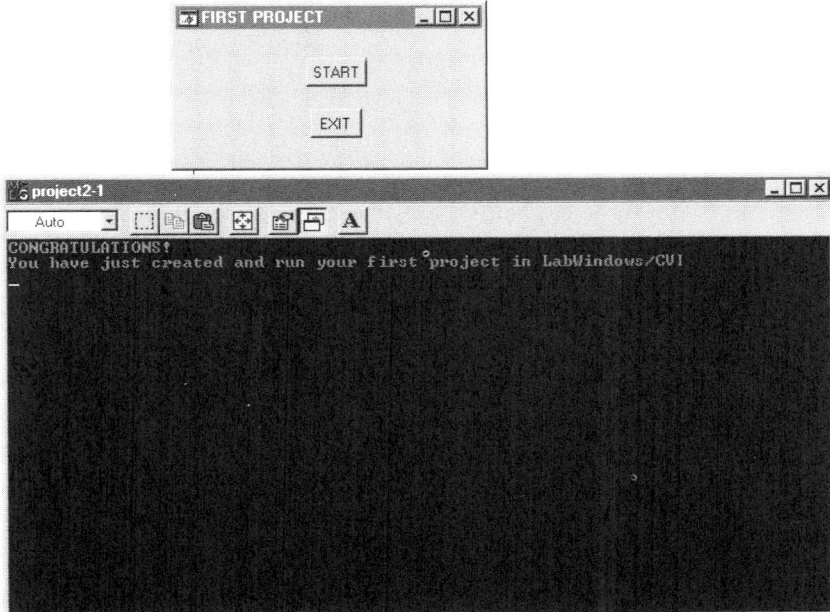

Figure B–12
Run with Output Directed to Windows Console

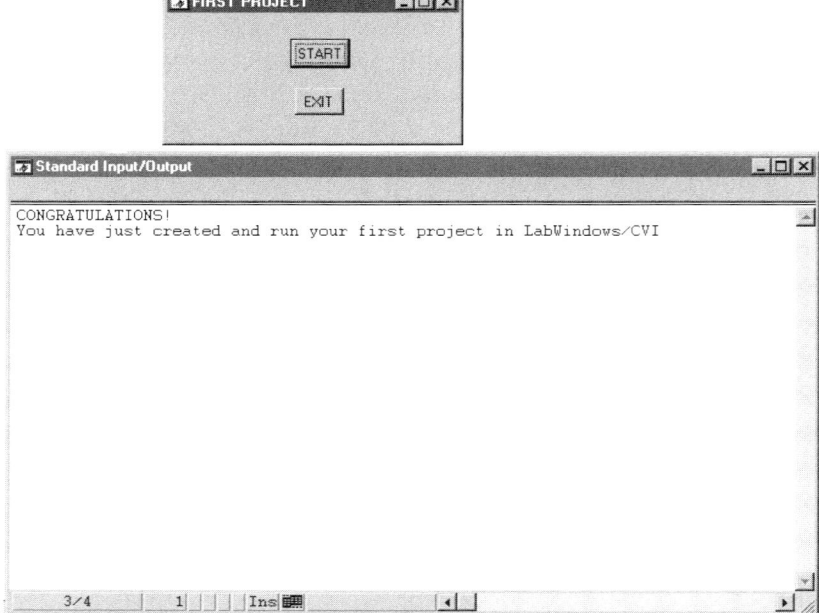

Figure B–13
Run with Output Directed to Standard Output

When the **Instrument Driver Support Only** box is checked in the GUI shown in Figure B–11, the project links to only a smaller set of *CVI* libraries. On the PCs, the **Stand-Alone Executables** and **DLLs** are created when this command is enabled and do not use the *CVI* Run-Time Engine DLL, cvirte.dll, which is very large compared to the instrsup.dll that is used instead. On Solaris 1 and 2 workstations, the DLL is called libinstrsup.so. On HP-UX, it is called libinstrsup.sl.

Note: This command is not available in Windows 3.1.

The **Version Info...** button brings up the **Version Info** dialog box as shown in Figure B–14. In the **File Version** box and the **Product Version** box, enter the number in the form: x,x,x,x, where x represents a number between 0 and 255.

Your company's name can be entered in the **Company Name** box and copyright information can be entered for your protection in the **Legal Copyright** box.

Clicking on the **More...** button brings up the **Version Info Sub-Dialog** box, shown in Figure B–15, in which you can enter additional information, such as comments and type of build. The **OK** button on the **Version Info** box returns you to the previous menu.

Appendix B • Project Window Environment 487

Figure B–14
Version Info Dialog Box

Figure B–15
Version Info Sub-Dialog Box

Using **LoadExternalModule** group in the dialog box shown Figure B–11 lets you add files that are loaded in the project at run-time to the executable via the **Add Files to Executable** button and its call to *LoadExternalModule*.

The **Help** button describes how to use *LoadExternalModule* and the use of the **Add Files to Executable** button.

Compile File

The **Compile File** command shown on Figure B–10 on the **Build** menu allows you to compile the file before you execute the project. If there are any compilation/build errors, they are shown in the **Build Errors** window.

Mark File For Compilation

The **Mark File For Compilation** command forces *CVI* to compile the source file on the next build. When a source file is opened, changed, and closed without recompiling, c appears next to the file name in the **Project** window. The file can be recompiled by clicking on the c in front of the file name in the **Project** window list.

Mark All For Compilation

The **Mark All For Compilation** command will force *CVI* to recompile all source files in the project at the next build.

External Compiler Support

The **External Compiler Support** command is used for building your executables or DLLs in one of the compilers supported by *CVI*. The compilers supported are given in Appendix A. This topic is discussed in Chapter 10, "External Compiler Support."

Create Distribution Kit

The **Create Distribution Kit** command enables you to make a set of disks from which you can install the executable program on another machine. The process is automatic by following the instruction of the **Create Distribution Disk** menu. The created executable is copied onto the formatted floppies or

Appendix B • Project Window Environment

on the hard disk. The **Install** and **Uninstall** icons are copied onto the floppies to facilitate creating the stand-alone executable and for removing it later (if needed). This is discussed in more detail in Chapter 8, "Creating Stand-Alone Executables and Distribution Disks."

Note: This command is only available in Windows.

Using the Run Menu

The **Run** menu contains the menu commands to run the project, continue and terminate execution during debugging, and set breakpoints. The **Run** command from the **Project** window menu is shown in Figure B–16, and the sub-menu commands are explained here.

Debug Project

The **Debug Project** command compiles and builds the project's executable or DLL. If the executable project is loaded, **Debug** followed by the project name appended by _dbg.exe is displayed as a sub-menu item. This com-

Figure B–16
Project Window Run Menu

mand runs the project for the currently selected configuration. If you have the DLL project loaded, the **Debug** command runs the executable specified by the **Select External Process...** command that is explained below. The shortcut for the **Debug Project** command is <Shift-F5>.

Continue

The **Continue** command resumes the program execution when the program is stopped at a breakpoint. The shortcut for this command is <F5>.

Terminate Execution

The **Terminate Execution** command is used for terminating the program when it is run in debug mode and is discussed in detail in Chapter 5, "Source Editor and Debugging Techniques." The shortcut for this command is <Ctrl-F12>.

Break at First Statement

The **Break at First Statement** command stops the program at the first executable statement, which is usually in the *main* function in C language.

Breakpoints...

The **Breakpoints...** command opens a **Breakpoints** dialog box listing all the breakpoints in the source file as shown in Figure B–17.
  The user has various options to add, delete, or edit project breakpoints as shown by the command buttons of the dialog box. Breakpoints are discussed in more detail in Chapter 5, "Source Editor and Debugging Techniques." The shortcut for the **Breakpoints** dialog box is <Shift-F9>.

Select External Process

Select External Process is only accessible when the **Target Type** is set to **Dynamic Link Library** in the **Build** menu. This command allows you to specify a stand-alone executable that uses your DLL. A dialog box appears, as shown in Figure B–18, when this command is selected and asks you to enter the file name along with the complete path and the command line arguments to an ex-

Figure B–17
Breakpoints Dialog Box

ternal program. The **Run Project** item in the **Run** menu changes to **Run filename.exe** where **filename** is the file name of the external program.

Execute

The **Execute** command launches the executable for the active configuration without attaching the debugger to the executable. The executable must be created using the **Create** menu item from the **Build** menu before using this command. This command is only active (un-dimmed) when the **Target Type** is **Executable.** The shortcut for this command is <Ctrl-F5>.

Figure B–18
Select External Process Dialog Box

Threads...

To use **Threads...**, you have to run the program using the debugger (breakpoints for example). When the program stops at the break point, select **Threads...** from the **Run** menu. The **Threads...** command displays the dialog box in the program being debugged, as shown in Figure B–19.

Use this dialog box to select a thread and select **View**. *CVI* displays the local variables and the current source position for the selected thread in the variable display in the source window. To view the information for the currently selected thread, use the **Up Call Stack**, **Down Call Stack**, and **Call Trace** commands from the **Run** menu in the **Source Editor** window.

Remember the following shortcuts:

Command	Shortcut
Debug	<Shift-F5>
Continue	<F5>
Terminate Execution	<Ctrl-F12>
Execute	<Ctrl-F5>
Breakpoints	<Shift-F9>

Using the Instrument Menu

The **Instrument** menu deals with how to **Load, Unload,** and **Edit** the instrument drivers, which are sets of high-level functions with function panels to control and communicate with equipment. An instrument driver can also be

Figure B–19
Threads Dialog Box

Appendix B • Project Window Environment

a software utility, like the **Programmer's Toolbox...** shown in Figure B–20. From this menu you load, unload, and edit the instrument drivers.

Load

The **Load** command allows you to load files ".fp" extensions from a dialog box. The selected file name appears in the **Instrument Menu.** Only the instrument driver programs and the function panels have the .fp extension and can be loaded. The **Programmer's Toolbox...** is shown loaded on the **Instrument** menu in Figure B–20.

Unload

The **Unload** command allows you to unload the selected instruments from a dialog box that appears when you select this command. You can select as many instrument drivers as you want to unload. When you click on **OK,** the selected instrument names are removed from the **Instrument Menu.**

Edit

You can use the **Edit** command to edit an instrument driver program using the **Source Editor** window. This advanced command will not be discussed in this text.

Details of creating and using instrument drivers are not part of this text but are covered in *LabWindows/CVI Instrument Driver Developer's Guide*.

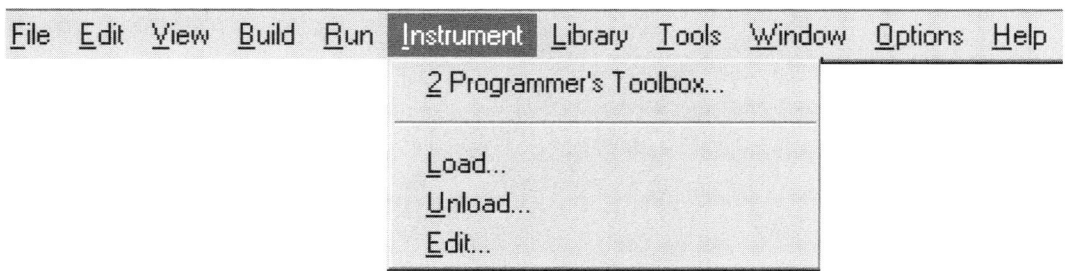

Figure B–20
Instrument Menu

Using the Library Menu

The **Library** menu is a comprehensive topic in *CVI* with numerous functions. Familiarizing yourself with this menu will give you a deep understanding of *CVI's built-in* functions and will reduce your development time considerably. You can access the function panels, which include a template of the function structure and an online context-sensitive help feature, enabling you to fill in function arguments in your source code. This feature can save you from thumbing through the seven manuals that come with *CVI* and typing in the information. The objective is to show you the use of these functions as encountered in the code when building the projects. Details explaining the function prototypes and their arguments and use are given when used in the code. The **Library** menu is shown in Figure B–21, and each of the commands on the sub-menu is defined.

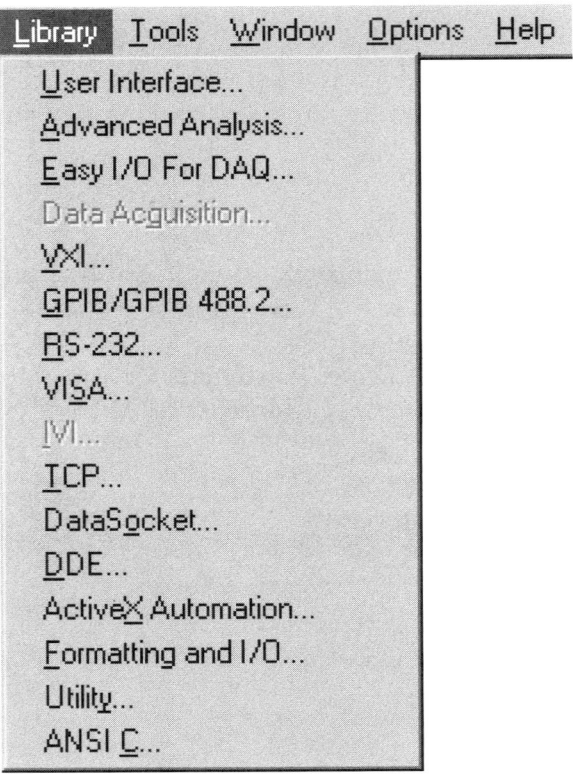

Figure B–21
Project Window Library Menu

Appendix B • Project Window Environment

User Interface

The **User Interface** library consists of functions that deal with the graphical user interface: graphs, strip-charts, menus, pop-up menus, controls, mouse and cursor controls, functions to control the appearance and behavior of the GUI, and many libraries, as shown in Figure B–22. These functions are covered in more detail in the *LabWindows/CVI User Interface Reference Manual*, or in On-line Help which lists all the functions alphabetically.

Advanced Analysis

The **Advanced Analysis** library consists of the functions pertaining to mathematical, statistical, engineering, analysis, and other routines that have been pre-written so the user does not have to re-invent the wheel. Some of the advanced analysis library function groups are shown in Figure B–23.

Figure B–22
User Interface Libraries

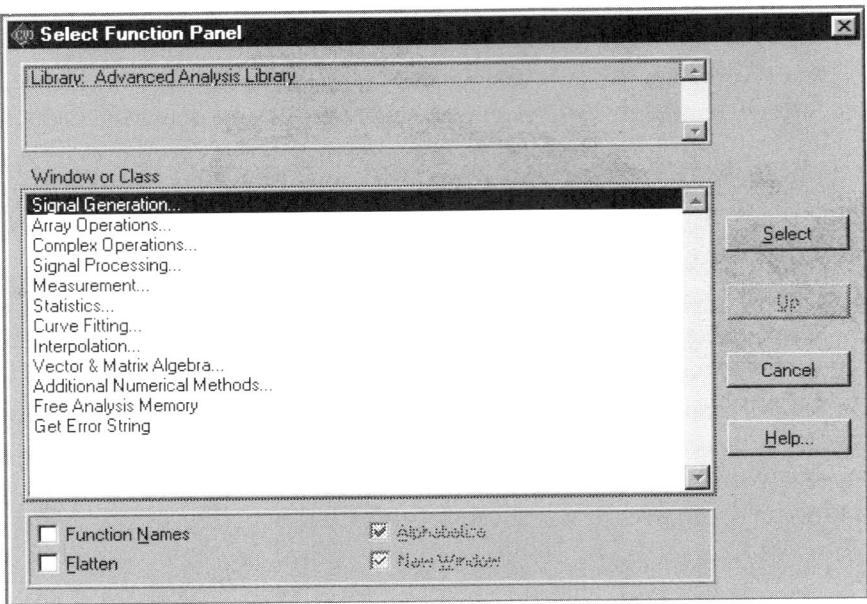

Figure B–23
Advanced Analysis Libraries

Easy I/O for Data Acquisition

The **Easy I/O for Data Acquisition** library consists of simpler **DAQ** library functions that are easier to use than the full **DAQ** library functions. The high-level function structure is shown in Figure B–24. Details of using the **DAQ** functions are given in the manuals that come with the National Instruments data acquisition hardware. This function exists for Windows only.

Data Acquisition...

The **Data Acquisition** library functions to control the National Instruments data acquisition boards are included in this menu.

VXI...

The VME **VXI** (**V**ersa-Modular Eurocard e**X**tensions for **I**nstrumentation) library functions enable the user to communicate and control the VXI-based instruments. This library comes with the drivers for the **VXI** equipment. These functions are shown in Figure B–25 by functionality. By clicking on

Appendix B • Project Window Environment

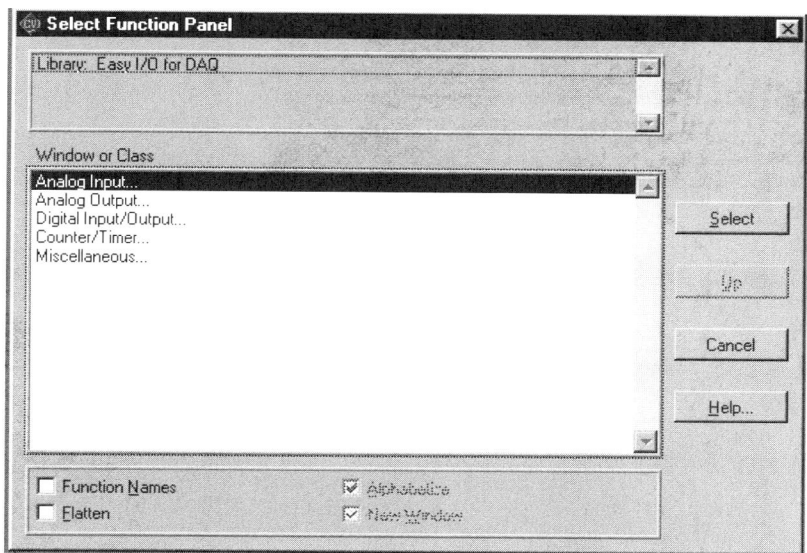

Figure B–24
Easy I/O for DAQ Menu

each of these items, you can experiment with the **VXI** library functions available. The **VXI**-based interface will not be covered in this text but in a forthcoming advanced book.

GPIB/GPIB 488.2

The **General Purpose Interface Bus** (GPIB) library consists of functions for communicating with and controlling the GPIB-based instruments that use the IEEE-488 or 488.2 protocol. To use these functions, you must have the National Instruments GPIB hardware interface card or another manufacturer's GPIB card that is compatible with *CVI* and the software drivers. The high-level GPIB drivers are listed in Figure B–26. This topic was discussed in Chapter 11, "GPIB Communications."

RS-232...

The **RS-232** library consists of functions for communicating over the serial port(s) with devices supporting the RS-232 interface. Using the serial port functions, you can read and write data to the serial port(s), set the

Figure B–25
VXI Utilities Library

Figure B–26
GPIB Function Library

configuration of the port(s), read and write data to files through the serial port(s), and perform utility functions dealing with serial interface protocols. The list of library functions is shown in Figure B–27. More information is given in *LabWindows/CVI Standard Libraries Reference Manual* or in On-line Help and in Chapter 12, "RS-232 Serial Communication."

VISA

The VISA (**V**irtual **I**nstrumentation **S**oftware **A**rchitecture) library is a single interface library for communicating with VXI, GPIB, and RS-232 instruments. The high-level functionalities of VISA are shown in Figure B–28.

IVI

The IVI (**I**nterchangeable **V**irtual **I**nstruments) library gives the user a way to create the VXI plug and play instrument drivers. The IVI wizard creates the skeleton of an IVI driver that includes the source code and the function panels. The list of high-level functions for the IVI library is shown in

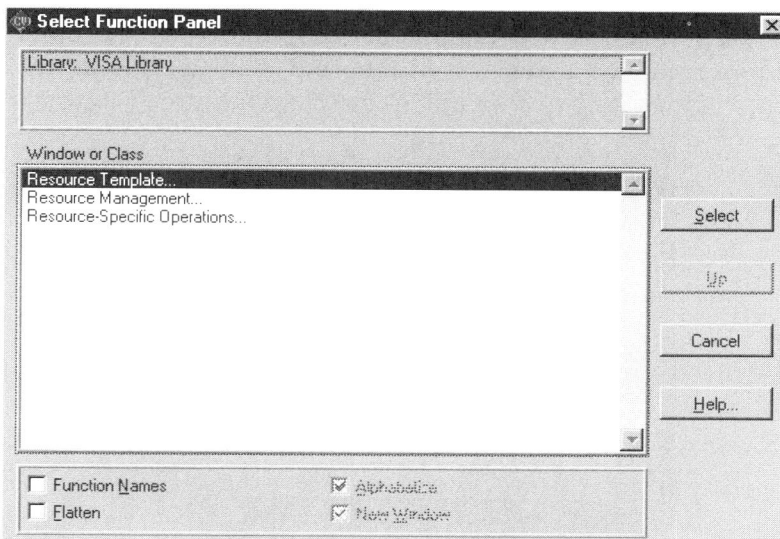

Figure B–27
RS-232 Serial Port Library

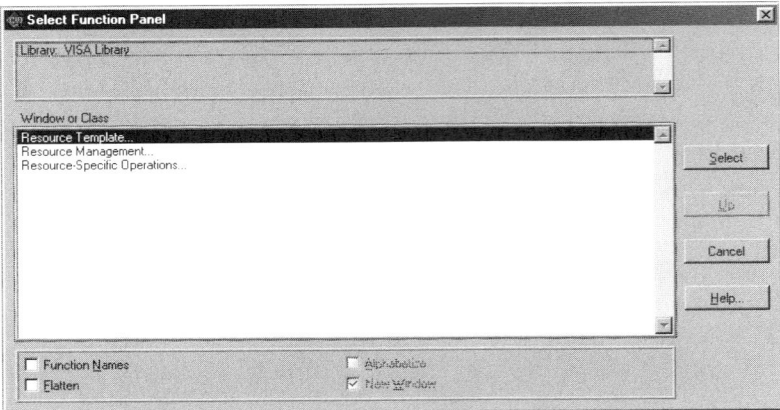

Figure B–28
VISA Library

Figure B–29. This material is covered in detail in *LabWindows/CVI Instrument Driver Developer's Guide.*

TCP

The TCP (**T**ransmission **C**ontrol **P**rotocol) library enables the user to communicate with other instruments or computers over the Ethernet. The top-level functionalites are shown in Figure B–30. See *LabWindows/CVI Standard Libraries Reference Manual* or in On-line Help for more information.

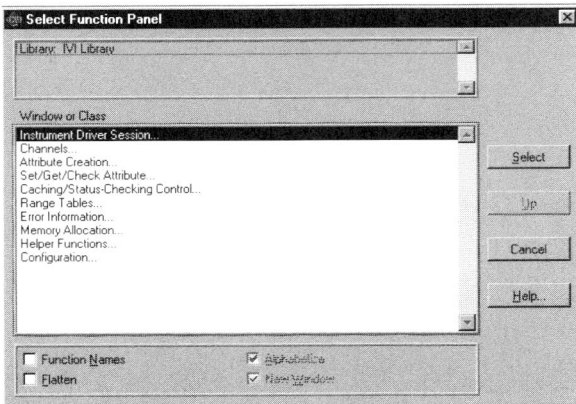

Figure B–29
IVI Library

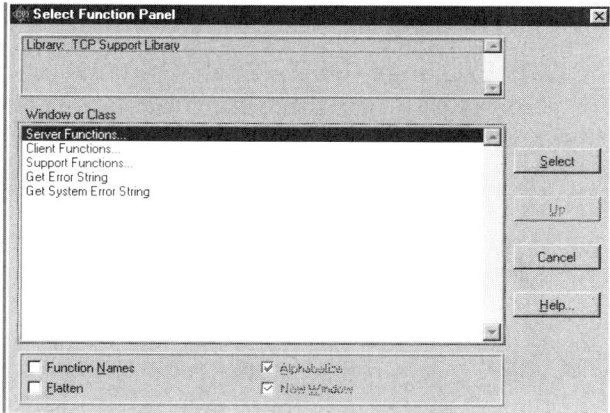

Figure B–30
TCP Support Library

DataSocket

The **DataSocket** library can be used to communicate with data from different sources, including HTTP (Hypertext Transfer Protocol), FTP (File Transfer Protocol), DSTP (Data Socket Transfer Protocol), and OPC OLE (Object Linking and Embedding (OLE) for Process Control) servers. The high-level **DataSocket** library functions are shown in Figure B–31. The **DataSocket** function will not be discussed in this text since it is an advanced topic.

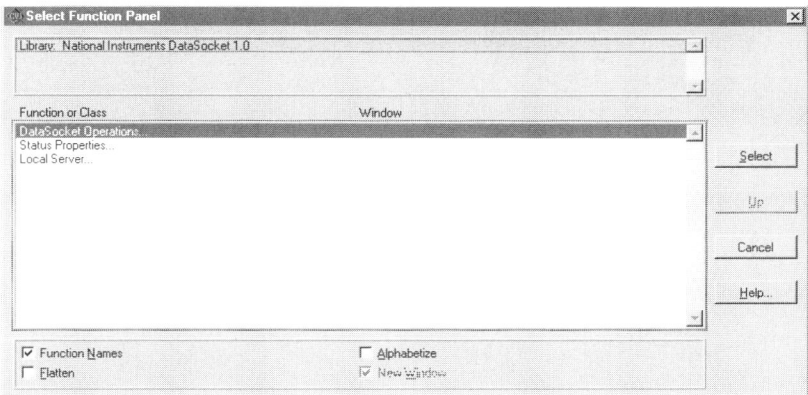

Figure B–31
Data Socket Library High-Level Functions

DDE

The **Dynamic Data Exchange** (DDE) library contains functions for creating a DDE client or a server. Among these functions are setting up the client or server, sending and receiving messages from the conversation partner, and creating hot links. The high-level functionalities are shown in Figure B–32. More information on this topic is available in Chapter 6, "DDE Library," in the *LabWindows/CVI Standard Libraries Reference Manual* or in Online Help. This command is applicable for Windows only.

X Property (Unix Only)

For *CVI* versions prior to 5.5, this menu item is visible when *CVI* is installed on Unix systems. The **X Client Property** library is for communicating among X clients for X Windows environment. More information on this topic is available in Chapter 9, "X Property Library," in *LabWindows/CVI Standard Libraries Reference Manual*.

ActiveX Automation (Windows Only)

The **Active X Automation** library is used with the ActiveX Automation Controller Wizard to control automation servers. You can use the wizard to

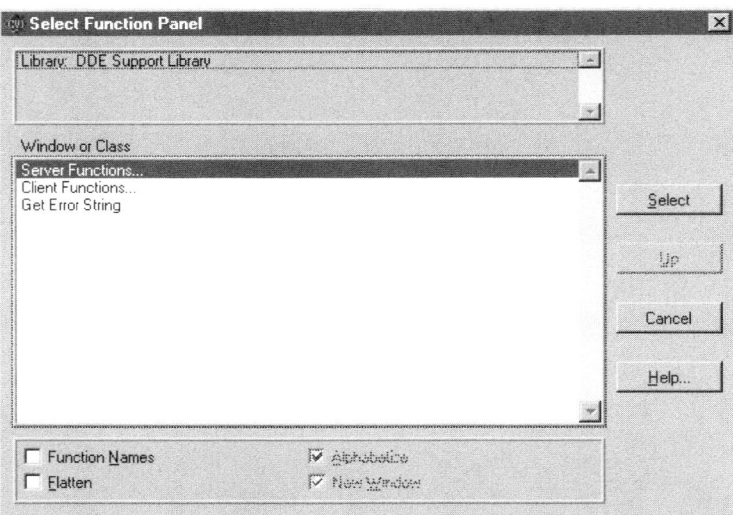

Figure B–32
DDE Support Library

browse an automation server and select the methods you want to use. It generates an instrument driver with C API and function panels for using the selected methods. The ActiveX Automation Library contains low-level functions that the generated instrument drivers use and the high-level functions interface that you can use to send and receive data and to communicate with the instrument driver functions.

The top-level menu is shown in Figure B–33. More details are available in Chapter 11, "ActiveX Automation Library," in *LabWindows/CVI Standard Libraries Reference Manual* or in Online Help.

Formatting and I/O

The **Formatting and I/O** library consists of functions for formatting the data in the source code and for inputting and outputting data to files. The top-level menu is shown in Figure B–34. More information on these functions is available in Chapter 2, "Formatting and I/O Library," in *LabWindows/CVI Standard Libraries Reference Manual* or in Online Help. A detailed description of *formatting* functions is given in Appendix D, "Formatting Functions."

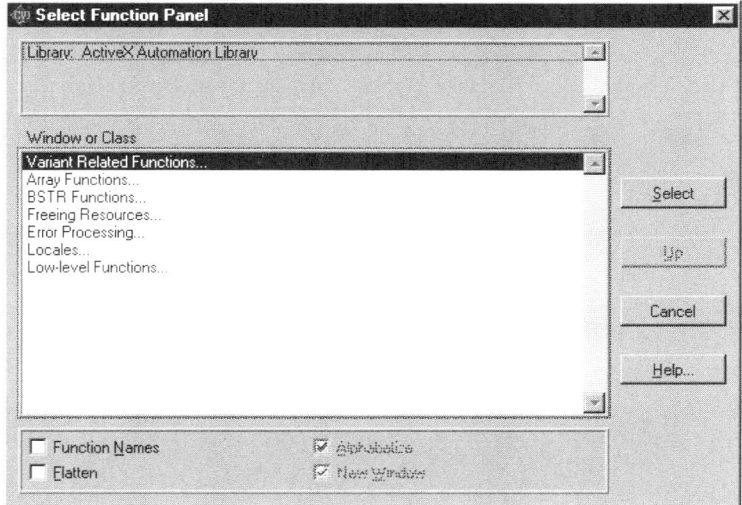

Figure B–33
ActiveX Automation Library

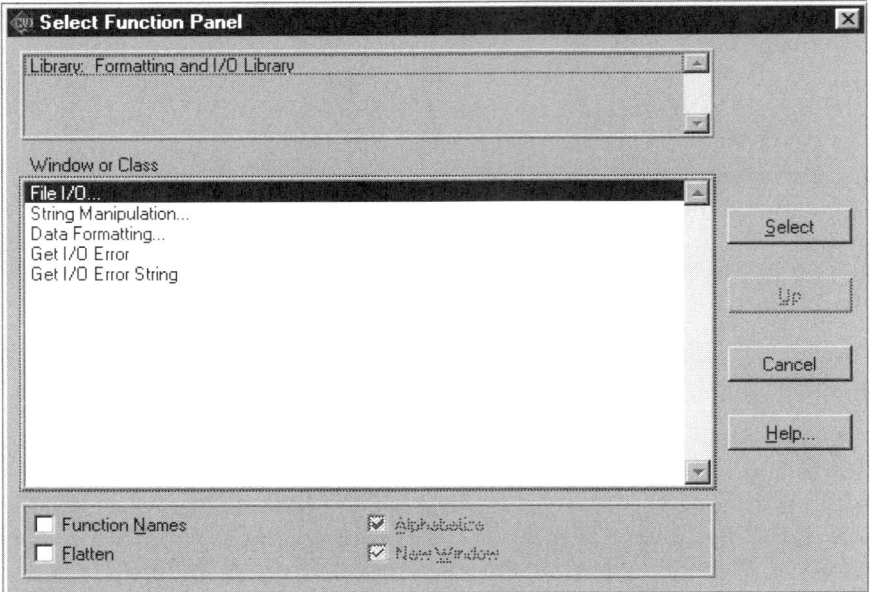

Figure B–34
Formatting and I/O Library

Utility

The **Utility** library contains miscellaneous functions that are not specific to any library category. A listing of these function categories is shown in Figure B–35. More details on this function are available in Chapter 8, "Utility Library," of *LabWindows/CVI Standard Libraries Reference Manual* or in Online Help.

ANSI C

American National Standards Institute (**ANSI**) C library contains the ANSI C functions that are part of *CVI*. The header files that contain the ANSI C libraries are shown in Figure B–36. Refer to Chapter 1, "ANSI C Library," of *LabWindows/CVI Standard Libraries Reference Manual* or in Online Help for more information.

Using the Tools Menu

The **Tools** menu commands are shown in Figure B–37.

Appendix B • Project Window Environment

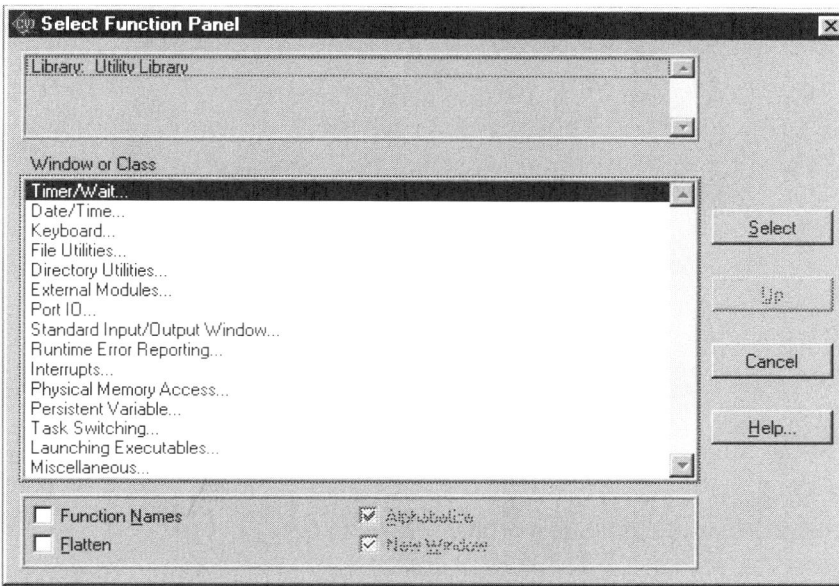

Figure B–35
Utility Library

Figure B–36
ANSI C Library

Appendix B • Project Window Environment

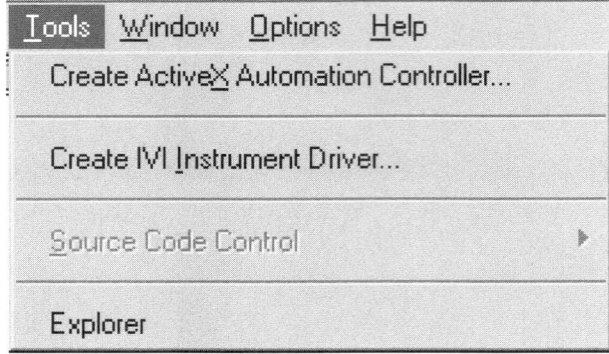

Figure B–37
Tools Menu

Create ActiveX Automation Controller (Windows Only)

The **Create ActiveX Automation Controller** function is available only in Windows. This command is used in conjunction with **ActiveX Automation Controller** to invoke a wizard to create a new instrument driver for an ActiveX Automation server. Figure B–38 shows the ActiveX Automation Server

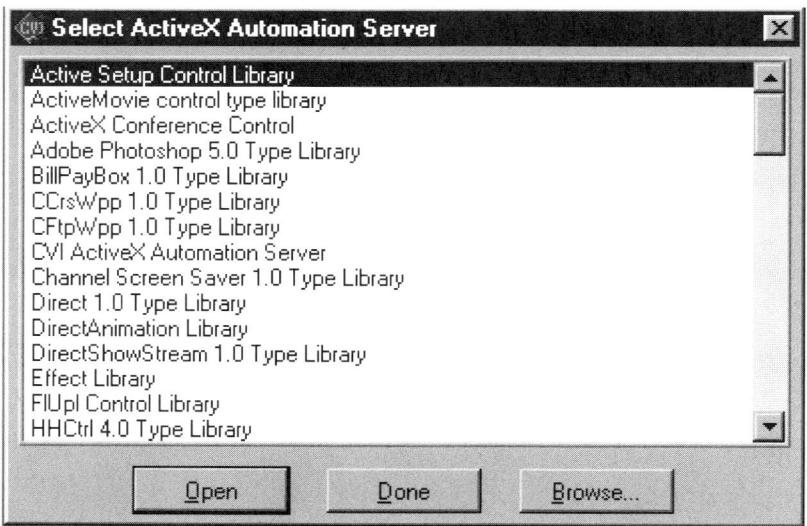

Figure B–38
ActiveX Automation Server Choices

Appendix B • Project Window Environment

choices. When you select this command, it takes a few seconds to go through the system registry list and display all the items in the registry. You can scroll in the window to select the appropriate servers. For more details, see Chapter 11, "ActiveX Automation Library," in *LabWindows/CVI Standard Libraries Reference Manual* or in Online Help.

Create IVI Instrument Driver

The **Create IVI Instrument Driver** function is available only in Windows. When this command is selected, it opens the Instrument Driver Development Wizard, as shown in Figure B–39. The wizard will help you create a function panel to control the instrument and generate a proper source file and include file.

When the **Next** button is selected, Figure B–40 appears, asking you to select the I/O type and the type of the instrument and guiding you through the steps of creating the instrument driver.

The procedure for creating instrument drivers is not part of this text and is referred to in an advanced book on *CVI*. Refer to *LabWindows/CVI Instrument Driver Developer's Guide*.

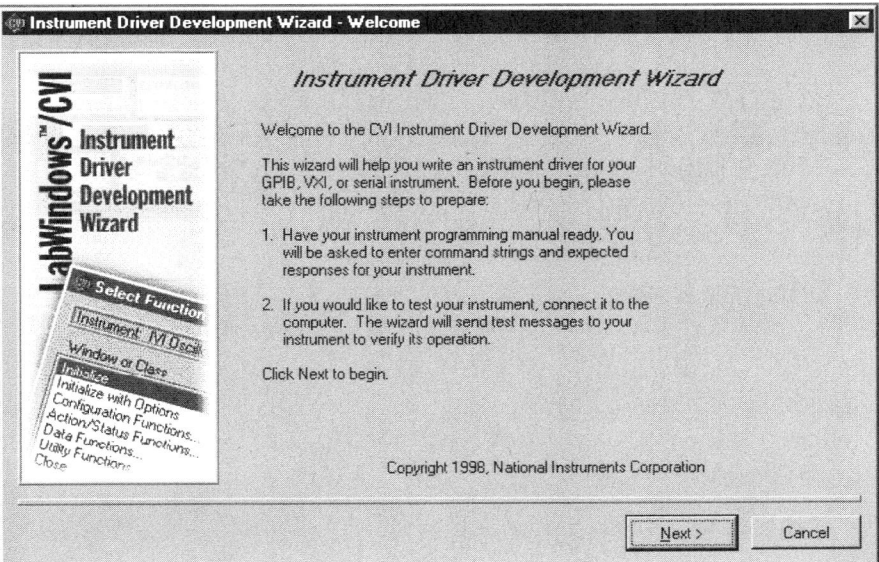

Figure B–39
Instrument Driver Development Wizard

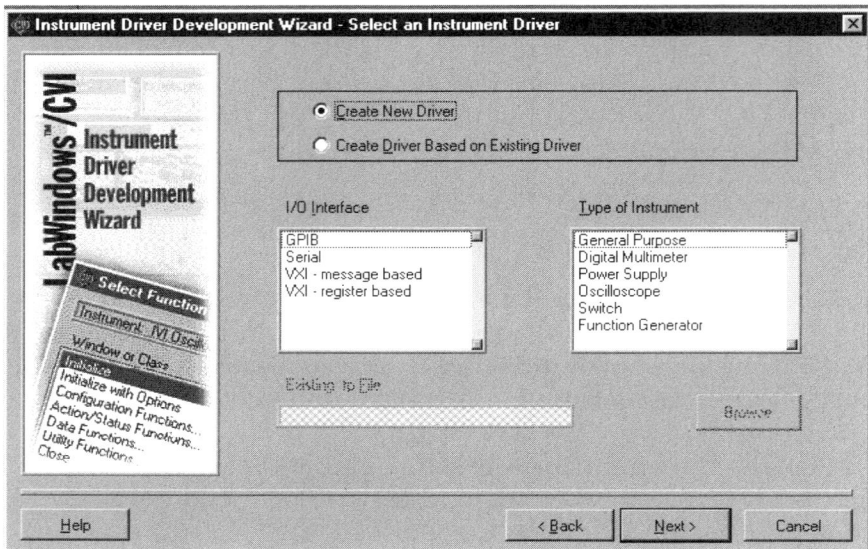

Figure B–40
Select Instrument Driver Menu

Source Code Control

The **Source Code Control** is used to attach your *CVI* project to a third party source code control system that implements the Standard Source Code Control Interface. The **Source Code Control** options must be configured using the **Options** dialog box of the **Project** window. After the **Options** are configured, you will be able to control the source code in your *CVI* project. For a comprehensive discussion on this topic, see *LabWindows/CVI User Manual*.

Using the Window Menu

The **Window** menu is shown in Figure B–41. This menu has a variety of features for window control and program execution functions used during debugging.

Cascade Windows

The **Cascade Windows** command is the same as the one in Windows, which arranges the open windows like a cascade or a waterfall. The windows' title bars are visible so you can select any window by clicking on the window title bar.

Appendix B • Project Window Environment

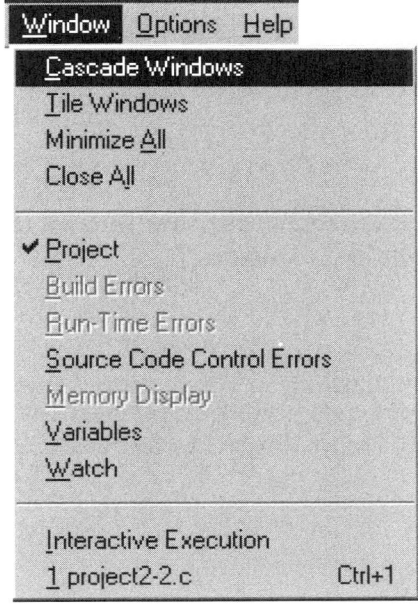

Figure B–41
Window Menu

Tile Windows

The **Tile Windows** command is the same as the one in Windows and arranges open windows next to each other like a floor tile. All open windows fit in smaller size next to each other on the computer desktop.

Minimize All

The **Minimize All** command is used in Windows only. This command hides all the *CVI* windows, including the **Project** window and any **User Interface Library** panels that are currently running. To re-size the minimized window, click on the *CVI* in the Windows task bar.

Close All

The **Close All** command closes all the *CVI* windows except the **Project** window and the GUIs used by the program during execution.

Project

The **Project** command brings the **Project** window to the front.

Build Errors

The **Build Errors** command brings the **Build Errors** window to the front if there are any compile or link errors.

Run-Time Errors

The **Run-Time Errors** command brings the **Run-Time Errors** window to the front if there are any errors which are found during the execution of the program.

Source Code Control Errors

The **Source Code Control Errors** command displays the warning and errors returned by the source code control systems when the **Source Code Control** command is executed from the **Tools** menu.

Memory Display

The **Memory Display** command opens the **Memory Display** window for viewing and editing the memory of the program. While the program is suspended during debugging, you are given the choice to edit and view the data in hexadecimal, decimal, single-precision floating point, double-precision floating point, or ASCII formats by selecting from the **Memory Display** dialog box shown in Figure B–42. You must check the **Edit Mode** box before you can modify the process' memory.

Variables

The **Variables** command brings the **Variable** window to the front. When executing a *CVI* project, it may become necessary to check the variables while debugging the program. This capability is discussed in Chapter 5, "Source Editor and Debugging Techniques."

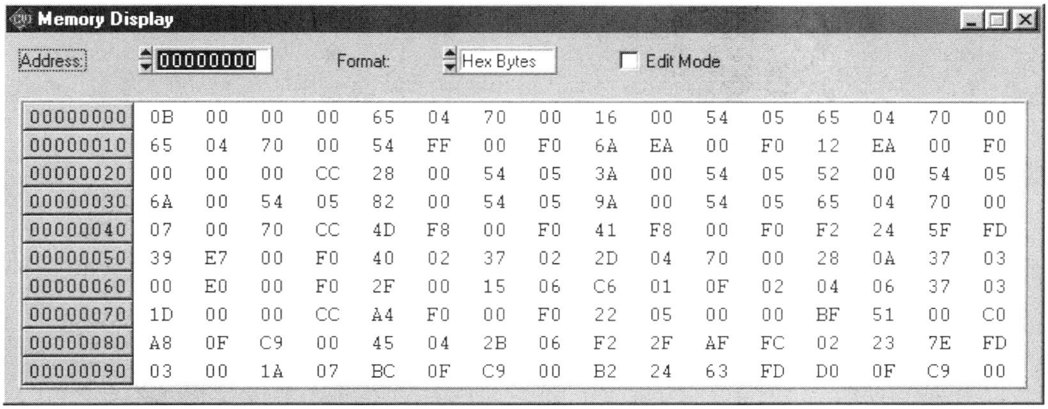

Figure B-42
Memory Display Window

Watch

The **Watch** command brings the **Watch** window to the front. This is a useful feature when debugging the software and wanting to observe the value of a variable or an expression. This feature was discussed in Chapter 5, "Source Editor and Debugging Techniques."

The following sub-menu commands appear on the **Window** menu dynamically if they are activated:

- **Array/Strings Display,**
- All open **User Interface** files,
- All open **Function Panels,**
- All open **Function Trees,**
- All open **Help Editor** windows,
- All open source files are listed on the bottom of the **Window** menu.

Interactive Execution

The **Interactive Execution** command brings the **Interactive Execution Window** (IEW) to the front. This window allows you to execute an incomplete program when you want to test a function without including it in the *main* program. The function executing in IEW can access variables and data declared as global in the **Source Code Editor** window, which cannot access functions and data declared in the IEW. The **Interactive Execution** command is a useful debugging and testing feature of *CVI*.

Using the Options Menu

The **Options** menu enables the user to set up preferences in the *CVI* environment for compiling, running, and enabling or disabling project options. This menu is shown in Figure B–43.

Build Options

The **Build Options** command sets the compiler options by opening a dialog box with the selections shown in Figure B–44.

The **Build Options** command allows you to set the following *CVI* compiler options.

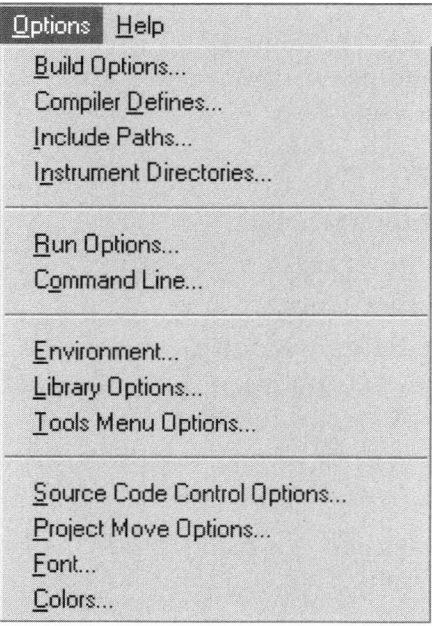

Figure B–43
Options Menu

Appendix B • Project Window Environment

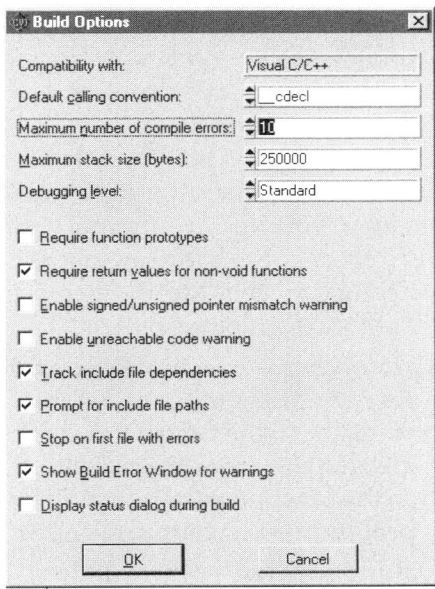

Figure B–44
Build Options

1. **Compatibility with**—This option is available in Windows only. Selecting this option displays the current compiler compatibility mode, you selected in the **Setup** procedure (in the **Select Compatibility Mode** screen during installation).

2. **Default calling convention**—This option is available in Windows only. Selecting this option sets the default calling convention to __cdecl unless you are using a Watcom compiler, which uses a stack-based calling convention.

3. **Maximum number of compiler errors**—Selecting this option sets the maximum number of compiler errors that should appear in the **Build Errors** window and stops the program and displays the errors before *CVI* finds more errors in the program.

4. **Maximum stack size (bytes)**—This option sets the maximum stack size in bytes required by your program for passing function arguments and storing automatic local variables. In most cases, the default value is sufficient unless you want to add many recursive function calls that can overflow the stack.

5. **Debugging level**—There are three debugging levels that can be set in the application. This option is active only when **Debug** is checked in the **Configuration** sub-menu under **Build**. The three debugging levels follow.

a) **No Run-Time Checking**—This option lets you set breakpoints in the source code, display variables, and watch windows but with no protection from run-time memory errors and without the use of the **Break on Library Errors** option.
b) **Standard**—This option allows you the same options as **No Run-time Checking** but allows you to set the **Break on Library Errors** option and have protection from run-time memory errors.
c) **Extended**—This option is the same as the **Standard** option but with added user protection to validate the beginning address passed to free dynamically allocated memory.

6. **Require function prototypes**—Selecting this option requires including the function's prototypes in your source code before the call to the function is made. A prototype must have the return type of the function and the data types of all arguments. If there are no arguments, the prototype must have the type `void` within the function parentheses. If you check this option and do not include the prototype(s), a compiler error will occur.

7. **Require return values for non-void functions**—Selecting this option generates a compiler warning if the function is of non-void type and does not end with a `return` statement that returns a value. This is not applicable for the `main` of the program and enables the programmer to set the function types correctly.

8. **Enable signed/unsigned pointer mismatch warning**—Selecting this option generates a warning at compile time for pointer assignments in which the left side and right side are not both signed or unsigned expressions.

9. **Enable unreachable code warning**—Selecting this option generates a compiler warning if some code is not reachable during execution. This option should always be set as it informs the programmer of a possible logic error in the code or of any redundant code.

10. **Track include file dependencies**—Checking this option keeps dependencies between the source files and the include files up-to-date.

11. **Prompt for include file paths**—Selecting this option prompts you to search for any header files that are listed in `#include` and that the compiler cannot find. This informs you that you have forgotten to include a required header file.

12. **Stop on first file with errors**—Selecting this option stops the compiler from finding errors in other files when it finds its first error, so one file at a time can be corrected for errors.

13. **Show Build Error Window for warnings**—Selecting this option shows the **Build Error** window when there are warnings in compilation. If this

option is disabled, the **Build Error** Window will not open and will not display the warnings even if warnings are generated by the compiler. It is advisable to enable this option.

14. **Display status dialog during build**—Selecting this option displays a dialog box during the build that shows the progression of the build with errors and warnings and percent complete. Unless you want to observe this information every time you build a project, this option should be disabled to make the project compile faster.

Compiler Defines

To avoid conflicts that arise due to function names and *typedef* between Windows Software Development Kit (SDK) and *CVI* libraries, the Windows SDK files must be included before the *CVI include* files. The *CVI* SDK files are in the path `cvi5\sdk\include` folder (directory), which is searched automatically by the *CVI* compiler. The user does not have to include the *CVI* SDK files in the project manually since *CodeBuilder* inserts it automatically.

If you use an external compiler to compile and link your source, you must use the SDK *include* files that come with the external compiler. If you use an external compiler to compile the source files but want to link in CVI, you need to use the *CVI* SDK *include* files.

Only `windows.h` must be included since it contains almost all the necessary *include* files, but including it increases the compilation time and memory usage substantially. To avoid this problem, the Microsoft macro /DWIN32/LEAN_AND_MEAN is used to reduce the memory usage and the compilation time. This macro includes only the more commonly used *include* files in `windows.h`. This selection is shown in Figure B–45.

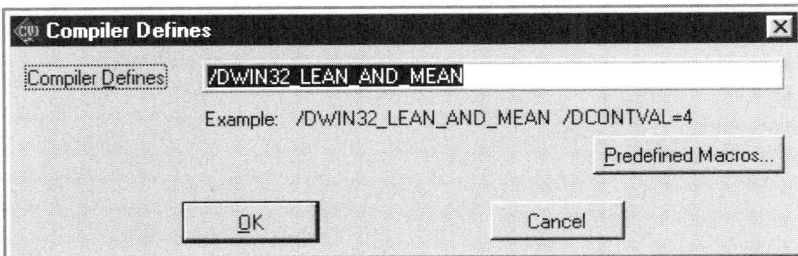

Figure B–45
Compiler Defines

Include Paths

The **Include Paths** command brings up a dialog box as shown in Figure B–46 in which you can list the header file paths that the compiler uses when searching for header files with simple path names. This dialog box consists of an upper box in which CVI saves the top list with the project file. CVI saves the bottom list from one session to another on the same machine regardless of the project.

Instrument Directories

In the **Instrument Directories,** you enter the directory name to search for the instrument drivers. The function panel .fp files of the dependent drivers store the names of the driver files on which they depend. The **Instrument Directories** dialog box is shown in Figure B–47. CVI loads the referenced instrument's (independent driver) .fp file. The hierarchy of the search for the instrument driver is as follows:

1. the folder (directory) of the referencing .fp file,
2. the directories listed in the Instrument Directories dialog box,
3. the sub-directories under the cvi5\toolslib folder (assuming cvi5 is where the program is installed),
4. the cvi5\instr directory.

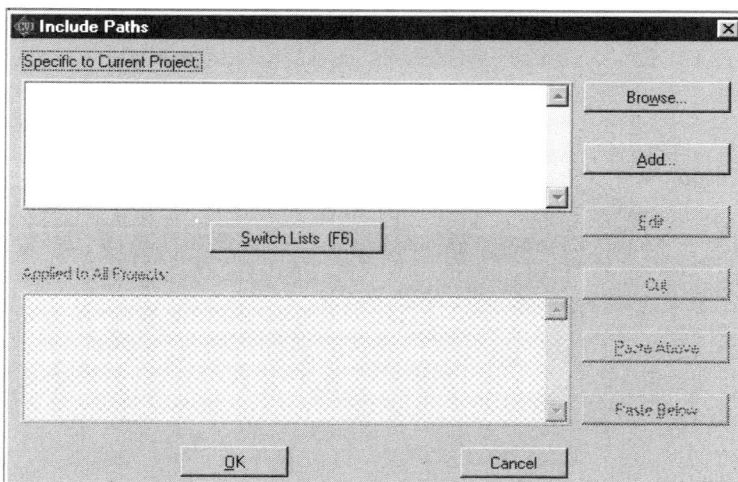

Figure B–46
Include Paths

Appendix B • Project Window Environment

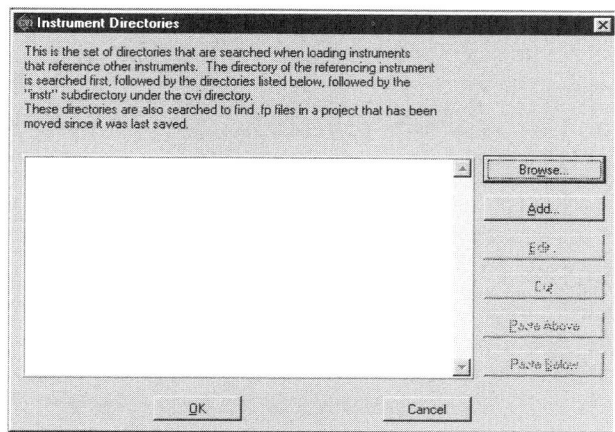

Figure B–47
Instrument Directories Dialog Box

Discussion of the creation of instrument drivers will not be included in this introductory book. For more information, refer to *LabWindows/CVI Instrument Driver Developer's Guide*.

Run Options

The **Run Options** command opens a dialog box, as shown in Figure B–48 to set the following run-time options.

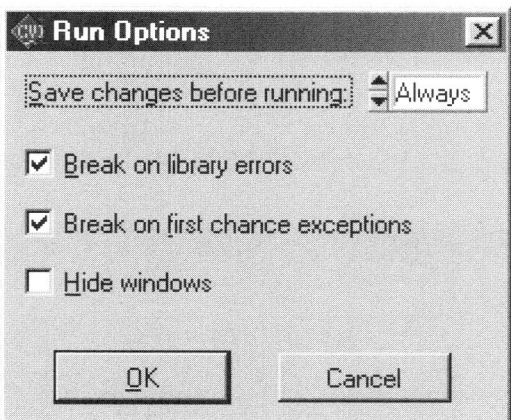

Figure B–48
Run Options

1. **Save changes before running**—The pull-down list box gives you three options: never save changes to modified files **(Never)**, ask before making changes **(Ask)**, and always save changes to modified files **(Always)**.
2. **Break on library errors**—This option sets a breakpoint at any function call to a *CVI* library function that results in an error.
3. **Break on first chance exceptions**—When this box is checked and an exception occurs in your program, *CVI* will suspend your program and will not allow your code to handle the exception.
4. **Hide windows**—This option causes all the *CVI* environment windows to be hidden until either a breakpoint occurs or the program terminates.

Command Line

The **Command Line** option invokes a dialog box, as shown in Figure B–49, in which you can enter the values for the arguments `argc` and `argv`, which are passed directly to the main program when the program is executed.

You can also pass command line arguments to a calling program when your project makes a DLL. You need to include the calling program's pathname and command line arguments using the **Select External Process** command in the **Run** menu.

Environment

The **Environment** option brings up a dialog box, as shown in Figure B–50, which has the following options.

1. **CVI environment sleep policy**—This option permits *CVI* to stay in sleep mode for a specified period of time while another event is being handled by the operating system. Other applications then have more processor time. Sleep mode may make *CVI* run slower, but you can specify how

Figure B–49
Command Line Dialog Box

Appendix B • Project Window Environment

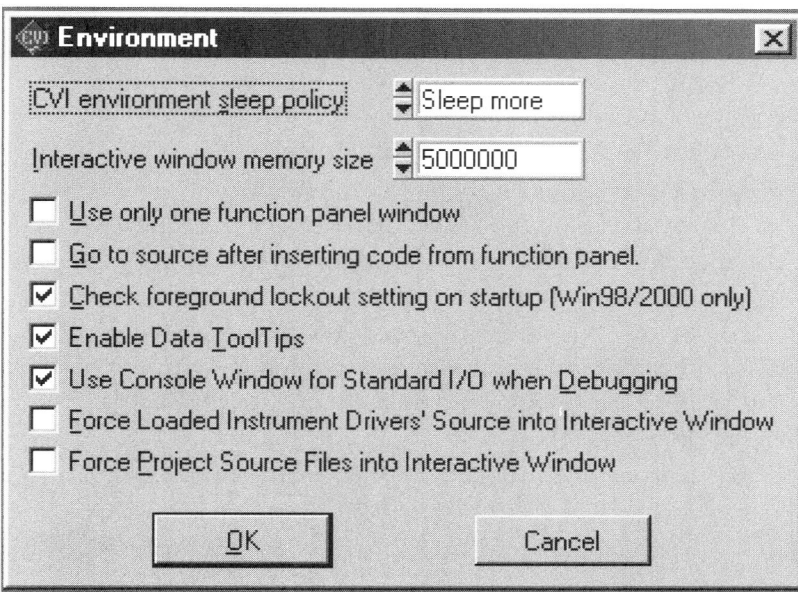

Figure B–50
Environment Options

much sleep time is ideal by experimenting with the items in the pull-down list box. The items in the list box give you a choice of the following:

- Do not sleep,
- Sleep some,
- Sleep more.

2. **Interactive window memory size**—This dialog box is used to set the memory allocation for executing function panels or code in the interactive window. Depending on your application, the default value may be sufficient but if *CVI* displays the message `Insufficient memory for interactive window`, you need to increase the allocation size.

3. **Use only one function panel window**—This option over-writes the current function panel window each time a new function panel window is selected.

4. **Go to source after inserting code from function panel**—Checking this box will automatically close the function panel after you have inserted source code from the function panel.

5. **Check foreground lockout setting on startup (Win98/2000 only)**—Windows 98 and Windows 2000 try to prevent the applications from

bringing themselves to the front. This is particularly the case when you are debugging your *CVI* program, and Windows 98 and 2000 disallow the application window to come to the front. When this box is checked, *CVI* will inform you on startup if the system settings will prevent the *CVI* windows from coming to the front.

6. **Enable Data ToolTips**—Check this box if you want *CVI* to display the values of variables and expressions when you move the cursor over these items in the **Source Code Editor** window while suspended during debugging.

7. **Use Console Window for Standard I/O when Debugging**—Check this box if you want to use the MS-DOS console window for displaying output strings and scanning input strings while debugging your project. This is enabled by default.

8. **Bring Debug Output Window to Front Whenever Modified**—Check this box to bring the **Debug Output** window to the front if you are using the output Utility Library *DebugPrintf* function as well as the Windows SDK *OutputDebugString* function. This window is used for debug string output and is accessible while your program is suspended.

9. **Force Loaded Instrument Driver's Source into Interactive Window**—Check this box if you run a function panel or interactive window code that loads the source file for an instrument driver (currently loaded in the **Instrument** menu), through *LoadExternalModule* or *LoadExternalModuleEx*. For further explanation of these library functions, see *LabWindows/CVI Standard Libraries Reference Manual* or in Online Help.

10. **Force Project Source Files into Interactive Window**—Check this box if you run a function panel or interactive window code that loads the source file for a currently loaded project through *LoadExternalModule* or *LoadExternalModuleEx*.

Library Options

The **Library Options** specify which *CVI* and user libraries are to be automatically loaded at the start of the project. The dialog box is displayed in Figure B–51.

The *CVI* libraries that you can select for running the project are listed at the top of the dialog box. See the section on library options in Chapter 3 of *LabWindows/CVI User Manual*.

Note: The libraries are loaded and shown on the **Library** menu only when *CVI* is launched the next time.

Appendix B • Project Window Environment

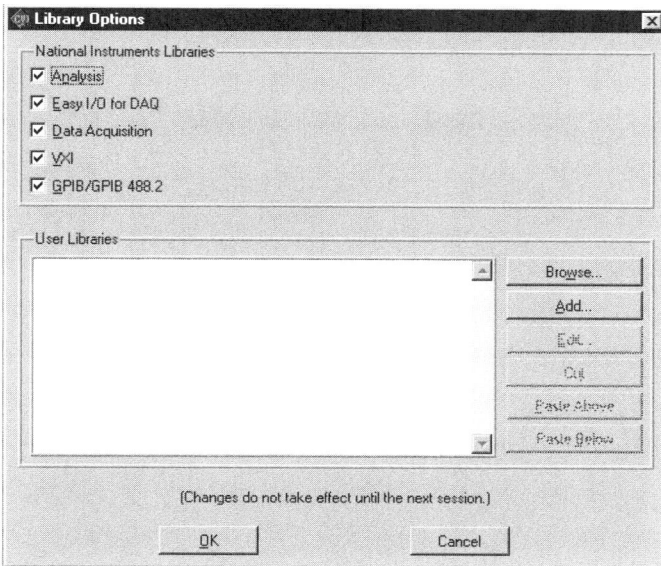

Figure B–51
Library Options

Tools Menu Options

Using the **Tools Menu Options,** you can add your own menu items to the **Tools** menu, as shown in Figure B–52.

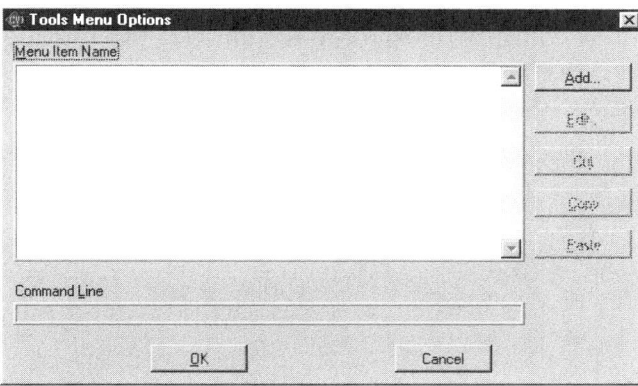

Figure B–52
Tools Menu Options

522   Appendix B • Project Window Environment

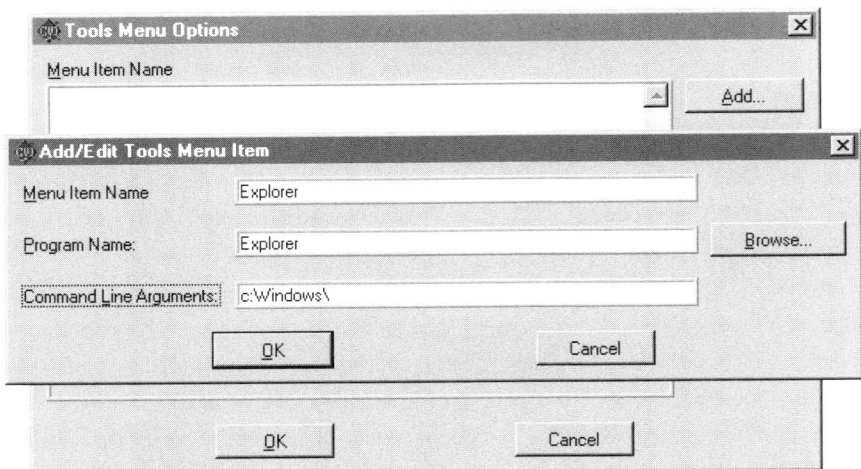

Figure B–53
Add/Edit Tools Menu Dialog box

By selecting the **Add** button from this dialog box, another dialog box will appear, as shown in Figure B–53. In this dialog box, enter the **Menu Item Name,** the **Program Name,** and the **Command Line Arguments** for the pathname of this menu item.

This menu item will show up on the **Tools** menu, as shown in Figure B–54, and whenever **Explorer** is selected from the **Tools** menu, the menu item will be executed.

Source Code Control Options

The **Source Code Control** was discussed in the "Using the Tools Menu" section above. When this option is selected, a dialog box, as shown in Figure B–55 is displayed.

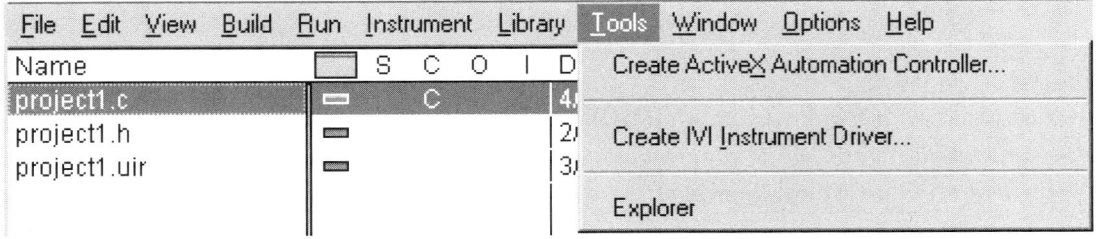

Figure B–54
Explorer Added to Tools Menu

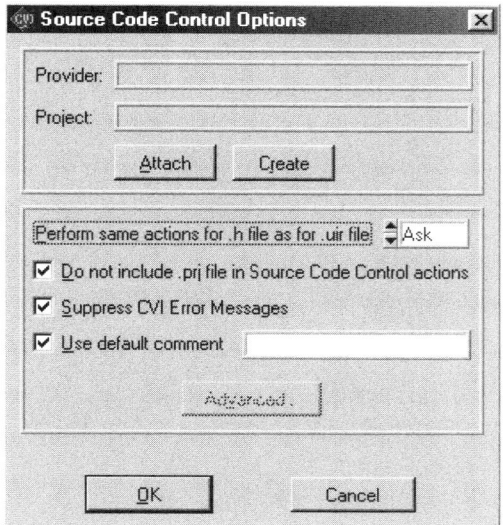

Figure B–55
Source Code Control Options

These options are explained in detail in Chapter 3 of the *LabWindows/CVI User Manual*.

Project Move Options

The **Project Move Options** command brings up a dialog box, as shown in Figure B–56, which, when selected, updates the pathnames, the include paths, the distribution kit, and the stand-alone executable if the project is moved to a different location.

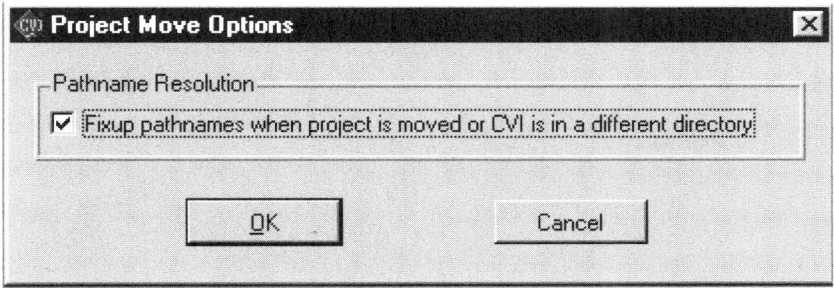

Figure B–56
Project Move Options

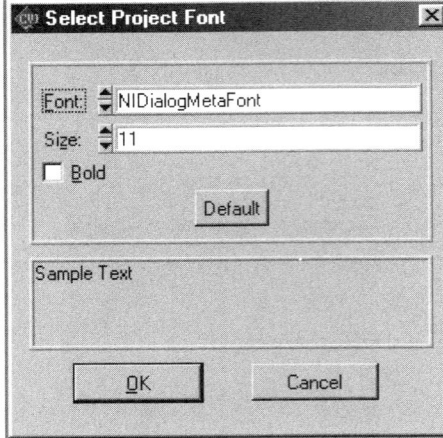

Figure B–57
Select Project Font

Font

The **Font** option enables you to select the fonts and font size for the text in the **Project** window. You can make your font selections in the **Select Project Font** dialog box, as shown in Figure B–57.

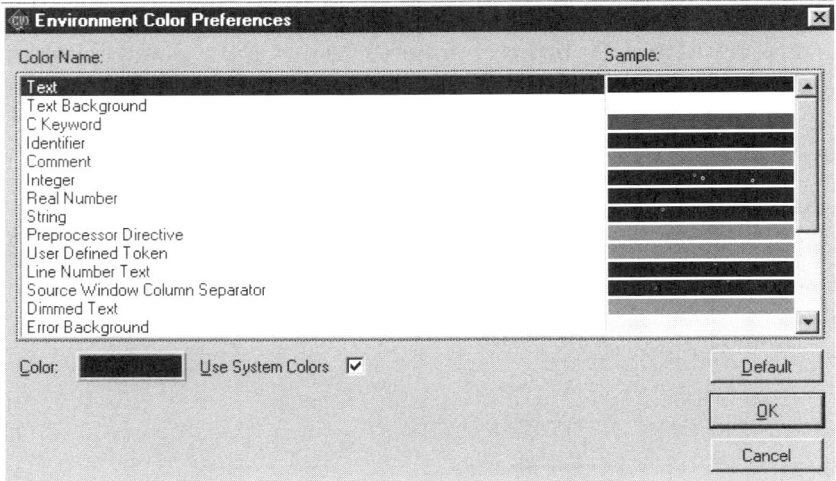

Figure B–58
Color Option

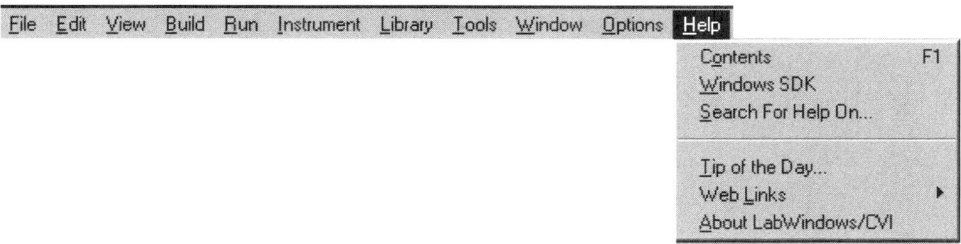

Figure B–59
Help Menu

Colors

The **Colors** option enables you to select the colors for the **Project, Source Code, Interactive Execution, Watch, Variable, String Display,** and **Array Display** windows. This option has no effect on the dialog boxes or function panels or on the **User Interface Editor** window. The dialog box for the selection of **Color** options is shown in Figure B–58.

Using the Help Menu

The **Help** menu provides with you online help with the various features of *CVI*. This menu is self-explanatory and uses the format of any other Window application. The **Help** menu is shown in Figure B–59.

Summary

In this appendix, the **Project** Window menu and sub-menu commands were explained step-by-step, including a description of each of the commands or a referral to the appropriate manual for further reading.

The purpose of these commands may not be clear to you at this point, but their use will become more apparent as you start using them when creating projects. At this time, just know what features are available from the **Project** Window and be able to use them as required when you create your projects.

User Interface Editor Environment

Chapter Highlights
- File Menu
- Edit Menu
- Create Menu
- View Menu
- Arrange Menu
- Code Menu
- Run Menu
- Library Menu
- Tools Menu
- Window Menu
- Options Menu
- Help Menu

This appendix gives a description of the **User Interface Editor** menu and sub-menu commands. You can refer to the menu commands as needed and do not need to read the complete appendix.

File Menu

The **File** menu in the **User Interface Editor** window shows the sub-menu commands, as shown in Figure C-1.

New, Open, Save, Save As, Close, and Exit *LabWindows/CVI*

The **New, Open, Save, Save As, Close,** and **Exit** *LabWindows/CVI* commands are similar to the **Project** window commands that were discussed in Appendix B, "Project Window Environment." The **Close** command is not on the **Project** Window **File** menu. In this menu, the **Close** command closes the opened user interface file.

Save Copy As

The **Save Copy As** command brings up a pop-up menu through which you can save the user interface file to another file name on the disk. The file name in the active window is not changed.

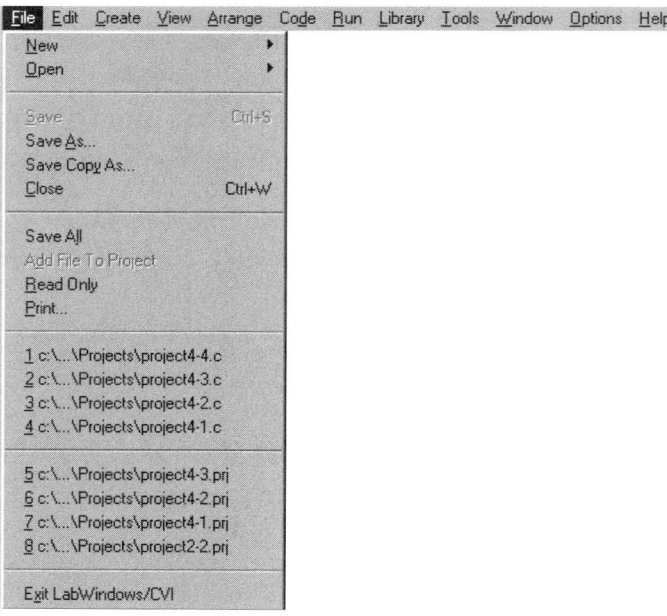

Figure C-1
User Interface Resource File Menu

Save All

The **Save All** command saves all the open files to directory location at which they were opened.

Add File to Project

The **Add File to Project** command adds the currently opened `.uir` file to the project list.

Read Only

The **Read Only** command prevents the user from making any changes to the currently opened file. The user interface file opens in read-only mode if the file's attribute on the disk is read-only. Otherwise, this command is disabled by default.

Print

The **Print** command opens a dialog box that gives you the choice of printing the entire GUI file or only the visible area to the default printer.

Edit Menu

The commands on the **Edit** menu are for editing panels, controls, and menu bars only. Figure C–2 shows the **Edit** menu.

Undo and Redo

The **Undo and Redo** commands are interdependent and can only be performed when there is something that can be edited. The **Undo** command returns the user interface resource file to its condition prior to the last change. The **Redo** command reverses the last **Undo** command. Notice that when you make a change to the user interface resource, the last change command appears with **Undo** or **Redo** on the **Edit** menu.

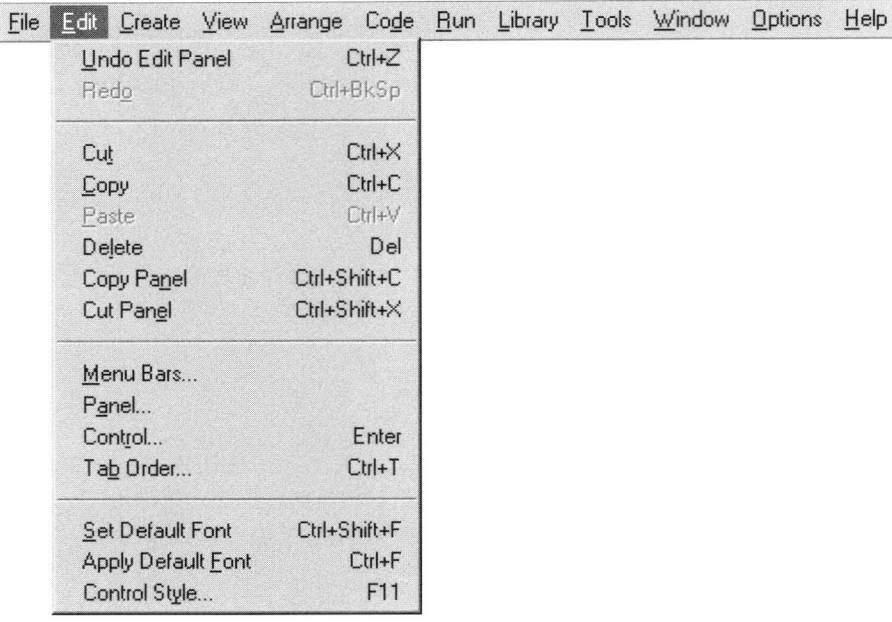

Figure C–2
User Interface Editor Edit Menu

Cut, Copy, and Paste

The **Cut, Copy,** and **Paste** commands are similar to any other Windows commands, but here you can cut, copy, and paste the objects placed on the panel or the menu bar. You can select a control, cut or copy it and place it with the **Paste** command.

You can select a single object or multiple objects by dragging the mouse around the items to be selected or by holding the <Shift> key and clicking on the objects individually. You can then cut, copy, and paste all these objects as a group.

Delete

The **Delete** command removes the objects from the panel without placing them on the clipboard, and therefore they cannot be restored with the **Paste** command. You can delete a single object or multiple objects by creating a frame around the objects with the mouse.

Cut Panel and Copy Panel

Using the **Cut Panel** and **Copy Panel** commands, you can remove the entire panel, placing it into the clipboard, and **Paste** it back to the same panel or on a different panel.

Panel

The **Panel** command opens the **Edit Panel** dialog box for the control object and was discussed in detail in Chapter 2, "Basics of Creating the Graphical User Interface."

Control

The **Control** command opens a dialog box for the active control, enabling you to edit its attributes.

Tab Order

The **Tab Order** command was discussed in Chapter 4, "Enhancing the User Interface."

Set Default Font

The **Set Default Font** command makes the font on the selected control/label become the default control/label font. All controls/labels created henceforth will use this font.

Apply Default Font

The **Apply Default Font** command sets the font of the currently selected control/label to the default control/label font. The default font from **Set Default Font** is used.

Control Style

The **Control Style** command changes the style of the selected control. To do so, click on the control you want to replace and select **Control Style**

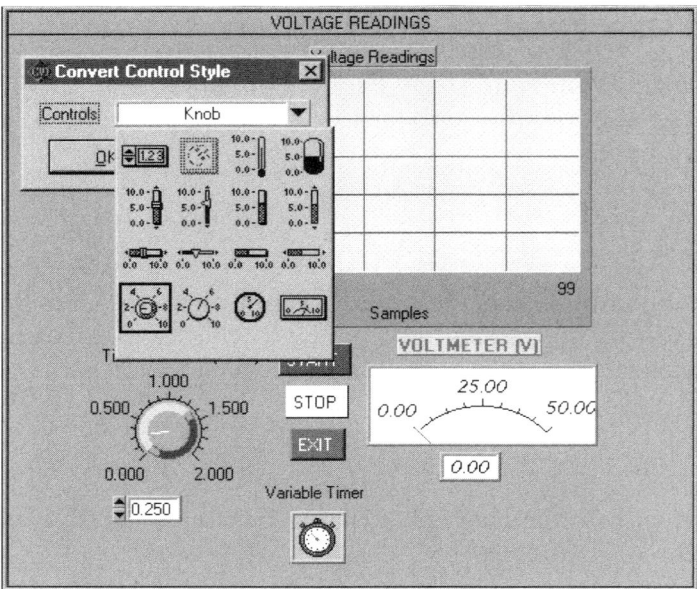

Figure C–3
Converting Control Style

(shortcut is <F11>). A dialog box is displayed in which you can select the control to replace. In Figure C–3, the knob is selected and **Control Style** selected from the **Edit** menu. You can replace the knob with any of the displayed numeric controls. If there are label/value pairs assigned to the replaced control, they are migrated unchanged to the new control.

Create Menu

You have used the features of the **Create** menu many times when creating the GUI. The best way to understand this menu is to go through all the items one by one and create GUIs using these controls, graphs, strip charts, gauges, thermometers, command buttons, binary switches, tables, and decoration boxes. There are some other useful and detailed features of this menu, and they are explained when encountered in the appropriate chapters. The **Create** menu is shown in Figure C–4.

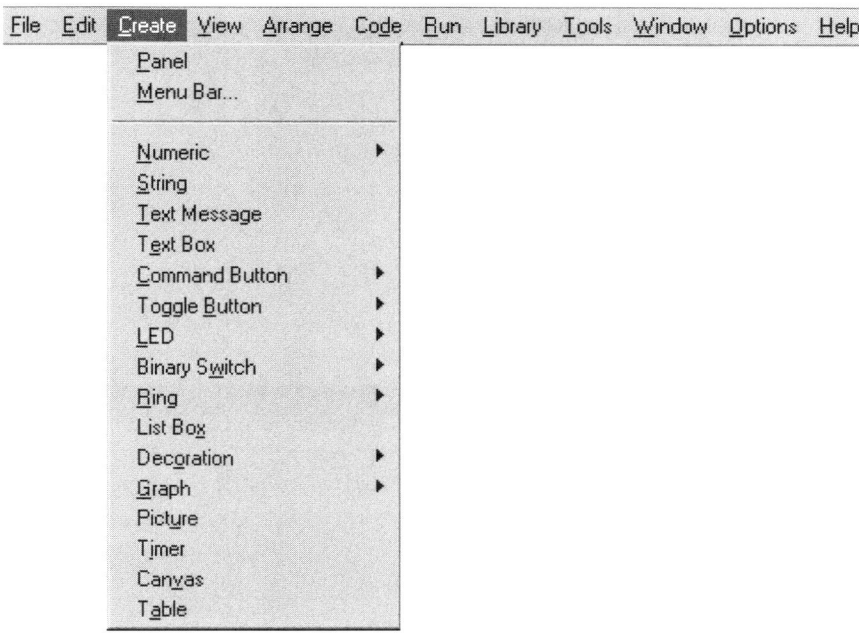

Figure C–4
User Interface Editor Create Menu

View Menu

The **View** menu is shown in Figure C–5. This menu is used for finding user interface objects, manipulating the panels to be displayed, and viewing the header file for the currently opened user interface.

Find UIR Objects

The **Find UIR Objects** command is used to find the GUI objects in the .uir files. By selecting this menu item, a dialog box opens as shown in Figure C–6.

Click on the **Search By** pull-down menu for a choice of various options for the search. The pull-down menu is shown in Figure C–7. You must know the type of control or menu item for which you are searching to be able to use this feature. The following items are valid search options.

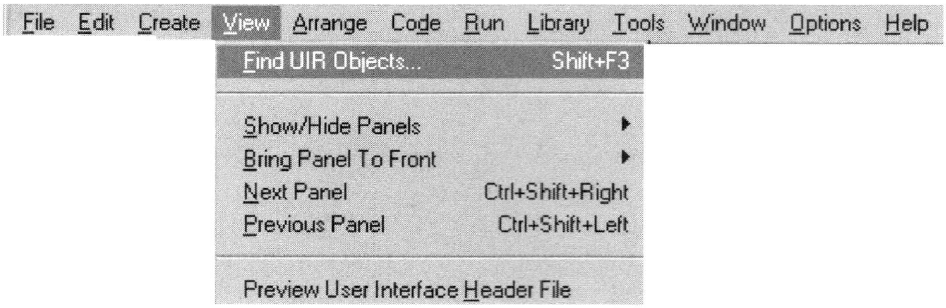

Figure C–5
User Interface Editor View Menu

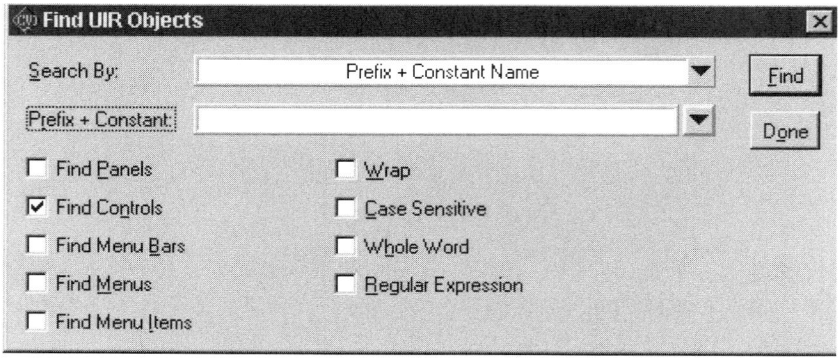

Figure C–6
Find UIR Objects Dialog Box

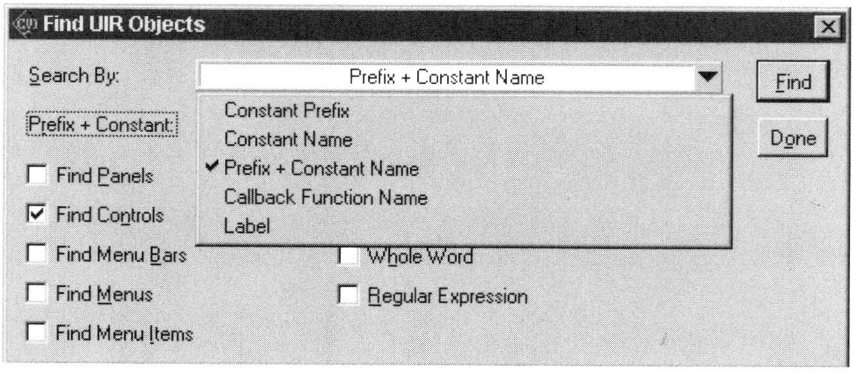

Figure C–7
Search By Ring Control

- **Constant Prefix**—This option is only valid for panels and menu bars.
- **Constant Name**—This option is valid for controls, menus, and menu items.
- **Prefix + Constant Name**—This option can be used for all items.
- **Callback Function Name**—This option is valid for all objects except menu bars.
- **Label**—This option is valid for all objects except menu bars.

The easiest way to search is to view a list of all the strings by clicking on the ring box arrow to the right of the **Prefix + Constant** box and checking the object to search as shown in Figure C–8. Only the correct matching strings will be displayed, depending on the search criterion used in the **Search By** box. You can then select **Find** to see the object on the user interface, as shown in Figure C–8.

If you check the **Wrap** box in Figure C–6, the search continues from the top of the file after the end is reached.

The **Case Sensitive** box, when checked, finds the exact match of uppercase and lowercase characters.

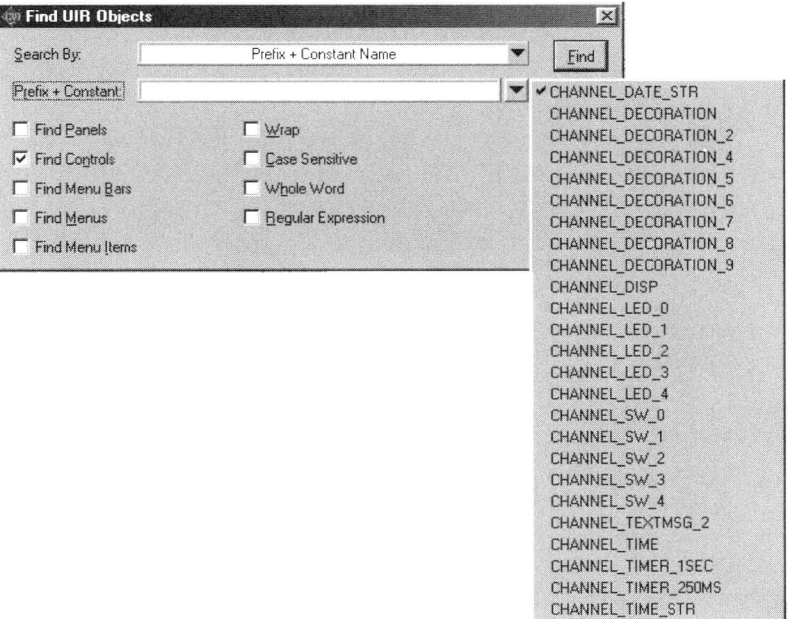

Figure C–8
Finding UIR Objects Using Prefix + Constant Box

The **Whole Word** box, as the name suggests, finds complete words, not part of the words segments.

The **Regular Expression** box can search using wildcard characters, such as ., *, +, ?, ^, $ or \. For a complete description of using the **Regular Expression,** refer to Table 4–1 in Chapter 4 of *LabWindows/CVI User Manual.*

Clicking on the **Find** button starts the search. When an item is found, Figure C–9 is displayed.

The **Find Prev** button searches backward from the present found object location.

The **Find Next** button searches forward from the present found object location.

The **Edit** button stops the search and opens the properties dialog of the control searched.

The **Stop** button ends the search.

Show/Hide Panels

The **Show/Hide Panels** command consists of a sub-menu as shown in Figure C–10. This command is useful when you have multiple panels in a project and you want to see or hide a particular panel. You can select panels by name by selecting **Panel List** and highlighting the panel you want to show or hide.

Bring Panel to Front

When the **Bring Panel to Front** command is selected, another sub-menu is displayed from which you can select the panel or again open a **Panel List** to bring the selected panel to the front.

Figure C–9
Successful Search Completion Dialog Box

Appendix C • User Interface Editor Environment

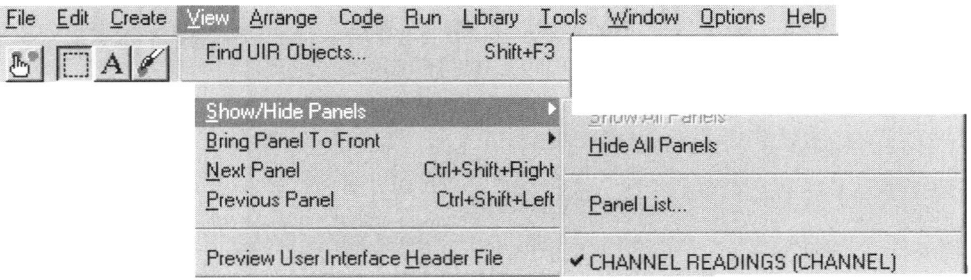

Figure C–10
Show/Hide Panel Sub-Menu

Next Panel

The **Next Panel** command brings the next panel to the front. This command is useful when you have multiple panels and you want to display them one by one for viewing or editing.

Previous Panel

The **Previous Panel** command brings the previous panel to the front. This serves the same purpose as the **Next Panel** command.

Preview User Interface Header File

The **Preview User Interface Header File** command brings up a list of the header (.h) files for the currently active GUI.

Arrange Menu

The **Arrange** menu is shown in Figure C–11. The purpose of this menu is to align objects in a certain way on the panel. Experimenting with the alignment tools in this menu will make you proficient in their use. Getting familiar with this menu will save you much frustration in trying to align objects manually on the panel. Arranging and aligning the objects is discussed in the "Starting the Project" section of Chapter 4, "Enhancing the User Interface."

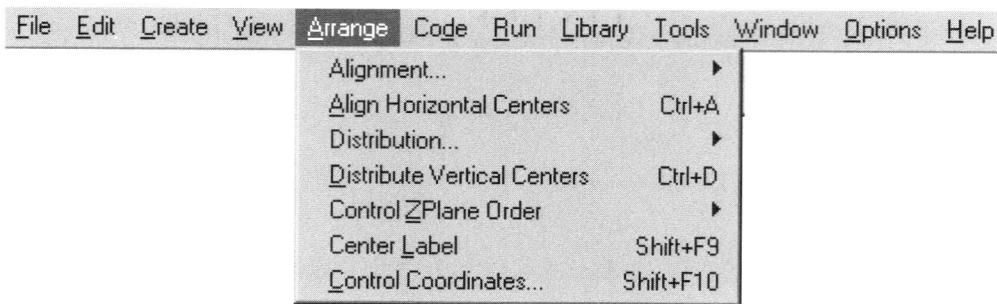

Figure C–11
Arrange Menu

Code Menu

The **Code Menu** is shown in Figure C–12. The **Code** menu was used in several previous chapters in which you created the skeleton code from the *GUI*.

Set Target File

The **Set Target File** command selects the target file for generating the source code from the user interface resource file. This opens a **Set Target File** dialog box as shown in Figure C–13 from which you can select the target file. Select the target file for the source code from this box and select **OK**, and the code will be added to the selected file.

Generate

The **Generate** command consists of sub-menu commands as shown in Figure C–14. In Chapter 2, "Basics of Creating the Graphical User Interface," you used the **All Code** command to generate the code from the user inter-

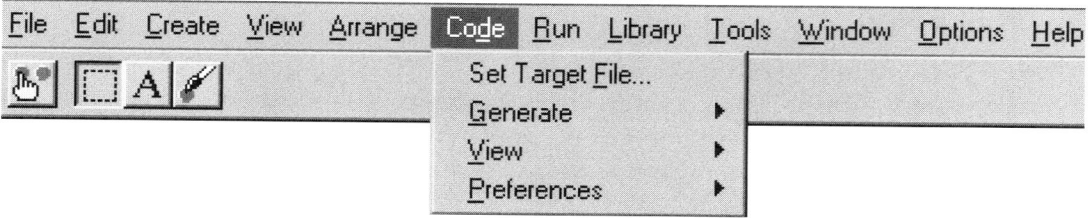

Figure C–12
Code Menu

Figure C–13
Set Target File Dialog Box

face resource file. Unless the option to **Always Append Code to End** is selected through the **Preference** menu item shown on this menu, *CodeBuilder* will insert the code at the cursor position in the source file. That is, if a source code file exists and more control(s) with callback functions were added to the user interface later, the code for the new control(s) will be inserted at the cursor position.

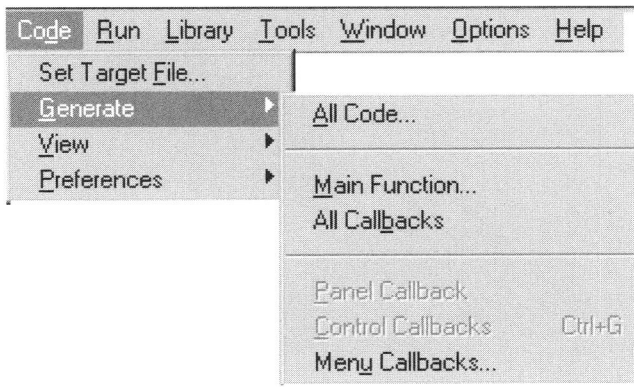

Figure C–14
Generate Sub-Menu

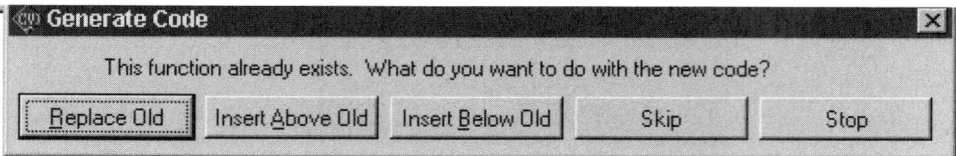

Figure C–15
Generate Code Dialog Box

In addition, in this case *CVI* will highlight the callback function name and prompt you by a **Generate Code** dialog box, shown in Figure C–15. You have an option of replacing the old function, inserting the new code above the old or below the old, skipping to the next function, or stopping. Make the appropriate choice by clicking on one of the buttons, and it will proceed through all the callback functions in your source file.

Main Function

The **Main Function** command creates the code for the main function and writes it to the target file. When the **Main Function** is selected from the submenu, the **Generate Main Function** dialog box is opened as shown in Figure C–16. You can select the panels that you wish to load in the main function. A default panel name is selected and is shown in the **Panel Variable Name** in

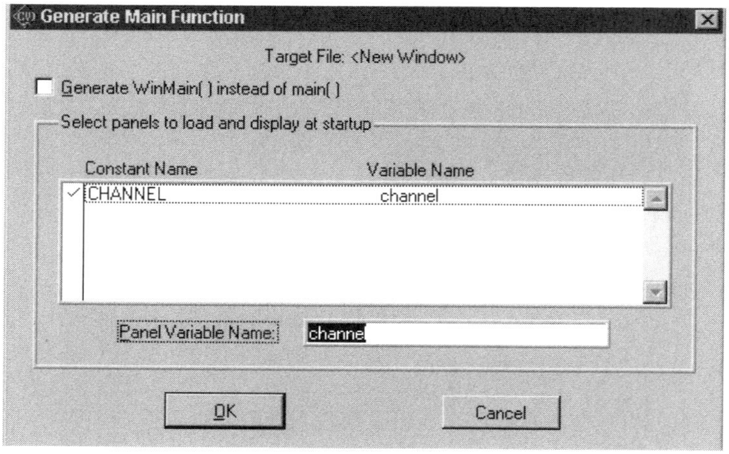

Figure C–16
Generate Main Function Dialog Box

this dialog box. *CodeBuilder* inserts the associated `#include` pre-processor directives, the variable declarations, and the *main* function in the target file.

All Callbacks

The **All Callbacks** command generates the code for all the callback functions and writes to the selected **Target File.** It also adds the `#include` statements that are required by these functions.

Panel Callback

The **Panel Callback** command is used when there is a callback function associated with the panel. Before you use this command, you must activate the panel by double-clicking on it. This command places the panel's callback skeleton code with the associated `#include` statements in the target file.

Control Callbacks

Using the **Control Callbacks** command allows you to create the skeleton code for the callback function associated with the selected control. This will place the code and any `#include` statements in the target file.

CVI uses a shortcut for the above operation. In the user interface resource file, right-click on the control object and select the **Generate Control Callback** command from the pop-up menu as shown in Figure C–17.

The use of this command was demonstrated in Chapter 4, "Enhancing the User Interface," through an example.

Menu Callbacks

The **Menu Callbacks** command produces the skeleton code for the menu and menu items that have callback functions assigned as their attributes. Clicking on this command button displays a dialog box as shown in Figure C–18. Select

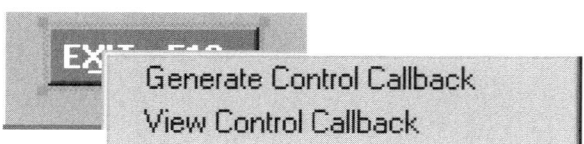

Figure C–17
Control Callbacks Shortcut

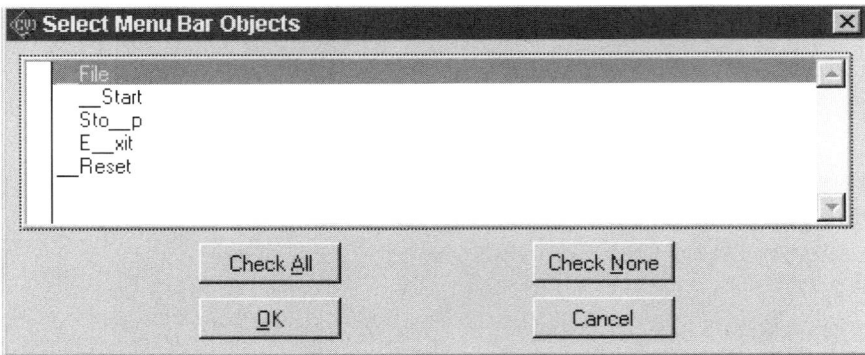

Figure C–18
Select Menu Bar Options

the menu item for which you want to generate the code and click **OK,** or if you want the code generated for all the menu items, click **Check All** and **OK.** The skeleton code with the include statements will be generated.

View

The **View** command is located on the **Code Menu,** as shown in Figure C–12 and is used for accessing a callback function code for either a **Panel Callback** or a **Control Callback.** The source code opens with the callback function name highlighted. The shortcut method for accessing a control callback is shown in Figure C–17; select **View Control Callback.**

This command searches for the selected function in all the source files in the project and all other open source files. If the function is found in a closed project, the file is opened automatically. This command is a great time-saver for locating a function in multiple source files in a project.

To return to the .uir file, select the **Find UI Object** command from the **View** menu in the **Source** window. You can also right-click on the callback function, and the panel shown in Figure C–19 will be displayed. Select the **Find UI Object** command from the pop-up menu or use <Ctrl-F> as a shortcut.

Preferences

Using the **Preferences** command from the **Code Menu** (Figure C–12), you can change the preferences for the default settings for *events* associated with control callback and panel callback functions. You can also select how the

Appendix C • User Interface Editor Environment

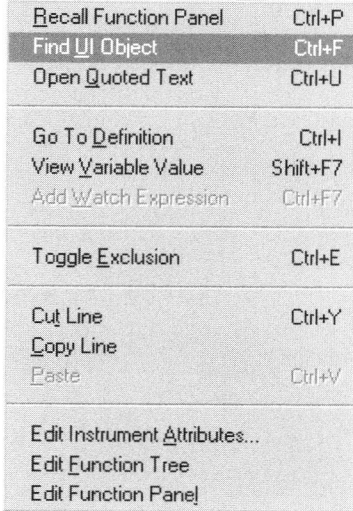

Figure C–19
Find UI Object Shortcut

generated code is to be added to the selected target file. The **Preference** menu item is shown in Figure C–20.

Default Panel Events/Default Control Events

To select which control callback events you want to use in your C code, select the **Default Panel Events** dialog box, shown in Figure C–21, or the **Default Control Events** dialog box, shown in Figure C–22.

These commands serve as an extra capability of *CVI* and allow you more control in the way the code is to be written based on your style and choice. You make your selections from the listed items and select **OK**. This selection gets included in the code. The timer events, however, always use the EVENT_TIMER_TICK and EVENT_DISCARD.

Always Append Code To End

The **Always Append Code To End** option adds code to the end of the source file. If you do not want the code at the end, place the cursor in the source code where you want the code added and do not check this option.

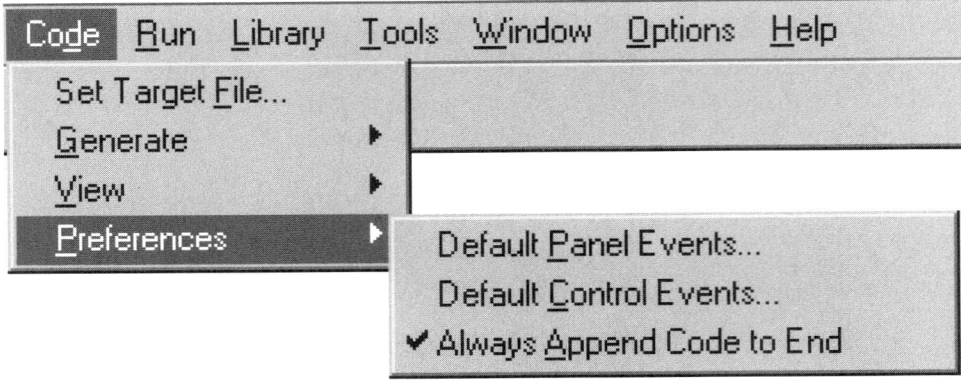

Figure C–20
Preferences Menu Item

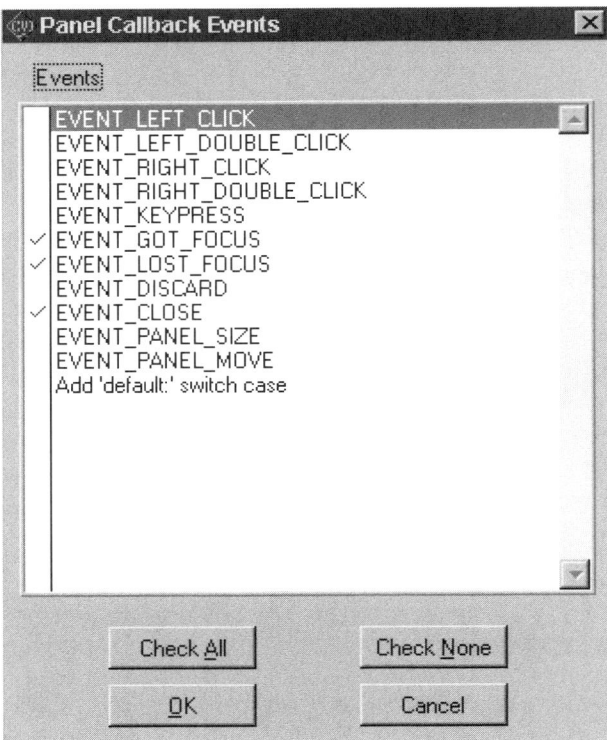

Figure C–21
Default Panel Events

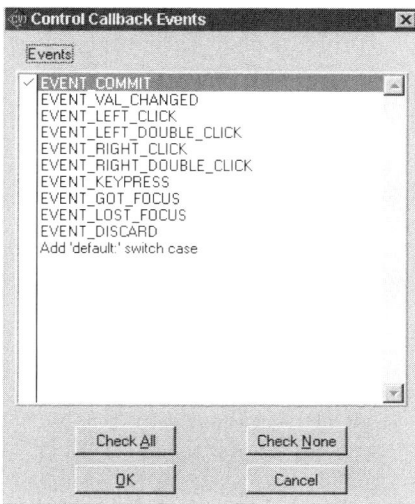

Figure C–22
Default Control Events

Run Menu

The **Run** menu shown in Figure C–23 contains some of the commands that appear in the **Source Code** and the **Project** windows. These menu items were discussed in Chapter 5, "Source Editor and Debugging Techniques." Some of the commands in this menu are very similar to the **Run** menu in the **Project** window and are covered in Appendix B, "Project Window Environment."

Figure C–23
Run Menu

Library Menu

The **Library** menu is explained in Appendix B, "Project Window Environment." You can also refer to *LabWindows/CVI User Manual* for more information.

Tools Menu

The **Tools** menu is explained in Appendix B, "Project Window Environment." You can refer to *LabWindows/CVI User Manual* for more information.

Window Menu

The **Window** menu is explained in Appendix B, "Project Window Environment." You can refer to *LabWindows/CVI User Manual* for more information.

Options Menu

The **Options** menu is shown in Figure C–24 and consists of the following menu items.

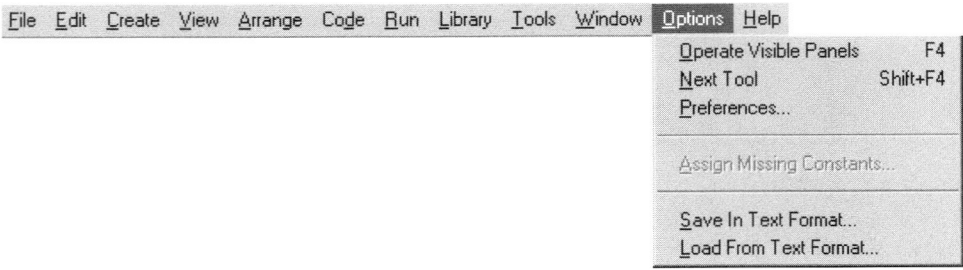

Figure C–24
Options Menu

Operate Visible Panels

The **Operate Visible Panels** command allows you to operate the controls as you would by using the **Operating Tool** shown in Chapter 3, "More Graphical User Interface." This icon shows the sequence of events generated on the top right corner of the panel (at a slower rate for visibility) when the control is selected during the application's execution.

Next Tool

The **Next Tool** command rotates you through all the tools shown on the **Tool Bar** of the **User Interface Editor** as explained in Chapter 3, "More Graphical User Interface." These commands can be selected one at a time by selecting <Shift-F4>.

Preferences

When **Preferences** command is selected, the **User Interface Editor Preferences** dialog box is displayed as shown in Figure C–25. Each of the dialog box options will be discussed.

Edit Color Preferences

The **Edit Color Preferences** group allows you to select the initial background color on the **User Interface Editor:** Windows. Click on the box to the right of **Initial editor background color:** and select the color of your choice from the pop-up color display panel.

Preferences for New Panels

The **Preferences for New Panels** group consists of the following choices.

- The **Resolution Adjustments** box lets you set the percentage to scale your panel and the objects on the panel when displayed on the screen. This is useful when the panels were created and viewed on monitors with differing resolutions. This attribute can also be set from the **Other Attribute** dialog box from the **Edit Panel** box, shown in Figure C–26.

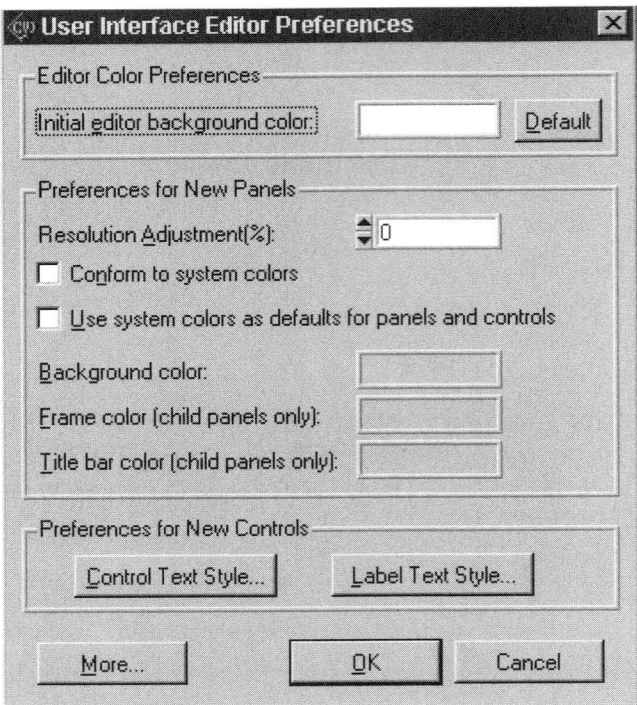

Figure C–25
User Interface Editor Preferences Dialog Box

- The **Conform to system colors** option box, when checked, forces the panels and controls to initially use the system colors on Windows, but the colors can be changed later on. This option can also be selected from **Other Attributes** from **Edit Panel** as shown in Figure C–26.
- When the **Use system colors as defaults for panels and controls** box is checked, the system colors are used as initial colors for panels and controls. With this option enabled, you cannot change the **Background Color, Frame Color,** and **Title bar color** boxes that are shown below this option on the dialog box. The **Frame Color** and the **Title bar color** are used for child panels only.

Preferences for New Controls

The **Preferences for New Controls** group, shown in Figure C–25, allows you to set the initial text attributes for the command buttons via the **Control Text Style** and the **Label Text Style** buttons. Setting the attributes here saves you

Appendix C • User Interface Editor Environment

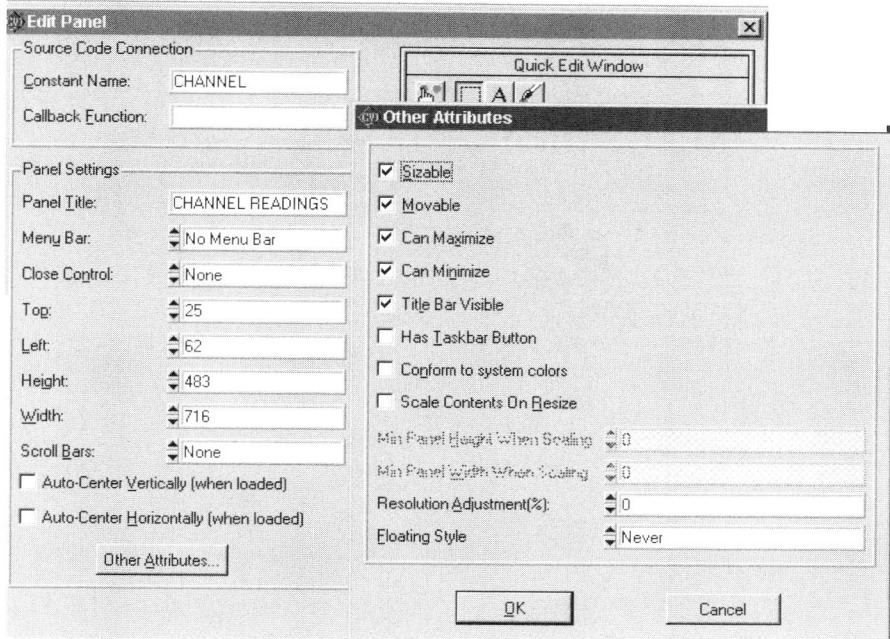

Figure C–26
Edit Panel/Other Attributes Dialog Box

from changing these attributes on every control later and gives you uniformity of the GUI design. The **Control Text Style** and the **Label Text Style** dialog boxes in which you can set your preferences are shown in Figures C–27 and C–28.

More

The **More** command button shown on the **User Interface Preferences** dialog box in Figure C–25 brings up the **Other User Interface Editor Preferences** dialog box as shown in Figure C–29. The dialog box consists of the following options.

- **Undo Preferences** group—Here you can set the number of changes that can be reversed for each file. The default number is 30. If you want the undo buffer to be empty after each file save, check the **Purge undo actions when saving file** box.

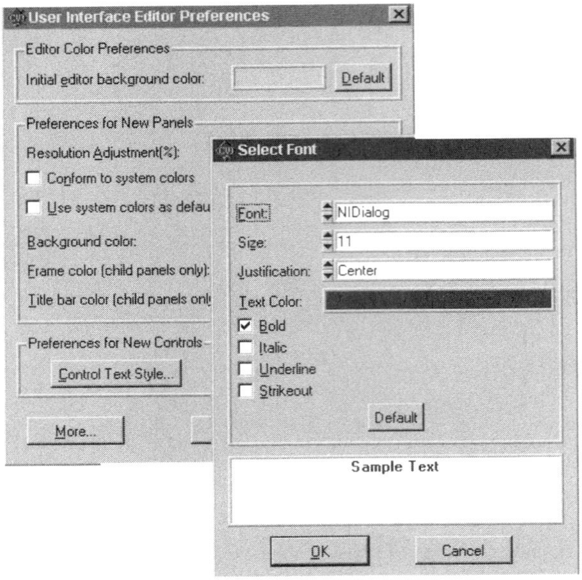

Figure C–27
Control Text Style Dialog Box

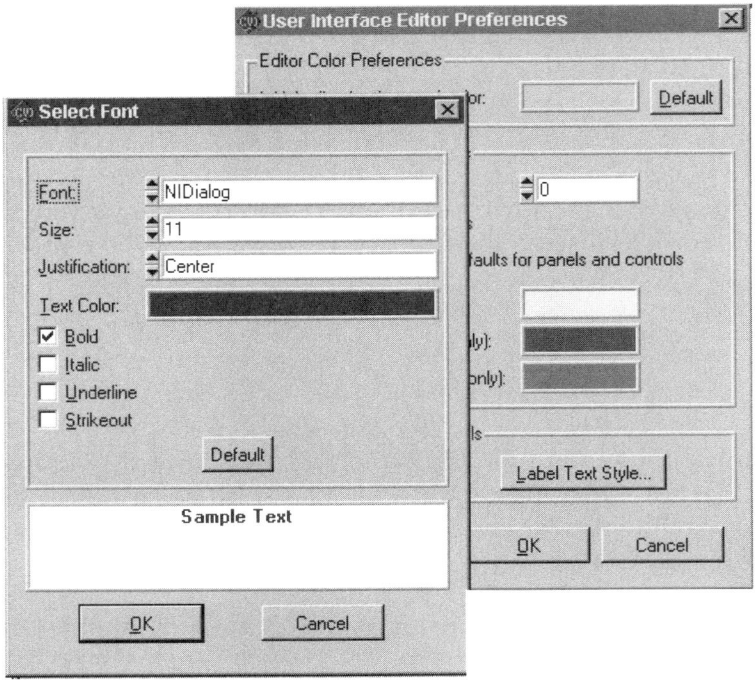

Figure C–28
Label Text Style Dialog Box

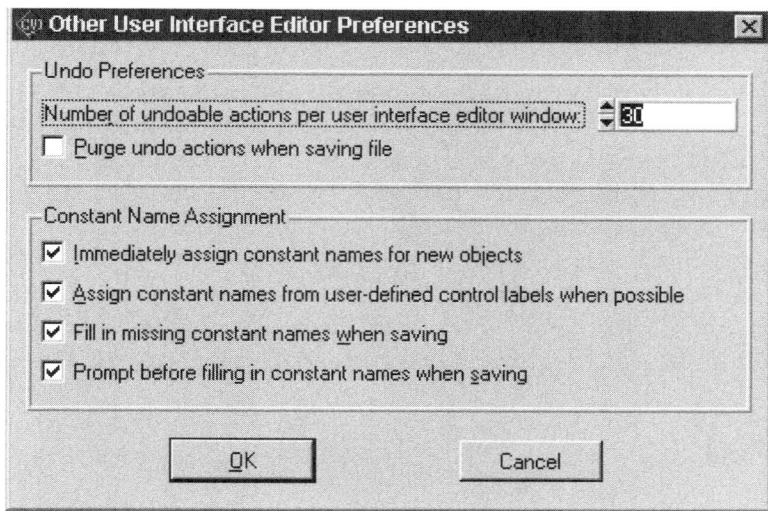

Figure C–29
Other User Interface Editor Preferences

- **Constant Name Assignment** group—This group consists of a number of options to facilitate the design of the GUI.
  - When you check the **Immediately assign constant name for new objects** box, the **User Interface Editor** generates constant names automatically when the control, panel, or the menu bar items are created. For the panels and controls, the constant names are generated immediately when they are created and can be modified as desired. The menu bar names are created when you exit the menu bar editor. Leave this option checked to facilitate the user interface design.
  - The **Assign constant names from the user-defined control labels when possible** box is only available if the **Immediately assign constant name for new objects** box is not selected, so it is kind of a superfluous option since the **User Interface Editor** by default creates its own constant name based on the control type. You can change the name later on, as always.
  - Checking the **Fill in missing constant names when saving** box fills in the constant names in the controls and panels if they are left blank. It is a good practice to leave this option selected.
  - The **Prompt before filling in constant names when saving** box gives you the option of confirming the number of items without constant names via a pop-up box.

Save in Text Format

The **Save in Text Format...** command is selected from the **Options** menu shown in Figure C–24. This command saves the contents of the **User Interface Editor** Window in ASCII text format for readability. A pop-up dialog box appears, asking you the path and the file name to use in saving the file. The .tui extension is automatically attached to this file. Do not save the user interface file as a .uir file since you may over-write your file and lose all of your work. The .tui listing gives the details of the .uir file controls in a readable ASCII format. Creating this file is useful when more information on the user interface is required. This file can be viewed with any text editor.

Load from Text

The **Load from Text Format** option is selected from the **Options** menu shown in Figure C–24. This command does the opposite of the **Save in Text Format** command. Open a new user interface file and select this command. A pop-up dialog box will prompt you for the file name you want to load. If you select a file with a .tui extension, an exact replica of the user interface file from which the .tui file was created will be displayed.

Help Menu

The **Help** menu enables the *CVI* online help so you can obtain more information on *CVI* features. This is the same menu as shown in the **Project** window, as discussed in Appendix B.

Appendix D • Formatting Functions

Figure D–1
Fmt Function Panel

```
n = Fmt (target_string, "%s<%d", source_integer);
```

In *Fmt* functions, the direction of conversion is always from right to left, which is shown by <, pointing in the direction of the data conversion. Consider this the tip of the arrow to indicate the direction of conversion. In the *Fmt* function, the *source* on the right side of the < is being converted to the *target* on the left side.

In looking at the above example, you converted an integer into a string. The %d on the right side of the < represents that an integer is being converted to a string, %s. The component in between the *source* and the *target* is called the *format string*. In the above example, the *format string* is %s<%d. The format string specifies the details of the type of all the data elements.

The return value *n* specifies the number of source items formatted into the target successfully. A -1 for *n* means an error in the formatting string and a -2 indicates an I/O error, meaning an error in the *format string*.

There is only one data element on the *target* side in the *Fmt* function, but it can have multiple *sources*.

```
char target_string[5];   //Define variable types
int source_integer, n;
source_integer= 99;
n = Fmt (target_string, "%s<%d", source_integer);
The result of this Fmt would be
target_string = "99";
```

Figure D-2
Fmt Function — Converting Integer to String

Look at a more complex example, shown in Figure D-2, which defines the variable types and how they are used in the *Fmt* function.

Let us look at another more practical example in Figure D-3.

In Figure D-4, you can see the one-to-one correspondence between the variables and the components in the *format specifiers*. This will give you a clear picture of how the *Fmt* function reacts to the variables and the *format specifiers*.

This example shows you how variables of different types can be mixed using the correct *format string* to form a more complex output string placed in *buf*.

Let us look at the *FmtFile* function by bringing up the function panel from **Library>>Formatting and I/O>>Data Formatting>>Formatting Functions>>FmtFile**. This panel is shown in Figure D-5. The prototype of this function follows.

```
n= FmtFile(fileHandle, format string, source1,source2,……..,
                                                     sourcen);
```

The *source* data is converted per the *format string* and written to the file specified by the *fileHandle*. The *fileHandle* is created using the *OpenFile*

```
char  *filename, *directory;
int fileNum, n;
filename="Test";   //File name is initialized to "Test"
fileNum = 5;   //Could be a test run number
directory ="c:\\cvi5\\samples\\class\\";   //Directory of the file
n= Fmt( buf, "%s<%s%s%d.out", directory, filename, fileNum);
The result will be
                    buf = "c:\\cvi5\\samples\\class\\Test5.out"
```

Figure D-3
Fmt Function — Converting Multiple Inputs to a String

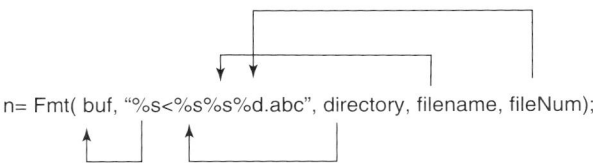

Figure D–4
Fmt Function—Variables and Format Specifiers Correspondence

function as shown in the "Using Files" section of Chapter 6. For example in the *WriteToFile* function in `project6.c`, the lines in Figure D–6 could have been replaced by the line in Figure D–7 to obtain the same output.

The last formatting function is *FmtOut*. This function does the same source formatting but displays the results to the **Standard Output** win-

Figure D–5
FmtFile Function Panel

```
//Create a file header with all the panel information
Fmt(buf, "%s<Operator: %s    Test #: %d    File Name: %s    Date: %s    Time:
        %s\n",              OperatorID, TestNum, WriteFileName, Date, Time);
//Write header to file
WriteFile (WriteFileHandle, buf, StringLength(buf));
```

Figure D–6
Writing to File Using *Fmt* Function

dow or the Console Window depending on your settings in the **Target Settings** option from the **Build** menu. The prototype of this function follows.

```
n= FmtOut( format string, source1,source2,......., sourcen);
```

Note there is no target string and the data is not written to any variable, just output to the **Standard Output** window or the Console Window. Let us look at the *FmtOut* function by bringing up the function panel from **Library>>Formatting and I/O>>Data Formatting>>Formatting Functions>> FmtOut,** as shown in Figure D–8.

Note that all the formatting functions use similar *format specifiers*. The difference between these functions is only to/from where they send/receive the *target* data.

Another item to note, the *target format specifier* (%s in the examples associated with *buf*) and/or < can be eliminated completely, in which case the function will assume the target to be a string by default (%s). If you want to omit the *target format specifier,* it is still better to leave < for visibility purposes.

```
n= FmtFile(WriteFileHandle, "%s<Operator: %s    Test #: %d    File Name: %s
       Date:    %s    Time: %s\n", OperatorID, TestNum, WriteFileName, Date, Time);
```

Figure D–7
Writing to File Using *FmtFile* Function

Appendix D • Formatting Functions

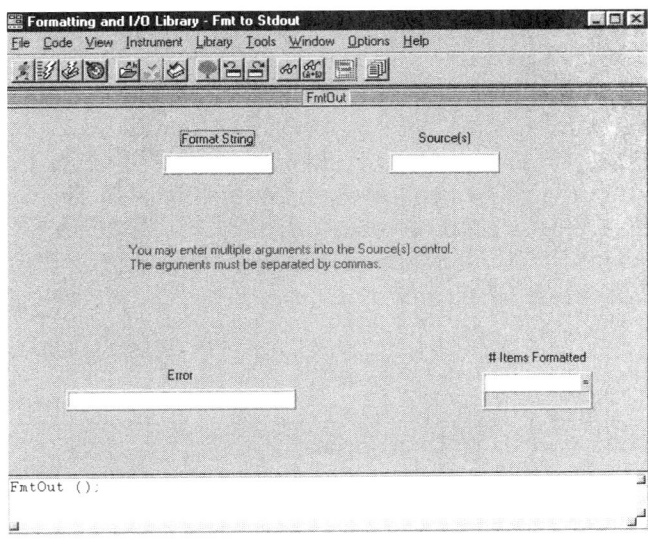

Figure D-8
FmtOut Function Panel

Format Code Specifiers

The *format specifiers* will now be explained to show you how they can be modified to give you more flexibility and control in manipulating the data elements. All *format strings* have a structure that is similar to

```
target_spec < source_spec_and_literals,
```

where `target_spec` is a *format specifier* that defines the nature of the target data and `source_spec_and_literals` consists of the *format specifier* and *literals* for the source data, showing the procedure to move the source data to the target.

Each *format specifier* has the following form.

```
% [rep]    formatcode  [[modifiers]]
```

The % represents the start of all *format specifiers*, `rep` indicates the number of times the format repeats with respect to the arguments. `Formatcode` indicates the data type, and `modifiers` represents optional additional codes that add more description to the data format and are always included in brackets.

Note: rep is not allowed when formatcode is a string.

This can be confusing but will become clear as you go through some examples of the *format specifiers*. *LabWindows/CVI Standard Libraries Reference Manual* explains the *formatcode* and the *modifiers* of the *Scan* and *Fmt* functions in detail and should be read for more information on this topic.

The *formatcode specifiers* are included in Table D–1 and have been taken from *LabWindows/CVI Standard Libraries Reference Manual*.

Table D–1 *Formatcode Specifier*

Formatcode	Data Type	Description
s	string	This character indicates that the corresponding argument is a string.
i	integer	This character represents that the corresponding argument is an integer. rep indicates an integer array of rep number of elements (if present).
x	integer (hexadecimal)	This character represents that the corresponding argument is an integer. If rep is present, the integer is an array of rep number of elements. This will convert the data to ASCII hexadecimal digits.
o	integer (octal)	This character indicates that the corresponding argument is an integer. rep indicates an integer array of rep number of elements. This will convert the data to ASCII octal digits.
d	integer (decimal)	This is identical to %i and is included for compatibility with the C *printf* function.
f	real number	This character represents that the corresponding argument is a real number. rep indicates a real number array of rep number of elements.
c	character	This character indicates that the corresponding argument is an integer with one significant byte. If rep is present, it means that the argument is an array of one-byte integers. There is no conversion to ASCII, and it is copied directly to the string.

Appendix D • Formatting Functions

Formatting Modifiers

Modifiers are optional codes that you can specify to give more information on the data item. You must enclose these *modifiers* in one set of brackets, even if you have a number of modifiers, and place the *modifiers* immediately after the format code.

A different set of *modifiers* operates with integers, another set with the floating points, and yet another for strings. Each of these *modifiers* is explained in the following tables. Table D–2 (taken from *LabWindows/CVI Standard Libraries Reference Manual*) lists the integer *modifiers* used with formatting strings %i, %d, %x, %o and %c. Although c is a character, it is considered part of the integer *modifiers*.

Table D–2 Integer Modifiers (%i, %d, %x, %o, %c)

Modifier	Purpose	Description
bn	Specify length	The *b* integer modifier species the length of the integer argument or the length of the integer arrays in bytes. Default is four-bytes, b2 is short integers, and b1 is single-byte integers.
in	Specify array offset	This modifier specifies the offset within an integer array, as a zero-based index into an array. The i modifier is valid if rep is used.
z	Treat string as integer array	The z integer modifier treats the data in the string as an integer array and is valid only if rep is present.
rn	Specify radix	The r integer modifier specifies the radix integer argument if the function converts the integer into string format. The radix that can be used are: 8 for octal, 10 for decimal (default), 16 for hexadecimal, 256 for eight-bit ASCII characters.
wn	Specify string size	The w integer modifier specifies the exact number of bytes in which to store the representation of the integer argument upon conversion to a string format. The default for n is 0, meaning the value can occupy whatever space it requires.

(*continued*)

Table D–2 Integer Modifiers (`%i`, `%d`, `%x`, `%o`, `%c`) (continued)

Modifier	Purpose	Description
pc	Specify padding	This modifier specifies a padding character, c, which fills the entire remaining field width to the left of the integer specified by the w character.
s	Specify as two's complement	The s integer modifier indicates that the function considers the integer argument a signed two's complement number. Be careful not to confuse this s with the format specifier.
u	Specify as unsigned	The u integer modifier indicates that the function considers the integer an unsigned integer.
onnn	Specify byte ordering	This modifier defines the byte ordering for raw data, to map it into the appropriate byte order of Intel format (PC) or Motorola (SPARC station) architecture. The bytes in Motorola format are in the reverse order of the Intel format.

Now let us look at the floating-point *modifiers*. These are the *modifiers* used with the `%f` *format specifier* and are shown in Table D–3, taken from *LabWindows/CVI Standard Libraries Reference Manual*.

Table D–3 Floating-Point Modifiers (`%f`)

Modifier	Purpose	Description
bn	Specify length	The b floating-point modifier species the length of the floating-point argument or the length of the floating-point arrays in bytes. Default is eight-bytes and single precision floating-point values are indicated by b4. Only 8 and 4 are valid values for n.
in	Specify array offset	This modifier specifies the offset within a floating-point array, as a zero-based index into an array. The i modifier is valid if rep is used.

(continued)

Appendix D • Formatting Functions

Table D–3 *Floating-Point Modifiers* (`%f`) (continued)

Modifier	Purpose	Description
z	Treat string as floating-point array	The z integer modifier treats the data in the string as a floating-point array. The z modifier is only valid if rep is present.
wn	Specify string size	The w floating-point modifier specifies the exact number of bytes in which to store string representation of the floating-point argument upon conversion to a string format. The default for n is 0, which means the value can occupy whatever space it requires.
pn	Specify precision	The p floating-point modifier specifies the precision of the floating-point number by indicating the number of places to the right of the decimal place. The default value is p6 if you do not specify anything.
en	Specify as scientific notation	The e floating-point modifier converts the value to scientific notation. If n is omitted, the default value of 3 is used.
f	Specify floating-point notation	The f floating-point modifier converts the value to a string format in floating-point notation.
t	Truncate	This modifier indicates that, in floating-point to integer transformations, the functions truncate instead of round the floating-point value.
r	Round	This modifier indicates that, in floating-point to integer transformations, the functions round instead of truncate the floating-point value. The default is truncation.

String *modifiers* are identified with the `%s` *format specifier* shown in Table D–4, which is taken from *LabWindows/CVI Standard Libraries Reference Manual*.

These descriptions of how and where to use the *modifiers* are overwhelming, and the best way to understand them is by way of examples, which follow in the next section.

Table D–4 String Modifiers (%s)

Modifier	Purpose	Description
i*n*	Specify array offset	The i string modifier gives the offset within a string. n is a zero-based index into the string indicating where to begin the processing.
a	Append	This modifier is used with the target format specifier and appends all converted data to the end of the target string after the ASCII NUL.
w*n*	Specify string size	When used with the source format specifier, this specifies the maximum number of bytes to be used from the string. The default of n = 0 means that the complete string is to be used. Any non-negative number to the size of the string length can be used. When used with the target format specifier, this specifies the exact number of bytes to store in the string, excluding the terminating ASCII NUL character. When n is 0 or is omitted, the function, by default, stores as many bytes as the source requests. When n is greater than the number of bytes available from the source, the remaining bytes are filled with ASCII NUL if you use the q modifier or blanks if no q modifier is used. When you use the w string modifier with the a modifier, n indicates the number of bytes to append to the string, excluding the terminating ASCII NUL.
q	Append NULs	When used with the target string in conjunction with the w modifier, the q string modifier specifies that the unfilled bytes at the end of the target string be filled with ASCII NUL instead of blanks.
t*n*	Teminate on character	When used with the source string, the t string modifier specifies that the source string terminates on the first occurrence of the character n, where n is the ASCII value of the character. When more than one t modifiers are used in the same specifier, the string terminates when any of the terminators occur.

(continued)

Appendix D • Formatting Functions

Table D–4 String Modifiers (%s) (continued)

Modifier	Purpose	Description
t-	Terminate when full	This is similar to n except that it specifies that there are no terminating characters. Reading of the source stops when the target is full or when the function has read the number of bytes specified with the w modifier.
t#	Terminate on number	This is equivalent to the t modifier with the ASCII values of the characters being +, - and 0 through 9. It stops the reading of the string when any of these characters are encountered.
jn	Specify characters to insert between each element of an array	CVI version 5.5 has introduced this modifier that allows you to specify characters that are inserted between each element when you use the repeat code to format multiple items.

Formatting Examples for the *Fmt* Function

Examples of the specific types of *Fmt* functions follow.

Integer to String

1. **Objective:** Convert an integer value to a string (see Figure D–9).

```
char target[10];      // This data type applies to all the examples below
int source;           // This data type applies to all the examples below
source =51;
Fmt ( target, "%s<%i", source);
```

Figure D–9
Integer to String

Result: target ="51"

Conversion: The *source* integer is converted to a string. `%i` is the format specifier for an integer (same as `%d`), while `%s` is the format for a string.

2. **Objective:** Convert an integer hexadecimal value into a string (see Figure D–10).

```
source =255 ;
Fmt ( target, "%s<%x", source);
```

Figure D–10
Integer Hexadecimal Value to String

Result: target ="ff"

Conversion: The *source* integer is converted into hexadecimal value as a string. `%x` is used for an integer hexadecimal *format specifier*.

3. **Objective:** Convert an integer octal value into a string (see Figure D–11).

```
source =255 ;
Fmt ( target, "%s<%o", source);
```

Figure D–11
Integer Octal Value to String

Result: target ="377"

Conversion: The *source* integer is converted into octal value as a string. `%o` represents an integer octal value.

4. **Objective:** Convert an unsigned integer value into a string (see Figure D–12).

```
source = -1 ;
Fmt ( target, "%s<%i[u]", source);
```

Figure D–12
Unsigned Integer Value to String

Result: target ="4294967295"

Conversion: The *source* integer is converted into a non-negative integer value as a string. The non-negative number for –1 is calculated as $2^{32} - 1$ for a thirty-two-bit integer. The `%u` is called an *unsigned integer modifier*.

Appendix D • Formatting Functions

Similarly, if the source in this example is –2, then the target= "4294967294", which is $2^{32} - 2$.

5. **Objective:** Convert an integer into a string with a specified width (see Figure D–13).

```
source =5678;
Fmt ( target, "%s<%i[w6]", source);
```

Figure D–13
Integer into String with Specified Width

Result: target = " 5678"

Conversion: The *source* integer is converted into a string with a field width of 6 and is right-justified. Therefore there are two blank spaces to the left of the string. The w modifier always represents the width of the field. It is followed by an integer that specifies the exact number of bytes in which to store the string representation after conversion.

6. **Objective:** Convert an integer into a string with a specified width and padding with character ^ (see Figure D–14).

```
source =5678;
Fmt ( target, "%s<%i[w6p^]", source);
```

Figure D–14
Integer into a String with Character Padding

Result: target ="^^5678"

Conversion: The *source* integer is converted into a string with a field width of 6 and is right-justified. The two spaces to the left of the string are padded with ^ characters to make the width of 6. The p followed by the padding character specifies the character to use for padding the string.

7. **Objective:** Convert an integer into a string with a specified width less than the source string's width (see Figure D–15).

```
source =5678;
Fmt ( target, "%s<%i[w2]", source);
```

Figure D–15
Integer to String with Specified Width Less than Source String's Width

Result: target ="*8"

Conversion: The *source* integer is converted into a string with a field width of 2, which is less than the source string length. Therefore a * is placed in the *target* string to indicate overflow, and only 2 characters, as specified, are shown in the target.

8. **Objective:** Convert to byte ordering when communicating with instruments that use Intel byte ordering (called *little endian*) and the Motorola byte ordering (called the *big endian*).

Suppose you are reading instrument data in Intel byte order in a string raw_buf and putting it in target con_buf. Then the code will look like Figure D–16.

```
short int raw_buf [100], con_buf [100];
Fmt(con_buf, "%100d<%100d[b2o01]", raw_buf);
```

Figure D–16
Reversing Byte Order

Conversion and Results: The %100d on the left side of < indicates that the data is to be converted into an integer array 100 bytes long, which is the dimension of con_buf. The *format specifier* to the right side of < converts the source short integer (indicated by b2) array raw_buf in the Intel byte order specified by o01, which means first byte 0, then byte 1.

Suppose you have an instrument that has Motorola-based byte ordering. This is specified as follows.

```
Fmt(con_buf, "%100d<%100d[b2o10]", raw_buf);
```

This is the same as shown in Figure D–16 but the o01, on the right side of < is now reversed to o10 to indicate that byte 1 is converted first and then byte 0. If you are using four bytes of data, then o0123 are in as-

Appendix D • Formatting Functions

cending order (Intel format) and o3210 are in descending order (Motorola format). The numbering shown after o are the byte numbers.

Short Integer to String

1. **Objective:** Convert a short integer into a string (see Figure D–17).

```
char target[20];      // This data type applies to all the examples below
short source;     // This data type applies to all the examples below
source = 6789;
Fmt ( target, "%s<%i[b2]", source);
```

Figure D–17
Short Integer to String

Result: target ="6789"

Conversion: The *source* short integer is placed into a *target* string. Note that [b2] represents a short integer. The modifier [b1] represents single-byte integers, converting only one byte to a string. So if you wanted to convert the source=67 to a string, then you will use the format shown in Figure D–18.

```
Fmt ( target, "%s<%i[b1]", source);
```

Figure D–18
Single-Byte Integer to String

In this case the target="67".

2. **Objective:** Convert a short-signed integer into a string (see Figure D–19).

```
source = -1;
Fmt ( target, "%s<%i[b2]", source);
```

Figure D–19
Short-Signed Integer to String

Result: target ="-1"

Conversion: The *source* short integer is placed into a *target* string with the negative sign.

3. **Objective:** Convert a short-signed integer into a string using the u modifiers (see Figure D–20).

```
source = -1;
Fmt ( target, "%s<%i[b2u]", source);
```

Figure D–20
Short-Signed Integer to String

Result: target ="65535"

Conversion: The *source* integer is converted into a non-negative integer value as a string. The non-negative number for –1 is calculated as $2^{16} - 1$ for a sixteen-bit integer. The u modifier considers the integer an unsigned integer.

4. **Objective:** Convert a short-signed integer into a string width of 6 using the w *modifiers* (see Figure D–21).

```
source =5678;
Fmt ( target, "%s<%i[b2w6]", source);
```

Figure D–21
Short-Signed Integer to String Using Width Modifier

Result: target = " 5678"

Conversion: The *source* integer is converted into a string with a field width of 6 and is right-justified. Therefore, two blank spaces appear to the left of the string. The value after the w specifies the right-justified width of the field.

5. **Objective:** Convert a short-signed integer into a string width of 6 using the w and p *modifiers* to add padding (see Figure D–22).

```
source =5678;
Fmt ( target, "%s<%i[w6p0]", source);
```

Figure D–22
Short-Signed Integer to String Using Width and Padding Modifiers

Result: target ="005678"

Conversion: The *source* integer is converted into a string with a field width of 6 and is right-justified. The two blank spaces to the left of the string are padded with 0 as specified by p0.

6. **Objective:** Convert a short integer into a string with a specified width less than the source string width (see Figure D–23).

```
source =5678901;
Fmt ( target, "%s<%i[b2w4]", source);
```

Figure D–23
Short Integer to String with Specified Width Less than Source String Width

Result: target ="*901"

Conversion: The *source* integer is converted into a string with a field width of 4, which is less than the source string length. Therefore a * is placed in the string as the left-most character of the specified width (4 in this example) to indicate overflow.

Inserting Characters Between Each Element of an Array

1. **Objective:** Insert specified characters between each element of an array (see Figure D–24).

```
#include <ansi_c.h>

static char buf[100];
static int ArrayOfIntegers[10],i;
for (i=0;i<10;i++)   ArrayOfIntegers[i]=i;

Fmt (buf, "%s<%10d[j3],-;", ArrayOfIntegers);
```

Figure D–24
Inserting Specified Character Between Each Element of Array

Result: buf="0,-;1,-;2,-;3,-;4,-;5,-;6,-;7,-;8,-;9,-;"

Conversion: The repeat code of 3 after the j modifier indicates that three characters following [j3] are to be inserted between each element of the array *ArrayOfIntegers*. Each element of *ArrayOfIntegers* is separated by a comma, a hyphen, and a semi-colon as specified by [j3],-;.

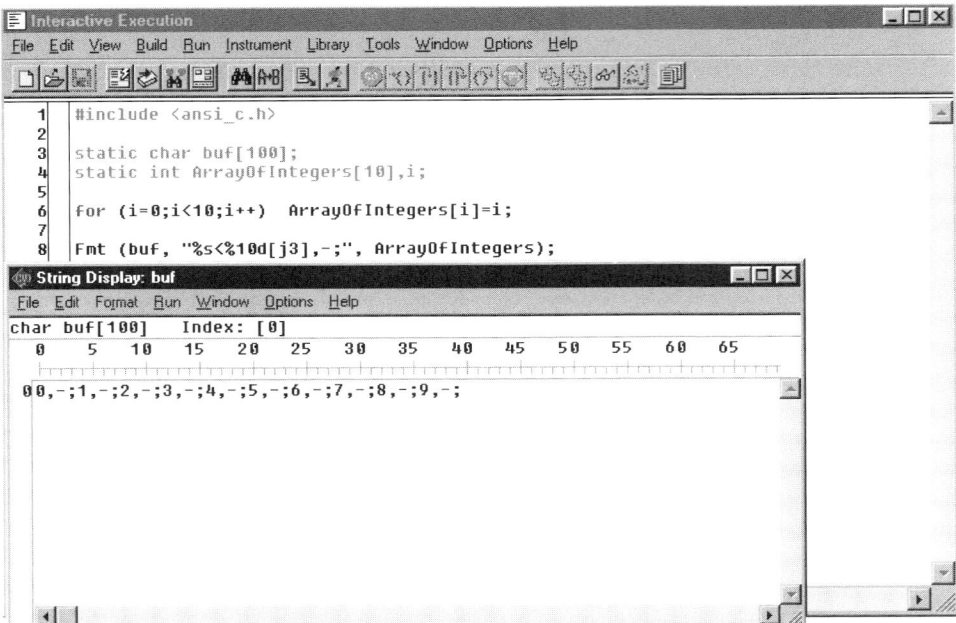

Figure D–25
Specified Character Insertion Demo Using Interactive Window

The Interactive Execution for this code fragment is shown in Figure D–25.

Real to String in Floating-Point Notation

1. **Objective:** Convert a floating-point number into a string (see Figure D–26).

```
char target[40];      // This data type applies to all the examples below
double y;             // This data type applies to all the examples below

y= 43.99876;
Fmt(target, "%s<%f", y);
```

Figure D–26
Floating-Point Number to String

Result: target="43.99876"

Appendix D • Formatting Functions

Conversion: The value of y is placed in the target string. %f is used for floating point values.

2. **Objective:** Convert a floating-point number into a string specifying the *precision modifier* (see Figure D–27).

```
y= 43.99876;
Fmt(target, "%s<%f[p3]", y);
```

Figure D–27
Floating-Point Number to String Using the Precision Modifier

Result: target="43.998"

Conversion: The value of y is placed in the target string with only three significant digits since p used with the %f represents precision. p followed by the value specifies the number of digits after the decimal point.

3. **Objective:** Convert a floating-point number into a string specifying the precision *modifier* with more precision digits than in the source (see Figure D–28).

```
y= 43.99876;
Fmt(target, "%s<%f[p10]", y);
```

Figure D–28
Floating-Point Number to String with More Precision Digits than in Source

Result: target="43.9987600000"

Conversion: The value of y is placed in the target string with ten significant digits. If there are not enough significant digits, as specified in p, zeros are added to the right side of the number to fill in the precision field.

4. **Objective:** Convert a floating point number into a string using the width *modifier* (see Figure D–29).

```
y= 43.99876;
Fmt(target, "%s<%f[w10]", y);
```

Figure D–29
Floating-Point Number to String Using Width Modifier

Result: target=" 43.99876"

Conversion: The value of y is placed in the target string with the same number of digits but is right justified and the field is filled with blanks on the left side of the string.

5. **Objective:** Convert a floating-point number into a string, specifying the width *modifier* less than the floating-point width (see Figure D–30).

```
y= 43.99876;
Fmt(target, "%s<%f[w4]", y);
```

Figure D–30
Floating-Point Number to String Specifying Width *Modifier* Less than Floating-Point Width

Result: target="43.9*"

Conversion: The value of y is placed in the target string with the width specified and the last character is * to indicate overflow.

6. **Objective:** Convert a very small floating point number into a string (see Figure D–31).

```
y= .00000000000013;
Fmt(target, "%s<%f", y);
```

Figure D–31
Small Floating-Point Number to String

Result: target="1.3e-13*"

Conversion: The value of y is placed in the target string in the scientific notation, which is forced by the *Fmt* command. The same is true for very large numbers.

Real to String in Scientific Notation

1. **Objective:** Convert a floating point number into a string using the scientific notation (see Figure D–32).

```
char target[40];      // This data type applies to all the examples below
double y;             // This data type applies to all the examples below

y= 32.459;
Fmt(target, "%s<%f[e]", y);
```

Figure D–32
Floating-Point Number into String With Scientific Notation

Result: target="3.2459e+001"

Conversion: The value of y is placed in the target string in the scientific notation. e is used for representing the scientific notation.

2. **Objective:** Convert a floating point number into a string using the scientific notation and specifying the precision modifier (see Figure D–33).

```
y= 32.459;
Fmt(target, "%s<%f[ep2]", y);
```

Figure D–33
Floating-Point Number to String Using Scientific Notation and Precision Modifier

Result: target="3.24e+001"

Conversion: The value of y is placed in the target string in the scientific notation. p2 represents two digits of precision.

3. **Objective:** Convert a floating point number into a string using the scientific notation and specifying the precision modifier and the number of digits in the exponent (see Figure D–34).

```
y= 32.459;
Fmt(target, "%s<%f[e2p4]", y);
```

Figure D–34
Floating-Point Number into String Using Scientific Notation Specifying Precision Modifier and Number of Digits in Exponent

Result: target="3.2459e+01"

Conversion: The value of y is placed in the target string in the scientific notation with four digits of precision and two digits in the exponent. The value after the e represents the number of digits in the exponent. The default of the exponent is 3 if the exponent modifier is omitted.

4. **Objective:** Convert a floating point number into a string using the scientific notation and specifying the precision and width modifiers and the number of digits in the exponent (see Figure D–35).

```
y= 32.459;
Fmt(target, "%s<%f[e2p4w16]", y);
```

Figure D–35
Floating-Point Number to String Using Scientific Notation and Specifying Precision and Width Modifiers plus Number of Digits in Exponent

Result: target=" 3.2459e+01"

Conversion: The value of y is placed in the target string in the scientific notation with four digits of precision and two digits in the exponent and justified to the right to allow for a field width of 16.

5. **Objective:** Convert a floating point number into a string using the scientific notation and specifying the width modifier that is too small (see Figure D–36).

```
y= 32.459;
Fmt(target, "%s<%f[ep2w6]", y);
```

Figure D–36
Floating-Point Number to String Using Scientific Notation and Specifying Too Small Width Modifier

Result: target="3.24e*"

Conversion: The value of y is placed in the target string in the scientific notation using a width of 6, which is too small to fit all the digits of the mantissa and exponent. The last character to the right is *, which indicates overflow because there is not enough width to display the exponents.

Integer and Real to String with Literals

Objective: Convert a combination of floating-point and integers with literal characters to a string (see Figure D–37).

```
char target[20];
int Machine=1;
Mem_size=4000;
Label= "COUNT";   // This assignment is for demo only, use
                      CopyString to assign "COUNT" to Label

Data_Val=93.456;

Fmt(target, '%s<:MACHINE%d:MEM%i:STORE '%s'; DATA
                  %f", Machine, Mem_size, Label, Data_Val);
```

Figure D–37
Floating-Point and Integers with Literal Characters to a String

Result:"MACHINE1;MEM4000;STORE 'COUNT'; DATA 93.456"

Appendix D • Formatting Functions

Conversion: All the *literals* and the values of the *format specifiers* are replaced by the values of the variables declared in this example. This method is very useful in communicating with instruments that use string commands like the GPIB instruments.

Concatenating Two Strings

Objective: Concatenate two strings using the *Fmt* function (see Figure D–38).

```
char FileName[260], *directory, *file;
directory="c:\\cvi5\\";
file="myfile";

Fmt (FileName, "%s<%s%s", directory, file);
```

Figure D–38
Concatenate Two Strings

Result: FileName="c:\cvi5\myfile"
Conversion: The first %s on the right of < takes the string value from directory, and the second %s the string value from file.

Appending to a String

Objective: Append characters to a string using the *Fmt* command (see Figure D–39).

```
char FileName[260], *directory, *file;
directory="c:\\cvi5\\";
file="myfile";

Fmt(FileName, "%s<%s", directory);   // FileName is " c:\cvi5\" at this line
Fmt(FileName, "%s[a]<%s", file);
```

Figure D–39
Appending Characters to a String

Result: FileName="c:\cvi5\myfile"
Conversion: The first *Fmt* command places the folder (directory) into the `FileName`, and the second *Fmt* appends the `myfile` string to `FileName` (which contains the folder (directory) from the first *Fmt* function).

Converting Decimal Integer to Hexadecimal String

Objective: Convert a decimal integer to a hexadecimal integer (see Figure D–40).

```
int source;
char  target[4];

source=255;  // Decimal integer
Fmt(target, "%s<%x", source);
```

Figure D–40
Decimal Integer to Hexadecimal Integer

Result: target = "ff"

Conversion: The *source* is converted to a hexadecimal string by assigning *source* integer the format string %x.

Scanning Functions

Like the **Formatting** functions, the **Scanning** functions play an important role in string and data manipulation. The **Scanning** functions work in the opposite direction as the **Formatting** functions.

The **Scanning** functions consist of *Scan*, *ScanFile*, and *ScanIn*. Scanning functions are used to break a single source item into various target items or to change the type of the data type from *source* to *target*.

These functions can be accessed from **Library>>Formatting and I/O>> Data Formatting>> Scanning.** You can choose *Scan*, *ScanFile*, and *ScanIn* from this menu. All the *Scanning* functions are similar in their actions; the only difference between these functions is where the data is read from and written to, just like the *Fmt*, *FmtFile*, and *FmtOut* functions.

The function panel for the *Scan* function is shown in Figure D–41. Its format follows.

```
n=Scan (source, formatstring, targetptr1,
                         targetptr2,....,targetptrn);
```

Scanning takes the source and converts it to target pointer(s) according to the *formatstring*.

Appendix D • Formatting Functions

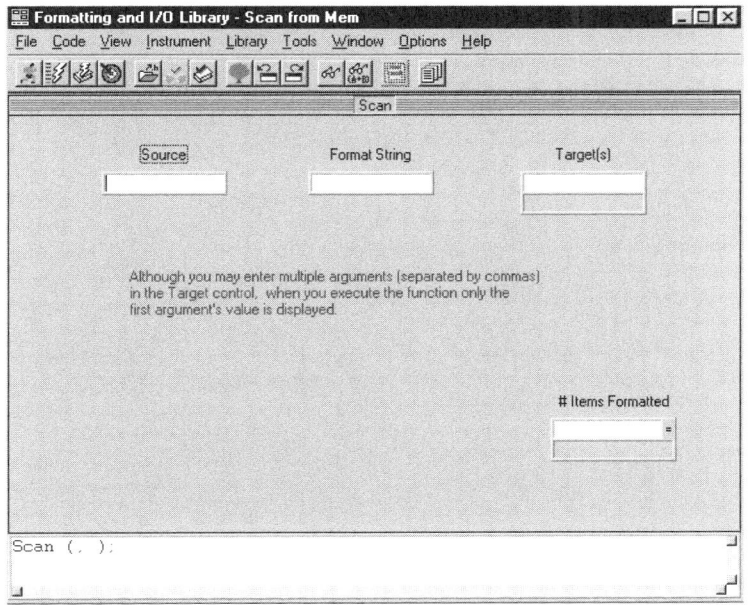

Figure D–41
Scan Function Panel

The function panel for the *ScanFile* function is shown in Figure D–42. Its format follows.

```
n=ScanFile (fileHandle, formatstring, targetptr1,
                    targetptr2,…..,targetptrn);
```

This function reads the data from the file specified in *fileHandle* and converts the *source* into the *target* pointer(s) per *formatstring*.

The last of the *Scanning* functions is *ScanIn*. The function panel for the *ScanIn* function is shown in Figure D–43. Its format follows.

```
n=ScanIn(formatstring, targetptr1,
                    targetptr2,…..,targetptrn);
```

This function reads the data from the **Standard Input** (the keyboard) and transforms the *source* data into the *target* pointer(s) per the *format string*.

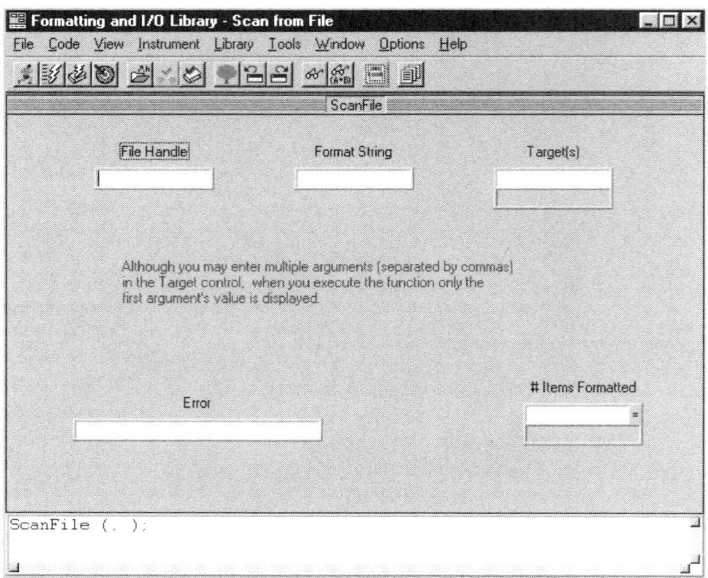

Figure D–42
ScanFile Function Panel

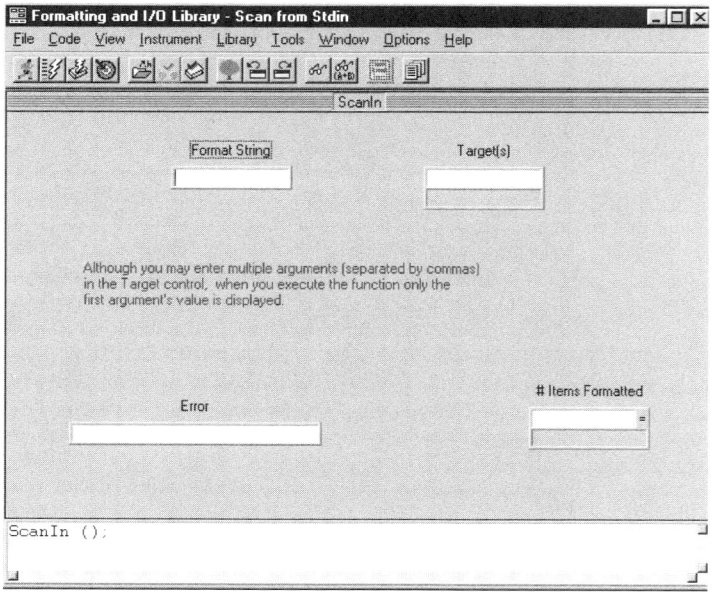

Figure D–43
ScanIn Function Panel

Scan Functions — Format String

The *format strings* and how they can be modified to give us more flexibility in the data separation will be demonstrated in this section. All *format strings* have a structure similar to

```
source_spec > target_spec_and_literals
```

where *source_spec* is a *format specifier* that defines the nature of the source parameter and *target_spec_and_literals* consists of the *format specifiers* and *literal* characters, which indicate how to divide the *source* data into the desired *target* of the *format strings*, shown as an example below.

```
%s > %f%d
```

Like in the *Fmt* function, the > serves as a reminder to show the direction of data transformation. Note that the direction of > is opposite to that of the *Fmt* function. The source is on the left side of > and the target on the right side. Again, remember that the direction of conversion is indicated by >, which looks like the tip of the arrow.

These *Fmt* and *Scan* functions are performing opposite functions but have many similarities, which you will notice as you read on. The *Fmt* function combines the source into one target, while *Scan* breaks it up into multiple targets. You will also notice that the source *format specifiers* are located to the left of > next to the source parameter. The target *format specifiers* are to the right of >, which is the side of the target parameters. Each *format specifier* has the following form.

```
%  [rep]   formatcode  [[modifiers]]
```

The % represents the start of all *format specifiers,* and `rep` indicates the number of times the format repeats with respect to the arguments. `formatcode` is the code character that indicates the data type, and *modifiers* are optional additional codes that add more attributes to the data format. *Modifiers* are always included in brackets.

Note: `rep` is not allowed when `formatcode` is a `s` or `l` (letter l) (string).

String to String

The *Scan* function in most cases does the opposite of the *Fmt* function. Whereas the *Fmt* function joins many sources into one target, the *Scan* function breaks the source into different pieces. Various capabilities of the *Scan* function can be demonstrated by examples.

1. **Objective:** Convert source string to target string (see Figure D–44).

```
char *source;     //This data type is used in all the examples below
char target[50];        // This data type is used in all the examples below

source= "MyName";

Scan(source, "%s>%s", target);
```

Figure D–44
Source String to Target String

Result: `"MyName"`

Conversion: This is a straightforward string to string conversion with no modifiers.

2. **Objective:** Leave blanks and tabs in the string (see Figure D–45).

```
source="         This        ";
Scan(source, "%s>%s[y]", target);
```

Figure D–45
Blanks and Tabs in String

Result: "   This"

Conversion: *Scan* skips leading spaces and tabs unless `y` is specified as the *modifier*.

3. **Objective:** Convert all characters up to the linefeed in string (see Figure D–46).

```
source="   This example is using:\n as a line feed ";
Scan(source, "%l[c]>%s", target);
```

Figure D–46
Converting All Characters Up to Line Feed in String

Result: "This example is using:"

Appendix D • Formatting Functions

Conversion: The %l[c] is used in the source to convert the characters in the source up to the linefeed character. \n is the new line character, so all characters up to it are copied.

4. **Objective:** Show an example of string conversion with white space character (see Figure D–47).

```
source=" This example ";
Scan(source, "%s>%s", target);
```

Figure D–47
String Conversion with White Space Character

Result: "This"

Conversion: *Scan* considers the white space character to be the terminator if the t modifier is not used. An example of using the t modifier follows.

5. **Objective:** Convert the string up to the terminator character (see Figure D–48).

```
source=" This string will stop, at the first comma ";
Scan(source, "%s>%s[t44]", target);
```

Figure D–48
Convert String Up to Terminator Character

Result: "This string will stop"

Conversion: *Scan* function uses the t modifier within the brackets to specify the ASCII number of the character at which the conversion stops (comma in this example). Suppose you wanted to stop at the first a in the word at, then you will specify [t97]. The ASCII value of a is 97. Then the target string will be "This string will stop,". Any C programming book would list the ASCII tables. More than one t modifier can be used, but the conversion to the target stops as soon as the first t modifier value is encountered.

6. **Objective:** Convert a string up to the number (see Figure D–49).

```
int target2;

source="abc789 ";
Scan(source, "%s>%s[t#]%d", target, target2);
```

Figure D–49
Convert String Up to a Number

Result: target="abc"   target2="789"

Conversion: The t# modifier is similar to the t modifier except the string terminates at the numbers 0 through 9 and +, and -. In this example, you have two targets: the first one stops at abc and the second at 789. This is a convenient way to break a string into its components of alpha and numeric characters.

7. **Objective:** Use string width modifiers (see Figure D–50).

```
source=" 1234567890123456789 ";
Scan(source, "%s>%s[w6]", target);
```

Figure D–50
Converting String with Modifiers

Result: "123456"

Conversion: Using *Scan*, you can extract a sub-string of a certain width by specifying the width of the string to be copied into the target by using the w modifier. The ASCII NUL is not counted in the string width. Just specify the width you want extracted into the target string.

If the width is zero, then all the bytes in the string are converted to the target. If you specify a width more than the source string, the remaining bytes are filled with blanks. If you use the q modifier, then the remaining bytes are filled with ASCII NULs.

8. **Objective:** Discard terminators from the source string (see Figure D–51).

```
char target2[100], target3[100];

source="abc, def; ghi; ";
Scan(source, "%s>%s[xt44]%s[xt59]%s[xt59]", target,
                                             target2, target3);
```

Figure D–51
Discarding Terminators from Source String

Result: target="abc" target2="def" target3="ghi"

Conversion: You can use the x modifier to discard the terminator, which in the first case is [t44], representing an ASCII value for ,. Then, discard the next two semi-colon terminators using [t59]. Using this *Scan* function, a complex string is broken into simple pieces.

Complementary to the [x] modifier, d modifier is applied to the *target specifier* and indicates that no target argument corresponds to the *target specifier*.

9. **Objective:** Convert selected bytes from the source (see Figure D–52).

```
source="abcdefghi";
Scan(source, "%s[i4]>%s", target);
```

Figure D–52
Converting Selected Bytes from Source

Result: target="efghi"

Conversion: The i modifier is placed next to the source string with a zero-based offset number to indicate from which byte of the source string the data will be transferred to the target. In this example byte 4 is e, so efghi is moved to the target string.

10. **Objective:** Append the source to the target string (see Figure D–53).

```
source="  abcdefghi";
Fmt(target, "%s<Example of appending strings:   ");
Scan(source, "%s[i4]>%s[a]", target);
```

Figure D–53
Appending Source to Target String

Result: "Example of appending strings: abcdefghi"

Conversion: To append a source string to the target, you use the a modifier in the target string. Remember that in the *Fmt* statement, %s< can be omitted without loss of meaning though it is recommended to at least use < for clarity.

Converting String to Integer and String

The examples in this section will show you how to break a string into an integer and a string.

1. **Objective:** Break an alphanumeric string to integer and alphabet (see Figure D–54).
   The same data type is to be used for the following examples.

```
char *source;
char target[40];
int numeric, number_converted;

source="9876Number";
number_converted=Scan(source, "%s>%i%s",&numeric,target);
```

Figure D–54
Breaking an Alpha-Numeric String

Result: numeric=9876   target="Number"   number_converted=2

Conversion: The source string is broken into an integer and a target string. Two items are successfully written to the target. If you put a 0 before the 9 in the source string, the integer will ignore the 0.

2. **Objective:** Show an example of erroneous integer converting (see Figure D–55).

```
source="Number9876";
number_converted=Scan(source, "%s>%s%i",target,&numeric);
```

Figure D–55
Erroneous Integer Converting

Result: target="Number9876"   numeric=6945060 (or some other erroneous number)
number_converted=1

Conversion: Trying to breakup the string this way will not work because the target sees the complete source string as one string and copies it into `target` leaving nothing for the numeric integer. Therefore, the number converted is only 1.

Appendix D • Formatting Functions

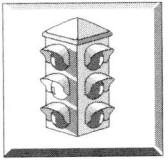

Caution: Never leave any spaces between the format specifiers in any of the Scan functions. Doing so may give erroneous results!

Converting Comma-Separated ASCII Numbers to Real Array

There may be times when you need to convert comma-separated ASCII numbers to an array of real numbers. You can use programming to achieve this, or you can use the *Scan* function to accomplish this very easily. Let us look at the example shown in Figure D–56.

```
char *source;
int return_value;
double array[7];

source="40.9, 56.9, 45.89, 1001.45, 99.99, 70.34, 40.01";
return_value=Scan(source, "%s>%7f[x]", array);
```

Figure D–56
Converting Comma-Separated ASCII Numbers to Real Numbers Array

Result: The array will contain seven elements.

array[0]=40.9 array[1]=56.9 array[2]=45.89
array[3]= 1001.45 array[4]= 99.99
array[5]= 70.34 array[6]= 40.01

The seven values of the *source* are converted to an array of real numbers, which you can access in your program by specifying the offset index into the array and obtain its value.

Scanning Non-NUL Terminated Strings

When you communicate with GPIB and RS-232 instruments, the data is not null-terminated. Understand that `gpib.h` (GPIB header file) contains the global variable *ibcnt*, which contains the actual number of bytes transferred during a GPIB instrument read. You can make use of this variable in the *Scan* function as shown in Figure D–57.

```
char source[20];
double result;
Scan(source, "%s[w*]>%f", ibcnt, &result);
```

Figure D–57
Null-Terminating Communication Data

This specifies the width of the string to be *ibcnt* and puts the NUL terminator in the last byte.

On a similar concept, you can convert a data file to an array using *CVI's* *FileToArray* function, which will not be discussed here but is used in project1-1. Look in Chapter 2 of *LabWindows/CVI Standard Libraries Reference Manual* for a description of this function.

Converting Integer Arrays to Real Arrays

If you want to convert an array of integers to an array of real numbers, the *Scan* function can do it for you very easily. The example given in Figure D–58 will show you how this is done.

```
int i_array[50];        // Array of 50 integer elements
double d_array[50];     //Array of 50 double elements

Scan(i_array, "%50i>%50f", d_array);
```

Figure D–58
Converting Array of Integers to Array of Floating-Point Numbers

Result: d_array contains the floating-point values of the integer i_array.

In the *Scan* function, 50 refers to the number of elements that you would like to convert.

Byte Swapping, with Integer Array to Real Array

Some instruments send back two-byte integers with the high byte followed by the low byte. At times, you would like to swap the bytes to be in the reverse order of that sent and, at the same time, write all the integer values to real values. The *Scan* function does it very conveniently as shown in Figure D–59.

```
int i_array[50];              // Array of 50 integer elements
double d_array[50];      //Array of 50 double elements

Scan(i_array, "%50i[o10]>%50f", d_array);
```

Figure D–59
Swapping Byte-Order and Integer to Real Values

Result: `d_array` contains the floating-point values of the integer `i_array` with byte-order swapped.

An important item to note, in the source, the o modifier changes the byte order from 01 to 10 meaning the high byte and low byte are switched around.

Converting Hexadecimal String to Decimal Integer

1. **Objective:** Convert a hexadecimal string to a decimal integer using the `%i` format string (see Figure D–60).

```
char *source;
int target;

source="3BE";   //Hexadecimal string
Scan(source, "%s>%i[r16]", &target);
```

Figure D–60
Hexadecimal String to Decimal Integer Using `%i` Format String

Result: target = 958;

Conversion: The source string is converted to decimal integer by giving target integer modifier a radix of 16.

2. **Objective:** Convert a hexadecimal string to a decimal integer using the `%x` format string (see Figure D–61).

```
char *source;
int target;

source="FF";   //Hexadecimal string
Scan(source, "%s>%x[r16]", &target);
```

Figure D–61
Hexadecimal String to Decimal Integer Using `%x` Format String

Result: target = 255;

Conversion: This accomplishes the same results as the previous example.

ScanFile Function

ScanFile reads the data from a file and converts the data as specified by the *formatcode* and the *modifiers*. All *formatcodes* and *modifiers* that apply to *Scan* functions can be used with the *ScanFile* functions. Let us look at an example of how *ScanFile* works.

The example shown in Figure D–62 will show you how to read the comma-separated numbers from a file, and if within array range, will convert all the elements to a real array value and discard the comma separators.

In the last line, the * is replaced by `NumberOfElements` to create an array of the correct number of elements to be read. The x modifier skips over the commas in the file.

```
#define ArraySize 100
int DataFileHandle, NumberOfElements;
double d_array[ArraySize];
//Open Test1.out file as read_only in ASCII format
DataFileHandle = OpenFile ("Test1.out", VAL_READ_ONLY,
                                                    VAL_OPEN_AS_IS,
VAL_ASCII);

//It is a good idea to always check the file for any opening errors.
if(DataFileHandle <0)
{
     //display message on Standard Output or console window
     FmtOut("Error on opening file: Test1.out\n");
}
//Read the number of elements into "NumberOfElements"
ScanFile(DataFileHandle, "%s>%i",&NumberOfElements);
if (NumberOfElements>ArraySize)
{
     //Display message and exit the program
     FmtOut("ArraySize exceeds the array dimensions\n");
     exit (-1);
}
 ScanFile(DataFileHandle, "%s>%*f[x]", NumberOfElements, d_array);
```

Figure D–62
Converting Comma-Separated from a File

Appendix D • Formatting Functions

```
char OperatorName [40];

FmtOut("Enter your name:   \n");
ScanIn("%1>%s[w40q]", OperatorName);
```

Figure D–63
Example of Using *FmtOut* and *Scan In*

ScanIn Function

The *ScanIn* function reads the data from the keyboard (the **Standard Input**) and converts it as specified in the *formatcode* and the *modifiers*. Let us look at an example that uses the *ScanIn* function.

Suppose you want to use a prompt for the operator's name and you want to save the name in a variable. You can write something like the code shown in Figure D–63.

In this example, the `%1`, as seen in the previous examples, is to accept characters up to the linefeed characters. As soon as the operator enters the <Return> after entering the name, or otherwise, the operator name is written to the variable `OperatorName`.

This example is just for demonstration purposes. It should not be used in such a way as to give the operator an opportunity of entering only a <Return> when a value is required. Some kind of error checking for the entered data should determine if an appropriate ASCII string has been entered. The error checking should determine that only ASCII characters are entered. No control characters or digits should be accepted. You may also want to limit the length of the `Operator Name` string so there is no chance of overflow.

Status Functions

There are two *Status* functions that are associated with the return information of the formatting and scanning functions. Let us discuss each one of these individually. The function *GetFmtErrNdx* returns the zero-based index of the formatting string in which the error occurred in the last *Fmt* or *Scan* call. The function panel is shown in Figure D–64.

This function has no argument and the return value is the index in which the error occurred in the *formatting string*. If Status is –1, then there is no

Appendix D • Formatting Functions

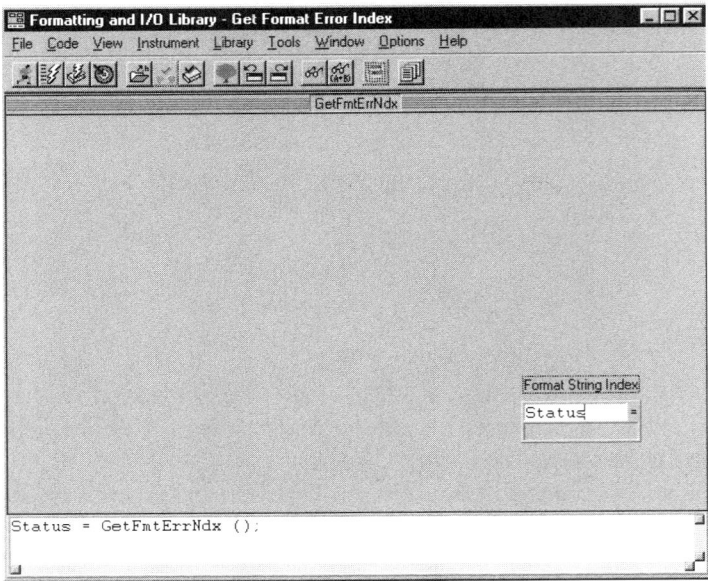

Figure D–64
GetFmtErrNdx() Function

error, else it returns the location of the error index. This function is useful for debugging the *Fmt* and *Scan* functions to find errors in the *formatcode*. An example, shown in Figure D–65, would clarify how this function is useful.

I have deliberately inserted an error in %i[r15] in the *target literal* by entering r15 instead of r16. When this code is executed, the pop-up panel shown in Figure D–66 is displayed.

If you select **Continue** from this pop-up panel, you can read the *Status* value as **3**, indicating the zero-based number 3 has an error. This is the fourth element in %i[r15], meaning the number 5 is erroneous.

```
char *source;
int target, Status;
source="3BE";   //Hexadecimal string
Scan(source, "%s>%i[r15]", &target);
Status=GetFmtErrNdx();
```

Figure D–65
Use of *GetFmtErrNdx* Function

Appendix D • Formatting Functions

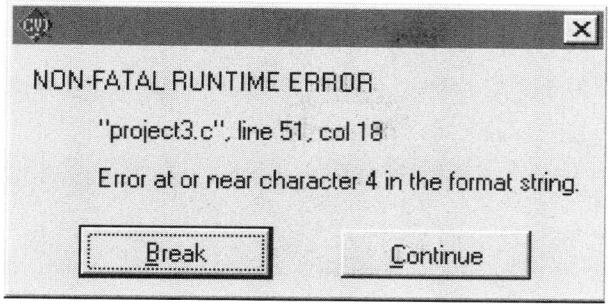

Figure D–66
Error Index

The other *Status* function is *NumFmtdBytes*. This library function was explained earlier and is repeated here. The function panel is shown in Figure D–67.

You have seen *NumFmtdBytes* used in previous chapters. It has no arguments and returns the number of bytes from the previous *Fmt* or *Scan* function. You do not have to physically count the bytes in a string in a *Fmt* or *Scan* function; just enter the *NumFmtdBytes* function where you would enter the bytes count in a function or output. This part has been explained previously.

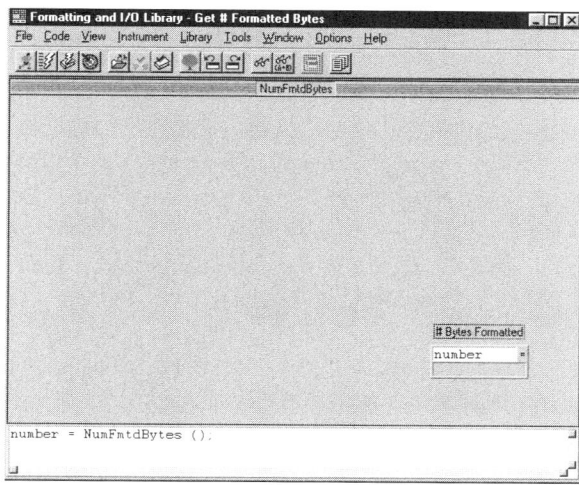

Figure D–67
NumFmtdBytes() Function

```
char target[100], Label[10];
float Data_Val;
int Machine=1, NumberofBytes, Mem_size;
Mem_size=4000;
Data_Val=93.456;
CopyString (Label, 0, "COUNT", 0, 6);

Fmt(target, "%s<:MACHINE%d:MEM%i:STORE '%s'; DATA %f", Machine,
                            Mem_size, Label, Data_Val);
NumberofBytes=NumFmtdBytes();
```

Figure D–68
Example of *NumFmtdBytes()*

This function is of great use when communicating with instruments via the RS-232 or the GPIB interfaces, which require a count of the number of bytes sent to or received from the instrument string. We will use the first example in the "Integer and Real to String with Literals" section as shown in Figure D–68.

The `NumberofBytes` will be 47. You can see that it can save you a lot of time because you don't have to manually count all the bytes in the string and you can avoid the potential mistakes of manual counting.

Using the Format Wizard

CVI comes with a utility for creating the *Fmt* and *Scan* function calls interactively, copying them on a clipboard, and pasting them into the code. This utility creates the formatting and scanning functions for you using the correct syntax based upon your inputs.

To access this utility from Windows, select **START>>National Instruments CVI>>Utility Programs>>Format Wizard** as shown in Figure D–69 below.

The **Format Wizard** main GUI is displayed as shown in Figure D–70. From this GUI, you have a choice of creating a *Fmt* or a *Scan* function call by selecting the **Function** switch. At the bottom of the GUI is the dialog box in which the structure of the function is created interactively as you make your selection. You can add the **Destination Name,** the **Sources** variables, the **Type,** and the **Format Modifiers** to complete the function call.

Appendix D • Formatting Functions

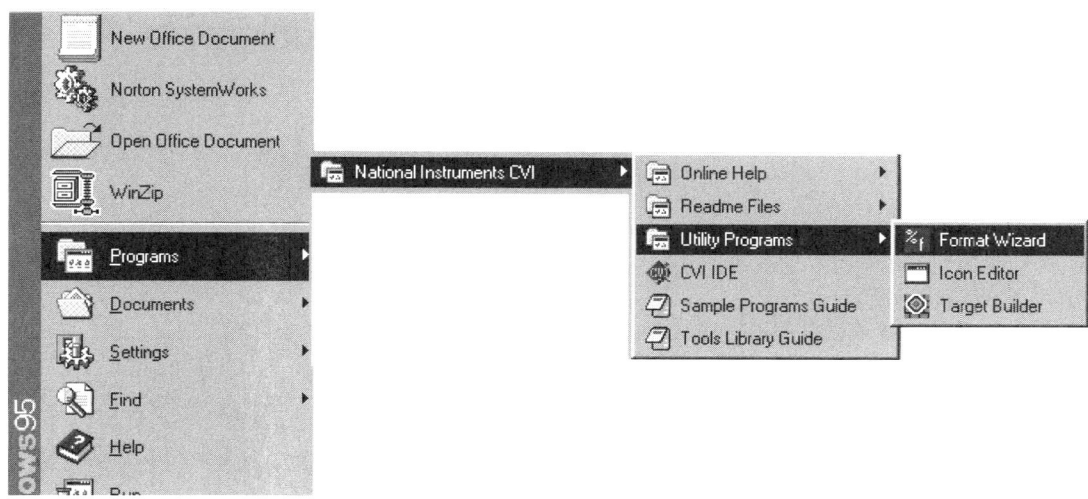

Figure D–69
Format Wizard Selection

The data types available from the pull-down **Type** menu are as shown in Figure D–71. When you select the data type, the appropriate *formatcode specifier* is inserted in the function call.

To understand the **Format Wizard,** let us look at an example. Select the **Function** switch to **Formatting** to create a *Fmt* function. Type in buffer for the **Destination Name** and leave the **Type** as **String.** Notice as the data is

Figure D–70
Format Wizard Main GUI

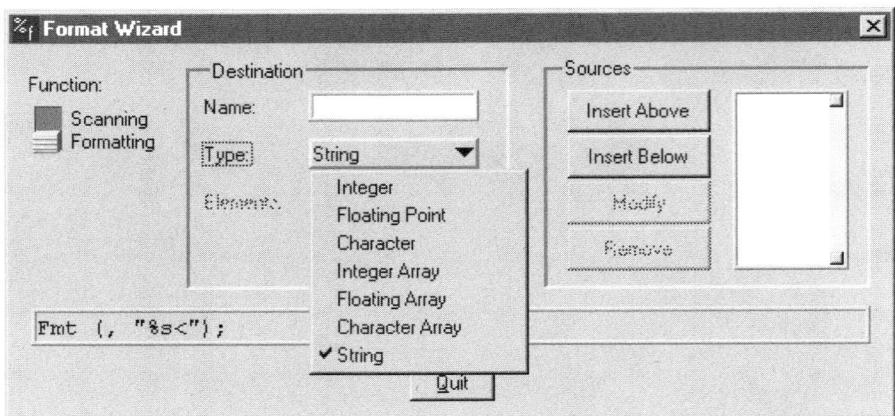

Figure D–71
Format Wizard Type Pull-Down Menu

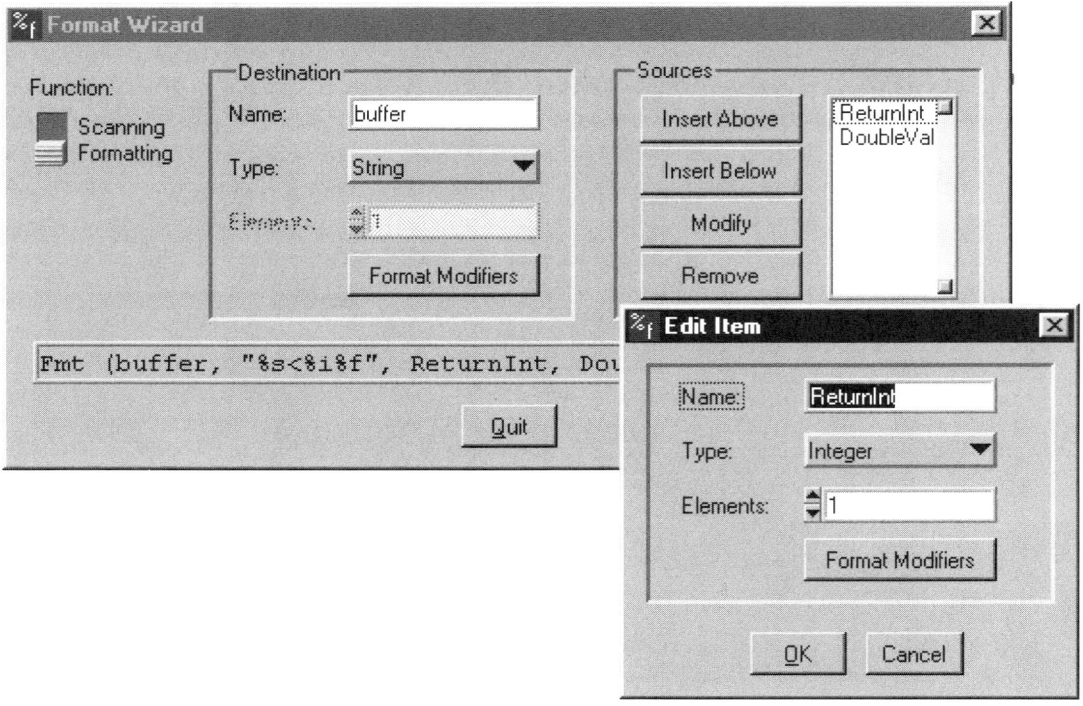

Figure D–72
Format Wizard Edit Item Dialog Box

Appendix D • Formatting Functions

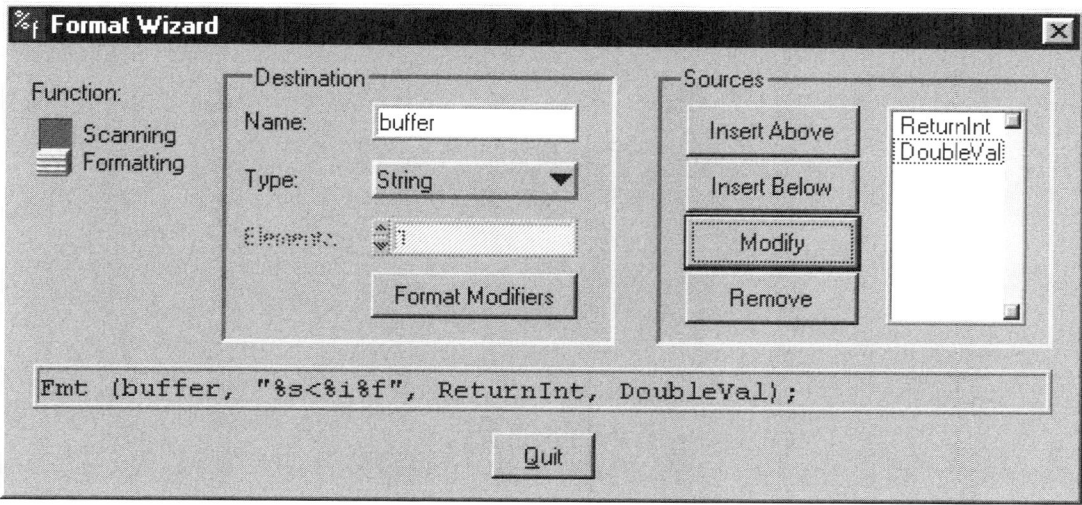

Figure D–73
Format Wizard Completed Function Call

entered, the *Fmt* function arguments and the *formatcode specifiers* are entered in the dialog box at the bottom. From the **Sources** group, select **Insert Above**. A pop-up dialog box as shown in Figure D–72 is displayed. Enter ReturnInt in the **Name** dialog box and select integer for **Type**. The **Elements** can be left as 1 unless an array is being entered, and you can enter the dimension of the array. Select **OK** to close this box.

Similarly, select **Insert Below** to insert a second source to this function and call it DoubleVal of **Type** float. Notice the function is now complete with the specified arguments and the correct *format specifiers* as displayed in Figure D–73. You can now copy and paste this function call in your code.

The **Format Modifiers** command button can be selected to add the *formatcode modifiers* if desired for both the **Destination** and **Sources**.

CVI DEMO PROGRAMS

Chapter Highlights
- Bmath Demo Version 1.0
- System Test Demo

Appendix E consists of two demo projects written in *CVI* and included here to show you the power and versatility of *CVI*.

The first demo project is a mathematics analysis tool, called **Bmath,** that analyzes functions parametrically. The second demo is a **System Test** project, more sophisticated than the Test Executive project you created in Chapter 7.

Bmath Demo Version 1.0

Introduction

Note: This is a demo version only. If you are interested in the full versions, please contact Yaakov Ben-Ami at ybenami@aol.com.

Welcome to the demo version of **Bmath,** a mathematics analysis tool designed to help anyone who needs to solve certain mathematical problems but does not want to spend a lot of time doing it.

What makes **Bmath** unique is its ease of use and its friendly and intuitive interfaces. It also has an online context-sensitive help system that will help you remember the functions of the different controls displayed on the user interface.

You can start the program installation by going to the bMath folder under **Projects** and clicking on Setup.exe. Follow on screen directions to create Bmath.exe. Once installed, click the Bmath.exe icon and the **Bmath—Main Menu** will be displayed as shown in Figure E–1. You will see a group of buttons; each will activate a mathematical utility.

The only utility available in this demo version is the functions analysis utility, which you can access by clicking on the **1D Function** as shown in Figure E–1. This utility allows you to analyze a function of up to ten variables, paramet-

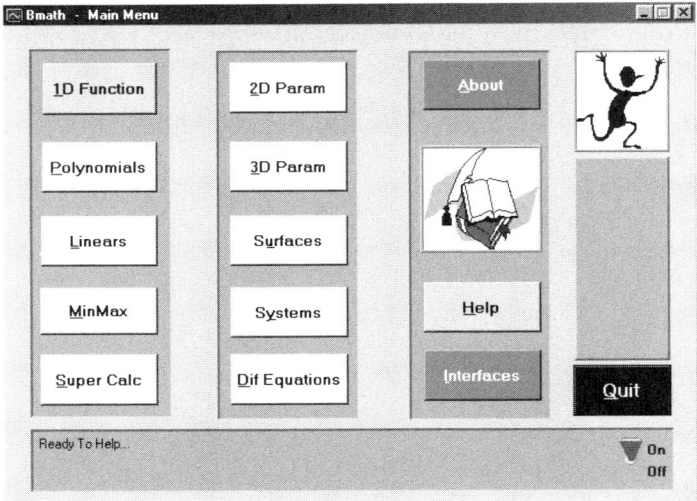

Figure E–1
Bmath — Main Menu

Appendix E • CVI Demo Programs

rically. The following sections will explain in detail the kinds of analyses you can perform using **Bmath** and how to do them. If you want to know more about the program's other features, please click the **About** button for contact information.

Loading/Editing Instructions

To start the functions utility, press the **1D Function** button. Doing that will display the **Functions** window as shown in Figure E–2.

To load an existing function from the menu bar, select **Options>>Load Function.** The selected function will display in the **Edit Function** dialog box. Load the function into memory by pressing the **Get** button. To enter a new function, identify the **Edit Function** window. In the **Edit Function** window, type a function of arguments a,b,c,...,j, and load the function into memory by pressing the **Get** button.

You can save any entered function to the file `generic.fun`. Later you can rename the file to anything that makes sense to you. However, make sure that the file has the extension `.fun`. To save a newly entered function into `generic.fun`, load it into memory first by pressing the **Get** button, and then from the menu bar, select **Options>>Save Function**.

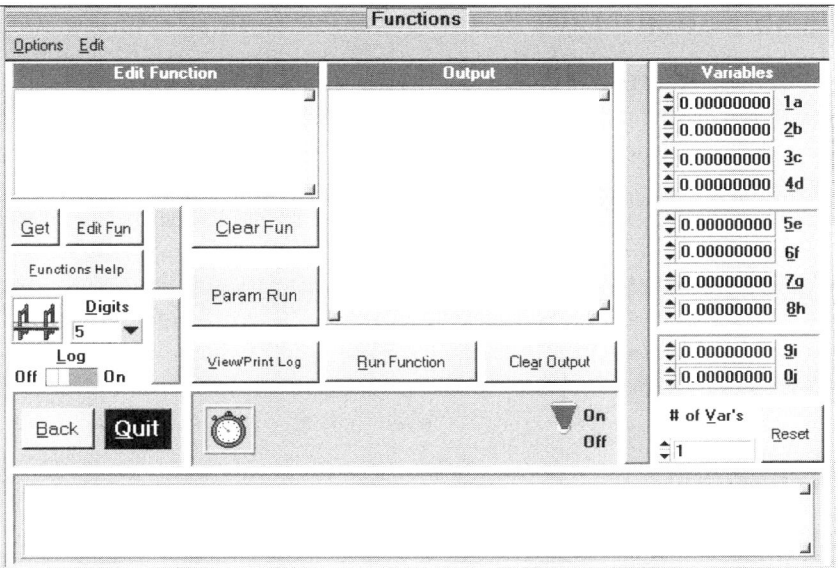

Figure E–2
Functions Window

To set the number of variables, use the **# of Var's** control. If, for example, you set the number of variables to two, the **a** and **b** numeric boxes, under **Variables,** will become enabled, and you can then enter the desired values. It is your responsibility to select the correct number of variables for each function that you load into memory. For example, if you choose a function of two variables, use **a** and **b** only as arguments, such as, `cos(a-b)`, `log(a) + b-exp(b)`, etc. Any excess argument will be assigned a value of 0.

After the function is loaded into memory and the number of variables is set, you can evaluate the function by pressing the **Run Function** button. Further analysis options are available through the **Parametric Run** utility. See details in the next section.

To enter a function expression correctly, you must observe algebraic precedence rules! The function's argument(s) must be enclosed within parentheses. Press the **Functions Help** button in the **Functions** window to see the available built-in functions. Look up any algebra or calculus text for argument ranges. Any violation will terminate a current task with an error message but will not bump you out of the **Bmath** program.

More help is available through the different utilities in the **Functions** window, the **Parametric Run** window, and the run time context-sensitive help utility.

Using Parametric Run

Clicking the **Param Run** button will display the **Parametric Run** window as displayed in Figure E–3. To use the parametric analysis utility, you need to load a function into memory. See the "Loading Editing Instructions" section.

You can execute a function evaluation at a selected number of points starting at the **Min** value and ending at **Max.** The numeric slide determines the function's parameter. To select a function parameter, move the slide's indicator to the parameter's name. For example, if you select **b**, the function will be treated as a one-dimensional function of the variable **b**. The other variables will be considered constants for this run.

Iterative Evaluation

From the **Parametric Run** window in Figure E–3, move the **Iter.** switch to the **On** position. You can then evaluate the function iteratively using the initial value of the parameter as shown in the **Variables** numeric controls in

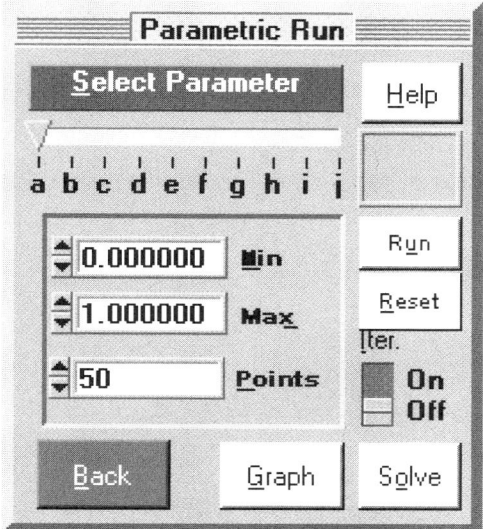

Figure E–3
Parametric Run Window

Figure E–2, and evaluate it a number of times as shown in the **Points** control. For example, if your function is cos(a), you will be evaluating the expression a(n+1) = cos(a(n)), where a(0) is picked up from the **Variables** numeric controls and the number of evaluations will be **Points** + 1.

Pressing the **Graph** button will activate the plotting utility explained later and shown in Figure E–4. Pressing the **Solve** button will activate the solver utility explained later and shown in Figure E–5.

Note: To activate the plotter and/or the solver utilities, you need to load a function into memory first.

Using the Plot Utility

To plot the graph of the loaded function, enter the **Min** and **Max** values in the **Parametric Run** window (Figure E–3) and press **Plot**. A graph of the loaded function will be drawn for the selected range.

Choose a **Line** or **Point** drawing style by clicking the **Style** switch in Figure E–4. You can control the number of points used to draw the graph through the **Points** control in the **Parametric Run** window. Once the graph is drawn, you can click on it to obtain the parameter and corresponding function values in the numeric controls **Fun.** and **Par.** at the bottom of the

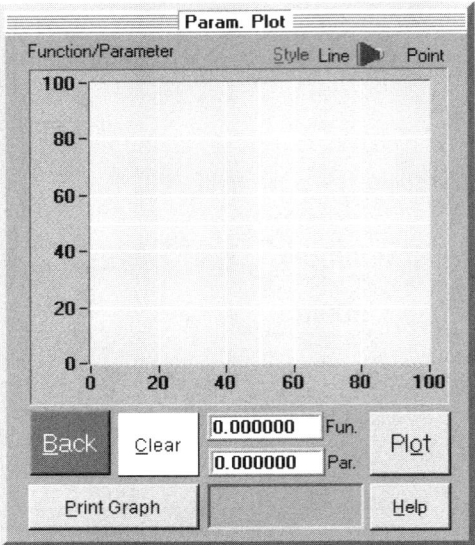

Figure E–4
Param. Plot Window

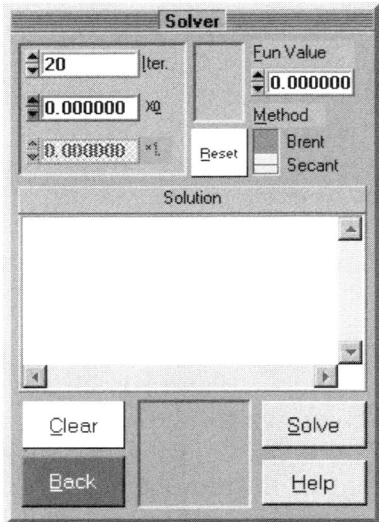

Figure E–5
Solver Window

window. The vertical axis represents the function value and the horizontal axis represents the parameter value. Pressing **Clear** will delete the graph and reset the coordinate axis. You can also print the graph by pressing the **Print Graph** button.

Using the Solver Utility

When you select the **Solve** button in Figure E–3, the **Solver** window is displayed as shown in Figure E–5.

The **Solver** finds a zero of the currently loaded function. It can solve for x in f(x) = 0 or f(x) = c, where c is a possible function value. The default value of c is 0.0, but you can change it through the **Fun Value** control. The desired number of iterations can be set through the **Iter.** control. Pressing the **Reset** button will set the controls to their default values and pressing the **Clear** button will clear the **Solution** box.

You choose either the **Brent** or **Secant** method by clicking the **Method** switch.

- **Brent Method**—You must supply the end points, through the X0 and X1 controls, of an interval that brackets the zero. The Brent method is explained in many advanced numerical analysis books, such as *Numerical Recipes in C* referenced in the bibliography.

- **Secant Method**—In this method, you must supply an initial guess for the zero through the X0 control. You will find an explanation of the Secant method in any numerical analysis book.

Note: You can use the **Plot** utility to get an idea of the approximate location of a zero.

System Test Demo

Introduction

System Test is a program designed to help you perform hardware testing of any system containing devices that are capable of interfacing with a PC. It is assumed that you know how to create the proper software to accomplish this objective.

The coding of the tests is done through the *DLL* TestsDll.dll. The *CVI* project that will enable you to do that is part of the CD-ROM included with this book. Make sure that all the DLL files are copied to the folder where the `Systest.exe` file is created by the installation. (See **System Test Start Up Procedure** below.) Be sure to read the `DllGuide.txt` file for detailed instructions regarding the test's coding and parameter interfaces.

The program was built around a set of basic tests, called molecules, each one of them designed to test a minimal, well-defined aspect of a particular system device. Once the coding of the molecules is done and the *DLL* is operational, **System Test** will allow you to construct test sequences using the molecules as building blocks.

The program will give you considerable flexibility in the design of the test sequences including:

1. conditional and statistical testing,
2. rearrangement of molecules within a given sequence,
3. a powerful variable interface that will allow you to communicate test parameter values to the molecules,
4. the ability to run individual molecules with a set of predefined default values,
5. extensive logging capabilities of test results,
6. an intuitive and user friendly GUI containing a run-time help utility.

System Test retrieves information from a set of text files contained in the `Systest` folder. The `System Test` folder on the CD-ROM is saved under **Projects>>Systest** and its sub-folders. Throughout this description, this will be referred to as the programs folder.

These files control the initial setup of program parameters, and you must be careful not to change their structure while editing them. If you do, **System Test** will not run correctly, but you will receive proper warning, indicating the source of the problem.

The use of the different files is described here.

1. `TestNames.ini`—Located in the programs folder, this is the main external file for test information. It contains the test names and the number of parameters of a certain type that a particular test (molecule) can receive as inputs or outputs.

    An example of a test section in `TestNames.ini` follows.

    ```
    [Sec1]
    ```
    TEST1 = "Device2 Test"
    Integers = 2

```
Doubles = 0
Longs = 0
Strings = 0
```

TEST1 is the name of the function in **System Test** that contains the calling code for the test named `Device2 Test`. The corresponding function name (the one you will code) in `TestDll.dll` is *ExTest1*. Read `DllGuide.txt` to learn how to code *ExTest1*. For this demo, the *ExTest1* function can receive two integers only as input and output. However, you can change the type and the number of variables to suit your testing needs.

Note: Every molecule can have ten integers, five doubles, five longs, and three strings. It's your responsibility to associate a meaningful test name with *TEST1*. It is a good idea to save a copy of this file under a different name.

2. `Vars.ini`—Located in the programs `ini` sub-folder, this file contains information about molecules' parameters. You access this information through **System Test.** See the "System Test Window (Main Window)" section for more details.

   A sample from `Vars.ini` follows.

```
[molec 1]
descrip = "Device2 Test\n"
integer 1 = "Integer 1: User Defined\n"
integer 2 = "Integer 2: User Defined\n"
Basic Test = "Device2 Test"
```

   Simply add the relevant information inside the quotation marks. You can create a `Vars.ini` template file, `Vars1.ini`, from within **System Test**, edit it, and rename it `Vars.ini`. `Vars1.ini` is not the name used by the program. The name was changed for safety reasons. After you are satisfied with the accuracy of the details, rename it to `Vars.ini`, which will be used by the program.

3. Default Values `.ini` files—Located in the programs **ini** sub-folder, these files contain default values for individual molecules. They can be created and edited from within **System Test.** The file name must match the relevant test name, e.g., `Device2 Test.ini`. See the "System Selective Testing Window" section for more details.

4. `bittest.bit`—Located in the programs folder, this file contains the test names of all the devices' built-in tests (BIT). The BITs are part of the general molecule list. This means that you have to code them in

TestsDll.dll, like any other molecule. You can create a `Template.bit` file from within **System Test,** edit it, and rename it `bittest.bit`. `Template.bit` is not the name used by the program. The name was changed for safety reasons. After you are satisfied with the accuracy of the details, rename it to `bittest.bit`, which will be used by the program.

5. Sequence files—Located in the programs `seq` sub-folder, these files contain names of molecules selected from within **System Test.** They are designed to test any desired aspect of your system. You can create them through the **Test Builder** utility in **System Test** or by using any text editor. Sequence files must have the extension `.seq`. Make sure the molecule names in a sequence file are correct.

6. Self Test Files (STF)—Located in the programs `seq` sub-folder, these files contain names of sequences you create through **System Test** or a text editor. They have a `.stf` extension and basically make up a sequence of sequences. They are named self test files because that is their customary usage; they are designed to test a more comprehensive aspect of your system, although theoretically you can accomplish the same objective by using regular sequences. It is mainly a convenience. Each STF file has the following section at the end of the file.

```
[Drop Out]
seq = -1
```

If `seq = 2` and `seq 2` in the `.stf` file fails, no further testing will be permitted unless you have editing permission. A -1 means no restrictions. The `.stf` files must be created outside the program by using any text editor. You can, however, create a `Template.stf` file from within **System Test,** edit it, and rename it anything meaningful to you. Make sure it has a `.stf` extension. Again, this is done for safety reasons. For more details, see the `.stf` section below.

System Test Start Up Procedure

1. You can start the installation of the program by clicking on `Setup.exe` in the `SysTest` folder. Follow on screen directions to create `SysTest.exe`. Once installed, double click the `SysTest.exe` icon in your installation folder. When the test program starts, the **Permissions** window shown in Figure E–6 will display on top of the **System Test** window (the **Main Window**) shown in Figure E–7, allowing you to enter the testing/editing codes.

Appendix E • CVI Demo Programs

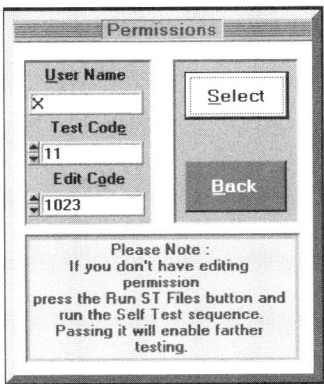

Figure E–6
Permissions Window

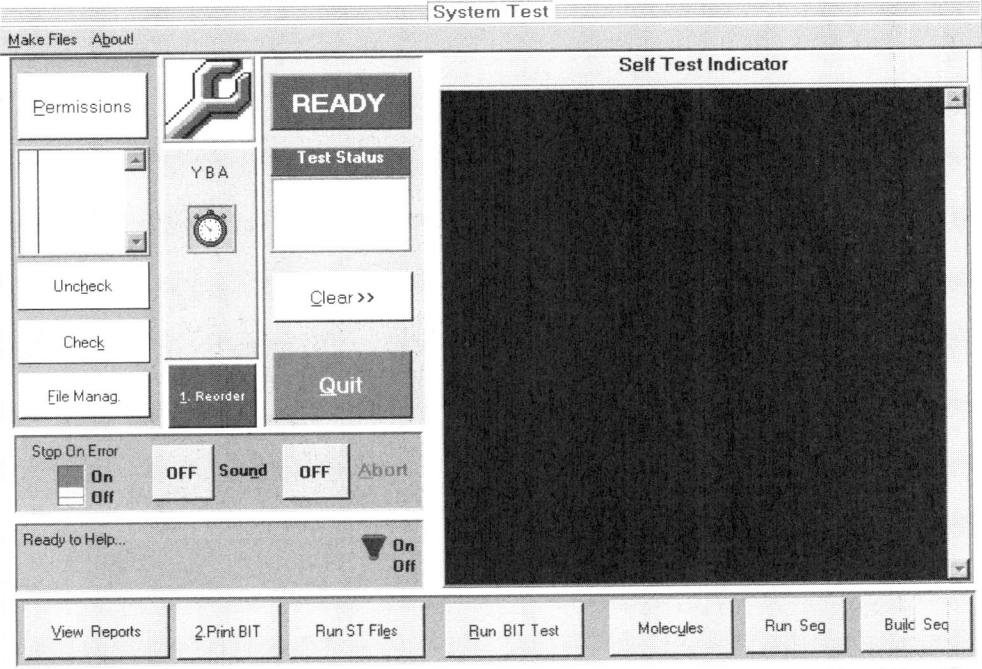

Figure E–7
System Test Window (Main Window)

2. If you entered the testing code, the only button available to you will be the **Run ST Files** button. Entering the editing code will enable all buttons. See the "How to Use Permission Levels" section. However, this feature is not included in the demo version. See the note following this list.
3. In the **System Self Tests** window shown in Figure E–8, press the **Load File** button and select the self test file that was designed to be the gate opener. If this test passes, farther testing will be enabled. Usually this test file checks if the test adapter is present and operational.
4. Press the **RUN ST Files** button.
5. If the test passes, all the testing buttons on the main window will be enabled.
6. If the test fails and you entered the **Test Code only,** the program will display a warning and you will not be allowed to continue with any testing.
7. If you entered the **Edit Code,** a warning will be displayed, but you will be permitted to continue with general testing.

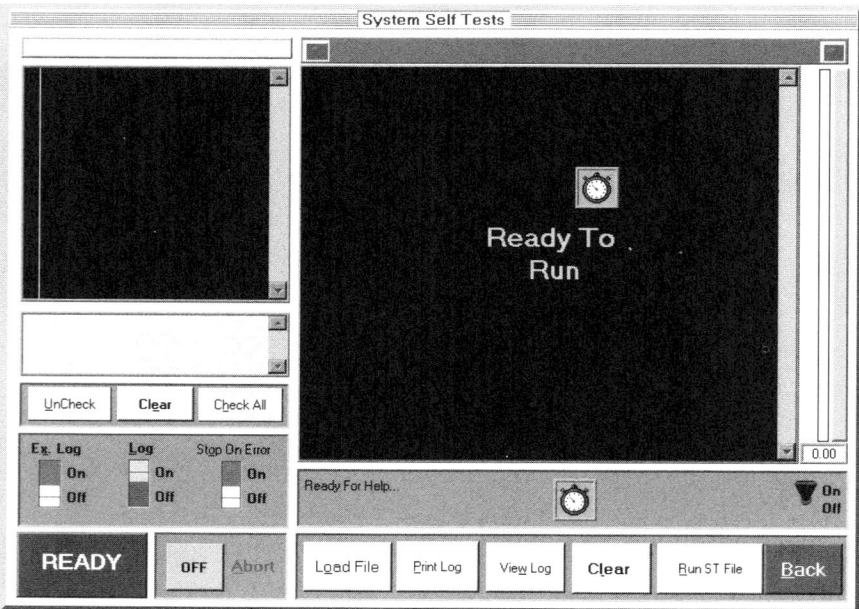

Figure E–8
System Self Tests Window

Note: In the demo version, the program will display the testing code `-11` and the editing code `-1023`. All you have to do is press the **Select** button. You can re-enter the above codes if, for some reason, they were modified or deleted from the **Permissions** dialog.

How to Use Permission Levels

Refer to Figure E–6 in which two access codes are available, **Test Code** and **Edit Code**. To run any test through the **System Test** program, you must have access to a **Test Code**. This code will allow you to run, in general, existing test sequence files with a `.seq` extension and existing self test files with a `.stf` extension. Having access to an **Edit Code** will allow you to do all the above. In addition, you will be allowed to create new test sequence files or edit existing ones.

System Test Window (Main Window)

Refer to Figure E–7 called the **System Test** or the **Main Window** from where you can launch all the program's utilities. Each button and its associated utility is explained, and if needed, the relevant procedure will be outlined. This window will enable you to run Built-In Tests (BIT) before any other type of testing is done. It will also let you enter the different codes you will need for testing and test editing, as was mentioned in the "System Test Start Up Procedure" section.

1. From the **System Test** window shown in Figure E–7, click the **Run BIT Test** button to display the **System Self Test** window (Figure E–8) and run a complete BIT of all the system's control devices present. If a BIT name is unchecked, it will not run. Note that while the BIT is running, you can click the **Abort** switch (in Figure E–8) to stop the run.
2. Press the **Uncheck** button to remove all check marks from the BIT names in the BIT list box.
3. Press the **Check** button to place check marks next to all the BIT names in the BIT list box.
4. The **Stop On Error** switch will stop the BIT run on first error. A pop-up will let you decide whether to continue the run or stop.
5. Press the **File Manag.** button (from Figure E–7) to access Windows Explorer.

6. The **Run ST File** button will bring up the **System Self Tests** window (Figure E–8). Once there you can run a collection of Self Test files. Press the **Molecules** button (from Figure E–7) to bring up the **System Selective Testing** window shown in Figure E–9 for single molecule and sequence testing.
7. Press the **Build Seq** button (from Figure E–7) to bring up the **Test Builder Window** shown in Figure E–10. Once there, you can build new test sequences or edit an existing one. Access to this window requires an editing permission code. Note: The **Test Builder Window** can be displayed in two modes. This step and step 8 describe these modes.
8. Press the **Run Seq** button (from Figure E–7) to bring up the **Test Builder** window again. Once there, you can run the existing test sequences. See the "Test Builder Window" section for details.
9. The menu bar will allow you to generate template .ini style files mentioned in the introduction. The following options are available under the **Make Files** menu.
    - **Make Vars.ini**—This option will allow you to create a template file, Vars1.ini, with all the test names and available variables for a par-

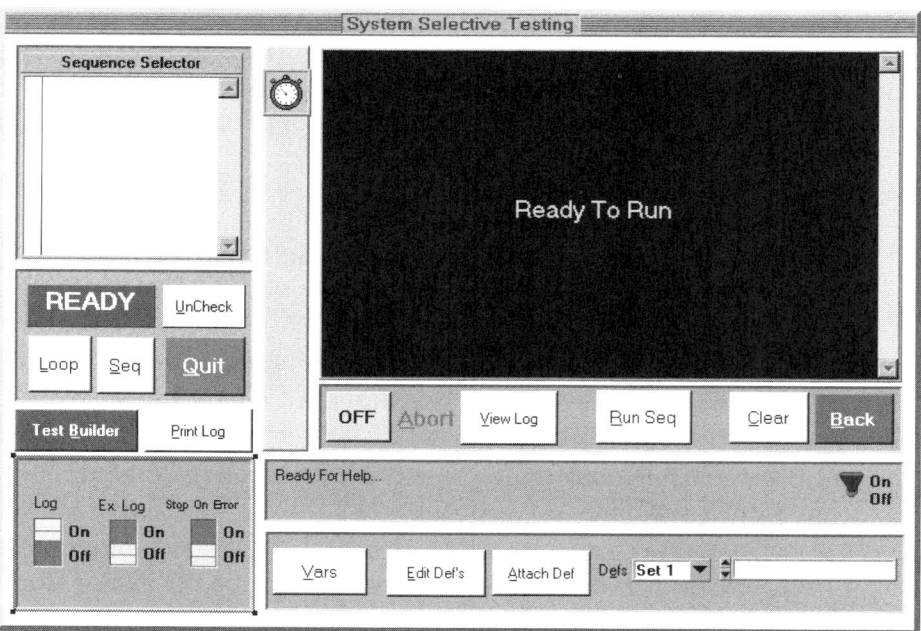

Figure E–9
System Selective Testing Window

Appendix E • CVI Demo Programs

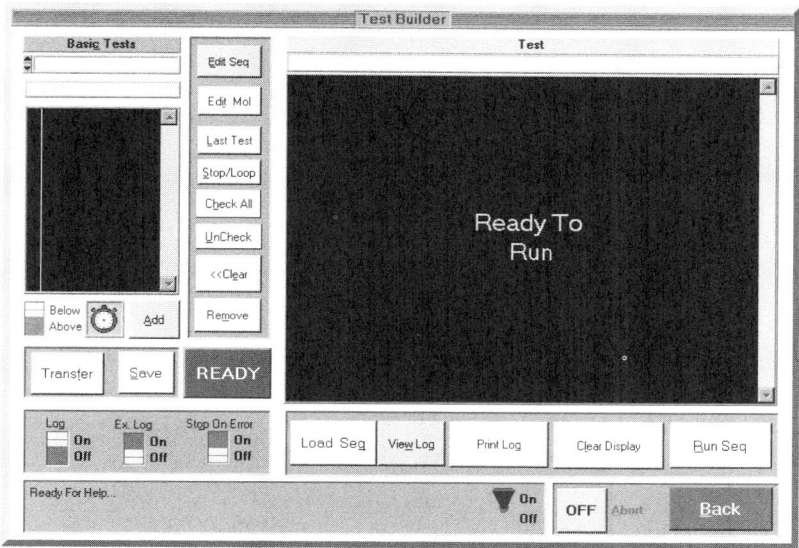

Figure E-10
Test Builder Window

ticular molecule. This file is used by the program to read a molecule and its variables and description. The information is displayed in both the **Vars** interface, described later in the "Vars Window" section, and in the **System Selective Testing** window. You can edit `Vars1.ini` and then copy it to `Vars.ini`. Make sure not to change the file's structure.

- Make Defaults ini—This option will allow you to create a variable's default values file for a particular molecule. The name of the file is the same as the molecule's name with a `.ini` extension. After the program creates the file, you can edit it and modify the variable's values to your liking. Make sure not to change the file's structure.
- Make STF File—This option will allow you to create the `Template.stf` file from within **System Test.** See an explanation in the introduction of this section. Make sure not to change the file's structure.
- Make BIT File—This option will allow you to create the `Template.stf` file from within **System Test.** See an explanation in the introduction of this section. Make sure not to change the file's structure.

10. Press the **Reorder** button from (Figure E–7) to display the **Molecules List Reordering** window as shown in Figure E–11. See an explanation of the utility in the next section.

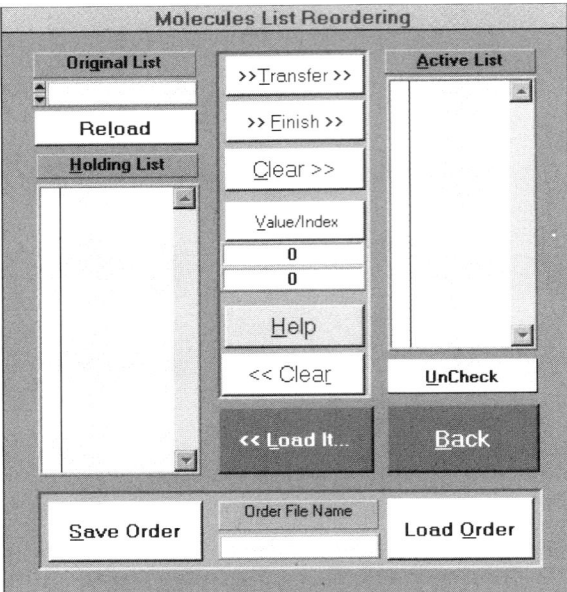

Figure E–11
Molecules List Reordering Window

Other controls and features are explained by the context-sensitive help utility in the **System Test** window.

How to Use the Molecules List Reordering Window

This utility will enable you to reorder the molecule names throughout **System Test** if a different order makes more sense for your testing needs. The program always starts with the **Original List** order. The **Original List** displays the names of all the molecules in the order they appear in the Test-Names.ini file. You can move a molecule name from the **Original List** to the **Active List** by clicking the **Transfer** button. However, make sure that you cleared the **Active List** first. Every time you transfer a name to the **Active List,** it will be removed from the **Original List**. That way you can create a newly ordered list of molecule names in the **Active List**. You don't have to transfer all the names to the **Active List**. By pressing the **Finish** button, all the names remaining in the **Original List** will be transferred to the **Active List** in their old order, after the names you have transferred manually. The reordering is complete when the **Original List** is empty and the **Active List** is full. At this point, you can save the new order into a file by typing the file name in the **Order File Name** box and pressing the **Save Order** button. If you don't enter a file name, the new order will be saved into the file `order.ord`.

Note: When you type in the order file name, do not add an extension to it. **System Test** will add the extension to the file name.

To load an existing order, press the **Load Order** button and select the desired order file. These files will have the extension `.ord`. To actually set the new order in **System Test,** you have to press the **Load It** button. You will notice that after pressing the **Load It** button, the **Holding List** gets populated with the newly ordered molecule names. This is merely an indicator that the new order is loaded into **System Test.** Press **Back** to return to the main window. **System Test** will now display the molecule names in the new order in all the relevant list controls.

System Self Tests Window

The **System Self Tests** window is displayed (Figure E–8) when you select the **Run ST Files** from the **System Test** window. This window will enable you to run the **Self Test Files** (STF).

1. Press the **Load File** button to select a self test file containing predetermined test sequences. Self Test files have a `.stf` extension.
2. Press the **Uncheck** button to remove all check marks from the test sequence names displayed in the **STF** list box in the upper left side of the **System Self Test** window. An unchecked sequence name will not execute when you are running the self test file containing that sequence.
3. Press the **Check All** button to place check marks next to all test sequence names displayed in the **STF** list box.
4. Click the **Ex. Log** switch to enable writing additional test information in the **STF** log file. Note: This option is not available in the demo version.
5. Click the **Log** switch to enable/disable generation of an **STF** log file.
6. The **Stop On Error** switch will stop the **STF** run on first error. A pop-up menu will let you decide whether to continue the run or stop.

7. Press the **Print Log** button to print the currently displayed self test file log.
8. Press the **View Log** button to view the currently displayed self test file log.
9. Press the **Run ST Files** button to run the currently displayed self test file. While the **STF** is running, you can click the **Abort** switch to stop the run.
10. Press the **Back** button to return to the **System Test** window (Main Window).

Other controls/features are explained by the context-sensitive help utility in the **System Self Tests** window.

System Selective Testing Window

This window is enabled from the **System Test** window by selecting the **Molecules** button. A box as shown in Figure E–9 is displayed that will enable you to run individual molecules for quick testing.

You will be able to create a one-time sequence and test molecules statistically.

1. Check molecule names in the **Sequence Selector** list box to run combinations of individual molecules.
2. Press the **Uncheck** button to remove all check marks from checked molecule names displayed in the **Sequence Selector** list box.
3. Press the **Loop** button to bring up the statistical testing window, titled **Loop/Select,** as shown in Figure E–12.
4. Press the **Seq** button to enable reordering (running order) of checked molecules. This will display the **Sequencer** window, as shown in Figure E–13.
5. Press the **Test Builder** button to bring up the **Test Builder** window shown in Figure E–10.
6. Press the **Print Log** button to print the log file containing results from the current molecule combination run.
7. Click the **Ex. Log** switch to enable writing additional comments to the selected log file type. This is not available in demo version.
8. Click the **Log** switch to enable/disable generation of a selected sequence log file.

Appendix E • CVI Demo Programs 617

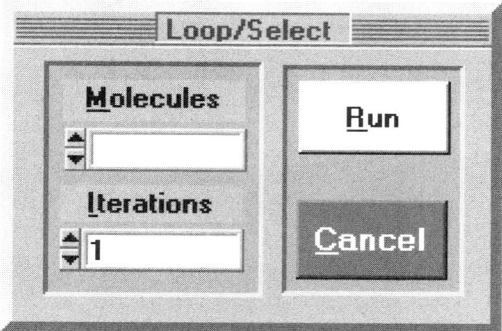

Figure E–12
Loop/Select Window

9. The **Stop On Error** switch will stop the selected sequence run on first error. A pop-up will let you decide whether to continue the run or stop.
10. Press the **Vars** button to bring up the variables interface shown in Figure E–14. Once there, you can set the parameters in the molecules participating in your selected sequence. For details see the **VarsWindow** section on page 622.

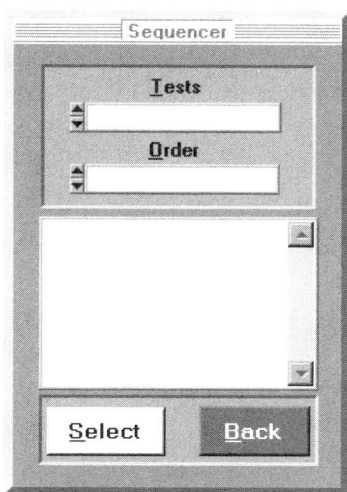

Figure E–13
Sequencer Window

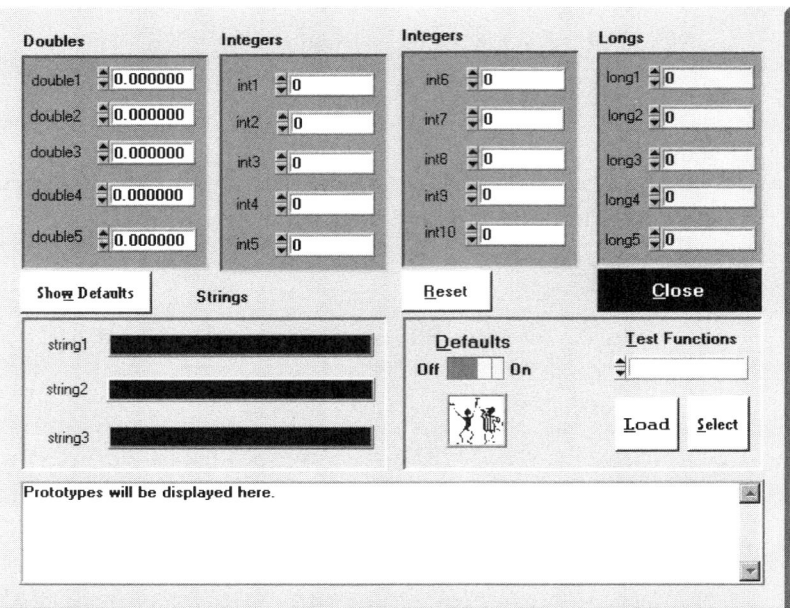

Figure E–14
Vars Window

11. Press the **Edit Def's** button to edit/view the default parameters for the molecule shown in the molecule ring box.
12. Press the **Attach Def** button to attach the default parameter's set number shown in the set numbers ring box.
13. Click the **Def's** ring box arrow to select a set of default parameter values for the molecule shown in the molecules ring box.
14. Press the **View Log** button to view the log file containing results from the current molecule combination run.
15. Press the **Run Seq** button to run the selected (checked) molecule combination. While the sequence is running, you can click the **Abort** button to stop the run.
16. Press the **Back** button to return to the **System Test** window (Main Window).

Other controls and features are explained by the context-sensitive help utility in the **System Selective Testing** window.

Loop/Select Window

The **Loop/Select** window (Figure E–12) is invoked when the **Loop** button is selected from the **System Selective Testing** window (Figure E–9).

1. Click the arrows in the **Molecules** ring box to select a molecule name for statistical testing.
2. Click the arrows in the **Iterations** ring box to select the number of times you want to run the selected molecule.
3. Press the **Run** button to start the test. While the loop is running, you can click the **Abort** switch to stop the run.
4. Press the **Cancel** button to return to the calling window.

Sequencer Window

The **Sequencer** window (shown in Figure E–13) is displayed when the **Seq** button is selected from the **System Selective Testing** window (Figure E–9).

1. Click the arrows in the **Tests** ring box to select the molecule name to be reordered.
2. Click the arrows in the **Order** ring box to set the run order for the molecule displayed in the **Tests** ring box.
3. Press the **Select** button to load the selected order into memory.
4. Press the **Back** button to return to the calling window.

Test Builder Window

This window, shown in Figure E–10, will enable you to build and edit test sequences. These sequences are stored in text files and can be called at any time. To access this window you will need an editing code (in the demo version).

1. Press the **Load Seq** button to display existing (previously created) test sequences. After selecting a test sequence, the participating molecules will be shown checked in the sequence list box.
2. Press the **Run Seq** button to run the loaded test sequence. While the sequence is running, pressing the **Abort** button will stop the run.
3. Press the **View Log Seq** button to view the log file containing the results from the current test sequence run.

4. Press the **Print Log** button to print the log file containing results from the current test sequence run.
5. Pressing the **Back** button will return you to the window prior to the **Test Builder** window.
6. To select a molecule for adding or transferring to a test sequence under construction, click the arrow in the **Basic Tests** ring box.
7. To edit the currently loaded test sequence, press the **Edit Seq** button.
8. To edit the parameters in a checked molecule name, press the **Edit Mol** button. This will bring up the **Vars** window (variables interface window) shown in Figure E-14.
9. Press the **Check All** button to place check marks next to all molecule names displayed in the test sequence list box.
10. Press the **Uncheck** button to remove all check marks from the molecule names displayed in the test sequence list box.
11. To remove a checked molecule name from the test sequence list box, press the **Remove** button.
12. To add the molecule name shown in the **Basic Tests** ring box to a sequence displayed in the test sequence list box, check a molecule, click the **Below/Above** switch, and then press the **Add** button.
13. To add a molecule name shown in the **Basic Tests** ring box to the end of a sequence displayed in the test sequence list box, press the **Transfer** button.
14. Press the **Save** button to load into memory any changes caused by pressing the **Edit Mol, Remove, Add,** or **Transfer** buttons.
15. Click the **Ex. Log** switch to enable writing additional comments to the selected log file type.
16. Click the **Log** switch to enable/disable generation of a selected sequence log file.
17. The **Stop On Error** switch will stop the selected sequence run on first error. A pop-up will let you decide whether to continue the run or stop.
18. Press the **Last Test** button to display the **Last Test** window as shown in Figure E-15.
19. Press the **Stop/Loop** button to display the **Stop/Loop** window shown in Figure E-16.

Other controls and features are explained by the context-sensitive help utility in the **Test Builder** window.

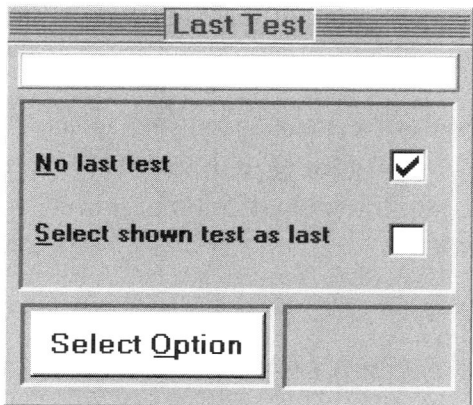

Figure E–15
Last Test Window

Last Test Window

The **Last Test** window (Figure E–15) will help you to determine the last molecule to be executed. No further testing will be allowed even if a sequence contains more molecule names later.

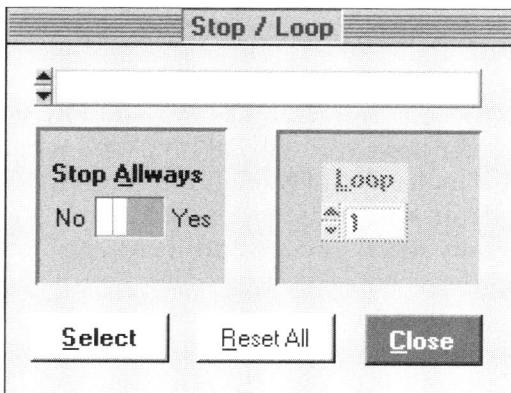

Figure E–16
Stop/Loop Window

1. You will see the name of the candidate molecule for **Last Test** in the string box. The name is picked up from the **Basic Tests** ring box in the **Test Builder** window.
2. Click one of the check boxes for your selected option.
3. Press the **Select Option** button to activate your selection.
4. The checked option will be in force from now on.

Stop/Loop Window

In the **Stop/Loop** window (Figure E–16), you can set an unconditional stop flag on any of the available molecules. The test sequence will stop after the selected molecule executes, and you will be asked whether or not to continue with the sequence execution.
Note: The **Loop** option is not available in the demo version.

1. Click the arrows in the ring box to select a molecule name.
2. Move the **Stop Always** switch to the desired position.
3. Press the **Select** button to activate your selection.
4. Press the **Close** button to return to the calling window.
5. Pressing the **Reset All** button will remove all stop flags from previously set molecules.

Vars Window

The **Vars** window is shown in Figure E–14. In this window, you will set the values you want to send to a selected molecule. The **Vars** window can be invoked from the **System Selective Testing** window or from the **Test Builder** window. In the first case, you will call the **Vars** window to set variable values for a selected molecule for a one time run only if you don't want to use the attached default set. In the second case, the **Vars** window will be invoked automatically when you are adding a variable's enabled molecule to a test sequence under construction or when you want to edit molecule variable values in an existing test sequence. See the "Test Builder Window" section for more details.

1. Click the arrows in the **Test Functions** ring box to select a molecule name for parameters editing.

2. Click the **Defaults** switch to the **Off** position to enable editing the selected molecule.
3. If the selected molecule accepts parameters, switching **Defaults** to **Off** will enable the **Select** button. Pressing the **Select** button will enable the entry boxes for the correct parameter types for the selected molecule. It will also display the descriptions of the available variables in the bottom window.
4. Pressing the **Select** button enables the **Load** button. After entering the desired values in the parameter boxes, press the **Load** button to load the selected values into memory.
5. When the **Vars** windows is invoked by the **Test Builder,** it displays with the **Select** button enabled and the **Default** switch invisible.
6. Press the **Reset** button to set all values in the parameter boxes to zero.
7. Press the **Show Defaults** button to view the default values of the molecule displayed in the Test Functions ring box.
8. Press the **Close** button to return to the calling window.

Bibliography

Appel, K., and J. McIntosh (Savannah River Technology Center) *LabWindows Controls Robotics for Radioactive Waste Tank.* National Instruments Corporation (User Solutions).

Axelson, J. *Serial Port Complete.* Wisconsin: Lakeview Research, 1998.

Clark, J.F. *SCXI Helps Test Aircraft Models for U.S. Government.* National Instruments Corporation (User Solutions).

Deitel, H.M., and P.J. Deitel. *C How to Program.* New Jersey: Prentice-Hall, Inc., 1994.

Harbinson, Samuel P. *C, A Reference Manual.* New Jersey: Prentice-Hall, Inc., 1995.

Hewlett-Packard Company HP 33120A Function Generator Arbitrary Waveform Generator User's Manual: Hewlett-Packard: Palo Alto, California, 1997 [Manual Part Number: 3312-90005].

Hewlett-Packard Company *HP 3478A Multimeter Operator's Manual:* Hewlett-Packard: Palo Alto, California, 1998 [Manual Part Number: 03478-900009].

Le Méur, G. *Using LabWindows to Test Microwave Components.* National Instruments Corporation (User Solutions).

Levin, I. *Detecting Intermittent Faults of 757 Aircraft Flap and Slat Controls During Thermal and Vibration Cycling.* National Instruments Corporation (User Solutions).

Mangin, R. *LabWindows-Based System Tests Cadillacs for Premium Sound.* National Instruments Corporation (User Solutions).

Mechaly, A. (IXEL) *LabWindows/CVI Helps Alcatel Ensure Clear Mobile Phone Reception.* National Instruments Corporation (User Solutions).

Mueller, S. *Upgrading and Repairing PCs.* Indiana: Que Corporation, 1998.

Mullet, K. and D. Sano, D. *Designing Visual Interfaces.* New Jersey: SunSoft Press, 1995.

National Instruments Corporation. Getting Started with LabWindows/CVI. National Instruments Corporation, 1999 [Part Number 320680E-01].

National Instruments Corporation. Getting Started with your Plug and Play GPIB Hardware and Software for Windows 95. National Instruments, 1998 [Part Number 321035D-01].

National Instruments Corporation. GPIB NI-488.2M Function Reference Manual for Win32. National Instruments Corporation, 1998 [Part Number 321038D-01].

National Instruments Corporation. GPIB User Manual for Windows 95 and Windows NT. National Instruments, 1998 [Part Number 321819A-01].

National Instruments Corporation. Hands-On Course, LabWindows/CVI Advanced. National Instruments, 1998 [Part Number 321057A-01].

National Instruments Corporation. Hands-On Course, LabWindows/CVI Basics. National Instruments, July 1996 [Part Number 320803D-01].

National Instruments Corporation. LabWindows/CVI Advanced Analysis Library Reference Manual. National Instruments, 1998 [Part Number 320686D-01].

National Instruments Corporation. LabWindows/CVI Basics I. National Instruments, 1998 [Part Number 320803E-01].

National Instruments Corporation. LabWindows/CVI Basics II. National Instruments, 1999 [Part Number 31057B-01].

Bibliography

National Instruments Corporation. *LabWindows/CVI Instrument Driver Developer's Guide*. National Instruments, 1999 [Part Number 320684 E-01].

National Instruments Corporation. *LabWindows/CVI Master Index*. National Instruments, 1998 [Part Number 320687D-01].

National Instruments Corporation. *NI-488.2 Function Reference Manual for Win32*. National Instruments, 1998 [Part Number 321038D-01].

National Instruments Corporation. *LabWindows/CVI Programmer's Reference Manual*. National Instruments, 1999 [Part Number 320685E-01].

National Instruments Corporation. *LabWindows/CVI Standard Libraries Reference Manual*. National Instruments, 1998 [Part Number 320682 D-01].

National Instruments Corporation. *LabWindows/CVI User Interface Reference Manual*. National Instruments, 1998 [Part Number 320683D-01].

National Instruments Corporation. *LabWindows/CVI User Manual*. National Instruments, 1999 [Part Number 320681E-01].

Perry, G. *C By Example*. Indiana: Que Corporation, 1993.

Petzold, C. *Programming Windows*. Redmond, Washington: Microsoft Press, 1999.

Press, W., S. Teukolsky, et al. *Numerical Recipes in C: The Art of Scientific Computing*. New York: Cambridge University Press, 1994.

Seyer, M. *RS-232 Made Easy: Connecting Computers, Printers, Terminals, and Modems*. New Jersey: Prentice Hall Computer Books, 1991.

Taylor, R. *LabWindows Goes Deep Sea Diving*. National Instruments Corporation (User Solutions).

Urias, R. *True Diagnostic Systems Uses LabWindows/CVI to Ease Development of Test Systems*. National Instruments Corporation (User Solutions).

Index

A

Active X
　Automation library functions, 502
　support, xxxv
Add/Edit Tools Menu Item
　Command Line Arguments, 522
　Menu Item Name, 522
　Program Name, 522
Add File to Project
　command, 529
　window list, 147
Add Files to Executable, 488
　command, 488
Add Files to Project
　command, 478
Advanced Analysis libraries, xxxv
　curve-fitting, xxxv
　dialog box, 496
　digital filters, integration and differentiation functions, xxxv
　linear equations, xxxv
　signal processing, xxxv
　statistical functions, xxxv
　time/frequency analysis, xxxv
Advanced Distribution Kit dialog box, 310
　Browse button, 310
　Command Line Arguments, 311
　Executable Filename, 311
　Help menu, 311
　Installation Name, 311
　Installation Script File, 310
　Program Group Name, 311
　Select button, 311
　Use Custom Script, 310
　Use Default, 311
Aligning objects
　Distribution…, 91
　Horizontal Centers, 91
　Vertical Centers, 91
Always Append Code to End, 343
　option, 539
American National Standards Institute. *See* ANSI.
analysis applications, xxxvii
analyze
　instrumentation data, 14
ANSI, xxxv, 2, 219, 554
　environment, xxxvii
ANSI C, xxxv, 3, 554
ANSI C library functions, 504
Apply Default Font
　command, 531
architecture
　open software, xxxvi, 3
Array Display window, 525
Array/Strings Display, 511

ASCII, 4, 8, 138, 182, 185, 186, 437
asynchronous transmission
  definition, 428
Automatic Skeleton Code Generation. *See CodeBuilder*.
Auto Save Project
  command, 476

B

Background Color, 547
Balancing Parentheses
  checking, 152
Baud Rate
  description, 436
bibliography, xxxvi, xxxv, 321, 394, 433, 605, 625–27
Binary Switch, 2
  creating, 93
Bmath
  1D Function button, 601
  Bmath.exe icon, 600
  Brent Method, 605
  Functions Help, 602
  Functions window, 602
  Graph button, 603
  installation
    Setup.exe, 600
  introduction, 600
  Iterative Evaluation, 602–3
  Loading/Editing Instructions, 601–2
  Main Menu, 600
  Param Run button, 602
  Parametric Run
    using, 602
  Parametric Run utility, 602
  Parametric Run window, 602, 603
  Plot button, 603
  Plot utility, 605
    using, 605
  Print Graph button, 605
  Run Function button, 602
  Secant Method, 605
  Solve button, 603
  Solver, 605
  Solver Utility
    using, 605
  Solver window, 605
Borland C++, xxxvi
Break at First Statement
  command, 490
Break on first chance exceptions
  option, 518
Break on library errors
  option, 514, 518
Breakpoint, 165
  Break at First Statement, 166

Breakpoints command, 490
Breakpoints Dialog Box, 167, 168, 490
  Add/Edit Item, 168
    Condition box, 168
    File box, 168
    Line box, 168
    Pass Count box, 168
  Delete All, 168
  Disable All, 168
  Enable All, 168
  Go to Line, 168
  Continue, 167
  Finish Function, 167
  Go To Cursor, 167
  programmatically setting
    Breakpoint, 165
  purpose, 165
  Step Into, 167
  Step Over, 167
  Terminate Execution, 167
  Toggle Breakpoint, 166
  using, 165–69
Build
  Compile File
    to compile program, 81
Build Errors
  box, 147
  command, 510
Build Errors in Next File, 147
Build Errors window, 488
Build menu, 31
"Build Menu" section, Appendix B, Project Window
        Environment, *28*
Build Options
  command, 512

C

C++ Builder, xxxvi
callback function, xxxvii, 51
  association, 16
  for controls, 52
  for panels, 53
  how to assign, 21
  number of times called, 17
  to terminate program, 24
  used for, 15
Callback Function and Generated Events, 51–53
Callback Function dialog box, 22
callback function names
  skeleton code from, 25
callbackData, 52, 53
calling convention
  for C
    __cdecl, 321
  standard
    __stdcall, 320

Index

Cascade Windows
   command, 508
check boxes, 2
child panels
   explanation, 20
CIC, 367, 370, 394
Clear Tag
   view/remove Tags, 154
ClearListCtrl, 251, 283, 297
Close All
   command, 509
CloseFile, 184, 225
CLRDTR, 465
CLRRTS, 465
Code menu, 27
Code Selection Modes
   Character Select Mode
      description, 150
      icon, 150
   Column Select Mode
      description, 150
      icon, 150
   Line Select Mode
      description, 150
      icon, 151
   Select Mode
      Beginning/End of Selection
         Viewing, 151
      using, 150–52
CodeBuilder, 25, 27, 28, 29, 33, 36, 45, 54, 56, 60, 81, 94, 106, 109, 110, 138, 196, 211, 240, 250, 251, 270, 271, 319, 328, 329, 331, 343, 345, 354, 515, 539
   cannot use, 319
CodeBuilder generated code
   analyzing, 33–35
Coloring Icon, 190
Coloring Tool, 267
   using, 92
Colors
   option, 525
ComBreak, 463
ComFromFile, 450, 451, 455
COM Port
   description, 436
Command Button, 2, 14
   changing attributes, 21
   how to create, 21
   label appearance, 22
   properties box, 21
Command Line
   command, 518
commit event, 39
   definition, 39
CompareStrings, 295
Compatibility With
   option, 513

compatible compilers, 309
   used when creating DLLs, 486
   used when creating stand-alone executables, 486
Compile File
   command, 488
Compiler Defines, 515
Compiler Options
   dialog box, 88
ComRd, 445
ComRdByte, 449
ComRdTerm, 449
ComSetEscape, 465
ComToFile, 452, 454, 455
ComWrt, 444, 445
ComWrtByte, 447
configuration
   options
      Debug, Release, 28
Configuration
   command, 482
ConfirmPopup, 199, 214, 215, 220
Confirm Pop-up Panel, 199–200
console window, 145, 183, 558
Constant Name
   controls
      size limitations, 22
   convention
      uppercase letters, 37
   Menu Bar Constant Name
      size limitations, 108
   Menu Bar Constant Prefix
      size limitations, 108
   panel
      limitations, 21
   prefixed by panel name, 22
Continue, 490
Control
   command, 522
   made active, 17
control box, 15
Control Callback, 541
control callback functions
   different from menu callback function, 110
control mode
   explanation, 39
   Hot, 18, 41
   Normal, 39
   types of, 39
   Validate, 18, 41
control objects
   operate on, 71
Control Style
   command, 531
Control Text Style, 549
Controller-In-Charge. *See* CIC
Copy
   command, 530

Copy Panel
   command, 531
CopyString, 222, 292
CorelDraw, 228
CRC, 455
Create ActiveX Automation Controller function, 506
Create Console Application, 485
Create Distribution Disk, 485
Create Distribution Kit, 305
   command, 488
Create Distribution Kit dialog box, 305
   Add Group button, 305
   Advanced button, 310
   Browse button, 306
   Build, 305, 310
   Cancel button, 310, 312
   Default button, 310, 312
   Delete button, 310
   Distribute Objects and Libraries for All Compilers, 307
   Edit File Group, 308
   Edit Group, 305
   File Groups, 307
   Install Language, 307
   Install Low-Level Support Driver, 307
   Install Program Manager Icons, 309
   Install Run-Time Engine, 306
   Installation Directory, 305
   kBytes Reserved on First Disk, 306
   Media Size, 306
   Replace Existing Files, 310
   Target Path, 305, 306
Create IVI Instrument Driver function, 507
Create menu, 20, 38
Create Object File
   Source Editor Option
      create object file, 84
Creating and using the Menu Bar, 102–9
Creating Distribution Disks, 305–14
Creating DLL Containing Callback Functions, 354–64
Creating DLLs with the user interface, 328–39
Creating DLLs without user interface, 319–28
Creating interactive measurement applications, 14
Creating object file for use in an external compiler, 342–54
Creating Stand-alone Executable, 303–305
   target settings, 304
Creating Stand-Alone Executables and Distribution Disks
   overview, 302–3
CTS lines, 437
CTS Mode, 437
   hardware handshaking, 437
Cut
   command, 530
Cut Panel
   command, 531

CVI, xxxv, 5
   acronym, 2
   add on tools
      Image Processing, 3
      PID, 3
      SQL Toolkit, 3
      Statistical Process Control, 3
      Test Executive, 3
      Test Stand, 3
   analysis routines, xxxv
   as an integrated ANSI environment, xxxv
   based on C programming language, xxxv
   code generation, xxxv
   data acquisition and instrumentation control software, xxxv
   data analysis tools, xxxv
   database manipulations, mathematical modeling, analysis, testing, 5
   definition, xxxvi
   external compiler support, xxxvi
   I/O libraries, xxxv
   industries used in, 3
   operating system support
      for version 5.5, xxxix
      prior to version 5.5, xxxix
         Sun Solaris, xxxix
         Unix, xxxix
         Windows 3.1, xxxix
   original version
      Disk Operating System (DOS), 14
   programming environment, xxxv
   prototyping tools, xxxv
   Starting *CVI*, 5–7
   Starting using Windows Start button, 468
   system requirements, xxxix–xl
      hard disk space required, xxxix
      monitor resolution, xxxix
      RAM, xxxix
   system requirements for SPARC systems
      hard disk space, xxxix
      main memory, xxxix
      swap space, xxxix
   use instrumentation libraries, xxxvi
   used by
      academia and research labs, xxxv
      Alcatel, 5
      Commercial Avionics Systems, 4
      Thomson-CSF in France, 5
   used for, xxxv, xxxvi
   used in
      High-Energy Heart Defibrillators, 5
   user interface tools, xxxv
   using GPS, 5
   What is, 2–3
   Where used, 3–5

Index

CVI 5.5 and later versions
  project
    compiled state saved, 31
CVI compatible compilers, 468
  Borland C++ 4.51 or higher, C++ Builder 4.0, 468
  Microsoft Visual C++ 2.2 or higher, 468
  Symantec C++ 7.2 or higher, 468
  Watcom C++ 10.5 or higher, 468
CVI environment
  current environment identification, 19
CVI IDE icon, 19
CVI installation, 468
  copying Projects folder, 469
CVI Manuals. See bibliography.
CVI Programming Model
  understanding, 15–18
CVI support libraries
  `cvirt.lib`, 342, 349, 359
  `cvisupp.lib`, 342, 349, 359
  `cviwmain.lib`, 342, 349, 359
CVI Utility Library, 307
Cyclic Redundancy Code. See CRC.

D

DAQ, xxxvi, 4. *See* data acquisition
data acquisition, xxxv, 2, 8, 36
Data Acquisition library functions, 496
  Easy I/O, 496
data analysis, 2, 4
Data Bits, 437
data collision, 368
Data Types
  changing, 38
DataSocket, xxxv
DataSocket library functions, 501
DateStr, 127, 211
DB-9, 431
  (Male) Connector Pin Locations, 431
  nine-pins connector, 431
DB-9 to DB-25 adapter, 431
DB-25, 433
  (Male) Connector Pin Locations, 431
  twenty-five-pins connector, 431
DCE
  used by, 429
Debug, 484
Debug configuration, 483
Debug Project
  command, 489
Debugger
  finding, 163
  String Display window
    Format
      ASCII, 175
      Binary, 175
      Decimal, 175
      Hexadecimal, 175
      Octal, 175
  String Display window
    Edit, 175
  using, 163–65
  Variables window
    current function variables, 172
    Edit/Add Watch Expression
      Add, 177
      Cancel, 177
      File box, 177
      Function box, 177
      Project/DLL box, 177
      Replace, 177
      Scope box, 177
      Update Break when value changes box, 177
      Update Display continuously box, 177
      Variable Expression box, 177
    global variables, 171
    Options
      Add Watch Expression, 176
    Run
      View Variable Value, 173
    View
      Close Variable, 173
      Expand Variable, 172
      Follow Pointer Chain, 173
      Retrace Pointer Chain, 173
  View
    Array Display window, 174
      Format for real numbers
        Fix Precision, 175
        Floating Point, 175
        Scientific, 175
      Options
        Reset Indices, 175
        for multi-dimensional arrays, 175
    String Display, 175
  view value, 172
Debugging Level
  option, 513
  settings, 163
Decoration box, 2
  Raised Box, 89
Default Calling Convention
  option, 513
Default Control Events
  settings, 25
Default Panel Events
  settings, 25
Default Value
  definition, 37
DefaultCtrl, 129, 215, 226, 280
Delay, 46
Delete
  command, 530
DeleteListItem, 247, 248, 257, 275, 276

Developing *CVI* application
  overview, 36
dials, 2
DirSelectPopup, 207
DirSelectPopup Panel, 207–8
DiscardUIObjectsForDLL, 329, 355
  similar to *destructor* in C++, 329
Display status dialog during build
  option, 515
DisplayComStatus, 462
DisplayPanel, 34, 63, 211, 286
distribution disk
  files included, 302
  installing on target machine, 312
DLL, 302, 303, 304, 305, 309, 317, 318, 319, 321, 322, 323, 324, 325, 328, 329, 335, 336, 340, 354, 355, 357, 482, 627
  advantages, 318
  concept of, 317
  creating, 328, 335
  disadvantages, 318
  using in project, 325–28
DLL_PROCESS_ATTACH, 327, 331
DLL_PROCESS_DETACH, 328, 332
DllEntryPoint, 329, 353
DllMain, 319, 352, 353
DLLMain, 319, 320, 321, 329, 331, 332, 336, 338
  entry point, 329
drag-and-drop, 24
DTE, 429
DTR, 465
DTR line, 437
Dynamic Data Exchange library functions, 502
Dynamic Link Libraries (*See* DLL)
  Overview, 318–19
Dynamic linking
  description, 318

E

Edit
  command, 493
Edit Command Button dialog box
  attributes, 45
Edit File Group dialog box, 308
  Add button, 308
  Relative Path, 308
Edit menu, 45, 529
Edit Menu Bar
  Callback Function, 104
  Constant Name, 104
  dialog box, 104
  Insert New Item, 104
  Insert Separator, 104
  Item, 104
  Left hierarchy button, 105
  Menu Bar Constant Prefix, 104
  Modifier Key, 104
  Right hierarchy button, 105
  View, 105
EIA serial communication standards, 432
  RS-232C, 432
  RS-422A, 432
  RS-432A, 432
  RS-485A, 432
  standard, 432
Electronics Industries Association. *See* EIA.
Enable Data Tool Tips
  view value in Debugger mode, 174
Enable Signed/Unsigned Pointer Mismatch Warning
  option, 514
Enable Unreachable Code Warning
  option, 514
End Of File (EOF)
  CVI indicator, 186
Environment Option, 518
  Check foreground lockout setting on startup
    option, 519
  CVI environment sleep policy
    option, 518
  Enable Data ToolTips
    option, 520
  Force Loaded Instrument Driver's Source into Interactive Window
    option, 520
  Force Project Source Files into Interactive Window
    option, 520
  Go to source after inserting code from function panel, 519
  Interactive Window Memory Size
    option, 519
  Use Console Window for Standard I/O when Debugging
    option, 520
  Use Only One Function Panel Window
    option, 519
error-checking
  CRC, 455
  Kermit, 455
  XModem, 455
  YModem, 455
  ZModem, 455
ErrorPrintf, 178
event, 51
  DOUBLE_CLICK, 110
  EVENT_CLOSE, 53, 97
  EVENT_COMMIT, 18, 28, 52
  EVENT_DISCARD, 52, 53, 543
  EVENT_GOT_FOCUS, 17, 52, 53
  EVENT_KEYPRESS, 52, 53
  EVENT_LEFT_CLICK, 18, 52, 53
  EVENT_LEFT_DOUBLE_CLICK, 52, 53
  EVENT_LOST_FOCUS, 18, 52, 53
  EVENT_PANEL_MOVE, 53
  EVENT_PANEL_SIZE, 53
  EVENT_RIGHT_CLICK, 52, 53
  EVENT_RIGHT_DOUBLE_CLICK, 52, 53

Index

EVENT_TIMER_TICK, 56, 543
EVENT_VAL_CHANGED, 18, 52
  value changed, 39
  waits for, 51
event-driven
  fundamental difference from sequential programs, 15
event-driven code
  CVI generates, 15
event-driven programming model
  explanation, 15
event generation, 29
event processing
  stopped by *QuitUserInterface* function, 16
events generated, 15, 51
  by *callback* function, 17
  controlled by *RunUserInterface* function, 15
Excel, 188, 217, 225, 228, 230, 233
Exclude File From Build command, 479
Exclude/Include Lines, 169–71
  purpose, 170
  Remove All Excluded Lines, 170
  Resolve All Excluded Lines
      Dialog Box, 170
    #endif, 170
    #if 0, 170
    Cancel, 170
    Comment, 170
    Delete, 170
    Options, 170
    Skip, 170
    Unexclude, 170
    Unexclude All, 170
  Toggle Exclusion, 170
executable
  create, 29
Execute
  command, 491
Exit *LabWindows/CVI*
  command, 477
Explorer, 230, 232, 297
external compiler
  creating object files
    DllMain
      use, 354
    WinMain
      use, 343
  creation of callbacks, 343
  DLL
    creating, 354
  Microsoft Visual C++
    using object file, 348
External Compiler Support, 345
  command, 488
  dialog box, 345
    ANSI C Library, 347
    CVI Libraries, 347
    Header File, 347
    Object File, 347
    Other Symbols, 347
    Using LoadExternalModule to Load Object and Static Library Files, 347
  using DLL, 354
External Compilers Supported
  Borland C++ 4.51 or higher, 342
  Borland C++ Builder 4.0, 342
  Microsoft Visual C++ 2.2 or higher, 342
  Symantec C++ 7.2 or higher, 342
  Watcom C++ 10.5 or higher, 342

F

File
  handle name, 182
  Open
    append file, 182
    file as is, 182
    for reading and writing, 182
    for reading only, 182
    for writing only, 182
    truncate file, 182
  OpenFile, 182
  Save As, 45
  write data, 182
File I/O
  meaning, 182
File menu, 19, 25, 473, 528
  create new file, 19
FileSelectPopup, 205, 228, 230
FileSelectPopup Panel, 205
FindLstn, 405
FindPattern, 292, 293
Find, Search and Replace Commands, 160–62
Find UI Object, 542
Find UIR Objects
  command, 533
Firewire. *See* IEEE-1394.
  high data transfer rate, 366
First *CVI* Project
  creating the, 18–33
FlushInQ, 460
FlushOutQ, 459
Fmt, 212, 554, 555, 556, 565, 574, 577, 578, 581, 585, 594, 595, 597
FmtFile, 554
  function panel, 557
Fmt Function examples, 565–78
FmtOut, 554
  function panel, 589
Font
  option, 524
Font Type
  metafonts as scalable, 189
  Scalable fonts
    purpose, 189
  use metafonts, 315

Format and Precision, 41
formatcode, 560, 590, 591, 592
    meaning, 560
formatcode modifiers, 591
formatcode specifiers, 556, 599
Format Code Specifiers, 559–60
format specifiers, 556, 558, 560, 568, 581, 597
    explanation, 559
format strings, 581
    structure, 555
Format Wizard
    starting, 594
    using, 594–97
Formatting and I/O library functions, 503
Formatting and Scanning Functions
    introduction, 554
Formatting Function examples
    appending two string, 577
    concatenating two string, 577
    converting decimal integer to hexadecimal string, 578
    inserting characters between each element of an array, 571
    integer and real to string with literals, 576
    integer to string, 565
    real string to floating point, 572
    Real to String in Scientific Notation, 574
    short integer to string, 569
Formatting Functions, 554
    Fmt, 554
    FmtFile, 556
    FmtOut, 557
    formatcode specifiers, 553
    format string
        description, 555
    modifiers, 553
    source
        meaning of, 555
    target
        meaning of, 555
Formatting Modifiers, 561–66
formatting string, 591
Frame Color, 548
    used for child panels, 548
function panel
    all open, 511
    data entered in template, 79
    definition, 67
    features, 79
    Find, 143
    Insert, 143
        function call, 81
    inserting in code, 68
    purpose, 2
    recalling, 68
    using, 143
    View, 143
Function Trees, 511

G

gauges, 2, 37
General Purpose Interface Bus. *See* GPIB.
Generate Control Callback, 541
    use, 126
Generate Prototype
    creating, 148
Generate WinMain() instead of main(), 343
generic *control* callback
    prototype, 51
GenericMessagePopup, 200, 230
Generic MessagePopup Panel, 200–3
generic *panel* callback
    prototype, 52
generic *PanelCallback*
    no control argument, 53
GetComStat, 462
GetCtrlAttribute, 99
GetCtrlIndex, 243, 244, 245, 255
GetCtrlVal, 66, 384
GetFileInfo, 187, 219
GetFirstFile, 272, 274, 283
GetFmtErrNdx, 591
GetIndexFromValue, 256
GetInQLen, 461
GetLabelFromIndex, 245, 247, 255
GETMAXCOM, 465
GetNextFile, 274
GetNumCheckedItems, 251
GetNumListItems, 244, 275, 276
GetOutQLen, 461
GetPanelAttribute, 314
GetProjectDir, 211, 271
GetRS232ErrorString, 439
GetSystemTime, 222
GetValueFromIndex, 244, 245, 247, 255
glue code, 318
GPIB, xxxv, 2, 5, 36, 365, 594
    addressing scheme, 370
        primary address, 370
        secondary address, 370
    board-level routines
        description, 369
    bus *extender*, 367
    cable, 367
    communication between devices, 367–69
    data transfer rate, 366
    definition, 366
    device drivers, 408, 409
    device-level routines
        description, 369
    Getting started with, 369
    limitations, 368
    linear configuration
        definition, 368

Index

star configuration
   definition, 368
   VXI chassis connected, 370
GPIB 488 functions
   ibcmd, 386
   ibcnt, 383
   ibcntl, 383
   ibdev, 381
   ibfind, 381
   ibln, 384
   ibonl, 381
   ibrd, 386
   ibwrt, 386
GPIB 488.2
   library routines, 394
GPIB-488.2 Functions
   FindLstn, 405
   Receive, 407
   Send, 406
   SendCmds, 407
   SendIFC, 394
GPIB-based device
   unique address, 367
GPIB box. *See* GPIB interface board.
GPIB bus
   limitations, 368
GPIB *bus extender*, 367
GPIB communication
   addressing scheme, 370
   introduction and history, 366–67
   using the NI-488 routines, 370–88
GPIB devices
   Controllers, 367
   Listeners, 367
   Talkers, 367
GPIB error
   iberr, 383
GPIB/GPIB 488.2 Library Function Panel, 381
GPIB IEEE-1394, 366
GPIB Instrument Drivers
   using, 408–26
GPIB interface board
   different addresses, 367
   installation, 367
GPIB status
   ibsta, 381
grandchild panels
   explanation, 20
graph, 2, 14
graph control, 74
   Graph, 74
graphical programming language, xxviii
Graphical User Interface, 2. *See* GUI
   results presentation, xxxv
GraphPopup, 222
GUI, 2, 4, 529
   definition used in text
      same as user interface resource, 19

ease of use, 14
pronounced, xxxv
user-defined, 15

H

handle
   definition, 33
Handshaking
   hardware, 437
   software, 437
Help Editor, 511
Help menu, 525
Hewlett-Packard Interface Bus. See HP-IB.
Hide Windows
   option, 518
HidePanel, 284
histogram, 14
HP 33120A Function Generator/
      Arbitrary Waveform
      Generator, 394
hp3478 device drivers
   hp3478a_close, 425
   hp3478a_config, 409
   hp3478a_init, 409
   hp3478a_measure, 425
   hp3478a_write_data, 425
hp3478 global error
   `hp3478a_err`, 409
HP-IB, 366
HS488
   high speed data transfer, 366

I

ibcmd, 386
ibcnt, 383
ibcntl, 383, 391, 393, 405
ibdev, 381, 384, 388
iberr, 383, 391, 393
ibfind, 381
ibln, 384
ibonl, 381, 387
ibrd, 386
ibsta, 381, 383, 391, 393
ibwrt, 386
Icon Editor, 484
   Starting, 117
IEC 625-1, 366. See also IEC Bus.
IEC Bus, 366
IEEE, 366
IEEE-1394, 366
IEEE-488, 366, 369, 381, 391
IEEE-488 routines, 369
IEEE-488.1, 371, 391

IEEE-488.1 standards, 394
IEEE-488.2, 369, 394
IEEE-488.2 routines, 369
IEEE-488.2 standard
    description, 394
    using, 394–408
IEEE-488.2-1992, 366
Include
    file, 473
Include File to Build
    command, 479
include files
    `windows.h`, 515
    `formatio.h`, 332
    `gpib.h`, 611
    `utility.h`, 332
    external compiler support
        `cvirte.h`, 33
    toolbar header
        `toolbar.h`, 116
    user interface libraries
        `userint.h`, 33
Include Paths
    command, 516
InitCVIRTE, 345
InitUIForDLL, 329, 331, 355, 357
    similar to *constructor* in C++, 329
Input Queue Length
    description, 438
Insert Construct
    inserting C construct, 156
    Toolbar icon, 81
Insert Include Statements Dialog Box, 147
Insert New Item, 108
Inserting C Constructs in Code, 156–8
InsertListItem, 246, 248, 255
InsertTextBoxLine, 129, 407
installation and setup. *See* Appendix A.
Install DataSocket Support, 307
    not available prior to CVI 5.5 versions, 307
Install NI Reports Support, 307
    not available prior to CVI 5.5 versions, 307
InstallPopup, 205
Institute of Electrical and Electronic Engineers. *See* IEEE.
instrumentation, xxxvii, 2, 4
Instrument Directories, 516
Instrument Directories dialog box, 516
instrument driver creation, xxxv
instrument driver function panels, 408
instrument drivers, 5
Instrument Driver Support Only, 486. *See also* "Chapter 8, Creating Stand-Alone Executables and Distribution Disks"
instrument module, 409
Interactive Execution Window, 511, 525
    purpose, 144

running, 146
using, 144
interface tools
    large set of tools, 14
International Electrotechnical Commission.
            See IEC.
IsListItemChecked, 254
IVI library functions, 499

K

knob control, 2, 37
    creating, 59

L

Label Appearance group, 43
    Label Raised box
        purpose, 44
    Size To Text box
        purpose, 44
Label Text Style, 548
LabVIEW, xxxvi
LabWindows/CVI. *See* CVI
laplink. *See* null-modem.
LaunchExecutable, 228, 297
LED
    creating, 89
Library Functions
    Breakpoint, 165
    ClearListCtrl, 251
    CloseFile, 184
    ComBreak, 463
    ComFromFile, 450
    CompareStrings, 295
    ComRd, 445
    ComRdByte, 449
    ComRdTerm, 449
    ComSetEscape, 465
    ComToFile, 452
    ComWrt, 444
    ComWrtByte, 447
    ConfirmPopup, 199
    CopyString, 222
    DateStr, 127
    DefaultCtrl, 129
    Delay, 46
    DeleteListItem, 247
    DirSelectPopup, 207
    DisplayComStatus, 462
    DisplayPanel, 34
    ErrorPrintf, 178
    FileSelectPopup, 205
    FindLstn, 405
    FindPattern, 292
    FlushInQ, 460
    FlushOutQ, 459

Index

Fmt, 212
FmtFile, 554
FmtOut, 554
GenericMesagePopup, 200
GetCtrlAttribute, 99
GetCtrlVal, 66
GetCtrlIndex, 243
GetFileInfo, 187
GetFirstFile, 272
GetFmtErrNdx, 591
GetIndexFromValue, 256
GetLabelFromIndex, 245
GetNextFile, 274
GetNumCheckedItems, 251
GetNumListItems, 240
GetOutQLen, 461
GetProjectDir, 117
GetRS232ErrorString, 443
GetSystemTime, 222
GetValueFromIndex, 245
HidePanel, 284
InsertListItem, 246
InsertTextBoxLine, 129
InstallPopup, 205
IsListItemChecked, 254
LaunchExecutable, 228
LoadExternalModule, 520
LoadExternalModuleEx, 520
LoadMenuBar, 112
LoadPanel, 286
LoadPanelEx, 329
MakeColor, 251
MakePathname, 217
MesagePopup, 200
MultiFileSelectPopup, 205–7
NumFmtdBytes, 219
OpenComConfig, 438
OpenFile, 182
PlotStripChart, 81
ProcessSystemEvents, 46
PromptPopup, 203
QuitUserInterface, 35
ReadFile, 185
ReadLine, 185
RecallPanelState, 281
Receive, 407
RemovePopup, 205
ResetTextBox, 129
ReturnRS232Err, 443
RunUserInterface, 34
SavePanelState, 279
Scan, 578
ScanFile, 578
ScanIn, 578
Send, 406
SendCmds, 407
SendIFC, 394

SetComTime, 438
SetCtrlAttribute, 95
SetCtrlVal, 49
SetCTSMode, 437
SetInputMode, 215
SetPanelPos, 286
SetXMode, 437
SplitPath, 293
Status, 591
StringLength, 219
Timer, 130
TimeStr, 127
Toolbar_Display, 119
Toolbar_InsertItem, 117
Toolbar_New, 117
WaveformGraphPopup, 208
WriteFile, 182
WriteLine, 186
XGraphPopup, 208
XModemConfig, 455
XModemReceive, 455
XModemSend, 455
XYGraphPopup, 208
YGraphPopup, 208
Library Options, 520
light emitting diode. *See* LED.
Line Icons, 161
   column, 153, 163
Line Numbers, 161
   View, 149
List Box, 2
   attributes, 237
      Label/Value Pairs, 237
      Label/Value Pairs dialog
         box, 238
         Label, 238
         Value, 238
   List Box
      Options, 237
   creating, 236
   Example 7-1, 238–48
   Example 7-2, 248–57
   introduction, 236–38
literals, 559
Load
   command, 493
LoadExternalModule, 302, 303, 305, 324, 347, 520
   using, 488
LoadExternalModuleEx, 520
LoadMenuBar, 112
LoadPanel, 33, 47, 314, 329
LoadPanelEx, 329

M

MakeColor, 251, 256
   creating color intensity levels, 251

MakePathname, 217
Manuals
 GPIB NI-488.2M Function Reference Manual for Win32, 369, 371, 381, 383, 408
 HP 33120A User's Manual, 394
 HP 3478A User's Manual, 424
 LabWindows/CVI Instrument Driver Developer's Guide, 476, 493, 500, 507, 517
 LabWindows/CVI Programmer Reference Manual, 305, 319, 324
 LabWindows/CVI Standard Libraries Reference Manual, xli, 131, 232, 369, 408, 443, 462, 499, 500, 502, 503, 504, 507, 520, 515, 560, 561, 562, 563, 588
 LabWindows/CVI User Interface Reference Manual, xli, 67, 99, 208, 295, 314, 495
 LabWindows/CVI User Manual, xli, 146, 175, 304, 307, 324, 508, 523, 546
 NI-VISA Programmer Reference Manual, 499
Manuals List. *See* bibliography
Mark All For Compilation
 command, 488
Mark File For Compilation
 command, 488
Maximum Number of Compiler Errors
 option, 513
Maximum Stack Size
 option, 513
Memory Display
 command, 510
Memory Display Window, 178–77
menu, xxxix, 16, 38, 39, 49
 synonymous to menu command.
Menu Bar, 473
 creation, 102
 description, 102
Menu Bar List, 102
 Copy, 104
 Create, 104
 Cut, 104
 Done, 104
 Edit
  Edit Menu Bar
   dialog box, 102
 Paste, 104
menu callback function
 different from control callback function, 110
 prototype, 110
 type, 110
menu command
 synonymous to menu, 102
menu item, 15, 31, 39
menu items
 same as
  sub-commands, 102
  submenu commands, 102

menus, xxxix, 2, 4
MessagePopup, 145, 146, 200, 220
MessagePopup Panel, 200
meters, 2, 37
Microsoft development tools, 14
Microsoft® Excel, 185
Microsoft platform, 14
Microsoft Visual Basic, 321
Microsoft Visual C++ 5.0, 342, 348, 358
Microsoft Visual C++, xxxvi
Minimize All
 command, 509
modal dialog boxes, 196. *See* popup panels
modifier, 590
Modifier Key, 104
 MenuKey (Ctrl), 104
modifiers, 561, 562, 563, 590
 meaning, 586
monitoring
 rocket engine controller, 8
Move Item Down
 command, 479
Move Item Up
 command, 479
MultiFileSelectPopup, 205
MultiFileSelectPopup Panel, 205–7
multi-port serial card, 429
multithreading, xxxv

N

National Instruments, xxxv, 18, 497, 626, 627
 conventions used in *CVI* manuals, xl
 CVI developed by, xxxv
 LabVIEW developed by, xxxvi
 RS-485 serial board, 433
 slogan
  "The Software is the Instrument," xxxvi
 TestBench, 262
New
 command, 473
new project
 create, 19
Next Build Error, 147
NI Spy, 391, 394
 purpose, 391
 starting, 391
NI Spy Utility, 391–94
No Sorting
 command, 481
Notepad, 185, 188, 217, 228, 230
null-modem, 433
numeric controls, 14
 creation, 37
 range values, 41
  maximum, 41
  minimum, 41

Index 641

thermometer, 191
used for, 37
NumFmtdBytes, 219, 593

O

Online Help
 manuals available Online, xl
OpenComConfig, 438, 441, 455
Open command, 475
OpenFile, 182, 183, 217, 556
OpenGL libraries, xxxv
Operating Tool, 16, 73, 547
Other Attribute dialog box, 547, 548
Output Queue Length
 description, 438

P

panel
 anatomy, 69–72
 Auto Center Horizontally, 70
 Auto Center Vertically, 70
 Child panel
  Attributes
 Frame Style, 70
  Frame Thickness, 71
   Beveled, 71
   Hidden, 71
   Outline, 71
   Raised, 71
   Raised Outline, 71
   Step, 71
  Size Title Bar Height to Font, 71
  Title Bar Thickness, 71
  Title Style, 72
   Bold, 72
   Fonts, 72
   Italics, 72
   Size, 72
   Strikeout, 72
   Text Color, 72
   Underline, 72
 command, 531
 create, 20
 creating new, 88
 definition, 20
 Edit, 88
 Height, 70
 Left, 70
 Menu Bar, 70
 Other Attributes
  Can Maximize, 71
  Can Minimize, 71
  Conform to System
   Colors, 71
  Has Task Bar Button, 71
  Scale Contents On Resize, 71
  Sizable, 70
  Title Bar Visible, 71
 Panel Title, 69
 pull-down submenus, 69
 Scroll Bars
  horizontal, 70
  vertical, 70
 Title, 88
 Top, 70
 Width, 70
Panel Callback, 541
panel handle(s)
 as static data variables, 33
Panel Title box, 37
parent/child/grandchild panel
 relationship, 20
Parentheses
 balancing, 152–53
Parity
 bit, 436
 description, 436
 Even, 436
 Mark, 436
 None, 436
 Odd, 436
 Space, 436
  description, 436
Paste
 command, 530
PC, xxxix, 4
PCs
 baud rate not supported, 433
PlotStripChart, 79, 80, 81, 82
Pop-up panel
 accessed from Library menu, 196
 working with, 196
Pop-up Panels, 196–210
 ConfirmPopup, 199
 DirSelectPopup, 207
 FileSelectPopup, 205
 GenericMessagePopup, 200
 InstallPopup, 205
 MessagePopup, 200
 MultiFileSelectPopup, 205
 PromptPopup, 203
 RemovePopup, 205
 WaveformGraphPopup, 208
 XGraphPopup, 208
 XYGraphPopup, 208
 YGraphPopup, 208
Powerpoint, 228
Preface, xxxv
Preference menu item, 539
Previous Build Error, 147
Print
 command, 477, 529

ProcessSystemEvents, 46, 295
    called by *RunUserInterface* function, 16
    normally not used, 16
programming environment, xxxvii, 2
Project
    adding file, 45
    command, 510
    creating new project, 88
    Open project, 7
Project Move Options, 523
Project window, 6, 7, 19, 25, 28, 46, 50, 56, 57, 63, 70,
        74, 76, 83, 84, 86, 88, 94, 106, 110, 116,
        138, 139, 142, 144, 147, 149, 173, 178,
        321, 328, 329, 335, 336, 343, 345, 355,
        358, 409, 471, 472, 473, 477, 479, 485,
        488, 489, 508, 509, 510, 525, 545
    adding files to, 196
    blank template, 6
    Build menu, 482
        building using supported compiler, 488
        compiling all files in the next build, 488
        compiling file, 488
        compiling file in the next build, 488
        creating distribution disks, 488
        creating executables/DLLs, 482
        selecting Debug or Release configurations, 483
        selecting the Target Type, 483
        setting the target, 484
        Target Settings dialog box, 484
    difference between windows, 139
    Edit menu, 477
        add files to project window list, 477
        exclude file from build, 479
        include file to build, 479
        move file one item down in list, 479
        move file one item up in list, 479
        remove file from build, 479
        select all files in project, 478
    File menu, 473
        automatically save project, 476
        exit CVI, 477
        files most recently closed, 477
        open a new file, 475
        open existing files, 475
        open existing Function Tree Editor window, 476
        open existing include file, 475
        open existing project file, 476
        open existing user interface file, 476
        open new Function Tree Editor window, 473
        open new include file, 473
        open new project file, 473
        open new source file, 473
        open new user interface file, 473
        print file, 477
        save all files, 476
        save command, 476
        save file as, 476

Instrument menu, 492
    edit instrument driver program, 493
    load instrument driver, 493
    unload instrument driver, 493
Library menu, 494
    Active X Automation library functions, 502
    Advanced Analysis library functions, 495
    ANSI C library functions, 504
    Data Acqusition library functions, 496
    DataSocket library functions, 501
    Dynamic Data Exchange library functions, 502
    Easy Data Acqusition library functions, 496
    Formatting and I/O library functions, 503
    GPIB library functions, 497
    IVI library functions, 499
    RS-232 library functions, 497
    TCP library functions, 500
    User Interface library functions, 495
    Utility library functions, 504
    VISA library functions, 499
    VXI library functions, 496
    X Client Property library functions, 502
Options menu, 512
    Build options
        current compiler compatiblity mode, 513
        Debugging Levels, 513
            Extended, 514
            No Run-time checking, 514
            Standard, 514
        default calling conventions, 513
        include Function Protypes, 514
        maximum compiler errors displayed, 513
        maximum stack size, 513
        pointer assignment warning, 514
        prompt for include files, 514
        require return value from non-void functions,
            514
        show build progression, 515
        show compiler warnings, 514
        stop on first file with errors, 514
        track include file dependencies, 514
        unreachable code warning, 514
    compiler defines settings, 515
    enter command line arguments, 518
    list paths to search header files, 516
    list paths to search path for instrument drivers,
        516
    Run options
        break on first chance exceptions, 518
        break on library errors, 518
        hide CVI environment window, 518
        save changes before running, 518
    set compiler options, 512
    setting adding menu items to Tools menu, 521
    setting Color options, 525
    setting environment, 518
        automatically close function panel, 519

Index 643

console window or Standard I/O during de-
    bugging, 520
display values of variables during debugging,
    520
foreground lockout, 519
IEW memory allocation or function panels, 519
load Instrument driver source into IEW, 520
load project source files into IEW, 520
overwrite function panel window, 519
sleep mode, 518
setting fonts option, 524
setting library options, 520
setting run-time options, 517
setting source code control options, 522
setting the project move options, 523
Run menu, 489
    break at first statement, 490
    compile and build and execute project, 489
    continue from breakpoint, 490
    Continue program execution, 490
    list of all breakpoints, 490
    run exectuable without debugger, 491
    Select External Process, 490
    terminate program execution, 490
    terminate program in Debug mode, 490
    Threads, 492
starting, 148
Tools menu, 504
    Create Active X Automation Controller, 506
    Create IVI Instrument Driver, 507
    Source Code Control, 508
View menu, 480
    do not sort files in list, 481
    show full dates of files in list, 481
    show full pathname, 481
    sort files by pathname, 481
    sort files in list alphabetically, 481
    sort files in list by date, 481
    sort files in list by file extension, 481
Window menu, 508
    all open Help editor windows displayed dynami-
        cally, 511
    All open source files displayed dynamically, 511
    Array/Strings displayed dynamically, 511
    bring Interactive Execution window
        to front, 511
    bring Variables window to front, 511
    bring Watch window to front, 511
    Build Errors
        window to front, 510
    Cascade Windows, 508
    Close All windows, 509
    display Source Code Control Errors, 510
    Function Panels displayed dynamically, 511
    Function Trees displayed dynamically, 511
    Minimize All windows, 509
    open Memory Display window, 510

Project window to front, 510
Run-time Errors
    window to front, 510
Tile Windows, 509
User Interface files displayed
    dynamically, 511
Project window, 409
Project window environment, 472
Project Window icons, 83–85
    meaning
        A icon, 85
        C icon, 84
        I icon, 84
        O icon, 84
        S icon, 84
        U icons, 85
Project window list
    remove file, 106
Project window menus, 472
Project6
    Overview, 188
Projects
    project1-1, 7, 16
    project2-1, 28
    project3-1, 56
    project3-2, 74
    project4-1, 94
    project4-2, 106
    project4-3, 116
    project4-4, 122
    project6, 188
    project7, 262
    project11-1, 371
    project11-2, 394
    project11-3, 408
    project12, 433
    SimpleDll, 329
    UseGUIExportDLL, 331, 336
    UseSimpleDLL, 327
Projects directory, 469
Prompt for Include File Paths
    option, 514
PromptPopup, 203, 215, 220
PromptPopup Panel, 203–4
protocol
    definition, 366
pull-down menu, 14, 39

Q

QuitUserInterface, 27, 60, 95, 97, 196, 211,
    232, 240, 250, 297, 329,
    343, 355
    application not terminated by, 15
    assigned to selected callback, 27
    used in conjunction with *RunUserInterface*, 15
    value passed to, 35

R

ReadFile, 185
ReadLine, 185, 228
Read Only
 command, 529
RecallPanelState, 281, 302
Receive, 407
Receive Data. *See* RD
Release configuration, 482
Remove File
 command, 479
RemovePopup, 205
RemovePopup Function, 205
Require Function Prototypes
 option, 514
Require Return Values for
   Non-void Functions
 option, 514
ResetTextBox, 128
ReturnRS232Err, 443
reuse, existing programs, xxxvi
Rings
 library functions used, 261
 similar to, 257
Ring controls 2, 14
 create, 258
 Horizontal Pointer Slide, 261
 Knob, 261
 Menu Ring, 258
 Meter, 261
 using, 257–61
 various, 258
 Vertical Level Slide, 261
RS-232, xxxvi, 4, 429
 communication with, 428
 Configuring, 435–42
 Configuring *XModem*, 455–57
 connector
  female type, 431
  male type, 431
 Connectors and Signals, 429–33
 Data Communication Equipment. *See* DCE
 Data Transmission Equipment. *See* DTE
 Error Checking Functions, 443
 multi-port serial cards, 429
 Reading a Data Byte, 448–49
 Reading Data Files, 452–55
 Reading Data using a termination
   Byte, 449–50
 Receiving Data Buffer, 445–46
 Receiving Data Packets, 457–58
 Sending a Data Byte, 447–48
 Sending and Receiving Data Packets, 455
 Sending Data Buffer, 444–45
 Sending Data Packets, 457
 utility functions, 459–64
  Break function, 463–64

 COM port Status
  obtaining, 462–63
 Flush Input Queue, 460–61
 Flush Output Queue, 459–60
 Queue Length, 461–62
 Writing Data Files, 450–52
RS-232 Input/Output Library
 read/write data, 444
RS-232 interface, 428
 data transfer
  asynchronous data transfer, 428
  synchronous, 428
RS-232 library functions, 497
 used with RS-485 boards, 433
RS-232 ports, 428
RS-422
 capabilities, 432
RS-422A
 better noise rejection ratio, 432
RS-485
 AT-Serial board, 465
 Communicating with, 465–66
 escapeCode, 465
 multi-drop usage, 432
RTE. *See* Run-Time Engine.
RTS, 465
RTS lines, 437
Run
 Execute
  to run project, 81
 icon, 81
 project1–1, 7
Run Options
 command, 517
Run-Time Engine, 302, 306
Run-Time Errors
 command, 510
 send debug output to, 178
RunUserInterface, 15, 27, 59, 95, 110, 232, 329, 355
 use, 15
 return value, 35
 responds to callback function events, 35

S

Save
 command, 476
Save All
 command, 476, 529
Save As
 command, 476
Save Changes Before Running
 option, 518
Save Copy As
 command, 528
SavePanelState, 279, 281, 302
Scan, 212, 554, 555, 578, 581, 582, 583, 584, 585, 587
 function panel, 578

Index

Scan Function examples, 581
  Byte swapping with integer array to real array, 588
  Comma Separated ASCII numbers to Real array, 587
  converting hexadecimal string to decimal integer, 589
  Converting integer arrays to real arrays, 588
  Scanning non-NUL terminated strings, 587
  String to Integer and String, 586
    erroneous integer conversion, 586
    separating alphanumeric string into integer and alphabet, 586
  String to String
    all characters up to linefeed in string, 582
    appending source to target string, 582
    converting selected bytes from source, 585
    converting string up to a number, 583
    converting string up to terminator character, 583
    discarding terminators from source string, 584
    leaving blanks and tabs in string, 582
    source string to target string, 582
    string conversion with white space character, 583
    string width modifiers, 584
ScanFile, 554, 578, 590
  function panel, 580
ScanIn, 554, 578, 591
  function panel, 580
Scanning Functions, 578
  Scan, 578
  ScanFile, 578
  ScanIn, 578, 603
SCPI, 366, 394
  hierarchical command tree structure, 366
  protocol, 366
scroll bars, 2
SDK
  used with external compiler, 515
Second *CVI* Project
  Creating the, 36–51
Select All
  command, 478
Select Compatibility Mode, 513
Select External Process, 518
Select External Process Dialog Box, 490
Select Project Font dialog box, 524
Send, 406
SendCmds, 407
SendIFC, 394
sequential programs
  description, 15
serial communication
  definition, 428
  Overview, 428–33
serial instrument control, 2

serial port communication
  advantages, 431
  disadvantages, 431
SetComTime, 438, 442
SetCtrlAttribute, 66, 95, 97, 98, 99
  library function attributes, 64
SetCtrlVal, 47, 332
SetCTSMode, 437, 441
Set Default Font
  command, 531
SetInputMode, 215
SetPanelAttribute, 314
SetPanelPos, 286, 314
SetPanelSize, 314
SETRTS, 466
SetSystemAttribute, 314
Setting and Clearing Tags, 153–54
Setting Line Numbers and Selecting Text, 149
SetXMode, 437, 441
SETXOFF, 466
SETXON, 466
Set XON/XOFF, 437
  software handshaking, 437
Show Build Error Window for warnings option, 514
Show Digital Display, 42
  purpose, 43
Show Full Dates
  command, 481
Show Full Pathnames
  command, 481
Show Inc/Dec, 42
skeleton code. *See CodeBuilder.*
sliders, 2, 37
Software Development Kit. *See* SDK.
Sort By Date
  command, 481
Sort By File Extension
  command, 481
Sort By Name
  command, 481
Sort By Pathname
  command, 481
Source Code, 101
Source Code Control, 508, 522
Source Code Control Errors
  command, 510
Source Code Editor, 19, 25, 28, 36, 81
Source Code window, 520, 525
Source Editor, 138, 472, 512
  Breakpoint
    insert, 139
  capability, 138
  Find Button Bar, 161, 162
    Find Next, 161
    Find Previous, 161
  Find Text, 160

Source Editor (*continued*)
  Line Icons
    defintion, 139
  Replace Button Bar, 162
    Replace, 160
    Replace All, 162
    Replace With, 162
  Replace Dialog Box, 161
  Run
    Activate Panels when Resuming, 170
  Tags
    insert, 139
  Toolbar, 140
    Debugging icons
      Add Watch Expression, 165
      Continue, 163
      Down Call Stack, 165
      Finish Function, 164
      Go To Cursor, 163
      Step Into command, 162
      Step Over, 164
      Terminate Execution, 164
      Up Call Stack, 165
      View Variable Value, 165
    Editing icons
      Compile File, 142
      File Save, 141
      Find, 142
      Find UI objects, 141
      Go To Definition, 141
      Insert Construct, 141
      New File, 141
      Open File, 141
      Recall Panel, 142
      Replace, 142
      Run Project, 142
      ProjectWindow icon, 142, 163
  Windows
    Memory Display, 178
    Variables, 171
    Watch, 176
Source Editor window, 329
  capability
    context sensitivity, 139
    difference between windows, 139
  Run menu
    Call Trace, 492
    Down Call Stack, 492
    Up Call Stack, 492
Source files
  Compare, 158–60
    comparing with, 158
    find next diference, 159
    ignore white space, 158
    match criteria, 159
    synchronize selections, 159
    synchronizing, 159
SPARC stations
  baud rate not supported, 436

SplitPath, 293, 452
stand-alone executable, 486
  error handling, 314–15
  files contained in, 302
  requirements, 301
Standard Commands for Programmable Instrumentation.
        See SCPI.
Standard Input, 579
  reading from keyboard, 185
Standard Input/Output, 145, 178, 485
Standard Output, 187, 327, 557
  writing to, 187
Start *bit*, 428
  use, 428
static linking
  description, 318
Status, 591, 592, 593
Status functions, 591
STDTR, 465
Stop bit, 428
  use, 428
Stop/Loop window, 622
Stop on First File with Errors
  option, 514
String Display window, 525
StringLength, 219
Strip Chart, 2, 14
  Attributes
    Edge Style, 76
    Grid Color, 76
    Label Appearance, 76
    Label Style, 76
    Points per Screen, 74
  Points X-axis
    Axis Name, 76
  Points Y-axis
    Axis Name, 76
    Maximum Value, 76
    Minimum Value, 76
  Scroll Mode
    Block, 74
    Continuous, 74
    Sweep, 76
  definition, 74
  different from graph control, 74
  traces
    multiple, 74
Sun, xxxix, 67
Sun SPARC, xxxix
Sun workstations, xl
switches, 14
Symantec C, xxxvi
synchronous transmission
  definition, 428
System Test
  Abort switch, 611, 616, 619
  Active List, 614
  Add button, 620
  Attach Def, 618

Index

Back button, 615, 616, 618, 619, 620
Basic Tests ring box, 620, 622
Below/Above switch, 620
Build Seq. button, 612
Cancel, 619
Check All button, 615, 620
Check button, 611
Close button, 622, 623
Def's ring box, 618
Defaults switch, 622, 623
Edit Code, 611
Edit Def's switch, 618
Edit Mol button, 620
Edit Seq button, 620
Ex. Log switch, 615, 616, 620
File Manag. button, 611
files
   `bittest.ini`, 607
   Default Values.`ini` files, 607
   Self Test Files, 608
   Sequence files, 608
   `TestNames.ini`, 606
   `Vars.ini`, 607
Finish button, 614
Holding List, 615
installation
   `Setup.exe`, 608
introduction, 605
Iterations ring box, 619
Last Test button, 620
Last Test molecule, 621
Last Test window, 620, 621
Load button, 623
Load File button, 610, 615
Load It button, 615
Load Order button, 615
Load Seq button, 619
Log switch, 615, 616, 620
Loop button, 616
Loop option
   not available in demo, 622
Loop/Select window, 616, 619
Make Files
   BIT File
      purpose, 613
   Defaults ini
      purpose, 613
   STF File
      purpose, 613
   `Vars.ini`
      purpose, 612
Molecules button, 612, 616
Molecules List Reordering window
   how to use, 613
Molecules List Reordering window, 613
Molecules ring box, 619
Order File Name, 615
Order ring box, 619
Original List, 614, 615

Permission levels
   how to use, 611
Permissions window, 609
Print Log button, 616, 620
Remove, 620
Remove button, 620
Reorder button, 613
Reset All button, 622
Reset button, 623
Run BIT Test button, 611
Run button, 611
Run Seq button, 612, 618, 619
Run ST Files button, 610, 612, 615, 616
Save button, 620
Save Order button, 615
Select button, 619, 622, 623
Select Option button, 622
Select ring box, 619
Self Test Files, 615
Seq button, 616
Sequence Selector list box, 616
Sequencer window, 616, 619
Show Defaults button, 623
start up procedure, 608
Stop Always switch, 622
Stop On Error switch, 611, 615, 617, 620
Stop/Loop button, 620
Stop/Loop window, 620, 622
System Selective Testing, 618, 622
System Selective Testing window, 612, 613, 619, 622
System Self Tests button, 615
System Self Tests window, 610, 612, 615, 616
System Test window, 614, 616, 618
System Test Window
   (Main Window), 609
SysTest.exe icon, 608
Test Builder, 619
Test Builder button, 616
Test Builder window, 612, 620, 622
Test Code, 610, 611
Test Functions ring box, 622
Tests ring box, 619
Transfer button, 614, 620
Uncheck button, 611, 615, 616, 620
Vars button, 617
Vars window, 617, 620, 622
   invoked by Test Builder, 622
View Log button, 616, 617, 618
View Log Seq button, 619
System Test Demo, 605

T

Tab Order. *See* Tab Order Creation.
   command, 531
   setting, 132
Tab Order Creation, 132–34
Tag, 153

Tag Scope
  scope of files
    setting Tag, 153
target format specifier, 558
Target Settings dialog box, 303, 484
  Application File, 484
  Application Icon File, 484
  Application Title, 484
  Create Console Application, 485
Target Settings for
            Dynamic Link Library
            dialog box, 322
  DLL Export Options, 324
  DLL File, 322
  Export What:, 324
  Help, 324
  Import Library Base Name, 322
  Import Library Choices, 322
  Instrument Driver Support
            Only, 322
  LoadExternal Module, 324
  Type Library, 323
  Version Info, 322
  Where to copy DLL, 322
target specifier, 585
Target Type
  command, 483
TCP/IP
  based devices, xxxvi
  communication, 2
TCP library functions, 500
Terminate Execution
  command, 490
Test Executive Project
  Designing, 262–70
Text Box, 2
  Adding, 122–32
  Control Appearance
            group, 122
  description, 122
  different from List Box, 236
  Label Appearance group, 122
  Scroll Bars, 122
  Visible Lines, 122
Text Color, 191
Text Message, 192
  creating, 94
Text Style
  attributes, 43
  proportional size fonts, 191
Text Style box, 43
  selecting
    Color, 43
    Fonts, 43
    Justification, 43
    Size, 43
thermometer, 37
Threads
  command, 492

Threads Dialog Box, 492
thunking, 318
thunking DLL, 319
Tick Style, 42
Tile Windows
  command, 509
Time Out
  description, 438
Timer, 130
  creating, 195
timer control
  adding, 57
  usage, 56
timer knobs, 18
TimeStr, 127
Title Bar, 472
  on Project window, 6
Title bar color, 548
  used for child panels, 548
Toggle Tag
  adding, removing Tag, 153
Tool Bar, 2, 16, 72, 547
  created programmatically, 115
  created using toolbar function library, 115
  creating, 115–22
  definition, 115
  description, 72–73
  Editing Tool Bar
    Coloring Tool
      use, 73
    Editing Tool
      use, 73
    known as, 73
    Operating Tool, 73
    Text Editing Tool
      use, 73
  icons, 16
    visible, 122
  instrument driver
    `toolbar.fp`, 116
  Shortcut Tool Bar
    known as, 72
Toolbar_Display, 119
Toolbar_InsertItem, 117, 122
Toolbar_New, 117
Tools Menu Options, 521
  Add button, 522
Track include file dependencies
  option, 514
Transmit Data. *See* TD
typical configuration
  CVI environment, 19

U

UART, 428
Undo and Redo
  commands, 529

Index

Universal Asynchronous Receiver/Transmitter. See UART.
Unix, 19, 305, 324
  CVI versions prior to 5.5, 433
  functions
    aioread, 433
    aiowrite, 433
    kernel, 433
Unload
  command, 493
Use Console Window for Standard I/O when Debugging
  console window display, 178
user interface
  as integral part of *CVI*, 14
  create code
    generate, 27
    select target file, 27
  select from Project window, 19
  setting
    code preferences, 25, 33, 45
User Interface Editor, 14, 16, 19, 20, 25, 27, 36, 37, 55, 69, 102, 139, 472, 528, 551
  creating
    new, 88
  part of CVI environment, 14
  prominent features, 69
User Interface Editor window
  Arrange menu, 537
  Code menu, 538
    Generate, 538. *See also* Chapter 2, "Basics of Creating the Graphical User Interface."
      All Callbacks, 541
      All Code, 538
      Control Callbacks, 541
      Main Function, 540
      Menu Callbacks, 541
      Panel Callbacks, 541
    Generate Code dialog box, 540
    Preferences, 542
      Always Append Code To End option, 543
      Default Control Events, 543
      Default Panel Events, 543
      Default Panel Events dialog box, 543
    Set Target File, 538
    Set Target File dialog box, 538
    View menu, 542
  Create menu, 532
  difference between windows, 139
  Edit menu, 529
    apply default font, 531
    change style of control, 531
    Copy
      command, 530
    Copy Panel command, 531
    Cut
      command, 530
    Cut Panel command, 531
    deleting objects, 530
    open Edit Panel dialog box, 531
    open the control dialog box, 531
    Paste
      command, 530
    set default font, 531
    set Tab order, 531
    undo and redo commands, 529
  File menu, 528
    add file to project, 529
    Close. *See* Project window, File menu commands.
    Exit LabWindows/CVI. *See* Project window, File menu commands.
    New. *See* Project window, File menu command.
    Open. *See* Project window, File menu command.
    print file, 529
    read only mode, 529
    Save. *See* Project window, File menu commands.
    save all open files, 529
    Save As. *See* Project window, File menu commands.
    save user interface to another file name, 528
  Help menu, 552
  Library menu, 546. *See* Appendix B, " Project Window Environment."
  Options menu, 545
    Load from Text, 552
    Next Tool, 547
    Operate Visible Panels, 547
    Preferences, 547
      Edit Color Preferences, 547
        Initial editor background color, 547
      More, 549
        Other User Interface Editor Preferences dialog box, 549
          Constant Name Assignment group, 551
          Purge undo actions when saving file, 549
          Undo Preferences, 549
      Preferences for New Controls, 548
        Control Text Style, 548
        Label Text Style, 548
      Preferences for New Panels, 547
        Conform to system colors, 548
        Resolution Adjustments, 547
        Use system colors as defaults for panels and controls, 548
    Save in Text Format, 552
    User Interface Preferences dialog box, 547
  Run menu, 545
  Tools menu, 546
  View menu, 533
    Bring Panel to Front, 536
    Find UIR Objects
      Edit, 536

User Interface Editor window
  View menu
    Find UIR Objects (*continued*)
      Find Next, 536
      Find Prev, 536
      Stop, 536
    Find UIR Objects dialog box, 534
      Case-Sensitive, 535
      Find button, 536
      Regular Expression, 536
      Search By
        Callback Function Name, 535
        Constant Name, 535
        Label, 535
        Prefix + Constant Name, 535
        Prefix + Constant Prefix, 535
      Whole Word, 536
      Wrap, 535
    finding user interface objects, 533
    Next Panel, 537
    Preview User Interface
        Header File, 537
    Previous Panel, 537
    Show/Hide Panels, 536
    Show/Hide Panels command
        Panel List, 536
  Window menu, 546. *See* Appendix B,
      "Project Window Environment."
user interface events
  monitored by *RunUserInterface* function, 16
User Interface Object
  Find, 155
User Interface Objects
  finding (Controls), 154–55
User Interface Overview, 14–15
user interface resource, xxxvii, 19
  definition used in text
    same as GUI, 19
Using the GUI DLL, 336–39
Utility library functions, 504

V

Variables
  command, 510
Variables, Arrays, and Strings
  working with, 171–76
Variable window, 510, 525
Version Info, 486, 487
Version Info dialog box
  Company Name, 487
  File Version, 487
  Legal Copyright, 487
  Product Version, 487
Version Info sub-dialog box, 486, 487

View Control Callback, 542
  viewing callback function from control, 127
View menu, 533
virtual instrument, xxxvi, 2, 426
virtual instrumentation, xxxv, 408
Virtual Instrumentation Software Architecture. *See*
    VISA.
VISA, 365
VISA library functions, 499
VME (Versa-Modular Eurocard)
    e**X**tensions for **I**nstruments. *See* VXI.
VXI, xxxvi, 4, 5
VXIbus, 366
VXI library functions, 496
VXI*plug&play*, 310, 322

W

Watch
  command, 511
Watch window, 176–77, 511, 525
Watcom, 321
  calling convention, 318
WaveformGraphPopup, 208
Win32, 3
Win32 Application, 348
Win32 Interactive Control Utility, 388–91, 389
  caveats, 390
Windows, xxxvi, xxxix, 3, 5, 14, 33, 66, 102, 115, 305,
    318, 321, 331, 383, 443, 473, 476, 502,
    513, 515, 547, 548
  definition as used in text, xxxvi
  NT version 4.0 Service Pack 3, xxxvi
Windows 3, 14
Windows 3.1, 318, 324. *See* Windows.
Windows 95. *See* Windows.
Windows 98. *See* Windows.
Windows 2000. *See* Windows.
Windows applications, 318
Windows Explorer. *See* Explorer
Windows NT. *See* Windows
WinMain, 343, 345, 346
Word, 228
WordPad, 185, 217, 230, 233
WriteFile, 182, 218, 219
WriteLine, 186
  complementary to *ReadLine*, 186
WriteToFile, 215, 585

X

XGraphPopup, 208, 225
XModem
  configuration, 455
  description, 456

Index

different from ComFromFile,
 ComToFile, 457
functions, 455
protocol, 457
XModemConfig, 455, 456, 457
XModemReceive, 455, 457
XModemSend, 455, 457, 458
XOFF, 466
XON, 466
X Windows, 502
XYGraphPopup, 208

Y

YGraphPopup, 208, 225

The Author

Shahid F. Khalid has over 30 years of experience in software engineering. He is presently a software engineer at Boeing, Canoga Park, California, where he works on testing the embedded software for the rocket engine control unit on the Delta IV program. The testing of the rocket engine controller is performed using *LabWindows/CVI* communicating via the RS-232 serial port and the GPIB interface to control various instruments that command the engine and acquire the data. He has also created *LabWindows/CVI* applications for laser beam alignment, test equipment diagnostics software for VXI platforms, and database applications and has worked on other company proprietary projects.

Shahid has worked as a software contractor in aerospace and defense with companies like GTE Government Systems and Loral Electro-Optical Systems. He has worked at Logicon Incorporated, and at Jet Propulsion Laboratories in Pasadena, California. He has been a Unit Head of a software development team at Singer-Kearfott in New Jersey, where he was involved with the development of inertial navigation guidance test software for the US Navy Mark VI program. He has worked as an analog circuit design engineer and computer programmer at various companies in Kansas, Tennessee, and New Jersey.

Shahid has a Bachelor's degree in Physics and Mathematics as well as a Master's degree in Mathematics from Panjab University in Pakistan. He also has a Master's Degree in Electronic Engineering from Kansas University and a Master's Degree in Computer Science from Stevens Institute of Technology (New Jersey).

He is married and has two daughters and a son. After living in Kansas, Tennessee, and New Jersey, he now enjoys his home in Agoura Hills, California. His hobbies include photography, traveling, swimming, reading self-study books on computers and programming languages, and playing with the computer.

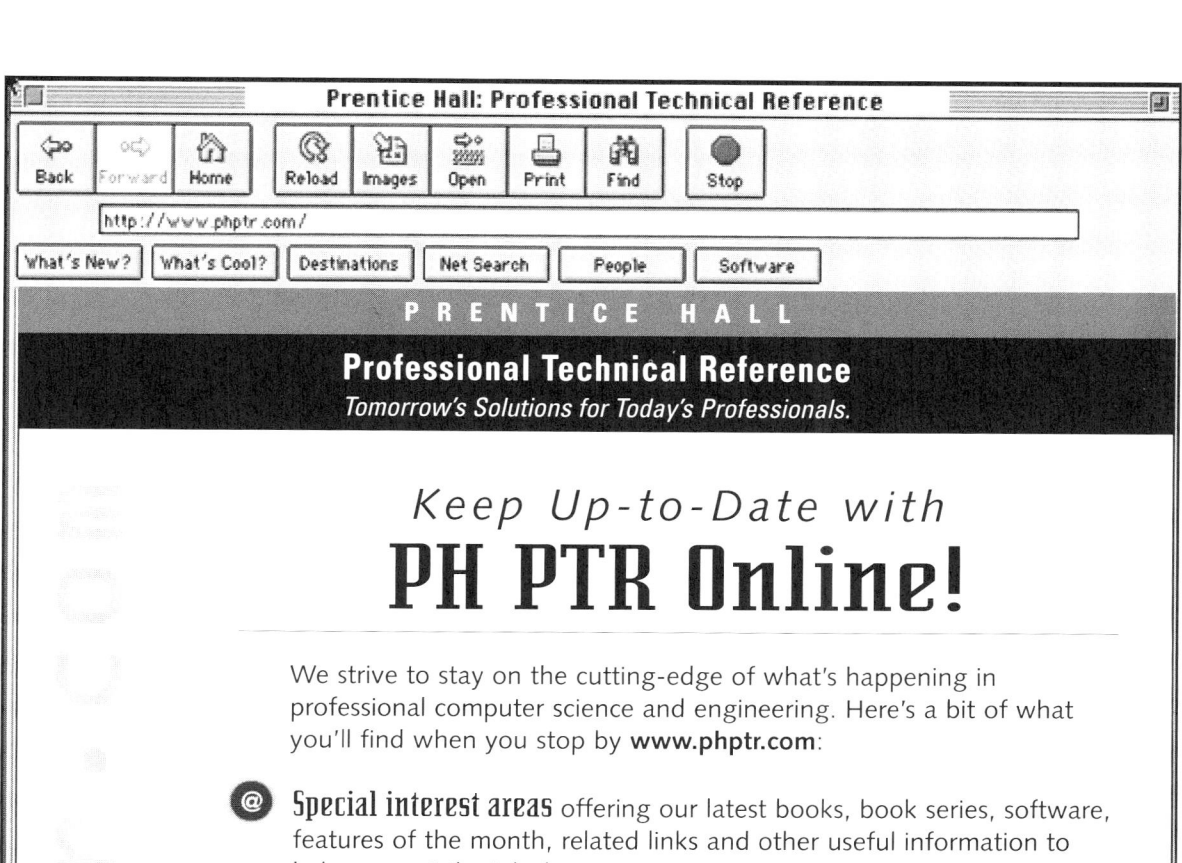

Keep Up-to-Date with
PH PTR Online!

We strive to stay on the cutting-edge of what's happening in professional computer science and engineering. Here's a bit of what you'll find when you stop by **www.phptr.com**:

Special interest areas offering our latest books, book series, software, features of the month, related links and other useful information to help you get the job done.

Deals, deals, deals! Come to our promotions section for the latest bargains offered to you exclusively from our retailers.

Need to find a bookstore? Chances are, there's a bookseller near you that carries a broad selection of PTR titles. Locate a Magnet bookstore near you at www.phptr.com.

What's New at PH PTR? We don't just publish books for the professional community, we're a part of it. Check out our convention schedule, join an author chat, get the latest reviews and press releases on topics of interest to you.

Subscribe Today! **Join PH PTR's monthly email newsletter!**

Want to be kept up-to-date on your area of interest? Choose a targeted category on our website, and we'll keep you informed of the latest PH PTR products, author events, reviews and conferences in your interest area.

Visit our mailroom to subscribe today! **http://www.phptr.com/mail_lists**

LICENSE AGREEMENT AND LIMITED WARRANTY

READ THE FOLLOWING TERMS AND CONDITIONS CAREFULLY BEFORE OPENING THIS SOFTWARE MEDIA PACKAGE. THIS LEGAL DOCUMENT IS AN AGREEMENT BETWEEN YOU AND PRENTICE-HALL, INC. (THE "COMPANY"). BY OPENING THIS SEALED SOFTWARE MEDIA PACKAGE, YOU ARE AGREEING TO BE BOUND BY THESE TERMS AND CONDITIONS. IF YOU DO NOT AGREE WITH THESE TERMS AND CONDITIONS, DO NOT OPEN THE SOFTWARE MEDIA PACKAGE. PROMPTLY RETURN THE UNOPENED SOFTWARE MEDIA PACKAGE AND ALL ACCOMPANYING ITEMS TO THE PLACE YOU OBTAINED THEM FOR A FULL REFUND OF ANY SUMS YOU HAVE PAID.

1. **GRANT OF LICENSE:** In consideration of your payment of the license fee, which is part of the price you paid for this product, and your agreement to abide by the terms and conditions of this Agreement, the Company grants to you a nonexclusive right to use and display the copy of the enclosed software program (hereinafter the "SOFTWARE") on a single computer (i.e., with a single CPU) at a single location so long as you comply with the terms of this Agreement. The Company reserves all rights not expressly granted to you under this Agreement.

2. **OWNERSHIP OF SOFTWARE:** You own only the magnetic or physical media (the enclosed software media) on which the SOFTWARE is recorded or fixed, but the Company retains all the rights, title, and ownership to the SOFTWARE recorded on the original software media copy(ies) and all subsequent copies of the SOFTWARE, regardless of the form or media on which the original or other copies may exist. This license is not a sale of the original SOFTWARE or any copy to you.

3. **COPY RESTRICTIONS:** This SOFTWARE and the accompanying printed materials and user manual (the "Documentation") are the subject of copyright. You may not copy the Documentation or the SOFTWARE, except that you may make a single copy of the SOFTWARE for backup or archival purposes only. You may be held legally responsible for any copying or copyright infringement which is caused or encouraged by your failure to abide by the terms of this restriction.

4. **USE RESTRICTIONS:** You may not network the SOFTWARE or otherwise use it on more than one computer or computer terminal at the same time. You may physically transfer the SOFTWARE from one computer to another provided that the SOFTWARE is used on only one computer at a time. You may not distribute copies of the SOFTWARE or Documentation to others. You may not reverse engineer, disassemble, decompile, modify, adapt, translate, or create derivative works based on the SOFTWARE or the Documentation without the prior written consent of the Company.

5. **TRANSFER RESTRICTIONS:** The enclosed SOFTWARE is licensed only to you and may not be transferred to any one else without the prior written consent of the Company. Any unauthorized transfer of the SOFTWARE shall result in the immediate termination of this Agreement.

6. **TERMINATION:** This license is effective until terminated. This license will terminate automatically without notice from the Company and become null and void if you fail to comply with any provisions or limitations of this license. Upon termination, you shall destroy the Documentation and all copies of the SOFTWARE. All provisions of this Agreement as to warranties, limitation of liability, remedies or damages, and our ownership rights shall survive termination.

7. **MISCELLANEOUS:** This Agreement shall be construed in accordance with the laws of the United States of America and the State of New York and shall benefit the Company, its affiliates, and assignees.

8. **LIMITED WARRANTY AND DISCLAIMER OF WARRANTY:** The Company warrants that the SOFTWARE, when properly used in accordance with the Documentation, will operate in substantial conformity with the description of the SOFTWARE set forth in the Documentation. The Company does not warrant that the SOFTWARE will meet your requirements or that the operation of the SOFTWARE will be uninterrupted or error-free. The Company warrants that the media on which the SOFTWARE is delivered shall be free from defects in materials and workmanship under normal use for a period of thirty (30) days from the date of your purchase. Your only remedy and the Company's only obligation under these limited warranties is, at the Company's option, return of the warranted item for a refund of any amounts paid by you or replacement of the item. Any replacement of SOFTWARE or media under the warranties shall not extend the original warranty period. The limited warranty set forth above shall not apply to any SOFTWARE which the Company determines in good faith has been subject to misuse, neglect, improper installation, repair, alteration, or dam-

age by you. EXCEPT FOR THE EXPRESSED WARRANTIES SET FORTH ABOVE, THE COMPANY DISCLAIMS ALL WARRANTIES, EXPRESS OR IMPLIED, INCLUDING WITHOUT LIMITATION, THE IMPLIED WARRANTIES OF MERCHANTABILITY AND FITNESS FOR A PARTICULAR PURPOSE. EXCEPT FOR THE EXPRESS WARRANTY SET FORTH ABOVE, THE COMPANY DOES NOT WARRANT, GUARANTEE, OR MAKE ANY REPRESENTATION REGARDING THE USE OR THE RESULTS OF THE USE OF THE SOFTWARE IN TERMS OF ITS CORRECTNESS, ACCURACY, RELIABILITY, CURRENTNESS, OR OTHERWISE.

IN NO EVENT, SHALL THE COMPANY OR ITS EMPLOYEES, AGENTS, SUPPLIERS, OR CONTRACTORS BE LIABLE FOR ANY INCIDENTAL, INDIRECT, SPECIAL, OR CONSEQUENTIAL DAMAGES ARISING OUT OF OR IN CONNECTION WITH THE LICENSE GRANTED UNDER THIS AGREEMENT, OR FOR LOSS OF USE, LOSS OF DATA, LOSS OF INCOME OR PROFIT, OR OTHER LOSSES, SUSTAINED AS A RESULT OF INJURY TO ANY PERSON, OR LOSS OF OR DAMAGE TO PROPERTY, OR CLAIMS OF THIRD PARTIES, EVEN IF THE COMPANY OR AN AUTHORIZED REPRESENTATIVE OF THE COMPANY HAS BEEN ADVISED OF THE POSSIBILITY OF SUCH DAMAGES. IN NO EVENT SHALL LIABILITY OF THE COMPANY FOR DAMAGES WITH RESPECT TO THE SOFTWARE EXCEED THE AMOUNTS ACTUALLY PAID BY YOU, IF ANY, FOR THE SOFTWARE.

SOME JURISDICTIONS DO NOT ALLOW THE LIMITATION OF IMPLIED WARRANTIES OR LIABILITY FOR INCIDENTAL, INDIRECT, SPECIAL, OR CONSEQUENTIAL DAMAGES, SO THE ABOVE LIMITATIONS MAY NOT ALWAYS APPLY. THE WARRANTIES IN THIS AGREEMENT GIVE YOU SPECIFIC LEGAL RIGHTS AND YOU MAY ALSO HAVE OTHER RIGHTS WHICH VARY IN ACCORDANCE WITH LOCAL LAW.

ACKNOWLEDGMENT

YOU ACKNOWLEDGE THAT YOU HAVE READ THIS AGREEMENT, UNDERSTAND IT, AND AGREE TO BE BOUND BY ITS TERMS AND CONDITIONS. YOU ALSO AGREE THAT THIS AGREEMENT IS THE COMPLETE AND EXCLUSIVE STATEMENT OF THE AGREEMENT BETWEEN YOU AND THE COMPANY AND SUPERSEDES ALL PROPOSALS OR PRIOR AGREEMENTS, ORAL, OR WRITTEN, AND ANY OTHER COMMUNICATIONS BETWEEN YOU AND THE COMPANY OR ANY REPRESENTATIVE OF THE COMPANY RELATING TO THE SUBJECT MATTER OF THIS AGREEMENT.

Should you have any questions concerning this Agreement or if you wish to contact the Company for any reason, please contact in writing at the address below.

Robin Short
Prentice Hall PTR
One Lake Street
Upper Saddle River, New Jersey 07458

About the CD

Welcome to the *LabWindows/CVI Programming for Beginners* CD. The software on this CD runs on your PC using Windows 95/98/NT 4.0 Service Pack 3 or the Windows 2000 operating system.

This CD contains an evaluation version of LabWindows/CVI 5.5 and all the project files used in this book. This CD also contains a trial version of System Test application and a mathematical application that analyzes the functions parametrically. Directions for installing the projects can be found in Appendix A of this book. All the projects on the CD have been run and tested thoroughly and neither the author nor Prentice Hall are liable for any errors or consequential damages occurring from the use of the software.

The "Projects" folder must be copied to the hard disk before running the projects. These projects have been tested to work with the full version of LabWindows/CVI 5.5. Most of these projects work with LabWindows/CVI 5.5 evaluation version, however due to the disabling of some of the menu commands the projects listed below cannot be run with the evaluation version. The exceptions and caveats of running the projects are listed below:

1. The menu commands to create distribution disks explained in Chapter 8, "Creating Stand-Alone Executables and Distribution Disks" cannot be used due to the disabling of the **Create Distribution Kit...** menu command.

2. The **External Compiler Support...** menu command explained in Chapter 10, "External Compiler Support" is disabled. "SimpleDLL.prj" and "GUIExportDLL.prj" projects will give a "File access permissions denied" error if you try to build the DLL.

3. All projects should be run as **Release** configuration (Execute) command. If you try to run in the **Debug** configuration you will get "File access permission denied" error.

4. Make sure that all the files copied from the CD-ROM to the hard disk have the **Read-only** and **Archive** attributes disabled including the files in "Input" and "Output" folders.

5. If the GPIB interface board is not installed before running project11-2 using the executable file (Application), the GUI will be displayed momentarily and the project will terminate.

Technical Support

Prentice Hall does not offer technical support for this software. If there is a problem with the media, however, you may obtain a replacement CD by emailing a description of the problem. Send your email to:

disk_exchange@prenhall.com